# THE HYPOTHALAMUS

*Research Publications:
Association for Research in Nervous and Mental Disease*

*Volume 56*

# The Hypothalamus

*Research Publications:
Association for Research in
Nervous and Mental Disease*

Volume 56

Editors:

**Seymour Reichlin, M.D., Ph.D.**
*Professor of Medicine
Tufts University
School of Medicine
Chief, Endocrinology Division
New England Medical Center Hospital
Boston, Massachusetts 02111*

**Ross J. Baldessarini, M.D.**
*Associate Professor of Psychiatry
Harvard Medical School-McLean
　Hospital
Assistant Director, Mailman
　Laboratory for Psychiatric Research
Boston, Massachusetts 02115*

**Joseph B. Martin, M.D., Ph.D.**
*Professor and Chairman
Departments of Neurology and Neurosurgery
Neurologist-in-Chief
Montreal Neurological Institute
McGill University
Montreal, Canada H3A 2B4*

Raven Press ▪ New York

Raven Press, 1140 Avenue of the Americas, New York,
New York 10036

© 1978 by Raven Press Books, Ltd. All rights reserved. This book is protected by copyright. No part of it may be reproduced, stored in a retrieval system, or transmitted, in any form or by any means, electronic, mechanical, photocopying, recording, or otherwise, without the prior written permission of the publisher.

Made in the United States of America

**Library of Congress Cataloging in Publication Data**
Main entry under title:

The Hypothalamus.

   (Research publications—Association for Research in Nervous and Mental Disease; v. 56)
   Includes bibliographical references and index.
   1. Hypothalamus. 2. Hypothalamo-hypophyseal system. 3. Hypothalamic hormones. 4. Neuroendocrinology. I. Reichlin, Seymour. II. Baldessarini, Ross J., 1937–  III. Martin, Joseph B. [DNLM: 1. Hypothalamus—Physiology. 2. Neurosecretion. W1 RE233P v. 56]
QP383.7.H96     599'.01'88     77-83691
ISBN 0-89004-167-9

## OFFICERS – 1976

SEYMOUR REICHLIN, M.D., Ph.D.
*President*
*New England Medical Center Hospital*
*171 Harrison Avenue*
*Boston, Massachusetts 02111*

ROSS J. BALDESSARINI, M.D.
*Vice President*
*Massachusetts General Hospital*
*Fruit Street*
*Boston, Massachusetts 02114*

JOSEPH B. MARTIN, M.D., Ph.D.
*Vice President*
*The Montreal General Hospital*
*1650 Cedar Avenue*
*Montreal H3G 1A4 Canada*

ROGER C. DUVOISIN, M.D.
*Secretary-Treasurer*
*Mount Sinai Medical Center*
*Fifth Avenue at 100th Street*
*New York, New York 10029*

## HONORARY TRUSTEES

CLARENCE C. HARE, M.D.   FRANCIS J. BRACELAND, M.D.   H. HOUSTON MERRITT, M.D.

## TRUSTEES

FRED PLUM, M.D., *Chairman*

SHERVERT H. FRAZIER, M.D.
ROBERT KATZMAN, M.D.
SEYMOUR S. KETY, M.D.

LAWRENCE C. KOLB, M.D.
ROBERT MICHELS, M.D.
CLARK T. RANDT, M.D.
LEWIS P. ROWLAND, M.D.

EDWARD SACHAR, M.D.
ALBERT J. STUNKARD, M.D.
MELVIN D. YAHR, M.D.

## COMMISSION – 1976

SEYMOUR REICHLIN, M.D., Ph.D.
*Chairman, Boston, Massachusetts*

ROSS J. BALDESSARINI, M.D.
*Boston, Massachusetts*

SEYMOUR KETY, M.D.
*Boston, Massachusetts*

FRED PLUM, M.D.
*New York, New York*

FLOYD BLOOM, M.D.
*La Jolla, California*

DOROTHY T. KRIEGER, M.D.
*New York, New York*

EDWARD SACHAR, M.D.
*Bronx, New York*

ROGER C. GUILLEMIN, M.D.
*La Jolla, California*

JOSEPH B. MARTIN, M.D.
*Montreal, Canada*

BERTA SCHARRER, M.D.
*Bronx, New York*

TOMAS HÖKFELT, M.D.
*Stockholm, Sweden*

RICHARD C. MICHAEL, M.D.
*Atlanta, Georgia*

SOLOMON H. SNYDER, M.D.
*Baltimore, Maryland*

## PROGRAM COMMITTEE

SEYMOUR REICHLIN, M.D., Ph.D.
*Chairman, Boston, Massachusetts*

ROSS J. BALDESSARINI, M.D.
*Boston, Massachusetts*

JOSEPH B. MARTIN, M.D., Ph.D.
*Montreal, Canada*

## COMMITTEE ON NOMINATIONS

CLARK T. RANDT, M.D.
*Chairman*

LAWRENCE C. KOLB, M.D.

MELVIN D. YAHR, M.D.

## COMMITTEE ON ADMISSIONS

HOWARD P. KREIGER, M.D.
*Chairman*

RONALD R. FIEVE, M.D.

DONALD H. SILBERBERG, M.D.

# Publisher's Note

*Titles marked with an asterisk (\*) are out of print in the original edition. Some out-of-print volumes are available in reprint editions from Hafner Publishing Company, 866 Third Avenue, New York. N.Y. 10022.*

      I. (1920) \*Acute Epidemic Encephalitis (Lethargic Encephalitis)
     II. (1921) \*Multiple Sclerosis (Disseminated Sclerosis)
    III. (1923) \*Heredity in Nervous and Mental Disease
    IV. (1924) \*The Human Cerebrospinal Fluid
     V. (1925) \*Schizophrenia (Dementia Praecox)
    VI. (1926) \*The Cerebellum
   VII. (1922) \*Epilepsy and the Convulsive State (Part I)
         (1929) \*Epilepsy and the Convulsive State (Part II)
  VIII. (1927) \*The Intracranial Pressure in Health and Disease
    IX. (1928) \*The Vegetative Nervous System
     X. (1929) \*Schizophrenia (Dementia Praecox) (Communication of Vol. V)
    XI. (1930) \*Manic-Depressive Psychosis
   XII. (1931) \*Infections of the Central Nervous System
  XIII. (1932) \*Localization of Function in the Cerebral Cortex
  XIV. (1933) \*The Biology of the Individual
   XV. (1934) \*Sensation: Its Mechanisms and Disturbances
  XVI. (1935) \*Tumors of the Nervous System
 XVII. (1936) \*The Pituitary Gland
XVIII. (1937) \*The Circulation of the Brain and Spinal Cord
  XIX. (1938) \*The Inter-relationship of Mind and Body
   XX. (1939) \*Hypothalamus and Central Levels of Autonomic Function
  XXI. (1940) \*The Disease of the Basal Ganglia
 XXII. (1941) \*The Role of Nutritional Deficiency in Nervous and Mental Disease
XXIII. (1942) \*Pain
XXIV. (1943) \*Trauma of the Central Nervous System
 XXV. (1944) \*Military Neuropsychiatry
XXVI. (1946) \*Epilepsy
XXVII. (1947) \*The Frontal Lobes
XXVIII. (1948) \*Multiple Sclerosis and the Demyelinating Diseases
XXIX. (1949) \*Life Stress and Bodily Disease
 XXX. (1950) \*Patterns of Organization in the Central Nervous System
XXXI. (1951) Psychiatric Treatment
XXXII. (1952) \*Metabolic and Toxic Diseases of the Nervous System
XXXIII. (1953) \*Genetics and the Inheritance of Integrated Neurological Psychiatric Patterns
XXXIV. (1954) \*Neurology and Psychiatry in Childhood
XXXV. (1955) Neurologic and Psychiatric Aspects of Disorders of Aging
XXXVI. (1956) \*The Brain and Human Behavior
XXXVII. (1957) \*The Effect of Pharmacologic Agents on the Nervous System
XXXVIII. (1958) \*Neuromuscular Disorders
XXXIX. (1959) \*Mental Retardation
   XL. (1960) Ultrastructure and Metabolism of the Nervous System
  XLI. (1961) Cerebrovascular Disease
 XLII. (1962) \*Disorders of Communication
XLIII. (1963) \*Endocrines and the Central Nervous System
XLIV. (1964) Infections of the Nervous System
 XLV. (1965) \*Sleep and Altered States of Consciousness
XLVI. (1968) Addictive States
XLVII. (1969) Social Psychiatry
XLVIII. (1970) Perception and Its Disorders
 XLIX. (1971) Immunological Disorders of the Nervous System
    50. (1972) Neurotransmitters
    51. (1973) Biological and Environmental Determinants of Early Development
    52. (1974) Aggression
    53. (1974) Brain Dysfunction in Metabolic Disorders
    54. (1975) Biology of the Major Psychoses: A Comparative Analysis
    55. (1976) The Basal Ganglia
    56. (1978) The Hypothalamus

# Preface

The hypothalamus has posed for many years a fascinating problem to investigators far out of proportion to its diminutive size. During the first three decades of this century, classical studies of clinical disorders and physiological experiments had shown that this region, weighing scarcely 4 g in adult humans, was associated with a wide variety of behavioral, autonomic, visceral, and endocrine functions. By 1939, knowledge of the hypothalamus had been defined sufficiently to justify a meeting of the Association for Research in Nervous and Mental Disorders. The 1940 volume summarizing this work proved to be a landmark and in turn, inspired many new studies. Many of the questions raised by workers of that era now appear to have been answered in a definitive fashion and entirely new concepts of nervous system function and neurohumoral control have now emerged. It is thus appropriate once again to consider the status of knowledge of the hypothalamus.

This is the first comprehensive publication on the hypothalamus under sponsorship of the Association for Research in Nervous and Mental Disorders since the historically important volume of 1940. Covered are major advances in neuroendocrine, behavioral, and physiological aspects of hypothalamic function over the last four decades. The concepts of neurosecretion and of neurohumoral control of the neurohypophysis put forth speculatively on the basis of nonspecific staining methods by Ernst and Berta Scharrer in 1940 have now been more fully established by the isolation and chemical characterization of the peptides of the neurohypophysis and by use of sophisticated immunohistochemical and electrophysiological methods. The portal vessel-chemotransmitter hypothesis of control of the anterior pituitary proposed by G. W. Harris on the basis of physiological studies has now been established through the identification of at least three hypothalamic releasing factors, TRH, LH RH, and somatostatin. Advances in endocrinology and pharmacology have made it possible to define feedback effects of hormones on the brain, and the actions of neurotransmitters in terms of molecular structure and modern receptor theory. The discovery of other neuropeptides such as the endorphins, enkephalins, substance P, and neurotensin, and their demonstration as secretory products of neurons has led to an explosive burst of understanding of cell communication within the nervous system with implications for neurobiology, behavior, and disease of the central nervous system. These advances have also permitted a new view of several classic systems of hypothalamic-pituitary control including those for growth hormone and gonadotropin regulation. Knowledge of pineal gland function has developed in a similarly explosive way.

In addition to providing a review of these basic aspects of hypothalamic function, this volume summarizes advances in the use of hypothalamic releasing factors in the evaluation of human disorder of the pituitary and places in a clinical context current knowledge of the physiological and pathological functioning of the vegetative and behavioral components of the hypothalamus.

*The Editors*

# Contents

1     Introduction
        *Seymour Reichlin*

15    The Endocrine Hypothalamus: Recent Anatomical Studies
        *Shirley A. Joseph and Karl M. Knigge*

49    Organization of LRF- and SRIF-Neurons in the Endocrine Hypothalamus
        *Karl M. Knigge, Shirley A. Joseph, and Gloria E. Hoffman*

69    Aminergic and Peptidergic Pathways in the Nervous System with Special Reference to the Hypothalamus
        *T. Hökfelt, R. Elde, K. Fuxe, O. Johansson, Hake Ljungdahl, M. Goldstein, R. Luft, S. Efendic, G. Nilsson, L. Terenius, D. Ganten, S. L. Jeffcoate, J. Rehfeld, S. Said, M. Perez de la Mora, L. Possani, R. Tapia, L. Teran, and R. Palacios*

137   The Magnocellular Neurosecretory System of the Mammalian Hypothalamus
        *Richard Defendini and Earl A. Zimmerman*

153   General Discussion

155   Biochemical and Physiological Correlates of Hypothalamic Peptides. The New Endocrinology of the Neuron
        *Roger Guillemin*

195   Biosynthesis, Packaging, Transport, and Release of Brain Peptides
        *Jeffrey F. McKelvy and Jacques Epelbaum*

213   General Discussion

217   Extrahypothalamic and Phylogenetic Distribution of Hypothalamic Peptides
        *Ivor M. D. Jackson*

233   Peptide Neurotransmitter Candidates in the Brain: Focus on Enkephalin, Angiotensin II, and Neurotensin
        *Solomon H. Snyder*

245 Peptide and Steroid Hormones and the Neural Mechanisms for Female Reproductive Behavior
*Donald W. Pfaff*

255 Steroid Hormone Action in the Neuroendocrine System: When is the Genome Involved?
*Bruce S. McEwen, Lewis C. Krey, and Victoria N. Luine*

269 Neurophysiological Organization of the Endocrine Hypothalamus
*Leo P. Renaud*

303 Pineal Gland as a Model of Neuroendocrine Control Systems
*David C. Klein*

329 Neuroendocrine Organization of Growth Hormone Regulation
*Joseph B. Martin, Paul Brazeau, Gloria S. Tannenbaum, John O. Willoughby, Jacques Epelbaum, L. Cass Terry, and Dominique Durand*

359 Hypothalamic Regulation of LH and FSH Secretion in the Rhesus Monkey
*E. Knobil and T. M. Plant*

373 Sleep-Related Endocrine Rhythms
*Robert M. Boyar*

387 Newer Understanding of Human Hypothalamic Pituitary Disease Obtained Through the Use of Synthetic Hypothalamic Hormones
*Lawrence A. Frohman*

415 Nonendocrine Diseases and Disorders of the Hypothalamus
*Fred Plum and Robert Van Uitert*

475 Subject Index

# Contributors

**Akira Arimura**
Department of Medicine
Veterans Administration Hospital
New Orleans, Louisiana 70146

**Robert M. Boyar**
Department of Internal Medicine
University of Texas Health Science
 Center at Dallas
Southwestern Medical School
Dallas, Texas 75235

**Paul Brazeau**
Department of Neurology
Montreal General Hospital Research
 Institute
McGill University
Montreal, Quebec, Canada

**Richard Defendini**
Division of Neuropathology
Department of Pathology
Columbia University College of
 Physicians and Surgeons
New York, New York 10032

**Dominique Durand**
Laboratorie de Physiologie
U.E.R. des Sciences
 Pharmaceutiques
Paris, France

**S. Efendic**
Department of Endocrinology
Karolinska Hospital
Stockholm, Sweden

**R. Elde**
Department of Histology
Karolinska Institute
Stockholm, Sweden

**Jaques Epelbaum**
Unite de Neurobiologie de
 l'INSERM 109
Paris, France

**Lawrence A. Frohman**
Division of Endocrinology and
 Metabolism
Department of Medicine
Michael Reese Medical Center
University of Chicago
Pritzker School of Medicine
Chicago, Illinois 60616

**K. Fuxe**
Department of Histology
Karolinska Institute
Stockholm, Sweden

**D. Ganten**
Department of Pharmacology
University of Heidelberg
Heidelberg, West Germany

**M. Goldstein**
Department of Psychiatry
New York University
School of Medicine
New York, New York 10016

**Roger Guillemin**
Laboratories for Neuroendocrinology
Salk Institute for Biological Studies
La Jolla, California 92037

**Gloria E. Hoffman**
Department of Anatomy
University of Rochester School of
 Medicine and Dentistry
Rochester, New York 14642

**T. Hökfelt**
Department of Histology
Karolinska Institute
Stockholm, Sweden

**Ivor M. D. Jackson**
Endocrine Division
Department of Medicine
New England Medical Center
Tufts University School of Medicine
Boston, Massachusetts 02111

## CONTRIBUTORS

**S. L. Jeffcoate**
National Institute of Biological
 Standards and Control
Holly Hill, Hampstead, London,
 England

**O. Johansson**
Department of Histology
Karolinska Institute
Stockholm, Sweden

**Shirley A. Joseph**
Department of Anatomy
University of Rochester School of
 Medicine and Dentistry
Rochester, New York 14642

**David C. Klein**
National Institutes of Health
Bethesda, Maryland

**Karl M. Knigge**
Department of Anatomy
University of Rochester School of
 Medicine and Dentistry
Rochester, New York 14620

**E. Knobil**
Department of Physiology
University of Pittsburgh
 School of Medicine
Pittsburgh, Pennsylvania 15261

**Lewis C. Krey**
The Rockefeller University
New York, New York 10021

**Åke Ljungdahl**
Department of Histology
Karolinska Institute
Stockholm, Sweden

**R. Luft**
Department of Endocrinology
Karolinska Hospital
Stockholm, Sweden

**Victoria N. Luine**
The Rockefeller University
New York, New York 10021

**Joseph B. Martin**
Departments of Neurology and
 Neurosurgery
Montreal Neurological Institute
McGill University
Montreal, Canada H3A 2B4

**Bruce S. McEwen**
The Rockefeller University
New York, New York 10021

**Jeffrey F. McKelvy**
Department of Biochemistry
University of Texas Health Science
 Center at Dallas
Southwestern Medical School
Dallas, Texas 75235

**G. Nilsson**
Department of Pharmacology
Karolinska Institute
Stockholm, Sweden

**R. Palacios**
Laboratorio de Immunohistocompatibilidad
Centro Hospitalario 20 Noviembre
Tlalpam, Mexico

**M. Perez de la Mora**
Departamento Biologia Experimental
Instituto de Biologia
Universidad Nacional Autonoma de
 Mexico
Mexico 20, D.F.

**Donald W. Pfaff**
The Rockefeller University
New York, New York 10021

**T. M. Plant**
Department of Physiology
University of Pittsburgh
 School of Medicine
Pittsburgh, Pennsylvania 15261

**Fred Plum**
Cerebrovascular Disease Research Center
Department of Neurology
New York Hospital-Cornell Medical
 Center
New York, New York 10021

# CONTRIBUTORS

**L. Possani**
Departamento Bioligia Experimental
Instituto de Biologia
Universidad Nacional Autonoma de
  Mexico
Mexico 20, D.F.

**J. Rehfeld**
Hospital Department of Medical
  Biochemistry
Aarhus University
Aarhus, Denmark

**Seymour Reichlin**
Department of Medicine
Tufts University School of Medicine
and
Endocrinology Division
New England Medical Center
Boston, Massachusetts 02111

**Leo P. Renaud**
Division of Neurology
Montreal General Hospital
Montreal, Quebec, Canada

**S. Said**
Department of Internal Medicine
University of Texas Health Science
  Center at Dallas
Southwestern Medical School
Dallas, Texas 75235

**Solomon H. Snyder**
Departments of Pharmacology and
  Experimental Therapeutics and
  Psychiatry and Behavioral Sciences
Johns Hopkins University
  School of Medicine
Baltimore, Maryland 20205

**Ludwig Sternberger**
Immunology Branch
Department of the Army
Aberdeen Proving Grounds
Ridgewood Arsenal, Maryland 21010

**Gloria S. Tannenbaum**
Department of Neurology
Montreal General Hospital
Research Institute
McGill University
Montreal, Quebec, Canada

**R. Tapia**
Departamento Biologia Experimental
Instituto de Biologia
Universidad Nacional Autonoma de
  Mexico
Mexico 20, D.F.

**L. Teran**
Departamento Biologia Molecular
Instituto Investigaciones
  Biomedicas
Universidad Nacional Autonoma de
  Mexico
Mexico 20, D.F.

**L. Terenius**
Department of Pharmacology and
  Toxicology
Uppsala University
Uppsala, Sweden

**L. Cass Terry**
Department of Neurology
Montreal General Hospital
  Research Institute
McGill University
Montreal, Quebec, Canada

**Robert Van Uitert**
Cerebrovascular Disease Research
  Center
Department of Neurology
New York Hospital-Cornell Medical
  Center
New York, New York 10021

**John O. Willoughby**
Department of Neurology
Montreal General Hospital
  Research Institute
McGill University
Montreal, Quebec, Canada

**Earl A. Zimmerman**
Department of Neurology
Columbia University College of
  Physicians and Surgeons
New York, New York 10032

*The Hypothalamus*, edited by S. Reichlin,
R. J. Baldessarini, and J. B. Martin,
Raven Press, New York, © 1978.

# Introduction

## Seymour Reichlin

The first meeting of the Association concerned with the hypothalamus took place in 1939, and the published volume of those proceedings (16) was a landmark in the development of neuroendocrinology. Today, we are compelled by the avalanche of new knowledge in the area of hypothalamic-pituitary regulation to survey the advances that have occurred since the early work of researchers such as Fulton, Bard, Long, Ranson, and Wislocki, all of whom contributed to the previous conference. For review of early history of hypothalamic research, see Anderson and Haymaker (1) and Meites et al. (39). More recent progress is reviewed in many monographs and articles (1,4,7,19,21,30–39,41,46–49,54–56).

In 1939 the Association honored Dr. Alfred Frölich, who had fled from Vienna before the outbreak of the Second World War. His report of the condition which came to be known as adiposogenital dystrophy or Fröhlich's syndrome was published in 1901 (14) and can also be looked upon as a landmark. This paper was printed in its entirety in the Association Proceedings (16), together with a photograph of his famous hypogonadal boy at the age of 14 (Fig. 1), who suffered from a tumor of the pituitary compressing the optic tract and hypothalamus. Incidentally, the tumor was removed by transsphenoidal hypophysectomy, the only safe procedure available at the time.

It was not an accident that Frölich reported this case; he and Harvey Cushing had together been students of Sir Charles Sherrington. But Frölich at first had no idea of the significance of his finding. He stated (15): "In looking back, it is amazing how little we knew in 1901. We had heard of acromegaly, and of Addison's disease, but at that time even epinephrine was an unknown entity related simply to what we called suprarenal extracts. . . . all we knew at the time was that the hypothalamus was an anatomical region lying beneath the thalamus. That is all we knew. Everything was in darkness like the gray substance, the tuber cinereum." Soon thereafter Erdheim, another in the line of great Viennese students of the hypothalamus, explained by careful pathological study that gonadal atrophy and obesity had arisen from a suprasellar lesion which caused damage to the hypothalamus and not directly to the pituitary. If we take a long leap forward from that time, we can conclude that this form of hypogonadism is due to a deficiency of the hypothalamic hypophysiotropic hormone, LHRH, a decapeptide

---

*Present address:* Endocrinology Division, New England Medical Center Hospital, Department of Medicine, Tufts University School of Medicine, Boston, Massachusetts 02111

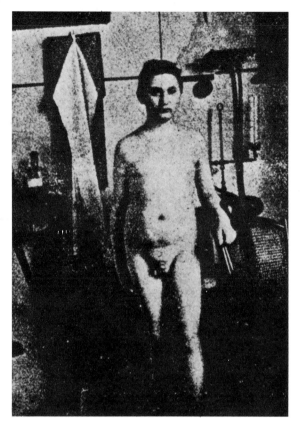

**FIG. 1.** The 14-year-old hypogonadal patient of Fröhlich first reported in 1901 and reprinted in *The Hypothalamus,* Proceedings of the Association for Research in Nervous and Mental Disease, December 20 and 21, 1939, New York.

secretion of specialized gonadotropin regulating neurons of the hypothalamus (see below). One can now, in fact, produce a form of hypogonadism by the use of antibodies directed against this molecule (Fig. 2).

Fröhlich's case is a splendid example of the way careful clinical study has predicted the nature of hypothalamic function and given the lead to anatomists and physiologists. Analysis of patients with tumors, injuries, or inflammations led to the early recognition that the hypothalamus was involved in regulation of visceral function, behavior, metabolism, and endocrine secretions. Well-established clinical syndromes recognized by 1939 include gastric ulceration due to basal brain damage, diencephalic epilepsy with its paroxysmal sympathetic discharge, hyperthermia and hypothermia, somnolence, manic states, anxiety and personality disturbance, sham rage, obesity, hypogonadism, and diabetes insipidus. Physiological and anatomical studies of these phenomena largely dominate discussions of the 1939 meeting as they dominated the literature of that time.

# INTRODUCTION

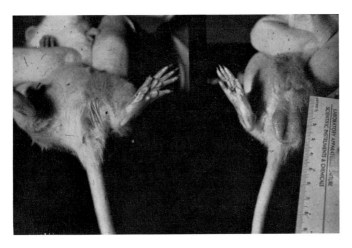

**FIG. 2.** Hypogonadism induced by immunoneutralization of endogenous luteinizing hormone releasing hormone (LHRH) in 2 litter mate male rats. One rat (*left*) was injected with a single dose of a potent antibody to LHRH on day 5 of life. The rat on the right was injected with normal rabbit serum. At 35 days of age, the rat treated with antibody shows virtually no testicular or scrotal development. This experiment illustrates that hypothalamic LHRH is needed to stimulate the pituitary for normal sexual development. Photograph courtesy of Barry Bercu.

In the 1939 meeting relatively little was said of the problem of neural control of the anterior pituitary, which will take up the bulk of the present meeting. Only one paper, that of U. Uotila in the 1939 meeting, dealt directly with the effects of pituitary stalk section. Although this worker noted loss of some pituitary functions in this and other publications (57,58), he did not, as pointed out by Brolin (5), differentiate clearly between neural and vascular components in the response. Dr. Chandler McC. Brooks, who is here with us today, emphasized in his 1939 paper (6) that the mechanism of adenohypophysial regulation was still obscure. This obscurity was to be cleared by the studies of Geoffrey Harris and Ernst Scharrer (2,13), both of whom died relatively young, but at a time that their far-seeing views had already gained widespread acceptance. Harris's contribution was to grasp the significance of the hypophysial-portal blood supply to the innervation-poor anterior pituitary, and to devise elegant and convincing experiments to prove that the vascular link was the essential conduit of information from hypothalamus to pituitary (25,26). Scharrer's contribution was the concept of neurosecretion, the idea that nerve cells could have a secretory function (51–53). Modern neuroendocrinology has married these two ideas.

Ernst Scharrer and his wife, Berta, who is with us today, summarized the first 6 years of their work on neurosecretion at the 1939 meeting in their contribution entitled "Secretory Cells Within the Hypothalamus" (52). As early as 1933, Ernst Scharrer had proposed that the structural peculiarities of the supraoptic and paraventricular nuclear cells were attributable to secretory activity. His reasons for thinking that this might be so include the striking anatomical variations as compared with other brain cells, their multinucleate character, the presence of protein-containing colloid-like vacuoles and granules, and an almost unique (for the central nervous system) intimate relationship between each cell and its surrounding capillary network. The Scharrers suggested that these nerve cells had a glandular function and introduced the term "diencephalic gland," which was resurrected as the "median eminence" gland more than 20 years later (46,47). The Scharrer hypothesis dominates current understanding of the endocrine hypothalamus and has been further refined into the concept of the neurosecretory neuron as the "neuroendocrine transducer" (60), which stands as the link between neural activity on the one hand and endocrine and metabolic regulation on the other (Fig. 3). The model of hypothalamic-pituitary control accepted by most workers today is outlined in Fig. 4.

Early work on neurosecretion (and, in fact, the first definition of this phenomenon) depended on histological methods and nonspecific stains such as the Gomori stain, first applied to pancreatic islet cells. Contemporary work on this system to be fully covered in this volume takes advantage of immunohistochemical methods using antibodies directed against the specific secretions of the neurohypophysis—oxytocin, vasopressin, and neurophysin. This approach has been further expanded to identify the histochemical localization of the hypothalamic releasing hormones and other brain peptides. Such studies have revealed a rich pattern of central peptidergic pathways.

Although the neurohypophysis was recognized to be involved in regulation of water balance, particularly through the pioneering physiological work of Verney (59) and Fisher and collaborators (12), neither of the neurohypophysial peptides had been isolated in 1939 (24). The studies of Van Dyke and collaborators (11) established the existence of the separate antidiuretic and oxytocic activity in neurohypophysial extracts. Finally, duVigneaud and colleagues established the structure of oxytocin in 1950 (43) and of vasopressin in 1954 (44), thus showing that it was possible to identify naturally occurring brain peptides. In 1955 these observations inspired the initiation of the search by Guillemin and collaborators (22) and Schally and collaborators (50) for the chemical structure of the hypothalamic hormones that regulate the secretion of the anterior pituitary gland. The early history of these efforts is reviewed by Saffran (49).

Although it was the first hypothalamic hormone to be identified by bioassay methods, corticotropin releasing factor (CRF) has still not been isolated in pure form; but three other hormones, thyrotropin releasing hormone (TRH), luteinizing hormone releasing hormone (LHRH), and somatostatin (growth hormone release inhibiting factor, SRIF), have been identified and synthesized, in 1969, 1971, and 1973, respectively (Table 1). At least five more such factors recognizable by bioassay await chemical identification (Table 2). The releasing hormones are not the only interesting pep-

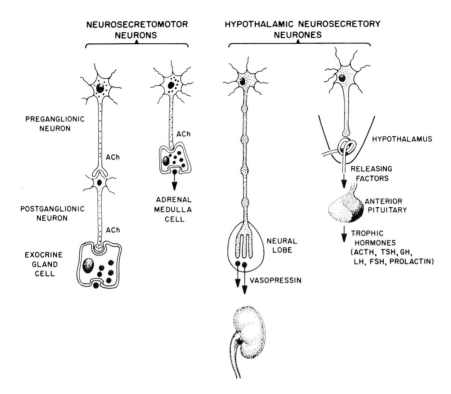

**FIG. 3.** Neural control over exocrine and endocrine gland function is exerted in a number of ways. Exocrine glands are innervated by postganglionic neurons which stimulate secretion through direct action of neurotransmitter on membrane receptors. The adrenal medulla is innervated by preganglionic sympathetic neurons of the nervous system which end on cell receptors. Neurons of the neurohypophysis and tuberohypophysial system are classified as neuroendocrine cells. The secretions are formed in the cell body, transported by axoplasmic flow to the nerve terminus, and released into the general circulation (in the case of vasopressin and oxytocin) and into the hypophysial-portal circulation (in the case of the hypophysiotrophic neurons). Because the hypothalamic neuron secretory product is peptide in nature, these neurons are termed peptidergic. In the terminology of Wurtman and Axelrod, neural cells which convert neural information to metabolic information are called "neuroendocrine transducer cells."

tides found in the hypothalamus. A partial list would include the endogenous morphine-like peptides, endorphins and enkaphalins, which appear to play a role in pain perception; angiotensin II, which is involved in central control of thirst, vasopressin release, and regulation of blood pressure; and other peptides such as substance P and neurotensin (8), whose function is as yet obscure (Table 3). The brain peptides are now viewed as being specialized neurotransmitter secretions of peptidergic neuron networks. By analogy with the supraopticohypophysial systems they are assumed to be formed in the cell bodies of neurons, transported by axoplasmic flow to nerve terminals, and released upon stimulation by propagated action potentials.

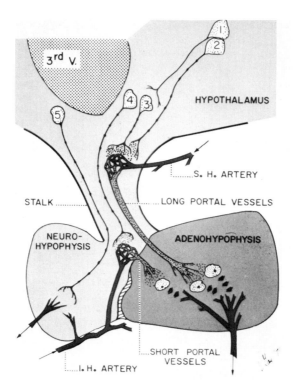

**FIG. 4.** The functional elements of control of the hypothalamic-pituitary unit. Two general types of neurons are involved in anterior pituitary regulation. One type, peptidergic neurons, forms the releasing hormones (shown here as No. 3, ending in the median eminence, and No. 4, ending in the pituitary stalk), in relation to the capillary plexus of the hypophysial-portal vessels. These neurons (corresponding to the designation "tubero-hypophysial" neurons) are neurosecretory, that is, they combine the function of excitable electrical tissue and secretory function and serve as "neuroendocrine transducers" (60) to convert neural information to hormonal information. Their function is analogous to the function of the classic supraoptico-hypophysial neuron (No. 5), the structure responsible for secretion of vasopressin. The second type of neuron is the link between the rest of the brain and the peptidergic neuron. These are largely monoaminergic and are believed to end on the cell body of the peptidergic neuron in a conventional manner (No. 1), or to end on the axon terminus of the peptidergic neuron (No. 2) in a manner termed by Schneider and McCann "axo-axonic." S.H.; superior hypophysial; I.H. inferior hypophysial; 3rd V., third ventricle (From 18a. Legend reprinted with permission from ref. 47a.)

Peptidergic neurons may provide systems of pharmacologically coded signals which produce specific effects by means other than the neurohistology of their efferent connections. For example, LHRH, the hypophysiotropic hormone that regulates the gonadotropic secretion of the pituitary, has been reported to increase sex behavior in female rats maintained on constant suboptimal levels of replacement sex steroids (40,42). Angiotensin II, which stimulates aldosterone secretion by the adrenal and has a direct pressor action on blood vessels, also exerts central effects on hypothalamic

TABLE 1. *Hypothalamic peptides of established structure and function (1976)*

| Name | Structure | Function |
|---|---|---|
| **Neurohypophysial hormones** | | |
| Vasopressin | Cys-Tyr-Phe-Glu-Arg-Cys-Pro-Arg-Gly-NH$_2$ with S—S bridge and NH$_2$ | Stimulates water resorption by the kidney |
| Oxytocin | Cys-Tyr-Ile-Glu-Arg-Cys-Pro-Leu-Gly-NH$_2$ with S—S bridge and NH$_2$ | Stimulates uterine contraction and "milk let-down" |
| **Hypophysiotropic hormones** | | |
| Thyrotropin-releasing hormone (TRH) | pyro-Glu-His-Pro-NH$_2$ | Releases TSH and prolactin[a] |
| Luteinizing hormone releasing hormone (LHRH, gonadotropin-releasing hormone, GnRH) | pyro-Glu-His-Trp-Ser-Tyr-Gly-Leu-Arg-Pro-Gly-NH$_2$ | Releases LH and FSH[b] |
| Somatostatin (growth hormone release inhibiting factor, SRIF) | H-Ala-Gly-Cys-Lys-Asn-Phe-Phe-Trp-Lys-Thr-Phe-Thr-Ser-Cys-OH with S—S bridge | Inhibits GH and TSH release; in extrahypothalamic sites has widespread distribution and and effects[c] |

[a] There is another prolactin stimulating factor in hypothalamic extracts in addition to TRH whose structure has not as yet been identified.
[b] There is overwhelming evidence that there is only one gonadotropic hormone releasing factor, but some believe that a separate FSH-regulating hormone has not been completely excluded.
[c] Somatostatin is found in extrahypothalamic brain and in the gastrointestinal tract (stomach, small and large intestine, pancreatic islet cells). In the GI tract somatostatin inhibits secretion of gastrin, insulin, and glucagon. Although not detectable in kidney, it inhibits renin secretion.

TABLE 2. *Hypophysiotropic hormones of established function but unknown structure (1976)*

| Name | Function |
|---|---|
| Corticotropin releasing factor (CRF) | Releases ACTH[a] |
| Growth hormone releasing factor (GRF, somatotropin releasing factor, SRF)[b] | Releases GH[b] |
| Prolactin releasing factor (PRF)[c] | Releases prolactin[c] |
| Prolactin inhibitory factor (PIF)[d] | Inhibits prolactin release[d] |
| Melanocyte stimulating hormone releasing factor (MSH-RF)[e] | Stimulates MSH release[e] |
| Melanocyte stimulating hormone inhibiting factor (MSH-IF)[e] | Inhibits MSH release[e] |

[a] A number of peptides will stimulate release of ACTH under certain conditions *in vivo* or *in vitro*. These include vasopressin and TRH. Neither of these peptides is believed to be an authentic CRF because of inconsistencies of response in several test systems.

[b] A number of amino acids and peptides will stimulate release of growth hormone under certain conditions *in vivo* or *in vitro*. These include glucagon. MSH, β-endorphin, neurotensin, substance P, TRH, LHRH, and a decapeptide isolated from porcine hypothalamic extracts that is chemically identical with the β chain of porcine hemoglobin. None is now believed to be an authentic GHRF because of inconsistencies of effects in various test systems, and because the GH release system is relatively easily affected by nonspecific factors.

[c] TRH, a peptide of established structure, is a potent prolactin releasing factor, has a role in maintenance of normal prolactin secretion, but is not the most potent releaser of prolactin found in hypothalamic extracts.

[d] Hypothalamic extracts contain two or more factors that inhibit prolactin release; one is dopamine, another GABA, and there may also be peptide(s).

[e] The status of knowledge of hypothalamic peptides that regulate MSH release is now in confusion because β-MSH, the most potent pigment regulating hormone of the pituitary, arises as an artifact from the breakdown of β-lipotropin. β-Lipotropin is an anterior pituitary hormone and may also be formed *in situ* in the hypothalamus. Classic methods for extraction of either hypothalamus or pituitary will cause breakdown of β-lipotropin with formation of β-MSH. In some but not all test systems, the tripeptide amide prolyl-leucyl-glycinamide is reported to inhibit release of bioassayable "MSH" and has been called MIF.

centers that control blood pressure and regulate vasopressin release (17,18, 45).

The newer insights into specialized peptidergic pathways are reminiscent of the recently disclosed biogenic amine neural pathways which largely arise in the mid-brain and are distributed to the hypothalamus, basal ganglia, forebrain, and spinal cord (Fig. 5). These systems, which include "dopaminergic," "serotoninergic," and "noradrenergic" components, disclosed by the brilliant histochemical work of the Swedish school of neuroanatomists (see 28), appear to exert important influences on the secretion of hypophysiotropic hormones on eating, drinking, the central control of body temperature, and displays of arousal and "affect," including rage and sexual behavior.

In the tradition of Herrick, neurologists have often sought understanding of complex functions of higher organisms by a study of the brain of simpler

TABLE 3. *Several hypothalamic peptides whose structure and action are known but whose physiological function is not fully understood*

| Name | Structure |
| --- | --- |
| Substance P[a] | Arg-Pro-Lys-Pro-Gln-Gln-Phe-Phe-Gly-Leu-Met-NH$_2$ |
| Neurotensin[b] | <Glu-Leu-Tyr-Glu-Asn-Lys-Pro-Arg-Arg-Pro-Tyr-Ile-Leu-CoOH |
| Angiotensin II[c] | Asp-Arg-Val-Tyr-Ile-His-Pro-Phe |
| leu-enkephalin[d] | Tyr-Gly-Gly-Phe-Leu-OH |
| meth-enkephalin[d] | Tyr-Gly-Gly-Phe-Met-OH |
| α-Endorphin[d] | Tyr-Gly-Gly-Phe-Met-Thr-Ser-Glx-Lys-Ser-Gln-Thr-Pro-Leu-Val-Thr-OH |
| β-Endorphin[d] | Tyr-Gly-Gly-Phe-Met-Thr-Ser-Glu-Lys-Ser-Gln-Thr-Pro-Leu-Val-Thr-Leu-Phe-Lys-Asn-Ala-Ile-Val-Lys-Asn-Ala-His-Lys-Gly-Gln-OH |
| γ-Endorphin[d] | Tyr-Gly-Gly-Phe-Met-Thr-Ser-Glu-Lys-Ser-Gln-Thr-Pro-Leu-Val-Thr-Leu-OH |
| Vasoactive inhibitory peptide[e] | His-Ser-Asp-Ala-Val-Phe-Thr-Asp-Asn-Tyr-Thr-Arg-Leu-Arg-Lys-Gln-Met-Ala-Val-Lys-Lys-Tyr-Leu-Asn-Ser-Ile-Leu-Asn-NH$_2$ |

[a] Substance P (permeability factor) was originally found in dorsal root ganglia and now has been shown to be distributed in both peripheral and central axon projections of sensory ganglia. This peptide is probably responsible for the wheal reaction in the triple response to skin damage. Its function in brain and sensation is unknown, but it is believed to interact with the endorphins. In large doses it has an intense sialagogic effect.

[b] Neurotensin is widely distributed in brain, and its identification in synaptosomes indicates a role as a neurotransmitter regulation of brain function. It causes a lowering of blood pressure and widespread increased capillary permeability and vasodilation.

[c] In the periphery, angiotensin II, a potent pressor agent, is formed by angiotensin I. Angiotensin II formed *in situ* in the hypothalamus, as in the periphery by converting enzymes, is believed to be involved in central regulation of blood pressure, drinking behavior, and ADH secretion.

[d] The endorphins and enkephalins make up a class of compounds that react with endogenous opiate receptors in various brain sites, including hypothalamus, amygdala dorsal root entry zone, and areas in ascending pain pathways. Their special function in the hypothalamus (as distinct from a role in pain perception) is unknown.

[e] VIP is found in the small intestine and in pancreatic islet cell tumors and more recently in hypothalamic extracts and in axonal endings in brain neurons. Although this peptide has many actions (produces hyperglycemia, acts as a hypotensive, stimulates water excretion by gut), its physiological role in either the brain or the extraneural tissues is unknown.

forms. It is of great interest, therefore, to note that the body of the sea anemone, a coelenterate, the simplest and most primitive class of animals with a nervous system, contains dopamine, serotonin, and norepinephrine, which appear by histochemical techniques to be contained in nerve cells. These substances are also present in the flatworm, a somewhat more advanced form. In higher forms, such as mammals, these primitive amines seem to have taken on new and sophisticated functions.

The peptidergic pathways also appear early in phylogenetic history. At least one of the hypothalamic peptides, TRH, has been identified in the brain of petromyzon which has no TSH, in amphioxus which has no pituitary, and in the invertebrate snail ganglia. Here is another example of a function (control of pituitary TSH and prolactin secretion) that has "co-opted" a

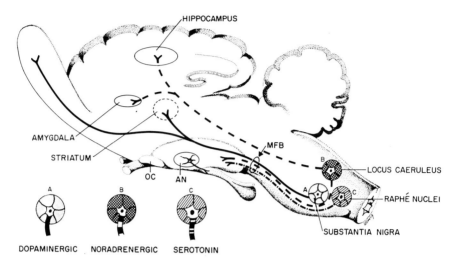

**FIG. 5.** Monoaminergic pathways in mammalian brain. The principal localization of the neurons containing norepinephrine, dopamine, and serotonin is in the mesencephalon and pons. Axons of these cells are distributed to widespread areas of cortex, limbic system, and striatum. The dopaminergic system of the arcuate is an exception to this general scheme of distribution. OC: optic chiasma; AN: arcuate nucleus; MFB: medial forebrain bundle. Reprinted with permission from ref. 31.

primitive molecule for its purposes (29,48). The primitive origins of the releasing factors and the unexpected finding that they are distributed outside of the hypothalamus may give us some clues to their evolutionary significance and their broader biological importance.

The technique of single-neuron recording of nerve cell activity in the hypothalamus has also been developed since the 1939 meeting. A number of workers, including John Green and Barry Cross (10), pioneered the study of the function of the supraoptic system with its relatively large cells (magnocellular neurons) in relation to physiologic stimuli. Investigation of the parvicellular (tuberohypophysial) system has been much more difficult, but several workers have finally been able to analyze its electrical activity; the electrophysiological organization of the hypothalamus will be reviewed in this volume.

Physiologically active peptides arising within the hypothalamus and other parts of the brain exert some of their effects on nerve cells (3). The mechanism of action of peptides on cells in general has been studied in extraordinary depth for the past two decades by pharmacologists and endocrinologists, and for the most part these mechanisms involve initial binding of the hormone to a cell membrane component (23,54). These insights can now be applied to the study of peptide hormone action in nerve cells.

As pointed out by Phillip Bard at the 1939 Hypothalamus meeting, patterned sexual function was dependent on the presence of sex steroids that induce the appearance of male and female mating behavior. To produce these effects steroids must act on target cells within the brain. Richard Michael, who is at this meeting as a commissioner, working together with Geoffrey Harris, was one of the early workers who showed that localized implants of estrogens into the brain could induce the same behavioral effects as systemic hormone therapy, thus indicating the existence of a restricted locus of steroid-sensitive neurons (27). These workers later showed that estrogenic hormones, marked with a radioactive tracer, also localized to certain hypothalamic neurons. The elucidation of steroid receptors in the brain, their anatomical localization, the nature of the binding to cells, and the subsequent mechanism of activation of neurone action can now be defined, as will be discussed in this volume using models of steroid hormone action adapted from peripheral systems such as the rat uterus or chick oviduct (9,20,41).

Although the pineal gland is not a part of the hypothalamus proper, it is being considered in this volume. In many species, the pineal gland interacts with hypothalamic elements to influence endocrine function. The circadian rhythms of the pineal in the rat appear to be regulated by the function of the suprachiasmatic nucleus of the hypothalamus, in turn controlled by the inferior accessory optic tract. In higher animals, the pineal gland is a much simpler model of neuroendocrine control than is the hypothalamus; the neurobiology of the pineal has been developed to a more advanced state than has the neurobiology of the hypothalamus. For these reasons, the student of the pineal has much to teach the student of the hypothalamus.

Because the manifestations of testicular and ovarian insufficiency are readily apparent clinically, hypogonadism as in Fröhlich's case was the first endocrine disease of hypothalamic origin to be recognized. Today, the development of radioimmunoassays has made it possible to study the moment-to-moment levels in the blood of all of the pituitary hormones and of the secretions of their target glands. These techniques have made it possible to apply classic neurophysiological methods such as localized lesions and electrical stimulation to disclose the anatomical pathways involved in pituitary regulation, and the factors that regulate its secretions. Extensive studies of each of the recognized tropic hormones of the anterior pituitary have been published (see 31 for review). Two model systems of anterior pituitary control have been selected for presentation in this volume. One involves a target gland with feedback loops, and the other a pituitary regulatory system that has no specific target gland. The regulation of sex hormone secretion in primates and the regulation of GH secretion, respectively, are the outstanding examples of this kind of control.

In the long-standing tradition of the ARNMD, the clinical relevance of basic research will be pointed out in this volume. The secretion of most of the pituitary hormones in man are regulated by intrinsic brain rhythms, which have importance for normal function and can also be used to inform the clinician when the system is abnormal. Several of the hypothalamic hormones have now been synthesized and introduced into clinical medicine. The use of these hormones has reinforced some concepts about the nature of hypothalamic-pituitary control in man, and the recognition of several syndromes of hypothalamic-pituitary disease.

Finally, the vegetative and behavioral manifestations of hypothalamic disease in man will be reviewed in the light of recent work on the organization of the hypothalamus for visceral and behavioral control. It is well to remember that this aspect of hypothalamic function relates to the most important and basic aspects of the physiological organization for survival.

To the audience, I once more say welcome, and express the conviction that you will find the following presentations intellectually thrilling.

## REFERENCES

1. Anderson, E., and Haymaker, W. (1974): Breakthroughs in hypothalamic and pituitary research. *Prog. Brain Res.*, 41:1-60.
2. Bajusz, E. (1965): Ernst A. Scharrer, editorial. *Neuroendocrinology*, 1:65-67.
3. Barker, J. L. (1976): Peptides: Roles in neuronal excitability. *Physiol. Rev.*, 56:435-452.
4. Blackwell, R. E., and Guillemin, R. (1973): Hypothalamic control of adenohypophysial secretions. *Annu. Rev. Physiol.*, 35:357-90.
5. Brolin, S. E. (1947): The importance of the stalk connection for the power of the anterior pituitary of the rat to react structurally upon ceasing thyroid function. *Acta Physiol. Scand.*, 14:233-244.
6. Brooks, C. McC. (1940): Relation of the hypothalamus to gonadotropic functions of the hypophysis. *Res. Publ. Assoc. Nerv. Ment. Dis.*, 20:525-550.
7. Brown, G. M., and Martin, J. B. (1973): Neuroendocrine relationships. In: *Progress in Neurology and Psychiatry*, edited by E. Spiegel, pp. 193-240. Grune & Stratton, New York.
8. Carraway, R., and Leeman, S. E. (1976): Characterization of radioimmunoassayable neurotensin in the rat. *J. Biol. Chem.*, 251:7045-7052.
9. Chan, L., and O'Malley, B. W. (1976): Mechanism of action of the sex steroid hormones. *N. Engl. J. Med.*, 294:1322-1328, 1372-1381, 1430-1437.
10. Cross, B. A., and Green, J. D. (1959): Activity of single neurones in the hypothalamus: Effect of osmotic and other stimuli. *J. Physiol. (Lond.)*, 148:554-569.
11. Van Dyke, H. B., Chow, B. F., Greep, R. O., and Rothen, A. (1941): The isolation of a protein from the pars neuralis of the ox pituitary with constant oxytocic, pressor and diuresis-inhibiting effects. *J. Pharmacol. Exp. Ther.*, 74:190-209.
12. Fisher, C., Ingram, W. R., and Ranson, S. W. (1935): Relation of hypothalamico-hypophyseal system to diabetes insipidus. *Arch. Neurol. Psychiatry*, 34:124-163.
13. Fortier, Claude (1972): Geoffrey Wingfield Harris, obituary. *Endocrinology*, 90:851-854.
14. Fröhlich, A. (1901): Ein Fall von Tumor der Hypophysis ceribri ohne Akromegalie. *Wien. Klm. Rundsch.*, 15:883-886, 906-908.
15. Fröhlich, A. (1940): Discussion of paper by Bailey, P., Tumors involving the hypothalamus and their clinical manifestations. In: *The Hypothalamus, Vol. XX*, publication of the ARNMD, p. 723. Hafner Publishing Co., New York.
16. Fulton, J. F. (Ed.) (1949): *Proceedings of the Association for Research in Nervous and Mental Disease*. Hafner Publishing Co., New York.
17. Ganten, D., Hutchinson, J. S., Schelling, P., Ganten, U., and Fischer, H. (1976): The iso-renin angiotensin systems in extrarenal tissue. *Clin. Exp. Pharmacol. Physiol.*, 3:103-126.

18. Ganten, D., Schelling, P., Vecsei, P., and Ganten, U. (1976): Iso-renin of extrarenal origin. *Am. J. Med.*, 60:760–772.
18a. Gay, V. (1972): *Fertile Steril.*, 26:51.
19. Gipsen, W. H., Greidanus, T. van W., Bohus, B., and deWied, D. (Eds.) (1975): Hormones, homeostasis and the brain. *Prog. Brain Res.*, 42:1–381.
20. Gorski, J., and Gannon, F. (1976): Current models of steroid hormone action: A critique. *Annu. Rev. Physiol.*, 38:425–450.
21. Greep, R. O. (1974): History of research on anterior hypophysial hormones. In: *Handbook of Physiology, Vol. IV, Part 2,* edited by R. O. Greep and E. B. Astwood, pp. 1–27. American Physiological Society, Washington, D.C.
22. Guillemin, R., and Rosenberg B. (1955): Humoral hypothalamic control of anterior pituitary: A study with combined tissue cultures. *Endocrinology*, 57:599–607.
23. Hardman, J. G. (1974): Cyclic nucleotides and hormone action. In: *Textbook of Endocrinology,* edited by R. H. Williams. W. B. Saunders Co., Philadelphia.
24. Hare, K. (1940): Water metabolism: Neurogenic factors. *Res. Publ. Assoc. Res. Nerv. Ment. Dis.*, 20:416–435.
25. Harris, G. W. (1955): *Neural Control of the Pituitary Gland.* Edward Arnold, London.
26. Harris, G. W. (1972): Humours and hormones: The Sir Henry Dale Lecture for 1971. *J. Endocrinol.*, 53:1–xxiii.
27. Harris, G. W., and Michael, R. P. (1958): Hypothalamic mechanisms and the control of sexual behaviour in the female cat. *J. Physiol.*, 142:26P (Abst).
28. Hökfelt, T., Fuxe, K., Goldstein, M., Johansson, O., Fraser, H., and Jeffcoate, S. (1975): Immunofluorescence mapping of central monoamines and releasing hormone (LHR) systems. In: *Anatomical Neuroendocrinology,* edited by W. Stumpf and L. Grant. S. Karger, Basel.
29. Jackson, I. M. D. (1977): Extrahypothalamic distribution of TRF, LRF and somatostatin and their function. In: *The Proceedings, Vth International Congress of Endocrinology, 1976* edited by V. H. T. James, pp. 62–66. Excerpta Media, Amsterdam.
30. Locke, W., and Schally, A. V. (Eds.) (1972): *The Hypothalamus and Pituitary in Health and Disease.* Charles C. Thomas, Springfield, Ill.
31. Martin, J. B., Reichlin, S., and Brown, G. M. (1977): *Clinical Neuroendocrinology.* F. A. Davis Co., Philadelphia.
32. Martini, L., and Ganong, W. F. (Eds.) (1966): *Neuroendocrinology, Vol. 1.* Academic Press, New York.
33. Martini, L., and Ganong, W. F. (Eds.) (1967): *Neuroendocrinology, Vol. 11.* Academic Press, New York.
34. Martini, L., and Ganong, W. F. (Eds.) (1969): *Frontiers in Neuroendocrinology, Vol. 1,* Oxford University Press, London.
35. Martini, L., and Ganong, W. F. (Eds.) (1971): *Frontiers in Neuroendocrinology, Vol. 2,* Oxford University Press, London.
36. Martini, L., and Ganong, W. F. (Eds.) (1973): *Frontiers in Neuroendocrinology, Vol. 3.* Oxford University Press, London.
37. Martini, L., and Ganong, W. F. (1976): *Frontiers in Neuroendocrinology, Vol. 4.* Raven Press, New York.
38. McCann, S. M., and Porter, J. C. (1969): Hypothalamic pituitary stimulating and inhibiting hormones. *Physiol. Rev.*, 49:240–284.
39. Meites, J., Donovan, B. T., and McCann, S. M. (Eds.) (1975): *Pioneers in Neuroendocrinology.* Plenum Press, New York.
40. Moss, R. L., and McCann, S. M. (1973): Induction of mating behaviour in rats by luteinizing hormone-releasing factor. *Science,* 181:177–179.
41. Naftolin, F., Ryan, K. J., and Davies, I. J. (Eds.) (1976): *Subcellular Mechanisms in Reproductive Neuroendocrinology.* Elsevier, Amsterdam.
42. Pfaff, D. W. (1973): Luteinizing hormone-releasing factor potentiates lordosis behaviour in hypophysectomized ovariectomized female rats. *Science,* 181:1148–1149.
43. Pierce, J. G., and duVigneaud, V. (1950): Preliminary studies on amino acid content of a high potency preparation of the oxytocic hormone of posterior lobe of the pituitary gland. *J. Biol. Chem.*, 182:359–366.
44. Popenoe, N. A., and duVigneaud, V. (1954): A partial sequence of amino acids in performic acid-oxidized vasopressin. *J. Biol. Chem.*, 206:353–360.

45. Ramsay, D. J., and Ganong, W. F. (1977): CNS regulation of salt and water intake. *Hosp. Pract.*, 12:63-69.
46. Reichlin, S. (1963): Medical progress, neuroendocrinology. *N. Engl. J. Med.*, 269:1182-1191, 1246-1250, 1296-1303.
47. Reichlin, S. (1974): Neuroendocrinology. In: *Textbook of Endocrinology, ed. 5*, edited by R. H. Williams, pp. 774-831. W. B. Saunders Co., Philadelphia.
47a. Reichlin, S. (1975): The control of anterior pituitary secretion. In: *Textbook of Medicine*, edited by P. B. Beeson and W. McDermott, p. 1671, W. B. Saunders Co., Philadelphia.
48. Reichlin, S., Saperstein, R., Jackson, I. M. D., Boyd, A. E., III, and Patel, Y. (1976): Hypothalamic hormones. *Annu. Rev. Physiol.*, 38:389-424.
49. Saffran, M. (1974): Chemistry of hypothalamic hypophysiotropic factors. In: *Handbook of Physiology, Vol. IV, Part 2*, edited by R. O. Greep and E. B. Astwood, pp. 563-586. American Physiological Society, Washington, D.C.
50. Schally, A. V., Arimura, A., and Kastin, A. J. (1973): Hypothalamic regulatory hormones. *Science*, 179:341-350.
51. Scharrer, E. (1966): Principles of neuroendocrine integration. Endocrines and the central nervous system. Assoc. Res. Nerv. Ment. Dis., 43:1-35.
52. Scharrer, E., and Scharrer, B. (1940): Secretory cells within the hypothalamus. In: *The Hypothalamus, Vol. XX*, Publication of the ARNMD, pp. 170-174. Hafner Publishing Co., New York.
53. Scharrer, E., and Scharrer, B. (1954): Hormones produced by neurosecretory cells. *Recent Prog. Horm. Res.*, 10:183-232.
54. Snyder, S. H., and Bennett, J. P., Jr. (1976): Neurotransmitter receptors in the brain: Biochemical identification. *Annu. Rev. Physiol.*, 38:153-175.
55. Swaab, D. F., and Schade, J. P. (Eds.) (1974): Integrative hypothalamic activity. *Prog. Brain Res.*, 41:1-516.
56. Szentogothai, J., Flerko, B., Mess, B., and Halasz, B. (1968): *Hypothalamic Control of the Anterior Pituitary*. Academiai Kiado, Budapest.
57. Uotila, U. U. (1939): On the role of the pituitary stalk in regulation of the anterior pituitary, with special reference to the thyrotropic hormone. *Endocrinology*, 25:605-614.
58. Uotila, U. U. (1940): The regulation of thyrotropic function by thyroxine after pituitary stalk section. *Endocrinology*, 26:129-135.
59. Verney, E. B. (1947): The antidiuretic hormone and the factors which determine its release. *Proc. R. Soc. Lond.* [Biol.], 135:25-106.
60. Wurtman, R. J. (1970): Neuroendocrine transducer cells in mammals. In: *The Neurosciences*, edited by F. O. Schmitt, pp. 530-538. The Rockefeller University Press, New York.

*The Hypothalamus*, edited by S. Reichlin,
R. J. Baldessarini, and J. B. Martin.
Raven Press, New York, © 1978.

# The Endocrine Hypothalamus: Recent Anatomical Studies

## *Shirley A. Joseph and Karl M. Knigge

Descriptive and experimental neuroanatomy of the central nervous system, as mirrored by the contributions in the 1939 symposium on the hypothalamus, depended upon a relatively small number of histological techniques. The Weigert procedure, with various modifications, served to map the major myelinated systems of the brain; experimental lesion studies depended largely upon the Marchi technique. An extensive catalog of methods for staining neurofibrils (Cajal and Bielschowsky) and cytoplasmic constituents formed the basis for detailed description of soma and processes. Earlier, the Golgi procedure had produced a monumental amount of information but was not in general use and, in fact, has not been used in studies of hypothalamic cytoarchitecture until relatively recent times. A Golgi study of neurons of the medio-basal hypothalamus by Szentagothai in 1962 (188) was considered an application of an "old-fashioned" technique.

The intervening years have witnessed the continued search and development of an impressive array of neuroanatomical techniques. The advent of the electron microscope signaled a new era in research, with recent advances including high-voltage and scanning microscopy, electron probe techniques, and membrane cleavage by freeze-fracture (20). At the light microscope level, the Falck-Hillarp fluorescence method (45) quickly unfolded a remarkable blueprint of the detailed anatomy of the monoaminergic systems of the brain. Introduction of controllable suppressive silver methods by Nauta and Gygax (132) has renewed experimental lesion studies. Knowledge of the fundamental property of axoplasmic flow has led to the development of tracing methods based upon anterograde movement of incorporated radiolabeled amino acids (58,100,112,209) and retrograde movement of substances such as horseradish peroxidase (101, 102) taken up by endocytotic activity in neuron terminals. Recognition of specific uptake (and reuptake) mechanisms form the basis of autoradiographic localization of radiolabeled transmitters (177), amino acids such as glycine, glutamate, and GABA (75), and hormones such as estrogen and corticosterone (118). Advances in immunology have made possible

---

*Department of Anatomy, University of Rochester School of Medicine and Dentistry, 601 Elmwood Avenue, Rochester, New York 14620

*15*

specific immunochemical procedures for localization of releasing hormones (7), neuropeptides of the supraopticohypophysial system (216), and enzymes involved in the biosynthesis of catecholamines (56). Classical neuroanatomical procedures have not, however, disappeared from the scene. The 100-year-old "reazione nera" of Golgi is experiencing a renaissance in neurobiological research (162). The application of computer technology (196) to Golgi-impregnated neurons provides enormous capability for analysis of neuronal geometry and circuit modeling. The development of EM procedures on Golgi-impregnated material (16,17) offers the bridge between analytical procedures at the light microscopic level and fine structural correlates at the EM level.

Many of these procedures are being applied to analysis of structure and function of the endocrine hypothalamus; some are yet to be applied. The following sections of this chapter present highlights of recent advances in anatomy of the hypothalamus with specific reference to those which appear to be related to neural control of the anterior pituitary gland.

## RETINOHYPOTHALAMIC PROJECTIONS AND THE SUPRACHIASMATIC NUCLEUS

In view of the commanding role of light in the neuroendocrine processes associated with the pineal and hypothalamic regulation of gonadal function, the question of retinohypothalamic connections continues to be of paramount interest and importance. Controversy regarding the existence of a retinohypothalamic projection based on degenerating nerve studies is reviewed by Haymaker et al. (132a), Hendrickson et al. (65), and Moore and Lenn (127). The more recent morphological evidence is quite convincing in support of a direct projection employing a variety of techniques which include the Nauta and Fink-Hemer methods for degenerating axons and terminals, 3-H amino acid autoradiography sulfide precipitation technique (at the light and EM level), and methods utilizing injections of horseradish peroxidase. Terminations of optic fibers have been demonstrated in the suprachiasmatic nucleus SCH by radioautography after intraocular injections of $^3$H-labeled amino acid in the opposum (125), guinea pig, rabbit and rat (65), ferret (190), cat (65,125), and in tree-shrew, hedgehog, galago, marmoset, and macaque (125). After enucleation or optic nerve section, degenerating terminals were seen in the region of the suprachiasmatic nucleus in lemon shark (60), fish (195), snake (134), duck (18,19), bird (63,119), duckbill platypus (29), hamster (146), rat (127,180), and in cat (65). The elegant studies of Mason and Lincoln (117), utilizing the method of cobalt sulfide precipitation combined with Timm's sulfide-silver method for intensification of heavy metals, demonstrated fiber projections from the optic chiasm into the posterior fifth of the suprachiasmatic as well as into the rostral part of the arcuate nucleus in the rat. At the EM level, using basically the same technique, Wenisch (210) observed a projection to the SCH only. There is now electrophysiological evidence to support these morphological findings of the neural pathway mediating visual input to the SCH. Sawaki (167) observed that photic stimulation or electrical stimuli when applied to the optic nerve in rat evoked a change in the mean firing rate in some suprachiasmatic neurons.

The hypothalamus is clearly involved in the circadian activity of many biological rhythms. Examples of some that have been studied extensively include pituitary ACTH and adrenal corticosterone secretion, pituitary prolactin, and pineal N-acetyltransferase, and melatonin.

The SCH has been implicated as a "biological clock" and neural pacemaker which regulates certain of these circadian oscillations; the cue driving its activity is environmental lighting. Bilateral destruction of this nucleus permanently disrupts the circadian rhythm in adrenocorticosterone secretion (126), drinking behavior and locomotor activity (181), female estrous cyclicity (182), sleep-wakefulness (74), and diurnal pineal N-acetyltransferase activity (128). With respect to pineal activity, the electrophysiological studies of Nishina et al. (133) reveal that neuronal activity of the SCH is augmented by visual stimulation and this increased activity, in turn, inhibits the sympathetic neurons which innervate the pineal gland. Additional studies which demonstrate the involvement of the SCH in neural control of the reproductive axis include the observations of Russak and Morin (156) that lesions in this nucleus in the golden hamster prevented testicular regression normally observed in short photoperiods. Immunocytochemical studies demonstrating that suprachiasmatic neurons contain neurophysin (217) and vasopressin (194) suggests that this may be one input into the circadian control of adrenocortical activity. In addition to the visual input to the SCH, serotonergic input from raphe nuclei is undoubtedly of considerable importance in modulating the efferent activity of this nucleus. Serotonin (5-hydroxytryptamine) (5-HT) levels and raphe nuclei are considered by some to be related to sleep-wakefulness cyclicity (80,81,114,120,147). In a recent monograph by Swanson and Cowan (187), efferents of the SCH have been traced by autoradiography to the ventromedial, dorsomedial, and arcuate nuclei of the hypothalamus. Although these intrinsic connections need further delineation, their role in neuroendocrine functions and biological rhythms may ultimately prove to be of great importance.

## THE PREOPTIC AND SEPTAL AREAS

The preoptic region has attracted much attention because of its role in the neural control of gonadotropin secretion (165). Extrahypothalamic as well as intrinsic connections of this region are being studied extensively. Anatomical and functional studies on one of its major afferents—the amygdalofugal fibers of the stria terminalis—have been presented by Raisman and Field (148,149), Field (46), Brown-Grant and Raisman (22), and Sawyer (164). Geometry of preoptic neurons with respect to striatal connections has been described carefully by Field and Sherlock (47) (Fig. 1). A unique finding in the analysis of Raisman and Field is the observation that there are twice as many synapses of nonstriatal origin on preoptic neurons in the female as in the male brain and that this sexual dimorphism

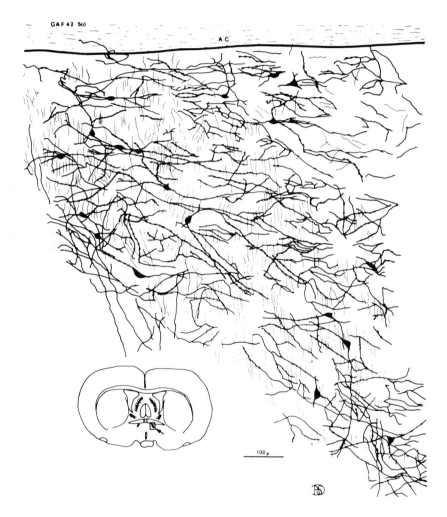

**FIG. 1.** A camera lucida drawing of all the neurons, axons, and dendrites seen in a single 100$\mu$ thick Golgi section taken in a coronal plane through the strial part of the preoptic area. Fine vertical lines indicate strial axons. The predominantly horizontal orientation of the dendrites (many of which are beaded) is clearly seen. *A.C.:* anterior commissure. Scale bar = 100$\mu$. Inset shows the position of the illustrated area (*arrow* and *rectangle*) on a drawing of a coronal section through the entire brain. From (47).

depends upon exposure to androgen during the first 10 postnatal days. Swanson (186) injected small amounts of $^3$H amino acid into the preoptic region and traced fiber connections which not only constitute the tuberoinfundibular tract but also appear to end in the periventricular nucleus and within the ventromedial and arcuate nuclei. Terminals within the mammillary complex (premammillary and supramammillary regions) were also demonstrable. Extrahypothalamic afferent projections were diffuse throughout many regions of the limbic system. Antidromic stimulation (44) shows that preoptic and anterior hypothalamic neurons project directly to the ventromedial and arcuate nuclei. By detailed autoradiographic analysis, Conrad and Pfaff (37) compared the efferent projections of the preoptic area with those from the anterior hypothalamic area and showed that the medial preoptic area and anterior hypothalamic neurons have different axonal trajectories throughout the hypothalamus; descending medial preoptic axons course in the medial portion of the medial forebrain bundle while anterior hypothalamic axons run in a bundle ventromedial to the fornix. Within the hypothalamus, the preoptic distribution of radiolabeled fibers was seen in the dorsomedial nucleus, ventrolateral portion of the ventromedial nucleus, ventral premammillary and supramammillary nuclei, and the dorsal regions of the septum. Fibers from the anterior hypothalamic area projected to the dorsomedial nucleus, capsule of the ventromedial nucleus, dorsal premammillary and supramammillary nuclei, and to the midlateral region of the septum. From the furthermost medial and periventricular zones in both the preoptic and anterior hypothalamic areas, axons arise that traverse the periventricular regions of the hypothalamus and project heavily to the arcuate nucleus.

## THE VENTROMEDIAL NUCLEUS

The size and complexity of the ventromedial nucleus (VMH) lends itself to extensive analysis. This area has been implicated in a variety of autonomic, neuroendocrine, and behavioral mechanisms (64) including the area of appetite control (183).

Studies over the past 10 years have expanded the role of the VMH to include regulation of certain hormones including growth hormone (9,10, 48,115,116), insulin (48,49,59,62,185), glucagon (49,50), thyrotropin (24), corticotropin (188), and gonadotropins (46,129,139,184).

In coronal sections, the ventromedial appears ovoid and consists of medium-dark staining cells. On the basis of cell density, this nucleus can be divided into two masses, the dorsomedial VMH, and the ventromedial VMH, separated by a cell-poor center. Unusual intranuclear and intracellular inclusions were seen by ultrastructural study in ventromedial cells by Taikeichi and Noda (189) and also, unlike most other hypothalamic neurons, ventromedial cell bodies are covered by a glial-type sheath. These workers also noted that the VMH contain both axosomatic and axodendritic type synapses and according to the size and type vesicles they contain, could be classified into two distinct types.

In the Golgi studies by Millhouse (123,124) the VMH dendrites were demonstrated to be long and generally unramified and to bear a number of spinous processes that extend beyond the confines of the nucleus into adjacent hypothalamic regions. He described three general axonal patterns. One type is characterized by having numerous collaterals that ramify extensively in VMH, another variety has few collaterals, and the third is devoid of collaterals. In this study as well as in others (33,40), a capsule can be demonstrated encircling the VMN. This capsule of axons, axon terminals, and dendrites is formed primarily of stria terminalis precommissural fibers. Other areas which contribute to the formation of this capsule are mammillary peduncle, medial forebrain bundle, and lateral hypothalamic fibers.

Efferent projections from the VMH are very extensive. With the Fink-Heimer method, Arees and Mayer (3) have shown degenerated axons extending from the VMH in a dorsal lateral and slightly rostral direction and terminating in the medial region of the lateral hypothalamus. Descending fibers from the VMH pass caudally in the periventricular system (123,124,188). The projection from the VMH into the medial forebrain bundle has also been reported (123,124). By autoradiography, Saper et al. (163) have shown additional connections with septum, medial preoptic area, amygdala, and with the central tegmental fields via the ventral supraoptic commissure. They also demonstrated short interconnections within the hypothalamus to the mammillary complex and anterior hypothalamic area.

## HYPOTHALAMIC MONOAMINES

The distribution of monoamine neurons in the central nervous system has been exhaustively examined with the Falck-Hillarp fluorescence technique (53,54,70,76,77,191). Several recent studies related to hypothalamic monoamine distribution will be noted.

The tuberoinfundibular dopamine neurons have generally been considered the only system originating in the diencephalon. With the development of a more sensitive fluorescence method employing glyoxylic acid (13,111), a second dopamine system has been identified by Bjorklund et al. (15). Cell bodies are located in the parafascicular thalamic nucleus and medial zona incerta (groups A11 and A13) and project to dorsal and anterior hypothalamic areas.

Using the glyoxylic acid fluorescence method, Sladek et al. (178,179) have demonstrated the presence of a significant population of bipolar cells in the subependymal layer along the third ventricle of the diencephalon of the rat. The perikarya of these cells are 10 $\mu$m in diameter, with one fluorescent process oriented toward the ventricle and another projecting into the adjacent neuropil of the periventricular stratum. Microspectrofluorometric analysis indicates the presence of dopamine in these cells.

One of the inherent problems of the formaldehyde as well as the glyoxylic acid-induced fluorescence method is the difficulty of precise identification of specific catecholamines, even with advanced instrumentation for spectral analysis. An approach with great potential is the immunofluorescence method introduced by Hokfelt et al. (56,72) using antibodies generated against the catecholamine-synthesizing enzymes (see Hokfelt et al., *this volume*); norepinephrine-containing neurons should be distinguished from dopaminergic cells by the presence of dopamine-$\beta$-hydroxylase. In general, initial applications of this method have complemented fluorescence with regard to dopamine and norepinephrine distribution in the hypothalamus. Cell bodies have been found in the medial preoptic and premammillary areas which exhibit no amine fluorescence but are immunoreactive when stained for dopadecarboxylase. An immediate reward of studies using antisera directed against the epinephrine-forming enzyme phenylethanolamine-$N$-methyl transferase (PNMT) has been the identification of epinephrine-containing neurons in the medulla and initial surveys of their ascending and descending distribution (71). In the hypothalamus, arcuate and paraventricular nuclei and periventricular and perifornical regions appear to be innervated by PNMT-containing neurons. It is surprising that PNMT immunofluorescence has not been reported for median eminence; this enzyme is present here in the highest concentration of any hypothalamic area examined biochemically by Saavedra et al. (160).

Kizer et al. (85) placed electrolytic lesions in the nigral A8–9–10 dopaminergic nuclei and noted a 40% decrease in dopamine content of the median eminence suggesting that these nuclei contribute dopaminergic projections to the median eminence.

The distribution of serotonin in the hypothalamus is less clearly detailed. The location of cell bodies in raphe nuclei of the brainstem and ventromedial reticular formation of the pons and mesencephalon and their ascending routes of projection have been adequately visualized by Falck-Hillarp fluorescence (55) (Fig. 2) but specific terminal distribution in the hypothalamus has been more difficult to identify. With fluorescence methods, the SCH is seen as a major serotonergic terminus. After various pharmacological treatments, lesser terminal distribution is seen in the septal and preoptic areas. This fluorescence map of serotonin distribution is at some variance with that presented by enzymatic isotopic assay of individual hypothalamic nuclei (23,158). The arcuate nucleus, for example, exhibits virtually no 5-HT fluorescence but contains more serotonin by chemical assay than does the SCH.

One of the most intriguing distributions of serotonin in the hypothalamus (and other areas of brain) is that which occurs to the cerebrospinal fluid.

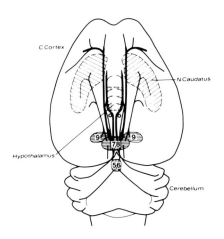

**FIG. 2.** Schematic illustration of subdivision of the ascending 5-HT pathways. A medial ascending 5-HT pathway exists which primarily innervates the hypothalamus and the preoptic area. This pathway originates from the mesencephalic raphe (B7, B8) and pontine raphe region (B5, B6). A lateral ascending 5-HT pathway also exists, which primarily innervates the cortical areas. This pathway originates mainly from the mesencephalic raphe cell groups (B7-B8). Also, a minor, far-lateral 5-HT pathway may exist which primarily innervates the extrapyramidal motor system. This pathway originates mainly from the mesencephalic raphe cell group (B7-B9). The cerebellum innervation originates mainly from the mesencephalic raphe (B7, B8) and possibly raphe cell groups (B5, B6) in the pons. From (55).

Richards et al. (151) and Richards and Tranzer (152,153) initially described the wide distribution of indoleamine nerve terminals which insinuate between ventricular ependymal cells and end freely in the cerebrospinal fluid (Fig. 3). Chan-Palay (32) has provided additional evidence for their

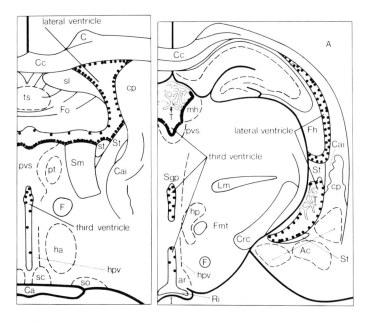

**FIG. 3.** Relative density of supraependymal yellow fluorescence in 2 frontal sections through the hypothalamus. *Crc:* Crus cerebri; *Ca:* Chiasma opticum; *F:* Columna fornicis; *Fmt:* Fasciculus mamillothalamicus; *Fo:* Fornix; *ha:* Nucleus anterior hypothalami; *Lm:* lemniscus medialis; *pt:* Nucleus paratenialis; *sc:* Nucleus suprachiasmaticus; *so:* Nucleus supraopticus; *ts:* Nucleus triangularis septi. From (111a).

serotonergic nature and Aghajanian and Gallager (1) have demonstrated that these are serotonergic projections of raphe origin. The functional significance of serotonin discharge from these terminals into the cerebrospinal fluid is unknown.

## MEDIAN EMINENCE

The need for continued and expanded research on the structure and function of the median eminence becomes more evident as each new contribution is presented on this remarkable window through which the many final pathways of the endocrine hypothalamus deliver their hormones to the pituitary gland (Fig. 4). General reviews have been presented by Kobayashi and Matsui (95) and by Knigge and Silverman (89); discussions of specific aspects of structure or function are found in Knigge et al. (91), Oksche et al. (137), Bjorklund et al. (15), Ajika and Hokfelt (2), and Gross et al. (61).

One of the more extraordinary features of the median eminence is the large number of biologically active molecules which are present here; it is probable that no other structure of equivalent volume in the central nervous system contains as many enzymes, neurotransmitters (established and putative), and hormones of both central and peripheral origin as does the median eminence. Not only are there a large number of different substances, but many are present in concentrations greater than anywhere in hypothalamus or brain: norepinephrine (39), tyrosine hydroxylase (84,159), phenylethanolamine $N$-methyltransferase (157,160), choline acetyltransferase (23), and histamine (24) to name a few. The volume of the median eminence of the rat hypothalamus is approximately 0.4 mm$^3$; of this, 60% is occupied by glia, tanycytes, the supraoptic-hypophysial tract (fibrous zone), a pituitary-portal capillary bed and an extensive perivascular space. Crowded into the remaining space are approximately $10^8$ axon profiles and terminals known or suspected to contain some 10 or more hypothalamic peptides, neurotransmitters (dopamine, norepinephrine, serotonin, acetylcholine, GABA), histamine, and a variety of other active substances. All are present in substantial concentrations. Exocytotic release of vesicle-bound material appears to occur randomly and without relationship to any morphologically visible synaptic mechanisms. The extracellular and perivascular space would appear to be a medium of remarkable composition. Although some regional topography is emerging, the phenomenon of diffusion occurring after release alone would suggest that large pools of nerve terminals and nonneuronal elements are bathed in an interstitial fluid containing a multitude of hormones and excitatory and inhibitory neurotransmitters. Neuropeptides such as luteinizing hormone-releasing factor (LRF), thyrotropin-releasing factor (TRF), and somatostatin are present in the median eminence for the purpose of being released into portal blood

# 24   ANATOMY OF THE ENDOCRINE HYPOTHALAMUS

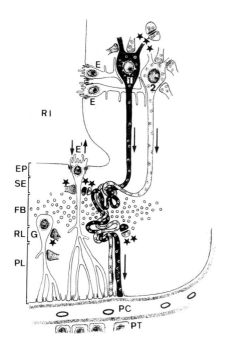

**FIG. 4.** Basic arrangement of neurons and ependymal cells in the region of the arcuate nucleus and in the median eminence. This diagram has been simplified, and some details are still hypothetical. The neurons of the arcuate nucleus have intimate contacts with two different populations of ependymal cells: 1. at the *central* level of the perikarya within the nuclear area, 2. at the *peripheral* level of the terminals within the median eminence. *1* aminergic and *2* peptidergic (releasing-factor producing) neurons of the arcuate nucleus. Axosomatic and axodendritic synapses (\*\*\*). Numerous presynaptic endings belong to ascending noradrenergic tracts. *E:* ependymal tanycytes form a link between the CSF and the somata of tuberal neurons. *Arrows* indicate the direction of transport of secretory materials within the axons. A number of tuberal axons run directly into the subependymal (*SE*) and reticular (*RL*) layers of the median eminence. Some fibers traverse the fiber layer (*FB*, with the crosssectioned supraopticohypophysial tract) where they may have increased contact area (exaggerated in the diagram). In the median eminence, long and branched ependymal tanycytes (*E'*) extend from the ependymal layer (*EP*) covering the infundibular recess (*RI*) to the outer surface of the median eminence (*PL*, palisade layer) that is covered by the primary capillaries of the portal circulation. *PT:* pars tuberalis. *G:* glial cell with ependyma-like processes and end-feet. Synaptoid contacts (\*) have been described between nerve endings of unknown origin (aminergic elements?) and ependymal or glial cells. Further, axoaxonal contacts (\*\*) occur in the median eminence. At the surface of the infundibular recess (*RI*) arrows indicate the possible direction of uptake or release of substances in the ependymal cells (*E'*). From (137).

and influencing adenohypophysial function; the role(s) of the various neurotransmitters present have not been clearly defined. Almost all neuronal models which have been proposed for the control of the releasing factor- (RF) producing neuron illustrate networks of synaptic relations on the soma or dendritic tree of these cells and the presence of transmitters in the median eminence is ignored. The preceding discussion regarding hormone and transmitter release from terminals in the median eminence does not take into consideration the possibility that different pools of hormone or transmitter exist and that terminals are anatomically organized as "terminal" and "nonterminal" dilatations as suggested by Cross et al. (38) for the supraoptic hypophysial system (Fig. 5). Release normally occurs only from a readily releasable pool which is contained in relatively few terminals. Studies

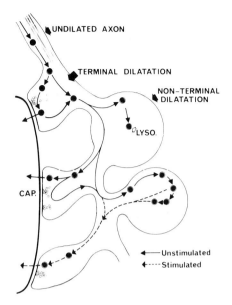

**FIG. 5.** Diagram of suggested movements of neurosecretory granules between readily releasable and storage pools in the magnocellular endocrine neuron. In the unstimulated gland, stored hormone in the nonterminal dilatations may be destroyed by lysosomes, but in stimulated conditions, some may be conveyed to the terminal dilatations abutting capillary membranes. From (38).

directed to this problem in median eminence should be both challenging and rewarding.

The term "tuberoinfundibular" tract is generally used to identify collectively all the projections or afferents to the median eminence. The term may require reexamination since it is already known that several afferent systems to the median eminence have their nuclei of origin well beyond the boundaries of the classical "tuberal" region of the hypothalamus.

The detailed anatomy of the monoaminergic innervation of the median eminence continues to be an area of great research interest and importance with regard to neural control of the pituitary gland. Fifteen years ago, Carlson et al. (31) reported that the external zone of the median eminence exhibited a prominent catecholamine fluorescence. This led to an immediate and sustained research interest in fluorescence histochemistry of this tissue. This fluorescence was soon shown to be neuronal, belonging in part to a projection from dopamine-containing perikarya in the arcuate-periventricular region (11,12,51,52,110). Microspectrofluorometric analysis together with a variety of experimental approaches led to the identification of norepinephrine in the internal zones of the median eminence (12,14,78) supporting earlier (104) and later (138) biochemical evidence for the presence of both amines.

A substantial literature has led to the current view that norepinephrine and dopamine are localized primarily in the internal and external zones of the median eminence respectively. The external zone of the median eminence is generally thought to be associated with events related to releasing hormones and pituitary control; the absence of noradrenergic input to this zone is considered disconcerting and puzzling in view of impressive evidence of noradrenergic involvement in neural control of several pituitary hormones.

The tuberoinfundibular dopamine system has been examined in depth and exhibits several notable features. A regional topography is apparent, with a heavy innervation in the lateral parts of the median eminence. Using a technique based on inhibition of catecholamine synthesis and estimation of the rate of disappearance of fluorescence, Fuxe et al. (57) have found that turnover of dopamine is more rapid in the medial than in the lateral parts of the median eminence; norepinephrine turnover in the subependymal layer was notably lower than that of dopamine (Fig. 6). Analysis of the tuberoinfundibular dopamine complex by Bjorklund et al. (14) also indicates regional differences within the median eminence, with separate groups of neurons in the arcuate and periventricular area.

These recent histofluorescence advances in monoamine distribution in the median eminence have as yet not been followed up by any sustained correlative electron microscopic studies such as those of Richards and Tranzer (153) (Fig. 7). The ability to identify specific monoamine terminals at the ultrastructural level in median eminence would represent a major advance. The technique of electron microscopic autoradiography of radiolabeled monoamines (69,75,99) appears to have some promise for median eminence studies (171,173).

The distribution of other monoamines in median eminence is less clear. Serotonin is of particular interest because of its emergence as a significant contributor to the regulation of several tropins (97,138,170). The median eminence of the rat contains 15 ng of serotonin per mg protein (158) and tryptophan hydroxylase, the enzyme necessary for its synthesis (87,143, 144). In spite of this significant concentration of serotonin in median eminence, histofluorescence studies beginning with Fuxe in 1965 have consistently failed to visualize it owing in some measure to the relative insensitivity of the Falck-Hillarp formaldehyde-induced fluorescence method. Pharmacological treatments which elevate intraneuronal 5-HT in other areas

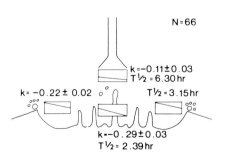

**FIG. 6.** Schematic representation of results from catecholamine (CA) turnover measurements in nerve terminals in the central part of the median eminence in normal male rat, based on microfluorimetric quantitation of changes in the formaldehyde-induced CA fluorescence following tyrosine hydroxylase inhibition produced by $\alpha$-methyl-$p$-tyrosine methylester (250 mg/kg i.p.). The disappearance of the fluorescence was exponential and the calculated T1/2 for the external layer was 2.39 h (medial part) and 3.15 h (lateral part) while T1/2 for the subependymal layer was 6.30 h. The k-values ± S.D. for the slope are also given. From (57).

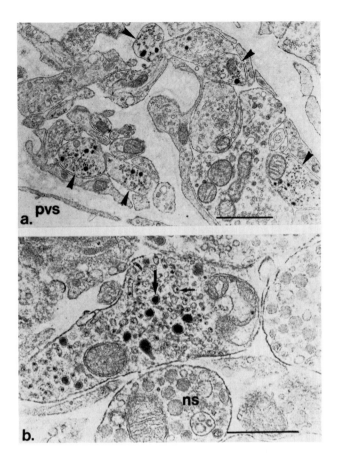

**FIG. 7.** Rat median eminence after fixation with chromate-dichromate buffered aldehydes and osmium. Note that several nerve terminals (arrowhead) contain small dense core vesicles (small arrow) and large dense core vesicles (large arrow) with a marked electron density while others, presumably neurosecretory neurons (ns), remain unreactive with this fixation method. pvs: perivascular space. **a:** 1 $\mu$m; ×16,500. **b:** 0.5 $\mu$m; ×42,000. From (153).

of the brain (145) have no effect on median eminence 5-HT. Calas (27,28) has presented light and electron microscopic evidence of $^3$H 5-HT uptake by some terminals in the median eminence of the duck, rat, and monkey; uptake was considered specific because low ($10^{-7}$M) concentrations of the $^3$H 5-HT were used in both *in vitro* as well as *in vivo* experiments in which tracer was introduced into the 3rd ventricle. 5-HT is accumulated by more than one mechanism and can be accumulated by nonserotonergic nerve endings (176). Only preliminary *in vitro* studies have been performed on the characteristics of the serotonin uptake processes in median eminence (90). Baumgarten and Lachenmayer (8) have also presented evidence of serotonergic nerve terminals in rat median eminence; serotonin neurotoxin (5,6 and 5,7-dihydroxytryptamine were administered intraventricularly) and degenerating terminals were subsequently seen in median eminence. Knigge et al. (90) have presented initial evidence for a nonneuronal localization of serotonin in median eminence; serotonin is present in tanycytes of this tissue and may be synthesized by them. Using a radiometric assay, Kizer et al. (87) were unable to detect any change in tryptophan hydroxylase activity in median eminence after castration, thyroidectomy, or adrenalectomy nor after treatment with testosterone, thyroxine, or dexamethasone. The tanycyte of median eminence and pinealocyte of the pineal gland are of similar embryonic origin, contain and/or may synthesize serotonin, and have similar synaptoid noradrenergic terminal relationships.

The technique of identifying neurons that project to the median eminence by antidromic stimulation pioneered by Markara et al. (113) and Sawaki and Yagi (168) has proven to be an outstanding procedure for both locating the perikarya of these projections as well as analyzing their synaptic organization (See Renaud, *this volume*). With this method, cells have been identified in various regions of the rat hypothalamus including the arcuate, ventromedial, suprachiasmatic nuclei, and periventricular regions; none were found in the supraoptic nucleus, medial preoptic, or anterior hypothalamic areas (214). These neurons are subject to recurrent inhibition and excitation (169,213), are influenced by projections from the preoptic region, amygdala, and hippocampus (82,215), are responsive to catecholamines (130), and have collaterals which project to several hypothalamic and extrahypothalamic sites (150). These neurophysiological methods must be combined with histochemical procedures for identifying the neurons being studied. Additional information on median eminence projections in the cat has been presented by Nojyo et al. (135) and in dog and sheep by Watkins (198).

A morphological study by Koritsanszky and Koves (98) supports electrophysiological data that medial preoptic neurons do not project to the median eminence. These workers placed lesions in the preoptic area and found no evidence of nerve terminal degeneration in the palisade (external) zone of median eminence. Other anatomical and physiological studies dealing with median eminence have presented an additional aspect of the neuronal organization of the basal hypothalamic region, namely the relatively few neurons which actually project to the median eminence. In a study of the effects of monosodium glutamate upon hypothalamic function, Holzwarth-McBride et al. (73) found that 85% or more of the neurons of the arcuate nucleus were destroyed without any significant effect upon volume or fine structure of the median eminence; dopaminergic neurons of the arcuate appeared to be least affected. These workers suggested that a very large proportion of arcuate neurons may be intrinsic to the nucleus or project to other areas of the brain. In subsequent studies, Hoffman (*unpublished data*) has noted further that immunocytochemically demonstrable LRF neurons of the arcuate region are also not significantly affected by monosodium glutamate. The fact that a large proportion of the neurons of the arcuate region are glutamate-sensitive, and that glutamic acid is the metabolic precursor of GABA, suggests that there may exist extensive neuronal circuits which utilize either the amino acid or GABA as transmitter. Support for the observations of Holzwarth-McBride is provided by the neurophysiological studies of Renaud (150). This investigator noted that less than 15% of medial hypothalamic neurons send axons to the median eminence as identified by antidromic stimulation.

Analysis of where nontuberoinfundibular neurons of the arcuate region project will obviously yield important information on the circuitry controlling the endocrine-motor neurons. At present, neurophysiological methods appear better suited to this type of discriminative analysis.

The use of systemically administered horseradish peroxidase may represent a useful procedure for identifying projections to the median

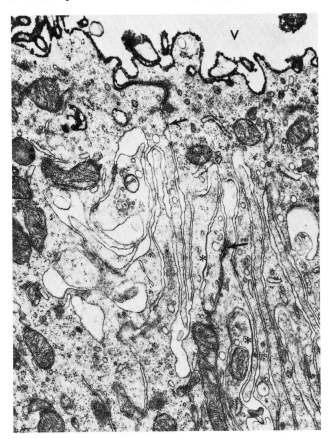

**FIG. 8.** Peroxidase, perfused through the cerebral ventricles (*V*), does not enter the extracellular clefts (*upper arrow*) between the special ependymal cells of the median eminence. Two gap junctions (*asterisks*) do not lie next to tight junctions (*lower arrow*). ×22,500. From (21).

eminence and other circumventricular organs. In a recent study by Broadwell and Brightman (26), this tracer was found to be incorporated by terminals in the median eminence of the mouse hypothalamus and transported to cell bodies of origin; they were found in the supraoptic, paraventricular and arcuate nuclei, the periventricular region, and in cells of the A1 and A5 catecholaminergic nuclei of the brainstem.

The role of tanycytes of the median eminence in a hypothesis of hormone delivery from cerebrospinal fluid to the pituitary portal capillaries has been reviewed by Knigge et al. (91,93). Several recent studies have provided additional information regarding these unique cells. The presence of tight junctions at their apical (ventricular) boundaries suggested that movement of substances from cerebrospinal fluid occurs by passage through tanycytes rather than between them. Careful studies with intraventricular horseradish peroxidase indicate clearly the barrier presented by these junctions (Fig. 8). Analysis of the junctions by freeze-fracture by Brightman et al. (21) suggests that they are rudimentary compared to those of choroid plexus or intestinal epithelium (Fig. 9). Nevertheless, tanycyte junctions impede the extracellular flow of horseradish peroxidase (MW 40,000), microperoxidase (MW 1,800), arginine (MW 174), and phenylalanine (MW 105) (202). Although excluded from intercellular movement by these tight junctions, endocytotic uptake of both horseradish peroxidase and ferritin has been observed (96,106,121,131). The uptake of peroxidase by tanycytes of

**FIG. 9.** Freeze-fracture preparation of the tight junctions between tanycytes of the median eminence. A junctional complex, consisting of tight junctions delineated by ridges and continuous gap junctions (*asterisks*), connects ependymal cells. Note large (*arrow*) and small (*arrowhead*) discontinuities. ×92,000. From (21).

median eminence of the quail appears to be influenced by neural afferents. The tanycytes of median eminence in this species appear unusually large and their ventriculoportal course readily apparent (Fig. 10). Following deafferentation of the basal hypothalamus of the quail, tanycytes of the posterior median eminence exhibit significantly greater uptake of peroxidase (94,136). The location of noradrenergic afferents in the inner (subependymal) layer of median eminence and their disconnection by deafferentation makes this monoaminergic input a likely candidate for the controller of tanycyte endocytotic activity. Autoradiographic studies of Calas (27) indicate that tanycytes may possess amino acid uptake mechanisms of different capacity. Intraventricular injection of $^3$H-pyroglutamic acid, histidine, or proline was followed by a marked concentration of radiolabeled grains of histadine in tanycytes, faint labeling with proline, and none with pyroglutamic acid.

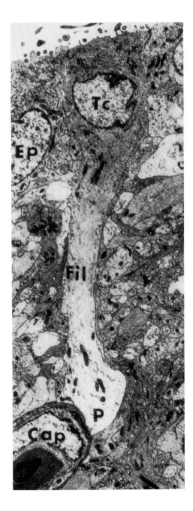

**FIG. 10.** A tanycyte (*Tc*) of median eminence of the quail, with a long basal process (*P*) terminating on the subependymal capillary (*Cap*) in the ventral third ventricular wall. *Fil:* filaments; *Ep:* ependymal cell. ×2,250. From (121).

Tanycytes which line the wall of the ventricular recess course through the bed of the arcuate nucleus. An unusual system of gap junctions between soma of the arcuate neuron and tanycyte process has been identified by del Cerro and Knigge (41). These junctions (Fig. 11) allow the passage of low molecular weight substance between the junctional partners; the implications of this with respect to possible transfer of releasing hormones to tanycytes is apparent.

The advent of the immunocytochemical procedures for releasing hor-

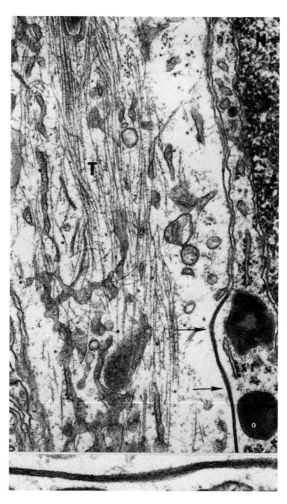

**FIG. 11.** A region of contact between neuron (*N*) and process of a tanycyte (*T*) in the arcuate nucleus of a 21-day-old rat. Part of the region of contact is occupied by a gap junction, seen in greater detail in the insert. ×26,100; insert: ×27,000. From (41).

mones offers an additional tool with which to examine the transport capacity of tanycytes. A number of studies (4,5,141) have specifically commented upon the absence of immunocytochemically demonstrable LRF in tanycytes of median eminence. A number of considerations suggest that caution should be exercised in interpreting this apparent failure to localize endogenous LRF in tanycytes. The same studies which have thus far failed to demonstrate LRF in tanycytes have also failed to visualize it in perikarya. LRF is present in many hypothalamic nuclei (139) and the studies of Hoffman et al. (67,68) and Knigge et al. (*this volume*) indicate that LRF is present in perikarya, probably in a structural form not recognized by all antisera. Further analysis may reveal that releasing hormones are present in tanycytes in a form not recognized by current immunocytochemical procedures nor by current antibodies because of an association of hormone with intracellular elements of the tanycyte.

The volume of the neural lobe of the rat pituitary gland is approximately equal to or slightly greater than the volume of median eminence. In the neural lobe, approximately 3 ng of neuropeptides are stored in $2 \times 10^{10}$ neurosecretory granules within terminals measuring 1300 to 1800 Å in diameter; the concentration of neuropeptide per granule is $1 \times 10^{-4}$ pg/granule (38). The granules in which LRF have been recognized immunocytochemically measure 650 Å in diameter (141), a volume one-tenth of the neural lobe granules. Assuming an equal packing density of LRF in the dense-core matrix of the smaller granules and reported values of 3 to 5 ng of LRF in median eminence, 3 to $5 \times 10^8$ granules would be required to package this amount of hormone. Even a cursory survey of literature which contains quantitative data on dense-core granule population in median eminence (73,154) suggests that the average number of dense-core vesicles per terminals is less than 100. The estimated number of LRF-containing terminals would therefore have to be about $10^6$, an unusually high proportion of the available terminals for only one of the many hormones and transmitters which presumably are also packaged in dense-core vesicles. A large amount of extragranular hormone may be present in the terminal or it may be present in some other compartment of the median eminence.

Clementi and Marini (34) and Coates (35,36) made initial observations on another unexplained feature of the median eminence and other circumventricular organs. These investigators described the presence of a variety of cells which are located in the cerebrospinal fluid associated generally with the modified ependyma overlying the circumventricular organs. Pauli et al. (140), Weindl and Schinko (203), and Scott et al. (171,172) have examined these cells extensively and determined that many have ultrastructural characteristics of neurons (Fig. 12). Processes ramify for long distances over the surface of the ventricular wall, perhaps interconnecting with neuron clusters over the several circumventricular organs. Other processes either terminate in the cerebrospinal fluid or penetrate into the

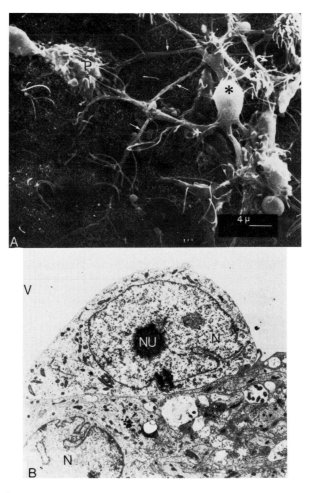

**FIG. 12.A.** Scanning electronmicrograph of primate infundibular recess demonstrating the presence of two morphologically distinct cell types. Type I (*) possesses numerous main branching processes from which secondary processes (*arrows*) extend at right angles and course horizontally over the ventricular lumen. The second cell type (type II) is significantly different in its surface structure and possesses a flattened, fluted process (P) which exhibits a ruffled appearance. ×2,760. **B:** Transmission electron microscopy of supraependymal type I cell in infundibular recess, possessing numerous ultrastructural criteria used to identify bona fide neurons. *N:* nucleus; *NU:* nucleolus; *V:* third ventricle. ×6,885. From (174).

underlying neuropil. The functional significance of this supraependymal network of neurons remains to be demonstrated.

The use of the organ-cultured median eminence as a model for functional studies have been described by Silverman et al. (166) and several studies relating to monoamine uptake and action reported (88,90,193). Another model of considerable interest is the one represented by transplants to the anterior chamber of the eye (42). Tissue placed here undergoes morphological changes similar to those occurring in organ-culture and is maintained for remarkably long periods of time (Fig. 13). LRF-containing neurons have been identified in monolayer culture of cells of the basal hypothalamus (92).

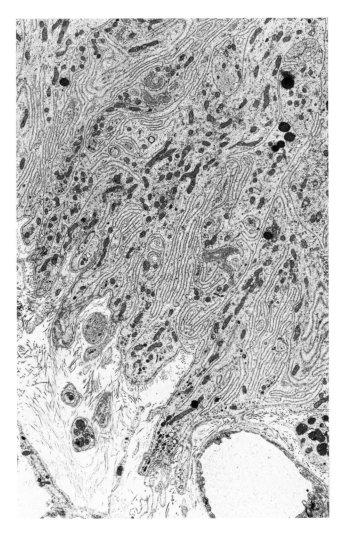

**FIG. 13.** External zone of a rat median eminence 41 days after transplantation to the anterior chamber of the eye. Densely packed glial (tanycyte) processes containing mitochondria and pleomorphic granulated vesicles terminate upon a perivascular connective tissue space. ×6,300. From (42).

## ORGANUM VASCULOSUM LAMINA TERMINALIS

An awareness and intense interest in the structure and neuroendocrine-related function of the organum vasculosum lamina terminalis (OVLT) resulted from the finding of immunoreactive LRF in this structure by Barry et al. (6). This original observation in the rat was quickly confirmed by other immunochemists and by radioimmunoassay of LRF in tissue from this region (86,208). Studies have been extended to include other species (27,203) and effects of such experimental manipulations as hypophysectomy (5) and deafferentation of the basal hypothalamus (206). Interest in this structure has been heightened further by reports indicating the presence of TRF and somatostatin (86), (Knigge et al. *this volume*) and by evidence that the OVLT may play a role in the central effects of angiotensin (142).

The original descriptions of vascular and cellular architecture of the OVLT were by Wislocki and co-workers (211,212) and Kuhlenbeck (103); it was referred to at that time as the supraoptic crest. Hofer (66) recognized that its structural characteristics warranted its inclusion in his catalog of circumventricular organs and he suggested the name by which we presently know this structure. Scattered studies during the next decade accumulated further information on its vasculature (43,83) and neuronal and ependymal histochemistry (30,107,108). Ultrastructural studies on the OVLT of the mouse and rat were initiated by Leveque et al. (109) and in the rabbit by Weindl et al. (204, 205). These studies have been continued in these and other species by Usui (192), Rohlich and Wenger (155), Wartenburg et al. (197), LeBeux (105), Weindl (199,200,201), Wenger (207), Weindl and Schinko (203), Schinko et al. (175), and Mikami (122).

From these studies emerged a picture of the OVLT as a structure with nearly classical attributes of the circumventricular organs bearing striking similarity to the median eminence. Although species differences exist with respect to ependymal and neuronal details, the basic architecture is fairly consistent and represents a brain region where neurosecretory neurons terminate upon the perivascular space of fenestrated capillaries and where a modified, nonciliated ependymal lining may functionally link the cerebrospinal fluid and the peripheral vasculature (Fig. 14).

The presence in the OVLT of LRF and other neuropeptides generally associated thus far with control of adenohypophyseal function raises intriguing questions with respect to neuropeptide functions as well as mechanisms of adenohypophysial control. Numerous studies on the angioarchitecture of the OVLT have failed to demonstrate a "portal" relationship between the OVLT capillary vasculature and the sinusoids of the adenohypophysis. If hormones such as LRF are released into capillaries of the OVLT, they may enter the systemic circulation and contribute an additional (presumably small) measure of control on gonadotropin secretion. Alternately, the architecture of the OVLT would allow delivery of neuropeptides to the cerebrospinal fluid (79,93).

# ANATOMY OF THE ENDOCRINE HYPOTHALAMUS

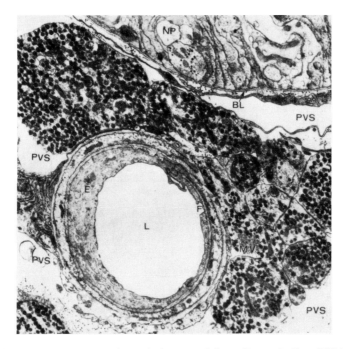

**FIG. 14.** Organum vasculosum of a male hamster 4 days after castration. Within the perivascular space (*PVS*) numerous neurosecretory axons are accumulated around a fenestrated (*F*) vessel. These axons are not surrounded by a basal lamina (*BL*). *E:* Endothelium; *L:* lumen; *MV:* microvesicles; *NP:* neurophil of parenchymal zone. ×7,500. From (203).

## REFERENCES

1. Aghajanian, G. K. and Gallager, D. W. (1975): Raphe origin of serotonin nerves terminating in the cerebral ventricles. *Brain Res.*, 88:221–231.
2. Ajika, K. and Hökfelt, T. (1975): Projections to the median eminence and the arcuate nucleus with special reference to monoamine systems: Effects of lesions. *Cell Tiss. Res.*, 158:15–35.
3. Arees, E. A. and Mayer, J. (1967): Anatomical connections between medial and lateral regions of the hypothalamus concerned with food intake. *Science*, 157:1574–1575.
4. Baker, B. L., Dermody, W. C., and Reel, J. R. (1974): Localization of luteinizing hormone-releasing hormone in the mammalian hypothalamus. *Am. J. Anat.*, 139:129–134.
5. Baker, B. L. and Dermody, W. C. (1976): Effect of hypophysectomy on immunocytochemically demonstrated gonadotropin-releasing hormone in rat brain. *Endocrinology*, 98:1116–1122.
6. Barry, J. and Dubois, M. P. (1973): Étude en immunofluorescence des structures hypothalamiques a competence gonadotrope. *Ann. Endocrinol.*, 34:735–742.
7. Barry, J., Dubois, M. P., and Poulan, P. (1973): LRF producing cells in the mammalian hypothalamus. *Z. Zellforsch.*, 146:351–366.
8. Baumgarten, H. G. and Lachenmayer, L. (1974): Indoleamine-containing nerve terminals in the rat median eminence. *Z. Zellforsch.*, 147:285–292.
9. Bernardis, L. L. and Frohman, L. A. (1970): Effects of lesion size in the ventromedial nucleus of the hypothalamus on growth hormone and insulin levels in weanling rats. *Neuroendocrinology*, 6:319–328.
10. Bernardis, L. L. and Frohman, L. A. (1971): Plasma growth hormone responses to electrical stimulation of the hypothalamus in the rat. *Neuroendocrinology*, 7:193–201.

11. Bjorklund, A., Enemar, A., and Falck, B. (1968): Monoamines in the hypothalamo-hypophyseal system of the mouse with special reference to the ontogenetic aspects. *Z. Zellforsch.,* 89:590–607.
12. Bjorklund, A., Falck, B., Hromek, F., Owman, C. H., and West, K. (1970): Identification and terminal distribution of the tuberohypophyseal monoamine fibre systems in the rat by means of stereotaxic and microspectrofluorimetric techniques. *Brain Res.,* 17:1–23.
13. Bjorklund, A., Lindvall, O., and Svensson, A. (1972): Mechanisms of fluorophore formation in the histochemical glyoxylic acid method for monoamines. *Histochemie,* 32:113–132.
14. Bjorklund, A., Moore, R. Y., Nobin, A., and Stenevi, U. (1973): The organization of tubero-hypophyseal and reticulo-infundibular catecholamine neuron systems in the rat brain. *Brain Res.,* 51:171–191.
15. Bjorklund, A., Falck, B., Nobin, A., and Stenevi, U. (1974): Organization of the dopamine and noradrenaline innervations of the median eminence-pituitary region in the rat. In: *Neurosecretion: The Final Neuroendocrine Pathway.* Sixth Intern. Congress of Neurosecretion, edited by F. Knowles and L. Vollrath, pp. 209–222. Springer-Verlag, Berlin.
16. Blacksted, T. W. (1965): Mapping of experimental axon degeneration by electron microscopy of Golgi preparations. *Z. Zellforsch.,* 67:819–834.
17. Blacksted, T. W. (1970): Electron microscopy of Golgi preparations for the study of neuronal relations. In: *Contemporary Research Methods in Neuroanatomy,* edited by W. J. H. Nauta and S. O. E. Ebbesson, pp. 186–216. Springer-Verlag, Heidelberg.
18. Bons, N. (1976): Retinohypothalamic pathway in the duck (*Anas Platyrkynchos*). *Cell Tiss. Res.,* 168(3):343–360.
19. Bons, N. and Assenmacher, I. (1973): Nouvelles recherches sur la voie nerveuse retino-hypothalamique chez les oiseaux. *C.R. Acad. Sci.,* 277:2529–2532.
20. Branton, D. (1966): Fracture faces of frozen membranes. *Proc. Natl. Acad. Sci.,* 55:1048–1056.
21. Brightman, M. W., Prescott, L., and Reese, T. S. (1975): Intercellular junctions of special ependyma. In: *Brain-Endocrine Interaction II. The Ventricular System in Neuroendocrine Mechanisms.* Second Internat. Symposium, edited by K. M. Knigge, D. E. Scott, H. Kobayashi, and S. Ishii, pp. 146–165. Karger, Basel.
22. Brown-Grant, K. and Raisman, G. (1972): Reproductive functions in the rat following selective destruction of afferent fibers to the hypothalamus from the limbic system. *Brain Res.,* 46:23–42.
23. Brownstein, M. (1975): Biogenic amine content of the hypothalamic nuclei. In: *Anatomical Neuroendocrinology,* edited by W. E. Stumpf, and L. D. Grant, pp. 393–396. Karger, Basel.
24. Brownstein, M. J., Saavedra, J. M., Palkovits, M., and Axelrod, J. (1974): Histamine content of hypothalamic nuclei of the rat. *Brain Res.,* 77:151–156.
25. Brownstein, M., Arimura, A., Sato, H., Schally, A. V., and Kizer, J. S. (1975): The regional distribution of somatostatin in the rat brain. *Endocrinology,* 96:1456–1461.
26. Broadwell, R. D. and Brightman, M. W. (1976): Entry of peroxidase into neurons of the central and peripheral nervous systems from extracerebral and cerebral blood. *J. Comp. Neur.,* 166:257–284.
27. Calas, A. (1975): The avian median eminence as a model for diversified neuroendocrine routes. In: *Brain-Endocrine Interaction II. The Ventricular System in Neuroendocrine Mechanisms.* Second Internat. Symposium, edited by K. M. Knigge, D. E. Scott, H. Kobayashi, and S. Ishii, pp. 54–69. Karger, Basel.
28. Calas, A., Alanso, G., Arnauld, E., and Vincent, J. D. (1974): Demonstration of indoleaminergic fibres in the median eminence of the duck, rat and monkey. *Nature,* 250:241–243.
29. Campbell, C. B. G. and Hayhow, W. R. (1972): Primary optic pathways in the duckbill platypus, *Ornithorynchus anatinus:* an experimental degeneration study. *J. Comp. Neurol.,* 145:195–208.
30. Campos, Ortega, J. A. and Ferres-Torres, R. (1965): Sobre el sustrato del organon vasculosum laminae terminalis de la rata albina. *Anales de Anatomia,* 14:381–409.
31. Carlsson, A., Falck, B., and Hillarp, N.-A. (1962): Cellular localization of brain monoamines. *Acta Physiol. Scand.,* (56, *suppl.*) 196:1–28.

32. Chan-Palay, V. (1976): Serotonin axons of the supra- and subependymal plexuses and in the leptomeninges: their roles in local alterations of CSF and vasomotor activity. *Brain Res.,* 102:103–130.
33. Chi, C. C. (1970): Afferent connections of the ventromedial nucleus in the rat. *Brain Res.,* 17:439–445.
34. Clementi, F. and Marini, D. (1972): The surface fine structure of the walls of the cerebral ventricles and of the choroid plexus in cat. *Z. Zellforsch.,* 123:82–95.
35. Coates, P. W. (1973): Supraependymal cells: Light and transmission electron microscopy extends scanning electron microscopic determination. *Brain Res.,* 57:502–507.
36. Coates, P. W. (1973): Supraependymal cells in the recesses of the monkey third ventricle (I). *Am. J. Anat.,* 136:533–539.
37. Conrad, L. C. A. and Pfaff, D. W. (1976): Efferents from medial basal forebrain and hypothalamus in the rat. I. An autoradiographic study of the medial preoptic area. *J. Comp. Neurol.,* 169:185–219.
38. Cross, B. A., Dybell, R. E. J., Dyer, R. G., Jones, C. W., Lincoln, D. W., Morris, J. F., and Pickering, B. T. (1975): Endocrine neurons. In: *Recent Progress in Hormone Research,* Vol. 31, edited by R. O. Greep, pp. 243–286. Academic Press, New York.
39. Cuello, A. C., Horn, A. S., MacKay, A. V. P., and Iversen, L. L. (1973): Catecholamines in the median eminence: new evidence for a major noradrenergic input. *Nature,* 243:465–467.
40. deOlmos, J. S. (1972): The amygdaloid projection field in the rat as studied with the cupric-silver method. In: *The Neurobiology of the Amygdala,* edited by B. E. Eleftheriou, pp. 145–204. Plenum Press, New York.
41. del Cerro, M. and Knigge, K. M. (1977): A system of gap junctions and some unusual glial-neuronal contacts in the rat arcuate nucleus. *Cell Tiss. Res., (in press.)*
42. Dellman, H. D. (1977): Ultrastructural of homografts of the rat median eminence into the anterior chamber of the eye. *Neuroendocrinology,* 22 (*in press.*)
43. Duvernoy, H. and Koritke, J. G. (1961): Sur la systematisation de la lame terminale (etude d'anatomie comparee). *Acta Anat.,* 47:378–379.
44. Dyer, R. G. and Cross, B. A. (1972): Antidromic identification of units in the preoptic and anterior hypothalamic areas projecting to one ventromedial and arcuate nuclei. *Brain Res.,* 43:254–258.
45. Falck, B., Hillarp, N. A., Thieme, G., and Thorp, A. (1962): Fluorescence of catecholamines and related compounds condensed with formaldehyde. *J. Histochem. Cytochem.,* 10:348–354.
46. Field, P. M. (1972): A quantitative ultrastructural analysis of the distribution of amygdaloid fibers in the preoptic area and the ventromedial nucleus. *Exp. Brain Res.,* 14:527–538.
47. Field, P. M. and Sherlock, D. A. (1975): Golgi studies of the sexually differentiated part of the preoptic area in the rat. In: *Golgi Centennial Symposium: Perspectives in Neurobiology,* edited by M. Santini, pp. 143–146. Raven Press, New York.
48. Frohman, L. A. and Bernardis, L. L. (1968): Growth hormone and insulin levels in weanling rats with ventromedial hypothalamic lesions. *Endocrinology,* 82:1125–1132.
49. Frohman, L. A. and Bernardis, L. L. (1971): Effects of hypothalamic stimulation on plasma glucose, insulin and glucagon levels. *Am. J. Physiol.,* 221:1596–1603.
50. Frohman, L. A., Bernardis, L. L., and Stachura, M. (1974): Factors modifying plasma insulin and glucose responses to ventromedial hypothalamic stimulation. *Metabolism,* 23:1047–1056.
51. Fuxe, K. (1963): Cellular localization of monoamines in the median eminence and in the infundibular stem of some animals. *Acta Physiol. Scand.,* 58:383–384.
52. Fuxe, K. (1964): Cellar localization of monoamines in the median eminence and the infundibular stem of some mammals. *Z. Zellforsch.,* 61:710–724.
53. Fuxe, K. and Hökfelt, T. (1970): Catecholamines in the hypothalamus and the pituitary gland. In: *Frontiers in Neuroendocrinology,* edited by W. Ganong and L. Martini, pp. 47–96. Oxford University Press, London.
54. Fuxe, K. and Hökfelt, T. (1970): Central monoaminergic system and hypothalamic function. In: *The Hypothalamus,* edited by L. Martini, M. Motta, and F. Fraschini, pp. 123–138. Academic Press, New York.

55. Fuxe, K. and Jonsson, G. (1974): Further mapping of central 5-hydroxytryptamine neurons: Studies with neurotoxic dihydroxytryptamines. *Adv. Biochem. Psychopharm.,* 10:1-2.
56. Fuxe, K., Goldstein, M., Hökfelt, T., Jonsson, G., and Lofstrom, A. (1974): New aspects in the catecholamine innervation of the hypothalamus and the limbic system. In: *Neurosecretion: The Final Neuroendocrine Pathway.* Sixth Internat. Congress of Neurosecretion, edited by F. Knowles and L. Vollrath, pp. 223-228. Springer-Verlag, Berlin.
57. Fuxe, K., Hökfelt, T., Jonsson, G., and Lofstrom, A. (1974): Aminergic mechanisms in neuroendocrine control. In: *Neurosecretion: The Final Neuroendocrine Pathway,* edited by F. Knowles and L. Vollrath, pp. 269-275. Springer-Verlag, Berlin.
58. Globus, A., Lux, H. D., and Schubert, P. (1968): Somadendritic spread of intracellularly injected triated glycine in cat spinal motor neurons. *Brain Res.,* 11:440-445.
59. Goldman, J. K., Bernardis, L. L., and Frohman, L. A. (1974): Food intake in hypothalamic obesity. *Am. J. Physiol.,* 227:88-91.
60. Graeber, R. C. and Ebbesson, S. O. E. (1972): Retinal projections in the lemon shark (*Negaprion brevirostris*). *Brain Behav., Evol.,* 5:461-477.
61. Gross, J. H., Knigge, K. M., and Sheridan, M. N. (1976): Fine structure of neurons of the arcuate nucleus and median eminence of the hypothalamus of the golden hamster following immolization. *Cell Tiss. Res.,* 168:385-397.
62. Hales, C. N. and Kennedy, G. C. (1964): Plasma glucose, nonesterified fatty acids and insulin concentrations in hypothalamic-hyperphagic rats. *Biochem. J.,* 90:620-624.
63. Hartwig, H. G. (1974): Electron microscopic evidence for a retinohypothalamic projection to the suprachiasmatic nucleus of passer demesticus. *Cell Tiss. Res.,* 153:89-99.
64. Haymaker, W., Anderson, E., and Nauta, W. J. H. (1969): *The Hypothalamus.* Charles C Thomas, Springfield, Illinois.
65. Hendrickson, A. E., Wagoner, N., and Cowan, W. (1972): An autoradiographic and electron microscope study of retinohypothalamic connections. *Z. Zellforsch.,* 135:1-26.
66. Höfer, H. (1958): Zur Morphologie der circumventricularen Organe des Zwischenhirnes der Saugetiere. *Vorh. Dtsch. Zool., Ges.,* M:202-251.
67. Hoffman, G. E. (1976): Immunocytochemical localization of luteinizing hormone-releasing hormone (LHRH) in murine and primate brain. *Anat. Rec.,* 184:429-430.
68. Hoffman, G. E., Knigge, K. M., Moynihan-McCourt, Melnyk, V., and Arimura, A. (1977): Luteinizing hormone releasing hormone (LHRH)-containing neuronal fields in mouse brain. *Neurosciences (in press.)*
69. Hökfelt, T. and Ljungdahl, A. (1971): Uptake of $^3$H noradrenaline and Y-$^3$H-aminobutyric acid in isolated tissues of the rat: An autographic and fluorescence microscopic study. *Brain Res.,* 34:87-102.
70. Hökfelt, T. and Fuxe, K. (1972): On the morphology and the neuroendocrine role of the hypothalamic catecholamine neurons. In: *Brain-Endocrine Interaction. Median Eminence: Structures and Functions,* edited by K. M. Knigge, D. E. Scott, and A. Weindl, pp. 228-265. Karger, Basel.
71. Hökfelt, T., Fuxe, K., Goldstein, M., and Johansson, O. (1974): Immunohistochemical evidence for the existence of adrenaline neurons in the rat brain. *Brain Res.,* 66:235-251.
72. Hökfelt, T., Fuxe, K., Goldstein, M., Johansson, O., Park, D., Fraser, H., and Jeffcoate, S. L. (1975): Immunofluorescence mapping of central monoamine and releasing hormone (LRH) systems. In: *Anatomical Neuroendocrinology,* edited by W. E. Stumpf and L. D. Grant, pp. 381-392. Karger, Basel.
73. Holzwarth-McBride, M. A., Hurst, E. M., and Knigge, K. M. (1976): Monosodium glutamate induced lesions of the arcuate nucleus. I. Endocrine deficiency and ultrastructure of the median eminence. *Anat. Rec.,* 186:185-205.
74. Ibuka, N. and H. Kawamura (1975): Loss of circadian rhythm in sleep-wakefulness cycle in the rat by suprachiasmatic nucleus lesions. *Brain Res.,* 96:76-81.
75. Iversen, L. L. and Schon, F. E. (1973): The use of autoradiographic techniques for the identification and mapping of transmitter-specific neurons in the CNS. In: *New Concepts in Neurotransmitter Regulation,* edited by A. J. Mandell, pp. 153-193. Plenum Press, New York.
76. Jacobowitz, D. M. (1975): Fluorescence microscopic mapping of CNS norepinephrine systems in the rat forebrain. In: *Anatomical Neuroendocrinology,* edited by W. E. Stumpf and L. D. Grant, pp. 368-380. Karger, Basel.

77. Jonsson, G. (1971): Quantitation of fluorescence of biogenic monoamines. *Prog. Histochem. Cytochem.,* 2:299–334.
78. Jonsson, G., Fuxe, K., and Hökfelt, T. (1972): On the catecholamine innervation of the hypothalamus, with special reference to the median eminence. *Brain Res.,* 40:271–281.
79. Joseph, S. A., Sorrentino, S., and Sundberg, D. K. (1975): Releasing hormones, LRF and TRF, in the cerebrospinal fluid of the third ventricle. In: *Brain-Endocrine Interaction II. The Ventricular System in Neuroendocrine Mechanisms.* Second Internat. Symposium, edited by K. M. Knigge, D. E. Scott, H. Kobayashi, and S. Ishii, pp. 306–312. Karger, Basel.
80. Jouvet, M. (1967): Mechanisms of the states of sleep, a neuropharmacological approach. *Res. Publ. Assoc. Res. Nerv. Ment. Dis.,* 45:86–126.
81. Jouvet, M. (1972): The role of monoamines and acetylcholine-containing neurons in the regulation of the sleep-waking cycle. *Ergeb. Physiol.,* 64:167–307.
82. Kawakami, M. and Sakuma, Y. (1977): Response of basomedial hypothalamic neurons to electrical and chemical stimulation in female rats. *Neuroendocrinology,* 22 *(in press).*
83. Kawakatsu, Y. (1961): Zur morphologie des ichlussplattenorgans (organum vasculosum laminae terminalis) einiger saugetiere. *Acta Anat. Nippon.,* 36:83–98.
84. Kizer, J. S., Muth, E., and Jacobowitz, D. M. (1976): The effect of bilateral lesions of the ventral noradrenergic bundle on endocrine-induced changes of tyrosine hydroxylase in the rat median eminence. *Endocrinology,* 98:886–893.
85. Kizer, J. S., Palkovits, M., and Brownstein, M. J. (1976): The projections of the A8, A9, and A10 dopaminergic cell bodies: Evidence for a nigral-hypothalamic-median eminence dopaminergic pathway. *Brain Res.,* 108:363–370.
86. Kizer, J. S., Palkovits, M., and Brownstein, M. J. (1976): Releasing factors in the circumventricular organs of the rat brain. *Endocrinology,* 98:311–317.
87. Kizer, J. S., Palkovits, M., Kopin, I. J., Saavedra, J., and Brownstein, M. J. (1976): Lack of effect of various endocrine manipulations on tryptophan hydroxylase activity of individual nuclei of the hypothalamus, limbic system and midbrain of the rat. *Endocrinology,* 98: 743–747.
88. Knigge, K. M. (1974): Role of the ventricular system in neuroendocrine processes. Initial studies on the role of catecholamines in transport of thyrotropin releasing factor. In: *Frontiers in Neurology and Neuroscience Research,* edited by P. Seeman and G. M. Brown, pp. 40–47. University of Toronto Press, Toronto, Ontario.
89. Knigge, K. M. and Silverman, A. J. (1975): Anatomy of the endocrine hypothalamus. In: *Handbook of Physiology,* sec. 7, vol. IV, edited by E. Knobil and W. H. Sawyer, pp. 1–32. American Physiological Society, Washington.
90. Knigge, K. M., Schock, D., and Sladek, J. R. Jr. (1975): Monoamines of median eminence. In: *Brain-Endocrine Interaction II. The Ventricular System in Neuroendocrine Mechanisms,* Second Internat. Symposium, edited by K. M. Knigge, D. E. Scott, H. Kobayashi, and S. Ishii, pp. 282–294. Karger, Basel.
91. Knigge, K. M., Joseph, S. A., Sladek, J. R., Notter, M. F., Morris, M., Sundberg, D. K., Holzwarth, M. A., Hoffman, G. E., and O'Brien, L. (1976): Uptake and transport activity of the median eminence of the hypothalamus. *Int. Rev. Cytol.,* 45:383–408.
92. Knigge, K. M., Hoffman, G., Scott, D. E., and Sladek, J. R., Jr. (1977): Identification of catecholamine and LHRH containing neurons in primary cultures of dispersed cells of the basal hypothalamus. *Brain Res.,* 120:393–405.
93. Knigge, M. K., Joseph, S. A., Hoffman, G., Morris, M., and Donofrio, R. (1977): Role of cerebrospinal fluid in neuroendocrine processes. Fifth Internat. Cong. Endocrinology, Hamburg, 1976. *Excerpta Medica (in press).*
94. Kobayashi, H. (1975): Absorption of cerebrospinal fluid by ependymal cells of the median eminence. In: *Brain-Endocrine Interaction II. The Ventricular System in Neuroendocrine Mechanisms.* Second Internat. Symposium, edited by K. M. Knigge, D. E. Scott, H. Kobayashi, and S. Ishii, pp. 109–122. Karger, Basel.
95. Kobayashi, H., Matsui, T., and Ishii, S. (1970): Functional electron microscopy of the hypothalamic median eminence. *Int. Rev. Cytol.,* 29:281–381.
96. Kobayashi, H., Wada, M., Uemura, H., and Ueck, M. (1972): Uptake of peroxidase from the third ventricle by ependymal cells of the median eminence. *Z. Zellforsch.,* 127: 545–551.

97. Kordon, C. (1976): New data on hormone-neurotransmitter interaction in gonadotropic regulation. In: *Neuroendocrine Regulation of Fertility,* edited by A. Kumar, pp. 180–185. S. Karger, Basel.
98. Koritsanszky, S. and Koves, K. (1976): Data on the absence of axon terminals of medial preoptic area neurons in the surface zone of the median eminence. *J. Neural Transm.,* 38:159–167.
99. Kramer, S. G., Potts, A. M., and Mangnall, Y. (1971): Dopamine: A retinal transmitter. II. Autoradiographic localization of $^3$H-dopamine in the retina. *Invest. Ophthalmol.,* 10:617–624.
100. Kreutzberg, G. W., Schubert, P., and Lux, H. D. (1975): Neuroplastic transport in axons and dendrites. In: *Golgi Centennial Symposium: Perspectives in Neurobiology,* edited by M. Santini, pp. 161–166. Raven Press, New York.
101. Kristensson, K. and Olsson, Y. (1971): Retrograde axonal transport of protein. *Brain Res.,* 29:363–365.
102. Kristensson, K., Olsson, Y. and Sjostrand, J. (1971): Axonal uptake and retrograde transport of exogenous proteins in the hypoglossal nerve. *Brain Res.,* 32:399–406.
103. Kuhlenbeck, H. (1955): Observations on some normal and pathologic histological variations of ventricular lining structures in the human brain. *Anat. Rec.,* 121:325.
104. Laverty, R. and Sharman, D. F. (1965): The estimation of small quantities of 3,4-dihydroxyphenylethylamine in tissues. *Br. J. Pharmacol.,* 24:538–548.
105. Le Beux, Y. J. (1972): An ultrastructural study of the neurosecretory cells of the medial vascular prechiasmatic gland. II. Nerve endings. *Z. Zellforsch.,* 127:439–461.
106. Leranth, C. and Schiebler, T. H. (1974): Uber die Aufnahme von Peroxidase aus dem 3. Ventrikel der Ratte; electronenmikroskopische Untersuchungen. *Brain Res.,* 67:1–11.
107. Leveque, T. F. and Hofkin, G. H. (1961): Demonstration of an alcohol-chloroform insoluble periodic acid-Schiff reactive substance in the hypothalamus of the rat. *Z. Zellforsch.,* 53:185–191.
108. Leveque, T. F. and Hofkin, G. H. (1962): A hypothalamic periventricular PAS substance and neuroendocrine mechanisms. *Anat. Rec.,* 142:252.
109. Leveque, T. F., Stutinsky, F., Porte, A. and Stoekel, M. W. (1967): Ultrastructure of the medial prechiasmatic gland in the rat and mouse. *Neuroendocrinology,* 2:56–63.
110. Lichtensteiger, W. and Langemann, H. (1966): Uptake of exogenous catecholamines by monoamine-containing neurons of the central nervous system: Uptake of catecholamines by arcuatoinfundibular neurons. *J. Pharmacol. Exp. Ther.,* 151:400–408.
111. Lindvall, O., Bjorklund, A., Hökfelt, T., and Ljungdahl, A. (1973): Application of the glyoxylic acid method to Vibratome sections for the improved visualization of central catecholamine neurons. *Histochemie,* 35:31–38.
111a. Lorez, H. P. and Richards, J. G. (1973): Distribution of indolealkylamine nerve terminals in the ventricles of the rat brain. *Z. Zellforsch.,* 144:511–522.
112. Lux, H. D., Schubert, P., Kreutzberg, G. W., and Globus, A. (1970): Exitation and axonal flow: autoradiographic study on motor neurons intracellularily injected with $^3$H-amino acid. *Exp. Brain Res.,* 10:197–204.
113. Makara, G. B., Harris, M. C., and Spyer, K. M. (1972): Identification and distribution of tuberoinfundibular neurones. *Brain Res.,* 40:283–290.
114. Marczynski, T. J. (1976): Serotonin in the central nervous system. In: *Chemical Transmission in Mammalian Central Nervous System,* edited by C. H. Hockman and D. Bieger, pp. 349–429. University Park Press, Baltimore.
115. Martin, J. B., Kantor, J., and Mead, P. (1973): Plasma GH responses to hypothalami, hippocampal and amygdalo-electrical stimulation. Effects of variation in stimulus parameters and treatment with a-methyl-*p*-tyrosine ($\alpha$MT). *Endocrinology,* 92:1354–1361.
116. Martin, J. B., Audet, J., and Saunders, A. (1975): Effects of somatostatin and hypothalamic ventromedial lesions on growth hormone release induced by morphine. *Endocrinology,* 96:839–847.
117. Mason, C. A. and D. W. Lincoln (1976): Visualization of the retino-hypothalamic projection in the rat by cobalt precipitation. *Cell Tiss. Res.,* 168(1):117–131.
118. McEwen, B. S. and Pfaff, D. W. (1973): Chemical and physiological approaches to neuroendocrine mechanisms: attempts at integration. In: *Frontiers in Neuroendocrinology,* edited by F. W. Ganong and L. Martini, pp. 267–335. Oxford University Press, New York.

119. Meier, R. E. (1973): Autoradiographic evidence for a direct retino-hypothalamic projection in the avian brain. *Brain Res.*, 53:417–421.
120. Michel, F. and Roffwarg, H. (1967): The split brain stem preparation: Effect in sleep wakefulness cycle. *Experientia*, 23:126–128.
121. Mikami, S. (1975): A correlative ultrastructural analysis of the ependymal cells of the third ventricle of Japanese Quail, Coturnix, coturnix japonica. In: *Brain-Endocrine Interaction II. The Ventricular System in Neuroendocrine Mechanisms*, Second Internat. Symposium, edited by K. M. Knigge, D. E. Scott, H. Kobayashi, and S. Ishii, pp. 80–93. Karger, Basel.
122. Mikami, S. (1976): Ultrastructure of the organum vasculosum of the lamina terminalis of the Japanese quail, *coturnix japonica*. *Cell Tissue Res.*, 172:227–243.
123. Millhouse, E. O. (1973): The organization of the ventromedial hypothalamic nucleus. *Brain Res.*, 55:71–87.
124. Millhouse, E. O. (1973): Certain ventromedial hypothalamic afferents. *Brain Res.*, 55:89–105.
125. Moore, R. Y. (1973): Retinohypothalamic projection in mammals: A comparative study. *Brain Res.*, 49:403–409.
126. Moore, R. Y. and Eichler, V. B. (1972): Loss of circadian adrenal corticosterone rhythm following suprachiasmatic lesions in the rat. *Brain Res.*, 42:201–206.
127. Moore, R. Y. and Lenn, N. J. (1972): A retino-hypothalamic projection in the rat. *J. Comp. Neurol.*, 146:1–14.
128. Moore, R. Y. and Klein, D. C. (1974): Visual pathways and the central neural control of a circadian rhythm in pineal serotonin n-acetyltransferase activity. *Brain Res.*, 71:17–34.
129. Moreshita, H., Kauamoto, M., Masuda, Y., Higuchi, F., Tomioka, M., Nagamachi, N., Mitani, H., Ozasa, T., and Adachi, H. (1974): Quantitative histological changes in the hypothalamic nuclei in prepuberal, puberal and postpuberal female rat. *Brain Res.*, 76:41–67.
130. Moss, R. L., Kelly, R. M., and Riskind, P. (1975): Tuberoinfundibular neurons: dopaminergic and norepinephrinergic sensitivity. *Brain Res.*, 29:265–277.
131. Nakai, Y. and Naito, N. (1975): Uptake and bidirectional transport of peroxidase injected into the blood and cerebrospinal fluid by ependymal cells of the median eminence. In: *Brain-Endocrine Interaction II. The Ventricular System in Neuroendocrine Mechanisms*. Second Internat. Symposium, edited by K. M. Knigge, D. E. Scott, H. Kobayashi and S. Ishii, pp. 94–108. Karger, Basel.
132. Nauta, W. J. H. and Gygax, P. A. (1951): Silver impregnation of degenerating axon terminals in the central nervous system. *Stain Technol.*, 26:5–11.
132a. Nauta, W. J. H. and Haymaker, W. (1969): Hypothalamic nuclei and fiber connections. In: *The Hypothalamus*, edited by W. Haymaker, E. Anderson and W. J. H. Nauta, pp. 136–208. Charles C Thomas, Springfield, Illinois.
133. Nishino, H., Kiyomi, K., and Brooks, C. M. (1976): The role of suprachiasmatic nuclei of the hypothalamus in the production of circadian rhythm. *Brain Res.*, 112:45–59.
134. Northcutt, R. G. and Butler, A. B. (1974): Retinal projections in the Northern water snake *Natrix sipedon sipedon*. *J. Morphol.*, 142:117–136.
135. Nojyo, Y., Ibata, Y., and Sand, Y. (1976): Demonstration of the tuberoinfundibular tract of the cat: Fluorescence histochemistry and electron microscopy. *Cell Tiss. Res.*, 168:289–301.
136. Nozaki, M., Kobayashi, H., Yanagisawa, M. and Bando, T. Monoamine fluorescence in the median eminence of the Japanese quail, *Coturnix Coturnix Japonica* following medial basal hypothalamic deafferentation. *Cell Tiss. Res.*, 164:425–434.
137. Oksche, A., Oehmke, H. J., and Hortwig, H. G. (1974): A concept of neuroendocrine cell complexes. In: *Neurosecretion: The Final Neuroendocrine Pathway*. Sixth Internat. Congress of Neurosecretion, edited by F. Knowles and L. Vollrath, pp. 154–164. Springer-Verlag, Berlin.
138. Palkovits, M., Brownstein, M., Saavedra, J. M., and Axelrod, J. (1974): Norepinephrine and dopamine content of hypothalamic nuclei. *Brain Res.*, 77:137–149.
139. Palkovits, M., Arimura, A., Brownstein, M., Schally, A. V., and Saavedra, J. M. (1974): Luteinizing hormone-releasing hormone LHRH content of the hypothalamic nuclei in rat. *Endocrinology*, 95:554–558.

140. Paull, W. K., Scott, D. E., and Boldosser, W. G. (1974): A cluster of supraependymal neurons located within the infundibular recess of the rat third ventricle. *Am. J. Anat.*, 140:129–133.
141. Pelletier, G. (1976): Immunohistochemical localization of hypothalamic hormones at the electron microscope level. In: *Hypothalamus and Endocrine Functions*, edited by F. Labrie, J. Meites and G. Pellitier, pp. 433–450. Plenum Press, New York.
142. Phillips, I. A. and Hoffman, W. E. (1976): Sensitive sites in the brain for the blood pressure and drinking responses to angiotensin II. In: *Actions of Angiotensin and Related Hormones*, edited by J. P. Buckley and C. Fenairo, pp. 325–356. Pergamon Press, New York.
143. Pickel, V. M., Joh, T. H., and Reis, D. J. (1976): Monoamine-synthesizing enzymes in central dopaminergic, noradrenergic and serotonergic neurons. Immunocytochemical localization by light and electron microscopy. *J. Histochem. Cytochem.*, 24:792–806.
144. Pickel, V. M., Shikimi, T., Joh, T. H., and Reis, D. J. (1975): Immunohistochemical localization of tryptophan hydroxylase in rat brain by light and electron microscopy. *Anat. Rec.*, 181:450.
145. Poitras, D. and Parent, A. (1975): A fluorescence microscopic study of the distribution of monoamines in the hypothalamus of the cat. *J. Morphol.*, 145:387–408.
146. Printz, R. H. and Hall, J. L. (1974): Evidence for a retinohypothalamic pathway in the golden hamster. *Anat. Rec.*, 179:57–66.
147. Pujol, J. F., Buguet, A., and Froment, J. L. (1971): The central metabolism of serotonin in the cat during insomnia: A biochemical and neurophysiological study after $p$-chlorophenylalanine or destruction of the raphe system. *Brain Res.*, 29:195–212.
148. Raisman, G. and Field, P. M. (1971): Sexual dimorphism in the preoptic area of the rat. *Science*, 173:731–733.
149. Raisman, G. and Field, P. M. (1973): A quantitative investigation of the development of collateral reinnervation after partial deafferentation of the septal nuclei. *Brain Res.*, 50:241–264.
150. Renaud, L. P. (1976): Tubero-infundibular neurons in the basomedial hypothalamus of the rat: Electrophysiological evidence for axon collaterals to hypothalamic and extrahypothalamic areas. *Brain Res.*, 105:59–72.
151. Richards, J. G., Lorez, H. P., and Tranzer, J. P. (1973): Indolealkylamine nerve terminals in cerebral ventricles: Identification by electron microscopy and fluorescence histochemistry. *Brain Res.*, 57:277–288.
152. Richards, J. G. and Tranzer, J. P. (1974): Ultrastructural evidence for the localization of an indoleamine in the supraependymal nerve from combined cytochemistry and pharmacology. *Experientia*, 30:287–289.
153. Richards, J. G. and Tranzer, J. P. (1974): The characterization of monoaminergic nerve terminals in the brain by fine structural chemistry. In: *Neurosecretion: The Final Neuroendocrine Pathway*, edited by F. Knowles and L. Vollrath, pp. 240–259. Springer-Verlag, Berlin.
154. Rinne, U. K. (1970): Experimental electron microscopic studies on the neurovascular link between the hypothalamus. In: *Aspects of Neuroendocrinology*, edited by W. Bargmann and B. Scharrer, pp. 220–228. Springer-Verlag, Berlin.
155. Rohlich, P. and Wenger, T. (1969): Elektronemikroskopische Untersuchungen am organum vasculosum laminae terminalis der ratte. *Z. Zellforsch.*, 102:483–506.
156. Rusak, B. and Morin, L. P. (1976): Testicular responses to photoperiod are blocked by lesions of the suprachiasmatic nuclei in golden hamsters. *Biol. Reprod.*, 15:366–374.
157. Saavedra, J. M. (1975): Localization of biogenic amine-synthesizing enzymes in discreet hypothalamic nuclei. In: *Anatomical Neuroendocrinology*, edited by W. E. Stumpf and L. D. Grant, pp. 397–400. Karger, Basel.
158. Saavedra, J. M., Palkovits, M., Brownstein, M. J., and Axelrod, J. (1974): Serotonin distribution in the nuclei of the rat hypothalamus and preoptic region. *Brain Res.*, 77:157–165.
159. Saavedra, J. M., Brownstein, M., Palkovits, M., Kizer, S., and Axelrod, J. (1974): Tyrosine hydroxylase and dopamine-$\beta$-hydroxylase: Distribution in the individual rat hypothalamic nuclei. *J. Neurochem.*, 23:869–871.
160. Saavedra, J. M., Palkovits, M., Brownstein, M., and Axelrod, J. (1974): Localization of phenylethanolamine $N$-methyl transferase in the rat brain nuclei. *Nature*, 248:695–696.

161. Saavedra, J. M., Palkovits, M., Brownstein, M., and Axelrod, J. (1974): Serotonin distribution in the nuclei of the rat hypothalamus and preoptic region. *Brain Res.*, 77:157–165.
162. Santini, M., ed. (1975): *Golgi Centennial Symposium: Perspectives in Neurobiology*, Raven Press, New York.
163. Saper, C. B., Swanson, L. W., and Cowan, W. M. (1976): The efferent connections of the ventromedial nucleus of the hypothalamus of the rat. *J. Comp. Neurol.*, 169:409–442.
164. Sawyer, C. H. (1972): Functions of the amygdala related to the feedback actions of gonadal steroid hormones. In: *The Neurobiology of the Amygdala*, edited by B. E. Elefteriou, pp. 745–762. Plenum Press, New York.
165. Sawyer, C. H. (1975): Some recent developments in brain-pituitary-ovarian physiology. *Neuroendocrinology* 17:97–124.
166. Silverman, A. J., Knigge, K. M., Ribas, J. L., and Sheridan, M. N. (1973): Transport capacity of median eminence. III. Amino acid and thyroxine transport of organ-cultured median eminence. *Neuroendocrinology*, 11:107–118.
167. Sawaki, Y. (1977): Retino hypothalamic projection: Electrophysiological evidence for the existence in female rats. *Brain Res.*, 120:336–341.
168. Sawaki, Y. and Yagi, K. (1973): Electrophysiological identification of cell bodies of the tubero-infundibular neurones in the rat. *J. Physiol.*, 230:75–85.
169. Sawaki, Y. and Yagi, K. (1976): Inhibition and facilitation of antidromically identified tubero-infundibular neurones following stimulation of the median eminence in the rat. *J. Physiol.*, 260:447–460.
170. Scapagnini, U., Moberg, C. P., VanLoon, G. R., de Groot, J., and Ganong, W. F. (1971): Relation of brain 5-hydroxytryptophan control to the diurnal variation in plasma corticosterone in the rat. *Neuroendocrinology*, 7:90–95.
171. Scott, D. E., Krobisch-Dudley, G. and Knigge, K. M. (1974): The ventricular system in neuroendocrine mechanisms. II. *In vivo* monoamine transport by ependyma of the median eminence. *Cell Tiss. Res.*, 154:1–16.
172. Scott, D. E., Krobisch-Dudley, G., Paull, W. K., Kozlowski, G. P., and Ribas, G. (1975): The primate median eminence. I. Correlative scanning transmission electron microscopy. *Cell Tiss. Res.*, 162:61–73.
173. Scott, D. E., Sladek, J. R. Jr., Knigge, K. M., Krobisch-Dudley, G., Kent, D. L., and Sladek, C. D. (1976): Localization of dopamine in the endocrine hypothalamus of the rat. *Cell Tiss. Res.*, 166:461–473.
174. Scott, D. E., Krobisch-Dudley, G., Paull, W. K. and Kozlowski, G. P. (1977): The ventricular system in neuroendocrine mechanisms. III. Suprapendymal neuronal networks in the primate brain. *Cell Tiss. Res.*, 179:235–254.
175. Schinko, I., Rohrschneider, I. and Wetstein, R. (1972): Electronenmikroskopische Untersuchungen am Subfornikalorgan der maus. *Z. Zellforsch.*, 123:277–294.
176. Shaskin, E. G. and Snyder, S. H. (1970): Kinetics of serotonin accumulation into slices from rat brain: Relationship to catecholamine uptake. *J. Pharm. Exp. Ther.*, 175:404–418.
177. Sidman, R. L. (1970): Autoradiographic methods and principles for study of the nervous system with thymidine-H³. In: *Contemporary Research Methods in Neuroanatomy*, edited by W. J. H. Nauta and E. O. Ebbesson, pp. 252–274. Springer-Verlag, New York.
178. Sladek, J. R., Jr., Cheung, Y., and Kent, D. (1976): New sites of monoamine localization in endocrine hypothalamus. *Anat. Rec.*, 181:481–482.
179. Sladek, J. R., Jr., and McNeill, T. H. (1977): Fluorescence histochemical identification of monoamine containing subependymal cells in mammalian and avian diencephalon. *J. Histochem. Cytochem. (in press)*.
180. Stanfield, B. and W. M. Cowan (1976): Evidence for a change in the retino-hypothalamic projection in the rat following early removal of one eye. *Brain Res.*, 104:129–136.
181. Stephan, F. K. and I. Zucker (1972): Circadian rhythms in drinking behavior and locomotor activity of rats are eliminated by hypothalamic lesions. *Proc. Nat. Acad. Sci. USA*, 69(6):1583–1586.
182. Stetson, M. H. and M. Watson-Whitmyre (1976): Nucleus suprachiasmaticus: the biological clock in the hamster. *Science*, 191:197–199.
183. Stevenson, J. A. F. (1969): Neural control of food and water intake. In: *The Hypothala-*

*mus,* edited by W. Haymaker, E. Anderson and W. J. H. Nauta, pp. 524–621. Charles C Thomas, Springfield, Illinois.
184. Stumpf, W. E. (1970): Estrogen-containing neurons and estrogen-neuron systems in the periventricular brain. *Am. J. Anat.,* 129:207–218.
185. Sutin, J. (1963): An electrophysiological study of the hypothalamic ventromedial nucleus in the cat. *Electroencephalogr. Clin. Neurophysiol.,* 15:786–795.
186. Swanson, L. W. (1976): An autoradiographic study of the efferent connections of the preoptic region in the rat. *J. Comp. Neurol.,* 167:227–256.
187. Swanson, L. W. and Cowan, W. M. (1975): The efferent connections of the suprachiasmatic nucleus of the hypothalamus. *J. Comp. Neurol.,* 160:1–12.
188. Szentágothai, J., Flerkó, B., Mess, B., and Halász, B. (1968): *Hypothalamic Control of the Anterior Pituitary.* Akademiai Kiado, Budapest.
189. Takeichi, M. and Noda, Y. (1974): Light and electron microscope studies of the ventromedial hypothalamic nucleus of the cat, with special reference to the fine structure of neurons and synapses. *Folia Psyciatr. Neurol. Jpn.,* 28:45–64.
190. Thorpe, P. A. (1975): The presence of a retinohypothalamic projection in the ferret. *Brain Res.,* 85:343–346.
191. Ungerstedt, U. (1971): Stereotaxic mapping of the monoamine pathways in the rat brain. *Acta Physiol. Scand. (Suppl.),* 367:1–48.
192. Usui, T. (1968): Electron microscopic studies on the ependymal cells of the organon vasculosum laminae terminalis in the adult rat. *Bull. Tokyo Med. Dent. Univ.,* 15:1–18.
193. Vaala, S. S. and Knigge, K. M. (1974): Transport capacity of median eminence: in vitro uptake of $^3$H-LRF. *Neuroendocrinology,* 15:147–157.
194. Vandesande, F., Dierick, K. and DeMey, J. (1975): Identification of vasopressin-neurophysin producing neurons of the rat suprachiasmatic nuclei. *Cell Tiss. Res.,* 156:381–390.
195. Vanegas, H. and Ebbesson, S. O. E. (1973): Retinal projections in the PEROH-like teleosti eugerres plumieri. *J. Comp. Neurol.,* 151:331–357.
196. Wann, D. F., Woolsey, T. A., Dierker, M. L. and Cowan, W. M. (1973): An on-line digital-computer system for the semiautomatic analysis of Golgi-impregnated neurons. *IEEE Trans. Biomed. Eng.,* BME-20:233–247.
197. Wartenberg, H., Hadziselimovic, F., and Seguchi, H. (1971): Experimentelle Untersuchungen uber die Passage der Blutgewebs-Schranke in den zircumventrikularen Organen des Meerschwein-Chengehirns. *Verh. Anat. Ges.,* 66:345–355.
198. Watkins, W. B. (1975): Neurosecretory neurons in the hypothalamus and median eminence of the dog and sheep as revealed by immunohistochemical methods. *Ann. N.Y. Acad. Sci.,* 248:134–152.
199. Weindl, A. (1969): Electron microscopic observations on the organum vasculosum of the lamina terminalis (OVLT) after intravenous injection of horseradish peroxidase. *Neurology,* 19:295.
200. Weindl, A. (1973): Neuroendocrine aspects of circumventricular organs. In: *Frontiers in Neuroendocrinology,* edited by W. F. Ganong and L. Martini, pp. 3–22. Oxford Univ. Press, London.
201. Weindl, A. (1974): Structural and functional investigations of the mammalian organum vasculosum of the lamina terminalis. *Ph.D. Thesis,* University of Rochester, Rochester, N.Y.
202. Weindl, A. and Joynt, R. J. (1972): The median eminence as a circumventricular organ. In: *Brain-Endocrine Interaction. Median Eminence: Structure and Function,* edited by K. M. Knigge, D. E. Scott, and A. Weindl, pp. 280–297. Karger, Basel.
203. Weindl, A. and Shinko, I. (1975): Vascular and ventricular neurosecretion in the organum vasculosum of the lamina terminalis of the golden hamster. In: *Brain-Endocrine Interaction II. The Ventricular System in Neuroendocrine Mechanisms.* Second Internat. Symposium, edited by K. M. Knigge, D. E. Scott, H. Kobayashi, and S. Ishii, pp. 190–203. Karger, Basel.
204. Weindl, A., Schwink, A., and Wetzstein, R. (1968): Der Feinbau des Gefassorgans der Lamina terminalis beim Kaninchen. II. Das neuronale und gliale Gewebe. *Z. Zellforsch.,* 85:552–600.
205. Weindl, A., Schwink, A., and Wetzstein, R. (1967): Der feinbau des gefaborgans der lamina terminalis beim kaninchen. I. Die gefaesse. *Z. Zellforsch.,* 79:1–48.
206. Weiner, R. I., Pattou, E., Kerdelhue, B., and Kordon, C. (1975): Differential effects of

hypothalamic deafferentiation upon luteinizing hormone-releasing hormone in the median eminence and organum vasculosum of the lamina terminalis. *Endocrinology,* 97:1597–1600.
207. Wenger, T. (1976): Ultrastructural changes in the vascular organ of the lamina terminalis following ovariectomy and hypophysectomy in the rat. *Brain Res.,* 101:95–102.
208. Wheaton, J. E., Krulich, L. and McCann, S. M. (1975): Localization of luteinizing hormone-releasing hormone in the preoptic area and hypothalamus of the rat using radio-immunoassay. *Endocrinology,* 97:30–38.
209. Weiss, P. A. (1969): Neuronal dynamics and neuroplasmic ("axonal") flow. In: *Cellular Dynamics of the Neuron,* edited by Samuel H. Barondes, pp. 3–34. Academic Press, New York.
210. Wenisch, H. J. (1976): Retino hypothalamic projection in the mouse: Electron microscopic and iontophoretic investigations of hypothalamic and optic centers. *Cell Tiss. Res.,* 167(4):547–561.
211. Wislocki, G. B. and King, L. S. (1936): The permeability of the hypophysis and hypothalamus to vital dyes, with a study of the hypophyseal vascular supply. *Am. J. Anat.,* 58:421–472.
212. Wislocki, G. B. (1940): Peculiarities of the cerebral blood vessels of the opossum: diencephalon, area postrema and retina. *Anat. Rec.,* 78:119–137.
213. Yagi, K. and Sawaki, Y. (1975): Recurrent inhibition and facilitation: demonstration in the tubero-infundibular system and effects of strychnine and picrotoxin. *Brain Res.,* 84:155–159.
214. Yagi, K. and Sawaki, Y. (1975): Recurrent neural circuits in the tubero-infundibular system. In: *Brain-Endocrine Interaction II. The Ventricular System in Neuroendocrine Mechanisms.* Second Internat. Symposium, edited by K. M. Knigge, D. E. Scott, H. Kobayashi, and S. Ishii, pp. 257–269. Karger, Basel.
215. Yagi, K. and Sawaki, Y. (1977): Medial preoptic nucleus: Inhibition and facilitation of spontaneous activity following stimulation of the median eminence in female rats. *Brain Res.,* 120:342–346.
216. Zimmerman, E. A., Hsu, K. C., Robinson, A. G., Carmel, P. W., Frantz, A. G., and Tannenbaum, M. (1973): Studies on neurophysin secreting neurons with immunoperoxidase techniques employing antibody to bovine neurophysin. I. Light microscopic findings in monkey and bovine tissues. *Endocrinology,* 92:931–940.
217. Zimmerman, E. A., Defendini, R., Sokol, H. W., and Robinson, A. G. (1975): The distribution of neurophysin-secreting pathways in mammillian brain. Light microscopic studies using the immunoperoxidase technique. In: *Abstracts of International Conference on Neurophysin Proteins: Carriers of Peptide Hormones. Ann. N.Y. Acad. Sci.,* 248:92–112.

*The Hypothalamus*, edited by S. Reichlin,
R. J. Baldessarini, and J. B. Martin,
Raven Press, New York, © 1978.

# Organization of LRF–and SRIF–Neurons in the Endocrine Hypothalamus

*Karl M. Knigge, Shirley A. Joseph, and Gloria E. Hoffman
In collaboration with Akira Arimura and Ludwig Sternberger

> *The problem of neurosecretion still offers a wide field for histological and cytological research, on the basis of which future investigation of the physiological action of the secretory products of the nerve cells will be conducted.*
>
> Ernst and Berta Scharrer (11)

The 1939 symposium of the Association for Research in Nervous and Mental Diseases represented an approximate midpoint in 20 years of struggle, for Ernst and Berta Scharrer, often against vehement opposition, to establish neurosecretion as a fundamental activity of neurons. In 1928 their first publication suggested a secretory function of nerve cells. In 1939, their suggestion that supraoptic neurons were the source of posterior lobe hormones was met with extraordinary coolness. At that time, the scientific community was reluctant to consider any possibility other than the view that antidiuretic hormone was formed by "particular cells in the neurohypophysis" (neural lobe) and that they were neurogenically stimulated by the supraoptic nuclei. The introduction by Bargmann of a chrom-alum hematoxylin histological procedure for "neurosecretory material" catalyzed rapidly the overwhelming evidence that supraoptic and paraventricular neurons were indeed engaged in the extraordinary activity of synthesizing hormones. Berta Scharrer (10) has provided an account of the struggle, disappointments, and triumph of those years.

The model of neurosecretion provided by the supraoptico-hypophysial system was adopted as the probable *modus operandi* of other hypothalamic neurons whose secretory activity was thought to be associated with control of adenohypophysial function. The last 30 years have witnessed remarkable achievements in neuroendocrinology—the accumulation of much fact as well as fiction—all without ever seeing the neurons in question or learning much about their anatomical organization. A variety of schema and terminologies have been introduced to conceptualize their organization and define

---

* Department of Anatomy, University of Rochester, School of Medicine, 601 Elmwood Avenue, Rochester, New York 14642.

the control systems they represent. They have accumulated a variety of names such as the parvicellular neurosecretory system, the hypothalamo-adenohypophysial system, release-forming and release-regulating system, and tuberoinfundibular neuronal system. Like the Scharrers in 1939, morphologists today have had to patiently await the development of the cytological technique with which to visualize these cells and bring reality to their form, location, and organization. Immunofluorescence was applied first by Barry and Dubois in 1973 (2) to demonstrate luteinizing hormone-releasing factor (LRF)-containing neurons in the preoptic region of the guinea pig. The last three years of research with immunocytochemical histology have yielded some startling insights into the architecture of these neuron systems. Although still in its infancy as a major cytochemical technique with respect to localization of brain hormones, it is already apparent that immunocytochemistry, alone or in combination with other procedures, represents a powerful research tool of the morphologist. Information gained thus far has already confirmed some of our concepts of neural control of the adenohypophysis; we have also been impressed that a considerable part of the apparent organization of neurohormone-producing cells extends beyond the classical tenets of neuroendocrinology.

## LRF AND SRIF NEURONS AND THEIR PROJECTIONS

The visualization of LRF-immunoreactive neurons by Barry and Dubois not only offered a preliminary indication of their topographical location in the brain, but also provided the first definitive evidence that this neurohormone was indeed localized in and probably synthesized by neurons. Following this important contribution, there appeared numerous reports on the immunocytochemical localization of LRF in brain of several species. These initial studies demonstrated LRF-immunoreactive material in nerve fibers and in presumed sites of terminal distribution such as median eminence and organum vasculosum of the lamina terminalis. Interestingly, there was a general failure to visualize LRF in neuron perikarya, and it was initially concluded that this indicated an absence of significant amounts of hormone in neuron cell bodies. Information based on radioimmunoassay of hormone extracted from various regions of the hypothalamus was, however, in conflict with this view; on theoretical grounds, the amount of radioimmunoassayable hormone present in the region of the arcuate nucleus should be demonstrable immunocytochemically, and much of this should be in perikarya.

Hoffman (3), using a Sorrentino F antiserum, demonstrated a pool of LRF-containing neurons in the retrochiasmatic-arcuate region. Subsequently, using Arimura 710 and 743 antisera, a second group of neurons was found in the preoptic-septal region (4,5,7). An analysis of the staining of hormone in neuron perikarya with different LRF antisera is presented later in this chapter. The topography of the LRF-containing neurons indicates a significant feature of their anatomical organization. LRF-containing neurons are not localized within the traditional nuclear groups of the hypothalamus and preoptic area. LRF neurons in the arcuate region are located in the lateral portions of the arcuate nucleus, extend into the internuclear regions rostral and lateral to the arcuate nucleus and below the ventromedial. Neurons of the preoptic-septal region (field II) are loosely scattered in the medial division of these areas as well as in the bed nucleus of the anterior commissure and the nucleus of the diagonal Band of Broca (Fig. 9). We have designated this as field I (Figs. 8, 9).

Figures 1 to 4 illustrate the cytological appearance of LRF-containing neurons of fields I and II as demonstrated by the immunocytochemical procedure using the unlabeled antibody enzyme method of Sternberger (13). Immunoreactive cells are 12 to 20 $\mu$m in diameter; fine granular reaction product generally fills the cytoplasm uniformly and reveals the negative image of a slightly eccentric nucleus. A single process, considered to be the axon, is seen frequently; shortly after leaving the perikaryon, it assumes a beaded appearance. Additional processes are rarely seen in field I suggesting that LRF may not be present in dendrites of these cells. Our studies with field I LRF neurons in tissue culture (8) reveal that the majority are bipolar. The neurons of field II are more clearly bipolar and exhibit a conspicuous vertical orientation of their processes. Projections from LRF fields I and II do not appear to form discrete bundles, but rather remain as diffuse fiber systems. Analysis of the projections are hampered also by the discontinuity of fiber staining; characteristically, only short, beaded segments are seen. The median eminence and organum vasculosum contain abundant immunoreactive material and are considered two major sites of terminal distribution.

Our impression is that the median eminence is a major terminus of field I neurons; fibers from these cells appear to project anteriorly and possibly project also the organum vasculosum. Field II neurons project to the organum vasculosum and appear to contribute to the median eminence; their ventral course is in the periventricular zone of the hypothalamus, close to the ependymal lining of the third ventricle. Fibers from both fields I and II can be seen for some distance radiating in directions which clearly suggest they are distributing to regions other than median eminence or organum vasculosum lamina terminalis (OVLT) (Fig. 10). Projections are made anteriorly into the septal region and diagonal Band of Broca, and dorsally to the anterior and periventricular thalamus. The ultimate destination of these paths are not known at present. In the guinea pig, significant concentrations of fibers (or terminals) are seen in the pretectal region and outer most strata of the superior colliculus. Silverman (12), in the same species using the Sorrentino F antisera, has described an additional projection into the ventral tegmentum.

In the several species examined thus far (rat, mouse, guinea pig, squirrel

**FIG. 1.** LRF neurons in field I of mouse hypothalamus, stained with Sorrentino F antiserum. ×750.

**FIG. 2.** LRF neuron in field II (preoptic-septal area) of mouse brain stained with Arimura 710 antiserum. ×750.

**FIG. 3.** LRF neurons in field II of mouse brain, stained with Arimura 743 antiserum. ×750.

**FIG. 4.** LRF neurons in tuberal nuclei of chicken hypothalamus, stained with F antiserum (preparation of Jeffrey Manasse). ×750.

**FIG. 5.** SRIF neurons in dog hypothalamus, stained with Joseph Charlie-8-antiserum. ×750.

**FIG. 6.** SRIF neuron in rat hypothalamus. ×750.

**FIG. 7.** SRIF neurons in the periventricular stratum of rat hypothalamus (preparation of Carol Bennett Clarke). ×750.

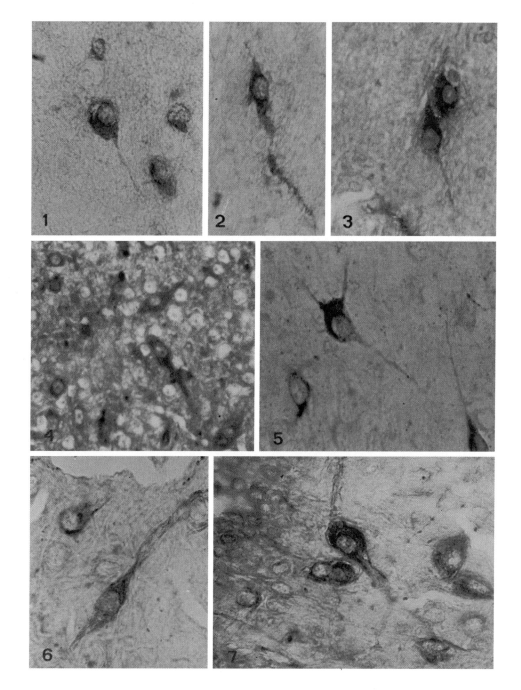

# ANATOMY OF THE ENDOCRINE HYPOTHALAMUS

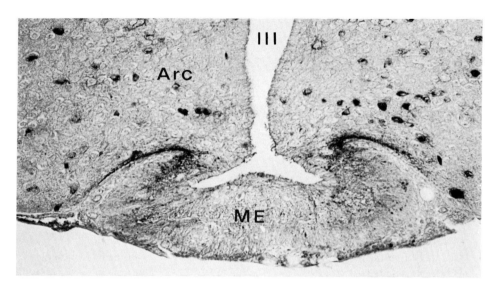

**FIG. 8.** Frontal section of mouse hypothalamus indicating the topography of neurons stained with F antiserum. This location (field I) extends through portions of the arcuate nucleus and lateral tuberal area beneath the ventromedial nucleus. *Arc:* arcuate nucleus; *ME:* median eminence; *III:* 3rd ventricle. ×277.

monkey, and dog) hypothalamic somatotropin release inhibiting factor (SRIF)-immunoreactive neurons are located in a fairly broad expanse of the periventricular stratum of the anterior hypothalamus as described by Parsons et al. (9) and Alpert et al. (1). Their proximity to the ventricular ependyma is remarkable; in the dog, loci of immunoreactive fibers and presumed terminals are seen between ependyma and in the 3rd ventricle (Fig.

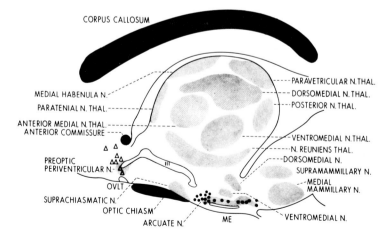

**FIG. 9.** Schematic midsagittal projection of LRF fields I (arcuate-retrochiasmatic area, *filled circles*) and II (preoptic-septal area, *open triangles*) of mouse brain.

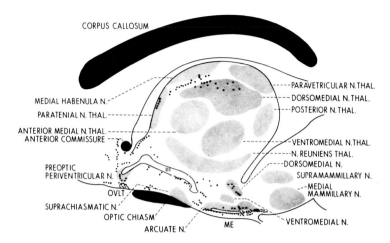

**FIG. 10.** Schematic midsagittal projection of fiber distribution from fields I and II of mouse brain.

11). In mouse, a consistent field of immunoreactive neurons is present in the dorsocaudal region of the interpeduncular nucleus (Fig. 12). Using the Joseph Charlie-8-antisera, cytological detail of immunoreactive neurons is adequate to make some initial comments on the structure of these cells

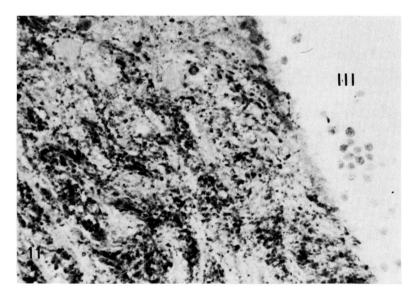

**FIG. 11.** Projections from anterior periventricular SRIF neurons along the ventricular wall of dog hypothalamus. Massive amounts of immunoreactive material are seen in this fiber system; projections to the ventricular wall are apparent. *III:* 3rd ventricle. ×675.

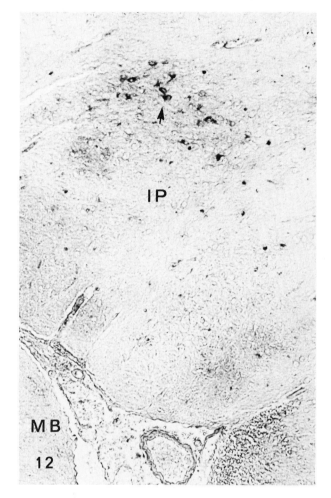

**FIG. 12.** Midsagittal section of mouse brain, indicating the position (*arrow*) of a field of SRIF cells in the dorsal region of the interpeduncular nucleus (IPN). *M:* mammillary body. ×206.

(Fig. 5–7). Immunoreactive antigen-antibody-PAP complex generally fills the cytoplasm in the form of irregular shaped clumps or large granules; the appearance in the perikaryon is distinctly different from that in immunoreactive axon segments or terminal distribution in median eminence. In dog, reaction product fills dendrites for 20–50 nm (Fig. 5), and the processes appear to extend for long distances (100–200 nm) without branching. In rat, reaction product is present in dendrites in such a way as to outline them for considerable distances (Fig. 6,7). In contrast to LRF-positive neurons which appear to be largely bipolar, a significant number of the SRIF neurons are multipolar. Axons appear to originate from the cell body at a relatively inconspicuous axon hillock and become narrow in diameter almost immediately. No cell or dendrite orientation within the SRIF field has been observed in preparations studied thus far.

Median eminence receives a massive SRIF input, especially in the rat, mouse, and dog (Fig. 13). This projection is derived, at least in part, from the anterior periventricular field of cells. In light of the limited number of SRIF cells, terminal arborization of each axon in the median eminence must be enormous. In dog, the median eminence projection of SRIF is conspicious

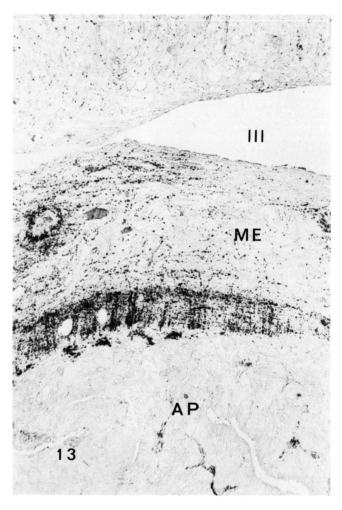

**FIG. 13.** Sagittal section of median eminence and pituitary gland of the dog. In this region a dense SRIF projection is present in the zona externa of the median eminence (ME). AP, adenohypophysis; III, 3rd ventricle. ×255.

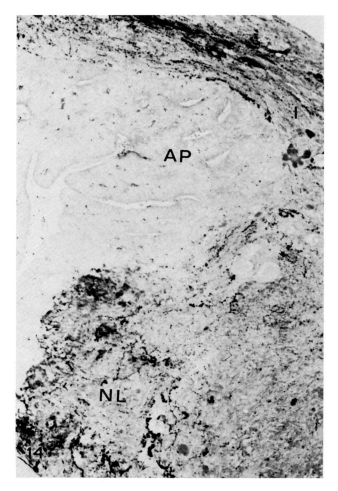

**FIG. 14.** Midsagittal section of dog median eminence and pituitary gland, illustrating the SRIF projection through the infundibulum (*I*) into the neural lobe (*NL*). A: adenohypophysis. ×370.

also because a significant portion continues into the neural lobe (Fig. 14). The organum vasculosum receives a detectable but relatively minor SRIF input. The ventromedial nucleus of the hypothalamus contains a conspicuous amount of SRIF immunoreactive reaction product (Fig. 15). In addition to beaded fiber profiles, fairly delicate immunoreactive granules are present around VMH neurons, especially in the posterior portion of the nucleus. Although our experience to date with respect to the morphological characteristics of terminal distribution of these projections is limited, their appearance in VMH strongly suggests that this nucleus receives afferent input from the anterior periventricular field. The projections of the interpeduncu-

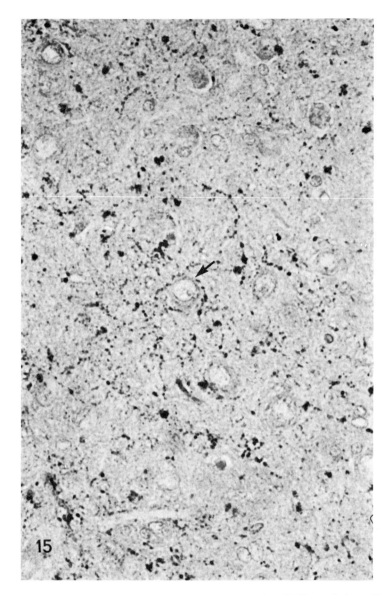

**FIG. 15.** High power photomicrograph of immunoreactive SRIF beaded profiles and terminal distribution around neurons (*arrow*) of the ventromedial nucleus of the hypothalamus of dog. The posterior portion of this nucleus appears to be especially innervated by projections from the periventricular field of SRIF neurons. ×800.

lar groups of SRIF neurons has been difficult to trace and no comment can be made at this time regarding their terminal distribution.

## IMMUNOSPECIFICITY AND NEURONS OF FIELD I

As noted above, perikarya of field I in mouse and rat hypothalamus were first seen with antiserum F and those in field II subsequently with antisera 710 and 743. Our initial interpretation of these results was to suggest that we were dealing with two anatomically, and perhaps functionally, separate fields of LRF-producing neurons (7). Further systematic study with antisera generated against a variety of LRF conjugates, as well as antisera of different "ages," has led us to consider the possibility that we may be visualizing two different neurohormones. Our analysis also suggests that we must consider the possibility that neurohormones are present in different physicochemical states in perikarya, axons, and terminals. Table 1 tabulates the staining of perikarya in fields I and II and fibers in median eminence and OVLT as demonstrated with antisera generated against different LRF-conjugates. Figure 16 schematically illustrates the possible sites of antigen (LRF)-antibody binding with four of these different antisera, based on the immunological indications that the three amino acids on each arm of the LRF molecule are the strongest determinants of its antigenicity. The site at which the bovine serum albumin (BSA) is conjugated is considered to be "blind" with respect to determining the conformation of the antibody. Figure 17 schematically illustrates the antigen-antibody relationships between LRF and these several antisera in perikarya where we propose that both terminal amino acids of the neurohormone are masked by their association

TABLE 1.
IMMUNOCYTOCHEMICALLY REACTIVE BRAIN SITES USING ANTIBODIES GENERATED WITH DIFFERENT LRF CONJUGATES

| LRF CONJUGATE | PERIKARYA | | MEDIAN EMINENCE | ORGANUM VASCULOSUM |
| --- | --- | --- | --- | --- |
| | FIELD I ARCUATE REGION | FIELD II PREOPTIC REGION | | |
| pGlu–His–Trp–Ser–Tyr–Gly–Leu–Arg–Pro–Gly–BSA (Arimura 710) | − | + | + | + |
| BSA–pGlu–His–Trp–Ser–Tyr–Gly–Leu–Arg–Pro–Gly·$NH_2$ (Arimura 743) | − | + | + | + |
| pGlu–His$^2$–Trp–Ser–Tyr–Gly–Leu–Arg–Pro–Gly·$NH_2$ BSA (Sorrentino F) | + | − | + | + |
| pGlu–His–Trp–Ser–Tyr$^5$–Gly–Leu–Arg–Pro–Gly·$NH_2$ BSA (Nett–Niswender 42) | − | − | + | + |
| pGlu–His–Trp–Ser–Tyr$^5$–Gly–Leu–Arg–Pro–Gly·$NH_2$ HRP (Sternberger 8) | − | + | + | + |

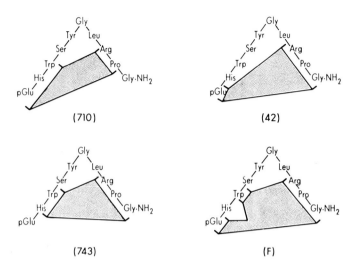

**FIG. 16.** Schematic representation of the binding sites on the LRF antigen used by antibodies in antisera generated with the several LRF-BSA conjugates shown in Table 1.

with a carrier or prohormone. In order to explain the failure of antisera 42 to stain any perikarya, we must suggest that the BSA conjugation is not specific for the $Tyr^5$ position, but includes a broader part of the molecule including the determinants Trp and His. Positive staining with this antisera in axons and median eminence suggests that at least the N-terminal has been

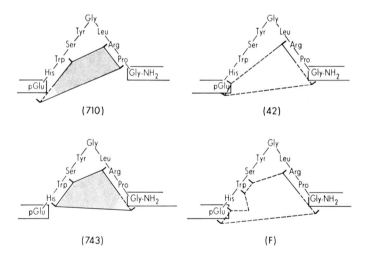

**FIG. 17.** Schematic representation of the antigen (LRF)-antibody relationship in perikarya. Both C- and N-terminals of the LRF molecule are represented as hidden and unavailable for antibody attachment because of their association with a prohormone. Under these conditions, sufficient binding affinity remains for antisera 710 and 743 to "stain;" insufficient binding affinity is present for antisera 42 and F.

freed of its carrier or released from its prohormone. Our current information does not provide any suggestion as to the state of the C-terminal after the neurohormone leaves the perikaryon. The ability of antisera B to stain perikarya of field II indicates a more specific conjugation of horseradish peroxidase (HRP) to the $Tyr^5$ position. The antibody generated with this conjugate should theoretically have the highest binding affinities for LRF of all those examined. Our experience suggests indeed that staining of field II perikarya is most intense with this antiserum.

Antisera 743 and 710 stain perikarya of field II while antiserum F does not; the reverse situation obtains in perikarya of field I. We suggest that the neurons of field II contain LRF while those of field I contain some other peptide sequence. The failure of antiserum F to bind to LRF *in perikarya* is attributed to the failure of the one remaining determinant site—$Trp^3$—to provide sufficient binding affinity for the antibody. Antiserum F, however, intensely stains perikarya of field I. We suggest that the primary structure of some other neurohormone is represented by a substitution of a different amino acid for histidine and that the resulting confirmation results in sufficient affinity of the F antibody to bind to it (Fig. 18). Antisera 743 and 710 do not attach to this molecule *in perikarya;* antisera 743 and 42 may potentially react after it has left the cell body.

Additional information is available to suggest that neurons of fields I and II are different. Table 2 describes the *in vitro* synthesis of "LRF" by hypothalamic blocks of tissue from these 2 areas. Synthesis is identified as the amount of radioimmunoassayable LRF (F antibody) present in the tissue after 1.5 hr incubation, as compared to a preincubation content. The large amount of LRF initially present in field I tissue is accounted for almost entirely by the hormone stored in the median eminence. After incubation, tissue containing field I neurons accumulated approximately 20 times more hormone than field II.

A qualitative difference with respect to biological activity is present also in hormone extracted from hypothalamic tissue of fields I and II. Table 3 documents a comparison of bio- and immunoassay of "LRF" from these 2 regions. Pieces of basal hypothalamus less median eminence (field I) and

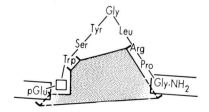

**FIG. 18.** Schematic representation of a primary structure of some other neurohormone present in perikarya of field I; the conformation of this molecule provides sufficient binding affinity for attachment and "staining" with F antiserum.

TABLE 2.
IN VITRO SYNTHESIS OF LRF BY FIELD I AND II OF RAT HYPOTHALAMUS[1]

| REGION | LRF CONTENT, ng. | | Δ |
|---|---|---|---|
| | pre-incubation | post-incubation | ng |
| FIELD I (+ M.E.) | 8.4 ± 1.1 | 15.7 ± 1.1 | 7.3 |
| FIELD II | 0.15 ± 0.08 | 0.53 ± 0.12 | 0.38 |
| CORTEX | N.D. | N.D. | |

1. Data of D.K. Sundberg

preoptic area (field II) were separately pooled and extracted with methanol and aliquots assayed for their LRF content as determined by radioimmunoassay and also by bioassay, measuring luteinizing hormone (LH) release from monolayer cultures of dispersed anterior pituitary cells. By radioimmunoassay, field I contained more than 10 times as much LRF; by bioassay, the LH-releasing ability of hormone extracted from field I was only twice as great as that of field II. The results suggest that, on an equimolar basis, hormone of field I has only one-fifth the "LRF"-activity of that in field II.

A number of interesting consequences result from these observations. If two neurohormones are involved, *both* are present in *median eminence* in a state where one or more of the antisera to LRF may demonstrate them immunocytochemically or quantitate them together as LRF. Both neurohormones are present in median eminence, the origin of one (LRF) being in the preoptic field II and the origin of the second being close to the median eminence in field I. Measurement of median eminence content of "LRF" is hence overestimated by the presence of the other cross-reacting molecule. Measurement of "hypothalamic" LRF content is similarly complicated if

TABLE 3.
LRF ACTIVITY IN FIELDS I AND II AS MEASURED BY RADIOIMMUNOASSAY AND BIOASSAY

| FIELD | LRF ACTIVITY, ng | |
|---|---|---|
| | RIA (E-Ab.) | BIOASSAY (LH RELEASE) |
| I | 1.2 | 0.89 |
| II | < 0.10 | 0.55 |
| ratio I:II | > 10:1 | 2:1 |

the retrochiasmatic-arcuate field I and/or median eminence are included in the tissue sample. Although our speculation proposes the presence of a neurohormone different from LRF in field I, "LRF-like activity" (as measured by radioimmunoassay) has been found in punches of the arcuate nucleus and in tissue extracts of hypothalamic pieces which do not include the preoptic area and median eminence. The salient tissue in this regard is the relative LRF activity on an equimolar basis between LRF and the neurohormone of field I. The proposed conformational change of this molecule would probably not completely prevent its recognition by "LRF" receptors in the pituitary. It is possible that our immunocytochemical studies have provided evidence for a partial structure and the locus of synthesis of a follicle-stimulating hormone (FSH)-releasing factor (RF).

Another example of a potentially "false positive" immunocytochemical reaction is seen in the staining for LRF in adenohypophysial cells. In both guinea pig and dog (Fig. 19), a population of cells (presumably gonadotrophs)

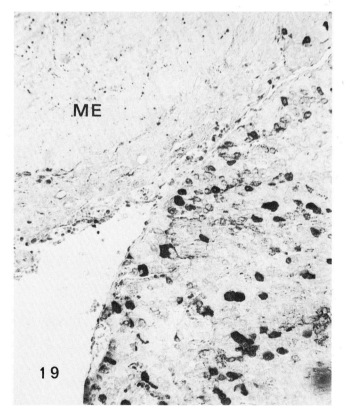

**FIG. 19.** Pituitary gland of dog illustrating intense LRF-immunoreactive response in a population of adenohypophysial cells. ME, median eminence. ×1375.

exhibit an intensive LRF-immunoreactive response. The results suggest a high concentration of membrane or intracellular receptor-bound LRF (14,15). The pituitary gland, however, like CNS tissue contains large amounts of enzyme which degrade LRF. If degradation involves peptide cleavage to yield octa- or nonapeptide fragments, these sequences may retain sufficient antigenic binding affinity on the C-terminal end of the molecule to be "seen" by antibodies, especially those of the F and 743 type.

## PEPTIDERGIC NEUROHORMONE SYSTEMS AND THE ENDOCRINE HYPOTHALAMUS

Our rapidly expanding knowledge of the anatomical organization of neuron systems associated with control of the adenohypophysis suggests that it may be of value to examine our variant terminologies, definitions, and concepts which have accumulated over the last several years. It is already clear that much of the terminology presently in use is neither appropriate to the specific issue of adenohypophysial control nor to the more general phenomenon of neurosecretion. Neurons which synthesize hormones associated with adenohypophysial control appear to be involved in other brain functions also; the same neurohormone may be released into portal blood in the median eminence, into cerebrospinal fluid, and may also act potentially as a neurotransmitter upon other neurons. Peptide-secreting neurons completely unrelated to adenohypophysial control appear to be present in extrahypothalamic sites. There may be present hormone-secreting neurons which project to one or more of the "windows of the brain" for delivery of their product to peripheral target organs other than adenohypophysis, kidney, or mammary gland. A meaningful classification of these divergent neuron systems is difficult to derive on the basis of function, location of neuronal fields, and details of synaptic relationships of distance the effector substance must travel to reach its site of action.

Sufficient direct and indirect evidence may be available to suggest that all the neurohormones associated with adenohypophysial control are peptides. For the benefit of at least neuroendocrine terminology, we propose that *all* neuronal systems which release a peptide as their effector substance be identified as the peptidergic neurohormone system. Further characterization is accomplished by identification of the specific peptide involved and also by the traditional neuroanatomical criteria of identifying origin and possible termination of the path involved. Table 4 illustrates the nomenclature of of this classification as applied to currently known or suggested projections of peptidergic neurohormone systems. The traditional "supraopticohypophysial system" becomes more carefully defined in terms of its components such as the vasopressin supraoptico-neural lobe projections, etc. The presence of SRIF in the neural lobe is described anatomically as an SRIF hypothalamo-neural lobe projection.

TABLE 4.

THE PEPTIDERGIC NEUROHORMONE SYSTEM

I. PEPTIDERGIC HYPOTHALAMO-NEURAL LOBE PROJECTIONS
   1. VASOPRESSIN SUPRAOPTICO-NEURAL LOBE
   2. OXYTOCIN PARAVENTRICULAR-NEURAL LOBE
   3. SRIF HYPOTHALAMO-NEURAL LOBE

II. PEPTIDERGIC HYPOTHALAMO-MEDIAN EMINENCE PROJECTIONS
   1. LRF FIELD I-MEDIAN EMINENCE
   2. LRF FIELD II-MEDIAN EMINENCE
   3. SRIF HYPOTHALAMO-MEDIAN EMINENCE
   4. TRF HYPOTHALAMO-MEDIAN EMINENCE
   5. VASOPRESSIN SUPRAOPTICO-MEDIAN EMINENCE

III. PEPTIDERGIC HYPOTHALAMO-OVLT PROJECTIONS
   1. LRF FIELD II-OVLT
   2. SRIF HYPOTHALAMO-OVLT

IV. PEPTIDERGIC HYPOTHALAMO-HYPOTHALAMIC PROJECTIONS
   1. SRIF HYPOTHALAMO-VENTROMEDIAL
   2. TRF HYPOTHALAMO-VENTRICULAR
   3. LRF HYPOTHALAMO-VENTRICULAR
   4. SRIF HYPOTHALAMO-VENTRICULAR

V. PEPTIDERGIC HYPOTHALAMO-MESENCEPHALIC
   1. LRF HYPOTHALAMO-TECTAL
   2. LRF HYPOTHALAMO-TEGMENTAL

A classification such as this is based upon anatomical organization and makes no inference as to the functional role of the projection in question even though its identification may include the name of a peptide known to act at the adenohypophysis or some other site. The SRIF hypothalamo-ventromedial projection which we describe need not implicate the ventromedial nucleus or SRIF at this site in either direct or indirect control of adenohypophysial growth hormone. The identification of a vasopressin supraoptico-median eminence projection makes no inference regarding a role of vasopressin in adenohypophysial function. A classification based upon nature of the effector substance and anatomical organization also makes no demands on the issue of whether the effector substance is released into a body fluid for extensive travel to its destination, whether it acts in a somewhat more restricted synaptoid junction or in the confined space of a true synaptic cleft. Further identification of these peptidergic systems as neurohormone or neurotransmitter projections would appear to only perpetuate a fruitless semantic argument. The term neurohormone is preferred by us to encompass all these projections; the original meaning of the word hormone as coined by Bayliss and Starling is hopefully broad enough to include the excitation which occurs at synaptoid junctions and synaptic clefts of peptidergic projections.

Finally, in our proposed nomenclature, the term neurosecretion is relieved of its onerous responsibility of being forced to represent difficult conceptualizations of neuronal structure and function instead of identifying the simple but remarkable phenomenon which it is, namely the secretion of any neuronal effector substance.

We (6) had earlier suggested an updated definition of the "diencephalic gland" of 1939, and proposed that the "endocrine hypothalamus" be that control subsystem containing, among other neural elements, the pool of specialized neurons whose secretory activity provides the neurohormones that participate in regulation of adenohypophysial function. Now that we have more definitive evidence that the neurohormones are indeed synthesized by neurons and have also proposed a classification in which these neurons are included, it is appropriate that our definition of the endocrine hypothalamus be "that control subsystem containing, among other neural elements, those projections of the peptidergic neurohormone system whose activity provides for the regulation of adenohypophysial function."

## ACKNOWLEDGMENTS

The authors express their appreciation for the valuable contributions made to these studies by Barbara Dolf, Terry Hayes, Vera Melnyk, Sally Poppelwell, Dolores Shock, and Julie Wilkins.

This project has been supported by NINCDS Program Project Grant NS-11642.

## REFERENCES

1. Alpert, L. C., Brawer, J. R., Patel, Y. C., and Reichlin, S. (1976): Somatostatinergic neurons in anterior hypothalamus: Immunohistochemical localization. *Endocrinology*, 98:255–258.
2. Barry, J. and Dubois, M. P. (1973): Étude en immunofluorescence des structures hypothalamiques a competence gonadotrope. *Ann. Éndocrinol. (Paris)*, 34:735–742.
3. Hoffman, G. E. (1976): Immunocytochemical localization of luteinizing hormone-releasing hormone (LHRH) in murine and primate brain. *Anat. Rec.* 184:429–430.
4. Hoffman, G. E., Moynihan-McCourt, J. A., and Knigge, K. M. (1976): Immunocytochemical localization of luteinizing-hormone-releasing hormone (LHRH). Differences with different antisera. *Neuroscience Abst.*, 90–28.
5. Hoffman, G. E., Knigge, K. M., Moynihan-McCourt, Melnyk, V., and Arimura, A. (1977): Luteinizing hormone releasing hormone (LHRH)-containing neuronal fields in mouse brain. *Neurosciences (in press)*.
6. Knigge, K. M. and Silverman, A. J. (1974): Anatomy of the endocrine hypothalamus. In: *Handbook of Physiology*, Sec. 7, Vol. IV, edited by E. Knobil and W. H. Sawyer. American Physiological Society, Washington, D.C.
7. Knigge, K. M., Joseph, S. A., Hoffman, G. E., Morris, M., and Donofrio, D. (1977): Role

of cerebrospinal fluid in neuroendocrine processes. V Int. Congress of Endocrinology, *Excerpta Medica (in press).*
8. Knigge, K. M., Hoffman, G. E., Scott, D. E., and Sladek, J. R., Jr. (1977): Identification of catecholamine and LHRH-containing neurons in primary cultures of dispersed cells of the basal hypothalamus. *Brain Res.,* 120:395–405.
9. Parsons, J., Erlandsen, S., Hegre, O., McEvoy, R. C., and Elde, R. (1976): Central and peripheral localization of somatostatin immunoenzyme immunocytochemical studies. *J. Histochem. Cytochem.,* 24:872–882.
10. Scharrer, B. (1975): Neurosecretion and its role in neuroendocrine regulation. In: *Pioneers in Neuroendocrinology,* Vol. 1, edited by J. Meites, B. T. Donovan, and S. M. McCann, pp. 257–265. Plenum Press, New York.
11. Scharrer, E. and Scharrer, B. (1939): Secretory cells within the hypothalamus. In: *The Hypothalamus and Central Levels of Autonomic Function.* Association for Research in Nervous and Mental Disease, Vol. XX. Hafner Publishing Co., New York.
12. Silverman, A. J. (1976): Distribution of luteinizing hormone-releasing hormone (LHRH) in the guinea pig brain. *Endocrinology,* 99:30–41.
13. Sternberger, L. A. (1974): *Immunocytochemistry.* Prentice-Hall, Englewood Cliffs, New Jersey.
14. Sternberger, L. A. and Petrali, J. P. (1975): Quantitative immunochemistry of pituitary receptors for luteinizing hormone-releasing hormone. *Cell Tiss. Res.,* 162:141–176.
15. Sternberger, L. A., Petrali, J. P., Joseph, S. A., Meyer, and Mills, (1977): Specificity of the immunocytochemical LH-RH receptor reaction. *Endocrinology (in press).*

*The Hypothalamus*, edited by S. Reichlin,
R. J. Baldessarini, and J. B. Martin,
Raven Press, New York, © 1978.

# Aminergic and Peptidergic Pathways in the Nervous System with Special Reference to the Hypothalamus

T. Hökfelt*, R. Elde, K. Fuxe, O. Johansson, Åke Ljungdahl, M. Goldstein, R. Luft, S. Efendic, G. Nilsson, L. Terenius, D. Ganten, S. L. Jeffcoate, J. Rehfeld, S. Said, M. Perez de la Mora, L. Possani, R. Tapia, L. Teran, and R. Palacios

The hypothalamus, as it lies at an important interface between brain and periphery, has through the years attracted interest from many points of view, for example, its rich content of biologically active substances such as putative neurotransmitters and neurohormones. Already in 1954 Vogt (332) in her study on the regional distribution of norepinephrine (NE) found particularly high concentrations of this amine in the hypothalamus, a localization which is true for many other transmitters. This structure also contains the classic neurosecretory systems described by the Scharrers and Bergman and others, with cell bodies in the magnocellular nuclei projecting to the neurohypophysis where their peptide hormones are released into the systemic circulation (see 18–20,36,193,194,296,316,318). Finally, the hypothalamus has provided an important source of the releasing and inhibitory hormones which have recently been identified and characterized, notably by Guillemin's and Schally's groups (6,44,58,233,295).

Against this background it is no wonder that the hypothalamus has also been extensively characterized with histochemical techniques. Shute and Lewis (302) mapped acetylcholinesterase-positive structures in the rat hypothalamus, and these studies were subsequently extended by Cottle and Silver (67). The first major study with the formaldehyde fluorescence method of Falck and Hillarp (95,96), dealing with the central nervous system, in fact, contained micrographs only from the hypothalamus (63). Subsequently, hypothalamic monoamine systems have become well known (2,23, 39–41,99–101,105,106,108,109,111,147,152–154,166,174,175,217,320).

The introduction of the classic immunohistochemical techniques of Coons and collaborators (66) into neurohistochemistry has resulted in a dramatic extension of the possibilities of identifying and mapping neuron systems on

---

* *Present address:* Department of Histology, Karolinska Institute, S-104 01, Stockholm, Sweden.

the basis of their content of specific macromolecules. This holds true for aminergic systems, where antibodies to the enzymes in the catecholamine (CA) and 5-hydroxytryptamine (5-HT, serotonin) synthesis are used (103, 104,114,119–121,132–134,148–154,158–161,180,269–272), for γ-aminobutyric acid (GABA)-containing neural systems with antibodies to the GABA-synthesizing enzyme, glutamate decarboxylase (GAD) (286,294, 340), and for a large number of peptides.

In the present chapter we will summarize findings on the distribution of CA systems and of several peptide-containing (and presumably "peptidergic") systems, based mainly on the indirect immunohistofluorescence technique. Here we will deal mainly with the hypothalamus, and for other areas of the nervous system the reader is referred to a recent review article (143). For the sake of completeness a few remarks will be made also on systems utilizing other transmitters such as acetylcholine (ACh), 5-HT, and other small molecules. Aspects falling outside the neurohistochemical field have intentionally been limited, but some pertinent biochemical studies on the distribution of neurotransmitters and peptides will be mentioned briefly, especially where histochemical findings are lacking. The classic anatomy of the hypothalamus is covered elsewhere in this volume in the chapter by Knigge (192), and for further details on this topic see Szentagothai et al. (321), Nauta and Haymaker (236), and Rethelyi and Halasz (283). For a broader survey of the hypothalamic hormones and their functional roles, see, in addition to other chapters of this book, the recent overview by Reichlin et al. (280). For further aspects on the pharmacology and physiology of peptides, see the articles by Barker (21), Moss (228), Nicoll (237), Pearse (257), Plotnikoff and Kastin (273), Prange et al. (276), and Renaud (281).

## METHODOLOGY

Immunohistochemistry is the basic technique used by us and in many other laboratories for mapping of enzymes and peptides in the hypothalamus and other areas of the nervous system. With this approach the high sensitivity and comparative specificity of the antigen-antibody reaction are applied to histological sections for subsequent examination in the light and electron microscope. The basic principles of this technique were described more than 30 years ago by Coons and collaborators (see 66). More recently advances in immunohistochemistry using peroxidase-labeled antibodies have offered possibilities to carry out studies both at the light and electron microscope level (see 235) with a higher sensitivity than available previously (see 317).

It must, however, be emphasized that the explosive development in the area of immunohistochemistry of the nervous system is mainly a direct effect of pioneering studies in other fields of research. Not until useful antigens had been isolated in pure form and their antigenic nature had been established was it possible to raise specific antisera useful in immunohistochemistry. In early work Gibb et al. (116) in 1967 established the antigenicity of dopamine-$\beta$-hydroxylase (DBH), the enzyme converting dopamine (DA) to NE, and later Goldstein and his collaborators isolated and prepared antisera to all four enzymes involved in CA synthesis (119–121 254; Fig. 4), i.e., in addition to DBH, tyrosine hydroxylase (TH), L-DOPA-decarboxylase (DDC), and phenylethanolamine-$N$-methyltransferase (PNMT). Roberts and colleagues (285,286) succeeded in purifying GAD, the GABA-synthesizing enzyme, to which antibodies were subsequently prepared. In similar developments, smaller peptides were isolated and their chemical structures were characterized, including the releasing or release-inhibiting factors, thyrotropin releasing hormone (TRH) (58,222), luteinizing hormone releasing hormone (LHRH) (6,295), and somatostatin (SOM) (44). Further examples are the peptides oxytocin (OXY) and vasopressin (VP) (8), substance P (SP) (64), originally isolated from equine brain and gut by von Euler and Gaddum (94), vasoactive intestinal polypeptide (VIP) (230), and the enkephalins (ENK), recently isolated and characterized by Hughes et al. (170). For references to studies of many peptides, see Table 2.

A brief description of the immunohistochemical technique used in our laboratory follows, but for information of immunohistochemical techniques in general, see Coons (66), Moriarty (227), Nairn (234), Nakane (235), and Sternberger (317); for detailed information of protocols used in other laboratories, see the references cited in relation to each substance.

## Antisera

Most of the specific antisera used in the present study were obtained from collaborating laboratories throughout the world. They were mostly raised in rabbits and have often been used in radioimmunoassays. In Tables 1

TABLE 1. *Characteristics of enzyme antisera*

| Antigen | Isolated from | Raised in | Reference |
| --- | --- | --- | --- |
| Tyrosine hydroxylase (TH) | Human pheochromocytoma | Rabbits | 254 |
| Aromatic L-amino acid decarboxylase (dopadecarboxylase) (DDC) | Beef adrenal gland | Rabbits | 119–121 |
| Dopamine-$\beta$-hydroxylase (DBH) | Beef adrenal gland | Rabbits | 119–121 |
| Phenylethanolamine-N-methyltransferase (PNMT) | Rat adrenal gland | Rabbits | 119–121 |
| Glutamic acid decarboxylase (GAD) | Mouse brain | Rabbits | 266 |

and 2 the types of antisera used, and their sources are given. The labeled antisera, i.e., fluorescein-isothiocyanate (FITC)-conjugated antibodies and the peroxidase-antiperoxidase (PAP) complex (317), were obtained from Statens Bakteriologiska Laboratorium (SBL, Stockholm, Sweden) and from Dr. Peter Biberfeld (Department of Pathology, Karolinska Hospital, Stockholm, Sweden), respectively. Control sera routinely used were preimmune serum (studies on enzymes) or a specific antiserum neutralized with the appropriate antigen (studies on peptides). In some cases normal rabbit serum was used as a control.

### Preparation of Tissues

For light microscopy rats were perfused with 4% formaldehyde (259) and the tissues were cut on a cryostat (Dittes, Heidelberg). For electron microscopy tissues fixed by perfusion with varying mixtures of formaldehyde and purified glutaraldehyde were cut in 40- to 75-$\mu$m thick sections on a Vibratome (116).

### Immunohistochemical Procedures

For immunofluorescence studies at the light microscopic level, tissue sections were incubated in a humid atmosphere at 37°C for 30 min with the specific antiserum in dilutions ranging from 1:10 to 1:160 and with the FITC-conjugated labeled antibodies diluted 1:4 under the same conditions. All sera contained 0.3% Triton X-100 (132). For electron microscopy the PAP technique (317) was used, and the Vibratome sections were incubated with the specific antisera diluted 1:500 or 1:1000.

### General Comments

Whereas the enzymes studied are large molecules which by themselves have antigenic properties, the peptides are all comparatively small substances. In most cases they have therefore been coupled with a carrier molecule such as bovine serum albumin, keyhole limpet hemocyanin, thyroglobulin, or $\alpha$-globulin (see Table 2). The antisera raised contained antibodies to these carrier proteins as well as to the peptides in question. Therefore, the antiserum is prepared before histologic reaction by absorption with the carrier protein to avoid unspecific staining. This step is especially important when studying peripheral tissues. In general, for immunohistochemistry the antisera are used in much higher concentrations than for radioimmunoassay. In immunofluorescence studies dilutions between 1:10 and 1:100 are mostly used, whereas the Sternberger technique (317) allows 100-fold higher dilutions. It may be pointed out that for specificity of the immunoreaction, high dilutions are, of course, preferable.

TABLE 2. Characteristics of peptide antisera

| Antigen | Structural characterization (ref.) | Conjugate | Raised in | Code No. | Source (ref.) |
|---|---|---|---|---|---|
| Luteinizing hormone releasing hormone (LHRH) | 6,295 | Synthetic LHRH/bovine serum albumin (BSA) | Rabbit | | 176,178 |
| Thyrotropin releasing hormone (TRH) | 58,233 | Synthetic TRH/human serum albumin | Rabbit | | 117,179 (see 34) |
| Somatostatin (SOM) (growth hormone release-inhibiting hormone) | 44 | (a) Synthetic SOM/human γ-globulin | Rabbit | 101,103 | 10 |
| | | (b) Synthetic SOM/thyroglobulin (TG) | Guinea pig | GP 2D | 25 |
| | | (c) Synthetic SOM/keyhole limpet hemocyanin (KLH) | Rabbit | R 141C | 85 |
| Substance P | 64 | Synthetic peptide/BSA | Rabbit | K 16, K 25 | 240 |
| Leucine-enkephalin | 170 | Synthetic peptide/TG | Guinea pig | GP 14B | 87 |
| Methionine-enkephalin | 170 | Synthetic peptide/KLH | Rabbit | R 6C | 146 |
| Angiotensin | 90,312 | Synthetic peptide/BSA | Rabbit | Denise | 112 |
| Vasoactive intestinal polypeptide | 230 | Synthetic peptide | Rabbit | | 292 |
| Gastrin | 125,126 | Synthetic peptide/BSA | Rabbit | 2720,4562 | 278 |
| Oxytocin | 81 | Synthetic peptide/TG | Rabbit | R 111 | 84 |
| Vasopressin (VP) | 81 | Synthetic peptide (lysine-VP)/TG | Guinea pig | GP 24 | 84 |
| Prolactin (PROL) | | Rat PROL 5A | Rabbit | | NIAMDD |

A major problem in immunohistochemistry is the poor penetration of the antibodies into the sections. We and other laboratories therefore use the detergent Triton X-100 as described by Hartman et al. (134) to solubilize membranes and enhance penetration.

A further problem with immunohistochemistry is the specificity of the immunoreaction and thus the interpretation of the results. Apart from unspecific binding, which usually can be revealed by using proper controls, cross-reaction(s) may occur between the antiserum and peptides structurally closely related to the antigen, against which the antiserum was raised. To exclude such cross-reactions fully appears to be impossible at present, but more exact information can be obtained by characterizing the binding properties of each antiserum with known peptides and their structural analogues. Thus, we know, for example, that our antiserum raised to methionine-ENK (met-ENK) cross-reacts about 10% with leucine-ENK (leu-ENK) but less than 0.1% with $\alpha$-, $\beta$-, and $\gamma$-endorphins (*to be published*). Our LHRH, TRH, SOM, OXY, VP, angiotensin (ANG), and VIP antisera have been tested to varying degrees as indicated in the original publications cited concerning their development. In the case of gastrin (GAS), a cross-reacting cholecystokinin-like peptide has recently been identified in the brain (72; J. F. Rehfeld, *personal communication*). Thus this peptide may be responsible for the distribution patterns described below. Against this background we refer in our original articles to the staining patterns in terms such as "SP-like immunoreactivity," "SP-positive," or "SP-immunoreactive material." In this chapter, however, we have for simplicity avoided such expressions.

## DISTRIBUTION PATTERNS

A brief survey of the histochemical distribution of the CAs and GABA (mainly as indicated by the distribution of their synthesizing enzymes), of some other putative transmitters, as well as of a number of peptides, will be given below. It will not be possible to include a full account on all systems studied and of the whole brain, or even of the whole hypothalamus. Schematic drawings of the distribution patterns at one hypothalamic level (Figs. 1, 2) and in the median eminence (Fig. 3) are presented.

## CATECHOLAMINES

The broad knowledge of the distribution of CA systems in the brain is based mainly on studies with the Falck-Hillarp technique and its recent modifications (35,95,96,166,218,336). More recently, possibilities of performing detailed "biochemical mapping" have become available due to refined technical improvements. The development of a slice-punch technique allowing sampling of extremely small parts of the brain (nuclei or even parts of nuclei) (249) combined with extremely sensitive assays has

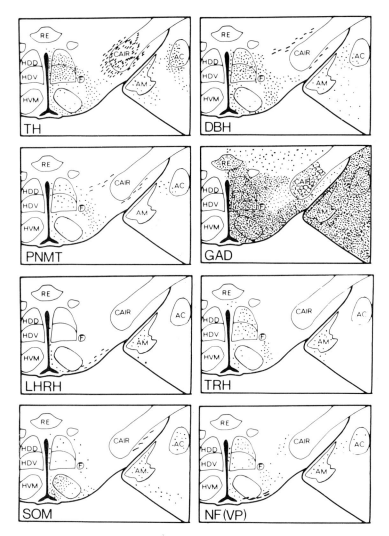

**FIG. 1.** Schematic drawing of the distribution of some enzymes and peptides in nerve terminals (*dots*) and axons (*dashes*) in the rat hypothalamus based on indirect immunofluorescence microscopy. All frontal sections show the same plane (A 4110 μm) in the König and Klippel atlas (195). Cell bodies have not been indicated. Nerve terminals in the median eminence are not included but are indicated in Fig. 3. For details on the distribution of the various substances see text. Abbreviations in the left corner indicate substance against which the antiserum was raised. TH, tyrosine hydroxylase; DBH, dopamine-β-hydroxylase; PNMT, phenylethanolamine-N-methyltransferase; GAD, glutamic acid decarboxylase; LHRH, luteinizing hormone releasing hormone; TRH, thyrotropin releasing hormone; SOM, somatostatin; NF (VP), neurophysin (vasopressin); AC, nucleus amygdaloideus centralis; AM, nucleus amygdaloideus medialis; CAIR, capsula interna, pars retrolenticularis; F, columna fornicis; HDD, nucleus dorsomedialis, pars dorsalis; HDV, nucleus dorsomedialis, pars ventralis; HVM, nucleus ventromedialis; RE, nucleus reuniens.

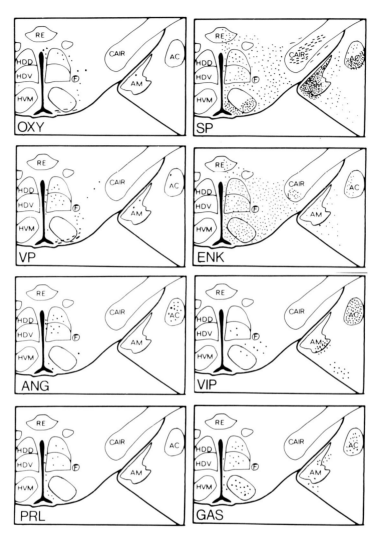

**FIG. 2.** Schematic drawing of the distribution of some peptides in nerve terminals (*dots*) and axons (*dashes*) in the rat hypothalamus based on indirect immunofluorescence microscopy (continuation of Fig. 1). For all details see legend to Fig. 1. OXY, oxytocin; SP, substance P; VP, vasopressin; ENK, enkephalin; ANG, angiotensin II; VIP, vasoactive intestinal peptide; PRL, prolactin; GAS, gastrin.

allowed a detailed quantitative biochemical mapping of CAs and their enzymes in the hypothalamic nuclei. Thus, the regional distributions of DA and NE (198,253), epinephrine (199,328), TH and DBH (199,289), PNMT (291), and 5-HT (29) have been mapped. The effect of hypothalamic deafferentation on these CAs and enzymes has recently been described (52,252). We refer to the original papers for details and to a recent review

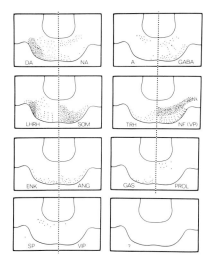

FIG. 3. Schematic drawing of the distribution of some catecholamines, GABA, and some peptides in nerve terminals in the rat median eminence (frontal section, midlevel). The drawings are based on formaldehyde-induced fluorescence technique and indirect immunofluorescence microscopy. The distribution of catecholamines and GABA is deduced from the distribution of their synthesizing enzymes. Some substances (DA, LHRH, SOM, TRH, ENK, ANG, and GAS) are present mainly in the external layer and thus potentially may be released into the portal vessels, whereas others (NA, A, PROL, and SP) are found mainly in the internal layer. For further details see text. DA, dopamine; NA, noradrenaline; A, adrenaline; GABA, γ-aminobutyric acid. For further abbreviations see Fig. 1.

article by Brownstein et al. (51). In the following we will mainly describe the hypothalamic CA neurons on the basis of immunohistochemical studies with antisera to the CA-synthesizing enzymes. As indicated in Fig. 4, epinephrine is synthesized from the amino acid tyrosine via L-dihydroxyphenylalanine (L-DOPA), DA, and NE. Each step is catalyzed by a specific enzyme. A DA neuron is characterized by the first two enzymes, TH and DDC; NE neurons have in addition DBH; and epinephrine neurons contain all four enzymes.

In the hypothalamus numerous TH- and DDC-positive cells but no cell bodies containing DBH or PNMT have been observed up to now (152, 153,161,320), indicating that CA cell bodies in this brain area mainly are dopaminergic. Some cells in the posterior hypothalamus (A 11 group) may, however, be noradrenergic (41).

### Hypothalamic DA Cell Bodies

The hypothalamic DA cell bodies may be considered to form two major systems, a *dorsal* and a *ventral* one, both of which generally follow the third ventricle (Fig. 5). The *dorsal DA system* is in direct continuity with the periaqueductal mesencephalic DA system (158,217), which surrounds the ventral aspects of the aqueductus cerebri Sylvii of the mesencephalon. It consists of small (diameter, 10 to 15 μm) cell bodies, which in turn appear to continue directly into the A 10 DA cell group. The most caudal part of the dorsal hypothalamic DA system, localized in the periventricular thalamic area medial to the fasciculus mammillo-thalamicus, is built up of large cells with their dendrites often oriented in the sagittal plane: these are the A 11 DA cell group (Fig. 6). Around the ventral aspect of the caudal part of the third ventricle, i.e., slightly ventral to and partly intermingling with the A 11

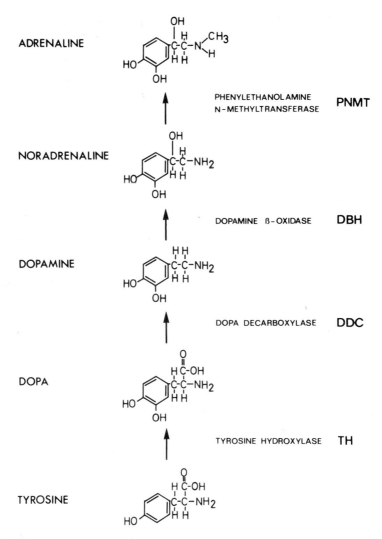

**FIG. 4.** Schematic presentation of catecholamine synthesis. Adrenaline (epinephrine) is synthesized from the amino acid tyrosine, via dihydroxyphenylalanine (DOPA), dopamine, and noradrenaline (norepinephrine). Each step is catalyzed by a specific enzyme.

group, numerous small, comparatively weakly TH-positive cell bodies are seen. In the rostral direction there is a direct continuation into the A 13 nucleus located in the zona incerta medially of the fasciculus mammillothalamicus (Fig. 7). However, cells do extend into the lateral parts of the zona incerta, into the medial forebrain bundle, and also ventrally to the fornix, continuing all the way down to the surface of the brain, thus joining the ventral system. These latter DA cells have not been assigned to any of the currently recognized hypothalamic DA cell groups.

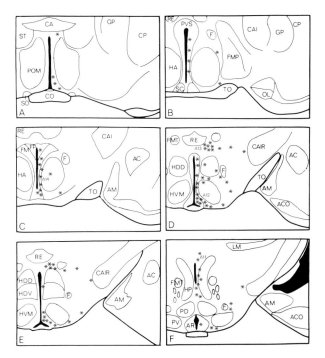

**FIG. 5.** Schematic drawing of the distribution of tyrosine hydroxylase-positive cell bodies (*asterisks*) at various levels of the rat hypothalamus. The frontal sections are drawn from König and Klippel at the following levels: A, A 6790 μm; B, A 6060 μm; C, A 5340 μm; D, A 4620 μm; E, A 4110 μm; F, 3430 μm. As indicated in the text all cell bodies probably represent dopamine cell bodies. A11 to A14 indicate name of cell body groups according to Dahlström and Fuxe (69). For further details see text. AC, nucleus amygdaloideus centralis; ACO, nucleus amygdaloideus corticalis; AM, nucleus amygdaloideus medialis; AR, nucleus arcuatus; CA, commissura anterior; CAI, capsula interna; CAIR, capsula interna, pars retrolenticularis; CO, chiasma opticum; CP, nucleus caudatus putamen; F, columna fornicis; FM, nucleus paraventricularis, pars magnocellularis; FMP, fasciculus medialis prosencephali; FMT, fasciculus mamillothalamicus; FP, nucleus paraventricularis, pars parvocellularis; GP, globus pallidus; HA, nucleus anterior; HDD, nucleus dorsomedialis, pars dorsalis; HDV, nucleus dorsomedialis, pars ventralis; HVM, nucleus ventromedialis; HP, nucleus posterior; LM, lemniscus medialis; OL, nucleus tractus olfactorii lateralis; PD, nucleus premamillaris dorsalis; POM, nucleus preopticus medialis; PV, nucleus premamillaris ventralis; PVS, nucleus periventricularis stellatocellularis; RE, nucleus reuniens; SC, nucleus suprachiasmaticus; SO, nucleus supraopticus; ST, nucleus interstitialis striae terminalis; TO, tractus opticus.

The *ventral DA system* is mainly confined to the arcuate nucleus (A 12 cell group) (Figs. 7, 14A, 15A). It may be mentioned that the ventrally located A 12 DA cells have a weaker TH-positive immunofluorescence than the ones lying more dorsally. The ventral system extends anteriorly all the way to the organum vasculosum laminae terminalis (OVLT), i.e., to the most rostral parts of the third ventricle. This group of DA cells has been termed A 14 (41) (Fig. 8A) but is anatomically not distinctly separated from

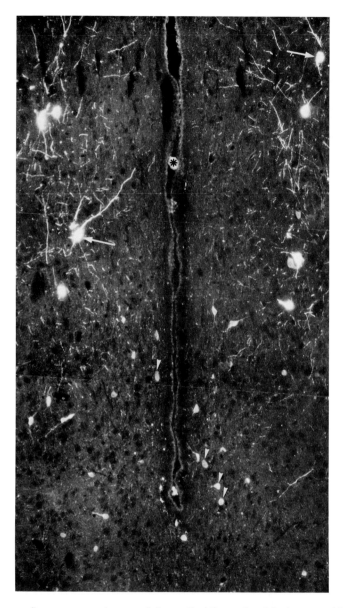

**FIG. 6.** Immunofluorescence micrograph (mount) of the periventricular area at the border between mes- and diencephalon after incubation with antiserum to tyrosine hydroxylase. Several large A11 cells with dendrites (*arrows*) are seen. Note also the small, weakly fluorescent cell bodies (*arrowheads*) surrounding the ventral tip of the ventricle (*asterisk*). ×105.

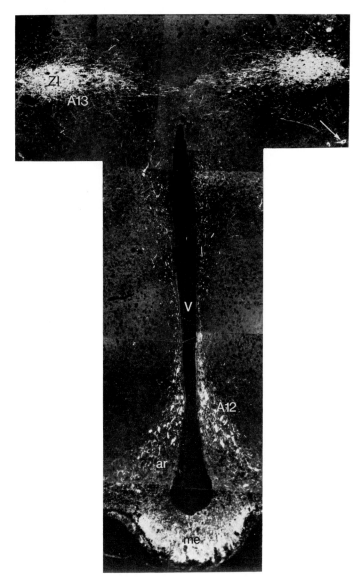

**FIG. 7.** Immunofluorescence micrograph (mount) of the basal hypothalamus, including the median eminence (me) and arcuate nucleus (ar), and of the zona incerta (ZI) after incubation with antiserum to tyrosine hydroxylase. The A13 and A12 cell groups are seen. Note the high density of catecholamine nerve terminals in the median eminence. Arrow points to some fluorescent cells in the dorsomedial nucleus. V, third ventricle. From Hökfelt et al. (158) ×65.

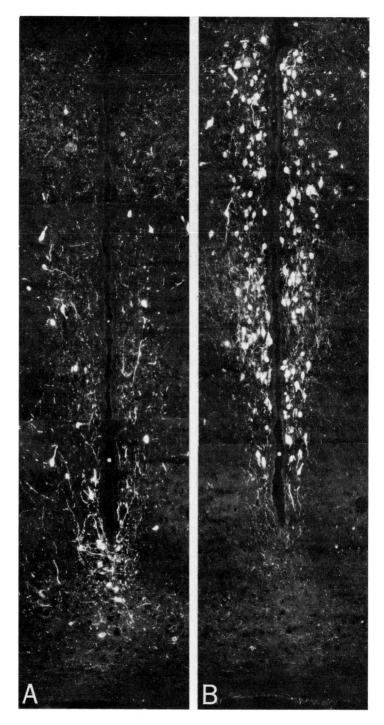

**FIG. 8.** Immunofluorescence micrographs (mounts) of consecutive sections of the anterior periventricular hypothalamic area after incubation with antisera to tyrosine hydroxylase **(A)** and somatostatin **(B)**. The dopamine cell bodies of the A14 cell group are present mainly around the basal parts of the ventricle **(A)**, whereas the numerous SOM-positive cell bodies tend to localize somewhat more dorsally **(B)**. Asterisks indicate third ventricle. ×90.

the A 12 group. It covers mainly the ventral half of the third ventricle, and the number of cells gradually decreases in the rostral direction. From the A 12 arcuate group cells extend dorsally in the periventricular area at the mid-hypothalamic level, and some cells are also present in the paraventricular, dorsomedial, and ventromedial hypothalamic nuclei. Cells also extend laterally from the arcuate nucleus in the basal hypothalamus along the surface of the brain (Fig. 9A). There are in the hypothalamus several cell bodies which are DDC but not TH positive. They are located mainly in the dorsal part of the suprachiasmatic nucleus and in the periventricular part of the posterior hypothalamus. Their localization agrees well with that of cells known both to accumulate exogenous, intraventricularly

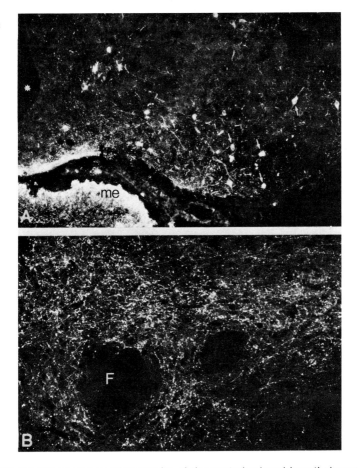

**FIG. 9.** Immunofluorescence micrographs of the posterior basal hypothalamus **(A)** and perifornical area **(B)** after incubation with antiserum to tyrosine hydroxylase **(A)** and phenylethanolamine-N-methyltransferase **(B)**. Many dopamine cell bodies (arrows) are located along the basal surface of the hypothalamus **(A)**. Note dense plexus of epinephrine nerve terminals in the perifornical area **(B)**. Asterisk indicates third ventricle. me, median eminence; F, fornix. ×105.

administered CAs (147) as well as to become strongly fluorescent after pharmacologic treatments designed to enhance CA fluorescence after the administration of L-DOPA (215). The latter finding indicates that these cells have the property of decarboxylating L-DOPA to DA. Since they do not seem to contain TH, the first and rate-limiting enzyme in CA synthesis, they are in all probability not DA or CA cells. They may, however, contain another type of cross-reacting decarboxylase, and the presence of 5-HT cells cannot be excluded (148). It is interesting to note that in the suprachiasmatic nucleus there are high concentrations of histamine (53).

### Hypothalamic Catecholamine Nerve Terminals

Defining the exact distribution of CA nerve terminal systems in the hypothalamus containing the three CAs, DA, NE, and epinephrine, on the basis of immunohistochemistry of normal animals is connected with considerable difficulties. The pattern described below should be considered preliminary and approximate because only the epinephrine neurons have a selective marker enzyme, PNMT, but in addition contain TH, DDC, and DBH. The last three enzymes are also present in NE neurons, whereas DA neurons contain TH and DDC. An identification must therefore rest on a quantitative and qualitative analysis and comparison of consecutive sections "stained" with antisera to TH, DBH, and PNMT, respectively (Fig. 10A-C). The occurrence of PNMT-positive nerve terminals indicates epinephrine neurons. More numerous or differently distributed DBH-positive, as compared to PNMT-positive, nerve terminals would suggest the presence of NE neurons. Finally, more numerous or differently distributed TH-positive, as compared to DBH- and PNMT-positive, nerve terminals probably represents DA nerve terminals. By combining lesion experiments, including deafferentations and 6-hydroxy-DA injections, with immunohistochemistry, we hope to obtain more exact information on the hypothalamic CA systems.

*Epinephrine Nerve Terminals*

PNMT-positive nerve terminals are present in highest concentrations in the dorsomedial and arcuate nuclei, in the parvocellular (medial) part of the paraventricular nucleus, in the perifornical area (Fig. 9B), and in the superficial basal hypothalamus. In the cranial direction PNMT-positive fibers are present in the periventricular region (Fig. 10A). Few or no PNMT fibers are found in the ventromedial nucleus, the median eminence (internal and subependymal layers contain some), the supraoptic and suprachiasmatic nuclei, the anterior hypothalamic nucleus, or the preoptic area. So far,

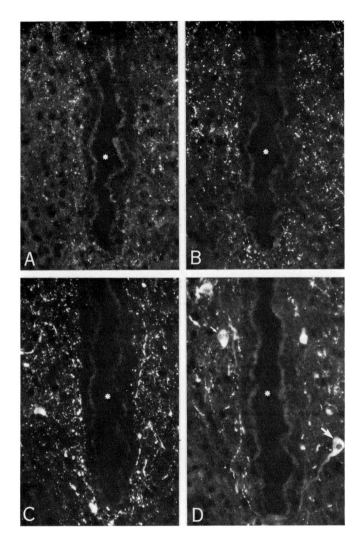

**FIG. 10.** Immunofluorescence micrographs of consecutive sections of the anterior hypothalamic periventricular area (ventral part) after incubation with antiserum to phenylethanolamine-$N$-methyltransferase **(A)**, dopamine-$\beta$-hydroxylase **(B)**, tyrosine hydroxylase **(C)**, and somatostatin **(D)**. Nerve terminals positive to all three catecholamine-synthesizing enzymes are found in the periventricular area **(A–C)**, but it is difficult to establish with certainty the quantitative relations between them. The presence of epinephrine nerve terminals is, however, certain **(A)**. It appears also as if there are more TH-positive nerve terminals than such positive to the two remaining enzymes, indicating the existence also of dopamine nerve terminals **(C)**. In this area SOM-positive cell bodies are present (*arrows*) **(D)**. Thus, epinephrine and possibly dopamine nerve endings may innervate SOM neurons. Asterisk indicates third ventricle. Magnification 90×.

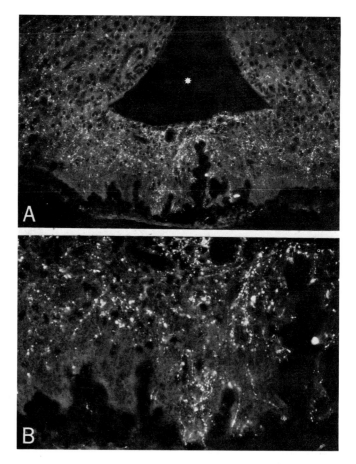

**FIG. 11.** Immunofluorescence micrographs of the basal hypothalamus including the median eminence after incubation with antiserum to dopamine-$\beta$-hydroxylase. A plexus of DBH-positive nerve terminals of moderate density is seen in the arcuate nucleus and in the internal layer of the median eminence. In the medial parts of the external layer fluorescent nerve terminals are also seen. **B:** A higher magnification of part of **A**. Asterisk indicates third ventricle. ×105 **(A)**; ×260 **(B)**.

PNMT-positive cell bodies have been identified only in the medulla oblongata in the lateral reticular nucleus (cranial part of A 1 cell group) and in the nucleus tractus solitarii area (cranial part of A 2 cell group). Thus, tentatively these cell groups can be considered to be the probable origin of the hypothalamic epinephrine nerve terminals. In contrast, Saavedra et al. (see 52) have reported that lesions of the A 1 and A 2 groups did not change the hypothalamic PNMT content and that a total deafferentation of hypothalamus reduced PNMT content by only 60%. The reason for this discrepancy is not clear, but there may, of course, be more PNMT-containing cell bodies present than hitherto assumed.

## Norepinephrine Nerve Terminals

Some NE nerve terminals have a very similar distribution to, and thus intermingle with, the epinephrine fibers. Thus, the dorsomedial nucleus, the arcuate nucleus (Fig. 11A) and the entire paraventricular nucleus, the perifornical area, the basal hypothalamus, including the most ventral parts of the ventromedial nucleus, and the retrochiasmatic area all contain numerous NE fibers together with the epinephrine nerves described above. In addition, the internal layer of the median eminence (Fig. 11A, B), the supraoptic nucleus, and the preoptic and periventricular nuclei contain moderate numbers of NE terminals. Sparse NE fibers are found in the external layer of the median eminence (mainly the medial part) (Fig. 11A, B), the preoptic area, the anterior hypothalamic nucleus, and the ventromedial and posterior hypothalamic nuclei.

## Dopamine Nerve Terminals

An extremely dense plexus of DA nerve terminals is found in the external layer of the median eminence (Figs. 7, 14A, 15A), mainly in its lateral parts extending into the pituitary stalk (Fig. 16A). This distribution was confirmed in an ultrastructural analysis of this area demonstrating that in the most superficial layer of the lateral external median eminence approximately one-third of all nerve endings are dopaminergic (2; see also 224 and Figs. 26, 29). The distribution of DA nerve terminals in other hypothalamic areas is so far only incompletely known. Very fine, weakly fluorescent TH-positive nerve terminals are present in the anterior hypothalamic nucleus, confirming the observations of Björklund et al. (39) obtained with the glyoxylic acid method. These authors described DA nerve terminals also in the preoptic suprachiasmatic nucleus, the preoptic medial nucleus, the periventricular area, and the zona incerta.

## Pathways

As indicated above it has generally been assumed that all NE and epinephrine nerve terminals have an extrahypothalamic origin and arise from cell bodies in the lower brainstem (7,326). It has also been thought that virtually all hypothalamic DA nerve terminals have their cell bodies in the hypothalamus. Although the A 12 group has been known to innervate mainly the median eminence (40,99,105) and the intermediate and posterior lobes of the pituitary (40), the projections of the remaining hypothalamic DA cell groups have not been known. Recently, Björklund et al. (39) have described an incerto-hypothalamic DA system consisting of a caudal part originating in the A 11 and A 13 groups and projecting to the dorsomedial nucleus and to the dorsal and anterior hypothalamic areas. The rostral part of this system is a build-up of the A 14 cells projecting into the medial preoptic area and the periventricular and suprachiasmatic preoptic nuclei. Some DA nerve terminals may have an extrahypothalamic origin, since Kizer et al. (190) recently have shown a 40% fall in DA in the median eminence after lesions in the A 8-A 9-A 10 region of the mesencephalon. These DA nerve terminals could possibly be located in the medial parts of the median eminence where high concentrations of TH-positive, but comparatively low concentrations of DBH-positive, nerve terminals are found (compare Fig. 11A with 14A).

## OTHER PUTATIVE TRANSMITTERS

### Acetylcholine

Comparatively little is known about the exact localization and distribution of cholinergic systems in the hypothalamus. This is partly due to the lack of a reliable and specific histochemical technique for acetylcholine neurons. Several histochemical studies on acetylcholinesterase (AChE) in the hypothalamus have been published (Fig. 12) (67,175,302,303), but difficulties in interpreting these results exist as discussed by Silver (304). Biochemical studies have revealed the presence of ACh (64a,98,210) and choline acetyltransferase (ChAc) (64a,135,210) in the hypothalamus. In a recent regional biochemical analysis of ChAc in the hypothalamus, Brownstein et al. (46) found in general a good correlation between the distribution of this enzyme and the cholinergic pathways previously described on the basis of histochemical staining of AChE (see above). The highest concentrations of ChAc were found in the median eminence. Recent developments in the field of histochemistry of cholinergic systems include a precipitation technique (59,184) as well as an immunohistochemical (225) approach to visualize ChAc and may make a more detailed mapping of cholinergic systems possible.

### γ-Aminobutyric Acid

The distribution of GABA neurons in various parts of the central nervous system has been described by Roberts and collaborators (see 286), most recently with the help of antibodies to glutamic acid decarboxylase and immunochemistry. Using an antiserum to GAD raised by Perez de la Mora et al. (266), we studied the distribution of GAD-positive nerve terminals in the hypothalamus. Extremely dense networks of GAD-positive fibers were observed in almost all hypothalamic nuclei (Fig. 13). The median eminence also contained GAD-positive nerve terminals (Fig. 13A) but in considerably lower concentrations. They were present in both the internal and external layers. Our findings agree well with the high levels of GABA found in the hypothalamus (see 286) and with the regional biochemical assay of GAD enzyme activity by Tappaz et al. (324). No GAD-positive cell bodies could be observed with our immunohistochemical technique. The

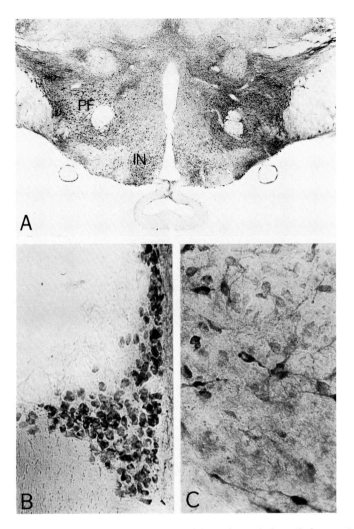

**FIG. 12.** Light micrographs of frontal sections of the guinea pig hypothalamus stained for histochemical demonstration of acetylcholinesterase. **A:** Numerous positive cells in the perifornical region (PF) and in the infundibular nucleus (IN). **B** and **C:** High magnifications of the lateral supraoptic nucleus and the infundibular nucleus, respectively. Note the varying intensity of the reaction. These micrographs were kindly supplied by Drs. Silver and Cottle and are from their work on the guinea pig hypothalamus (67). For further information see ref. 67. Magnification: 15× **(A)**, 105× **(B)**, and 240× **(C)**.

biochemical data of Brownstein et al. (52) on the effect of hypothalamic deafferentation on GAD activity indicate that the GABA nerve terminals in the median eminence have an intrahypothalamic origin, whereas part of the GABA fibers in other hypothalamic nuclei probably have their cell bodies outside the hypothalamus. GABA neurons can also be traced histochemically with autoradiographic techniques (167,171), but only limited reports of the hypothalamic distribution have appeared (171,222).

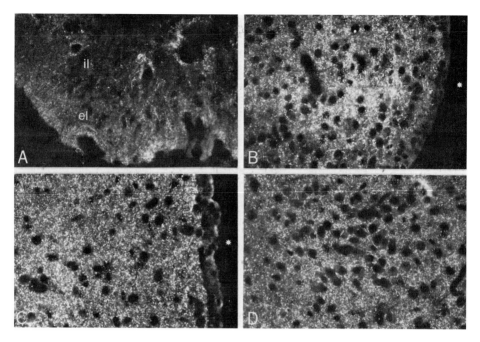

**FIG. 13.** Immunofluorescence micrographs of the median eminence **(A)**, the arcuate nucleus **(B)**, the dorsomedial nucleus **(C)**, and the ventromedial nucleus **(D)** after incubation with antiserum to glutamic acid decarboxylase. Dense networks of GABA nerve terminals are present in all hypothalamic nuclei. Note numerous GABA fibers in the internal (il) and external layer (el) of the median eminence **(A)**. Asterisks indicate third ventricle. ×85.

## Other Amino Acids

In addition to GABA, several other amino acids are transmitter candidates. They include glycine, glutamate, and aspartate (see 68 and 202). So far, autoradiography of accumulated labeled exogenous amino acids is the only histochemical technique applied in attempts to trace these neurons. Glycine neurons have been identified in the spinal cord (see 167), and preliminary findings on glutamate-accumulating neurons in the hippocampus have been reported (172). No histochemical information on the hypothalamic distribution of these amino acids is yet available.

## Histamine

Several attempts have been made to visualize histamine histochemically (see 38), including an o-phthaldialdehyde fluorescent derivative technique (83,181), but until now no satisfactory method has been developed by which histamine stores in brain can be visualized. For example, a successful technique with which to visualize histidine decarboxylase, the histamine-synthesizing enzyme, has not been reported. Several biochemical studies do, however, favor the existence of histamine neurons in the brain (see 62, 123, 297, and 315), although some histamine is probably also stored in mast cells. Histamine and histidine decarboxylase are present in synaptosomes and have a marked regional distribution, and changes in these markers can be obtained as a consequence of lesions in various places. The regional distribution within the hypothalamus has been reported by Brownstein et al. (53).

### 5-Hydroxytryptamine

The Falck-Hillarp technique is considerably less sensitive for 5-HT-containing neurons than for the CA neurons, and consequently our knowledge of the 5-HT pathways is incomplete. A review in this area has been published by Fuxe and Jonsson (111). More recently, Joh and collaborators (180) have purified tryptophan hydroxylase, the 5-HT-synthesizing enzyme, but so far mapping studies have been performed only to a limited extent. Since the second enzyme in the 5-HT synthesis, "5-hydroxytryptophan decarboxylase," is either the same enzyme (see 148) as or immunologically cross-reacts with the aromatic amino acid decarboxylating enzyme in CA neurons, there is also a possibility of mapping 5-HT neurons with antisera to DOPA decarboxylase (148).

According to the older histochemical data (7,69,100,101,326), the hypothalamic 5-HT nerve terminals all originate from cell bodies in the lower brainstem. However, the existence of 5-HT cell bodies in the hypothalamus has been claimed (313). Some DOPA decarboxylase-positive cells described above may represent 5-HT neurons.

Histochemically the highest numbers of 5-HT nerve terminals in the hypothalamus are found in the suprachiasmatic nucleus, but 5-HT terminals are spread all over the hypothalamus (100,101). They appear to arise mainly from the medial B7 and B8 cell groups of midbrain, with axons ascending in the medial subcortical 5-HT pathway (111).

The regional distribution of 5-HT in the hypothalamus has been reported by Saavedra et al. (290). After a total hypothalamic deafferentation, 5-HT levels decreased by approximately 60% in the island of hypothalamic tissue produced (52), supporting the view described above that some 5-HT cells may exist in the hypothalamus.

## DISTRIBUTION OF PEPTIDES

### Luteinizing Hormone Releasing Hormone

LHRH was structurally characterized as a decapeptide by Amoss et al. (6) and Schally et al. (295). The first immunohistochemical studies on

LHRH were published in 1973 by Barry et al. (32), Calas et al. (60), and Leonardelli et al. (214). Subsequently a large number of papers have appeared and the brains of both mammals (14a,15,16,24,27–33,57,60,75, 127,138,154,186,187,197,231,232a,263,299,300,306,337,345,348), including primates (25,26,55,240a), and lower species (3,72a,122,301) have been examined. Studies have also been carried out at the electron microscopic level (61,117,232,261,263,308). The immunohistochemical distribution agrees well with that found in regional radioimmunologic assays (9,37,49, 188,250,338; see also 224a) and in studies of subcellular fractions (22,277). We will deal with the LHRH distribution only to a limited extent since this peptide is discussed extensively by Knigge (192) in this volume and in other review articles (see 75, 186, 263, and 344).

## Cell Bodies

Difficulties were encountered initially in attempts to visualize LHRH-positive cell bodies, especially in the rat brain. However, immunologic studies of several species using various types of experimental variables, including treatment with drugs such as colchicine, methanol, melatonin, or barbiturates, have described numerous LHRH immunoreactive cell bodies (3,24–26,28–33,56,122,231,300,306,348). In rodents most of these cell bodies were found outside the basal hypothalamus in the pre- and suprachiasmatic regions. More recently, however, Hoffman et al., using an antiserum directed to a specific portion of the LHRH molecule, have described numerous LHRH-positive cell bodies in the arcuate nucleus of the rat (128, see 75). LHRH immunoreactive material has also been found in tanycytes by some authors (348).

## Nerve Terminals

The highest concentrations of LHRH-positive nerve terminals are found in the external layer of the median eminence (Figs. 14B, 15B). In the most rostral part the entire external layer is covered, but in more caudal parts the fibers are confined to the lateral aspects, with only few fibers in the medial parts. The ventral surface of the floor of the third ventricle is also covered by LHRH-positive nerve endings (Fig. 16B). The pituitary stalk contains only few LHRH nerve terminals (Fig. 16B). A dense plexus of LHRH-positive fibers is present in the OVLT close to the capillary network (Fig. 17A, B). The arcuate nucleus, the periventricular nucleus, and other hypothalamic nuclei contain only rare isolated nerve terminals (Fig. 14B). As pointed out already by Barry et al. (33), sparse LHRH-positive fibers are also present in extrahypothalamic areas such as the amygdaloid complex and the mesencephalon.

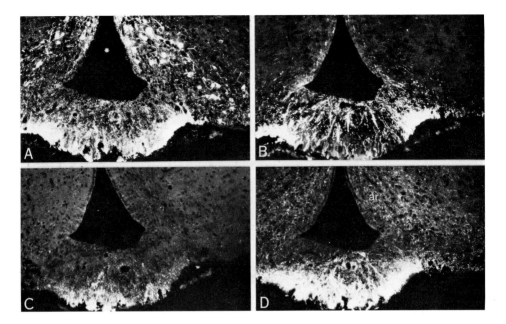

**FIG. 14.** Immunofluorescence micrographs of consecutive sections of the basal hypothalamus including the median eminence after incubation with antiserum to tyrosine hydroxylase **(A)**, luteinizing hormone releasing hormone **(B)**, thyrotropin releasing hormone **(C)**, and somatostatin **(D)**. LHRH **(B)**, TRH **(C)**, and SOM **(C)** nerve endings are all present in the median eminence but in different concentrations and with different distributions. LHRH-positive nerve endings are mainly present in the lateral parts, TRH-positive nerve endings in the medial parts, and SOM-positive nerve endings all over the median eminence. Note that SOM-positive nerve endings are present also in the arcuate nucleus. In this area there are also numerous TH-positive cell bodies and nerve endings mainly representing dopamine neurons **(A)**. Immunofluorescent cell bodies are not present after incubation with any of the antisera to the three peptides. Asterisk indicates third ventricle. ×75.

## Pathways

Barry et al. (33) suggested the existence of an LHRH pathway from cell bodies in the pre- and suprachiasmatic areas to the median eminence. In agreement with this, a complete deafferentation of the medial-basal hypothalamus in the rat causes a marked reduction (75 to 80%) of the LHRH content within the resulting island of tissue (45a,227a,252,337). The same operation does not change the LHRH content in the OVLT, however (49,227). These findings indicate that although the majority of LHRH nerve terminals in the median eminence originate outside the medial-basal hypothalamus, a small proportion may have their cell bodies in this area (see 75 and 138). The exact course of the axons giving rise to extrahypothalamic pathways is not known.

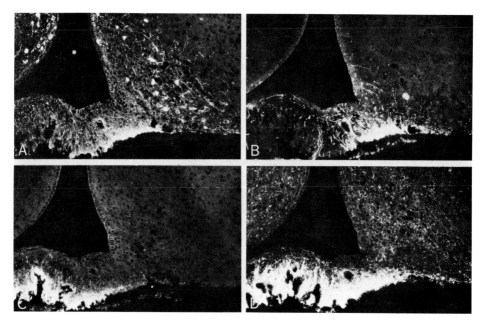

**FIG. 15.** Immunofluorescence micrographs of the basal hypothalamus including the median eminence at a level slightly caudal to the one shown in Fig. 14. The incubations have been performed with the same antisera as in Fig. 14, and the micrographs are shown in the same sequence. Note that LHRH-positive nerve endings are almost exclusively localized in the lateral parts of the median eminence at this level. Asterisk indicates third ventricle. ×75.

## Thyrotropin Releasing Hormone

TRH was isolated and structurally characterized as a tripeptide by Burgus et al. (58) and Nair et al. (233). By radioimmunoassay the distribution of TRH was found to be widespread throughout the nervous system (22,50, 54,173,204,240a,241,338). So far, few papers have been published on the immunohistochemical distribution of TRH (155,156).

### Cell Bodies

The TRH-positive cell bodies in the hypothalamus are mainly found in the dorsomedial nucleus (Fig. 18A) and the perifornical area.

### Nerve Terminals

The highest concentrations of TRH-positive nerve terminals are found in the medial parts of the external layer of the median eminence with high concentrations also in the pituitary stalk (Figs. 14C, 15C, 16C). In the dorsomedial nucleus, the perifornical region, and the parvocellular part of the

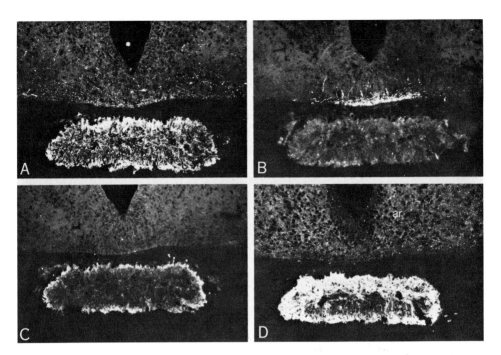

**FIG. 16.** Immunofluorescence micrographs of the posterior basal hypothalamus including the stalk. The incubations have been performed with the same antisera as in Fig. 14, and the micrographs are shown in the same sequence. Both TRH- and SOM-positive nerve endings are found in the stalk, whereas no LHRH-positive nerve terminals appear to enter the caudal parts of the stalk. Note LHRH-positive nerve terminals at the basal surface of the brain (*arrow*) and dense plexus of SOM-positive fibers in the arcuate nucleus (ar). Numerous dopamine fibers are also present in the stalk (**A**). Asterisk indicates third ventricle. ×75.

paraventricular nucleus, fibers are also seen. Moderate numbers of fluorescent fibers are found in the medial part of the ventromedial nucleus, in the periventricular area, and in the zona incerta. Scattered fibers are present in the ventral preoptic area, in the magnocellular part of the paraventricular nucleus, in the suprachiasmatic nucleus, and in the medial forebrain bundle. No fibers have so far been found in the OVLT. It should be emphasized that TRH-like immunoreactivity is also found in many extrahypothalamic areas including the spinal cord (50,155,156,173,241,339), but so far not outside the central nervous system.

### Pathways

Little is known of the TRH pathways in the brain. After a lesion mainly of the anterior periventricular hypothalamic area, all TRH-positive nerve terminals in the median eminence disappear (86). The extent of this lesion corresponds well with the location of the so-called thyrotropic area (98,124,

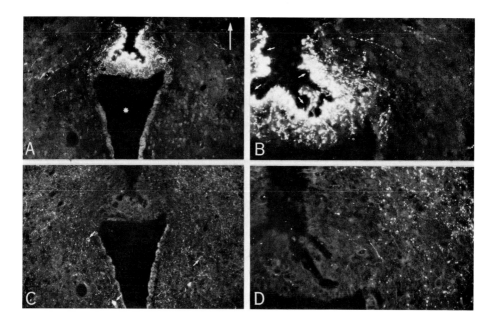

**FIG. 17.** Immunofluorescence micrographs of horizontal sections of the organum vasculosum laminae terminalis (OVLT) after incubation with antiserum to LHRH **(A,B)**, and tyrosine hydroxylase **(C)**, and dopamine-β-hydroxylase **(D)**. **A** and **C**, on one hand, and **B** and **D**, on the other hand, show consecutive sections. A dense plexus of LHRH-positive nerve endings is present in the OVLT close to the capillary plexus (*small arrows*). Numerous TH-positive and DBH-positive nerve endings are present in the surrounding brain tissue but do not penetrate into the OVLT **(C, D)**. Thus, the LHRH-positive nerve endings in the OVLT have a different relation to catecholamine nerve endings as compared to the situation in the median eminence (see Figs. 14, 15). Note TH-positive cell bodies (*arrows*) in the periventricular region. Asterisk indicates third ventricle and arrow points cranially. ×85.

130,200,279). The TRH-positive cell bodies so far discovered may partly be included in this lesion but have generally more caudal and lateral location. Thus, the lesion may also have affected axons running to the external layer of the medial eminence. After a total surgical isolation of the medial basal hypothalamus with the Halasz knife, the TRH concentration within the resulting island was reduced by 75% (54,252), indicating the existence of TRH cell bodies outside of this region. In agreement with this view, significant changes in several brain regions outside the lesion were not observed. It may be mentioned that TRH-positive nerve terminals in the ventral horns of the spinal cord disappear below the lesion after a total transection of the cord, indicating the existence of supraspinal descending TRH neuron systems (143).

### Somatostatin (Growth Hormone Release Inhibiting Hormone)

The existence of a growth hormone inhibiting factor was discovered by Krulich et al. (203) in the 1960s. In 1973 Brazeau et al. (44) isolated and

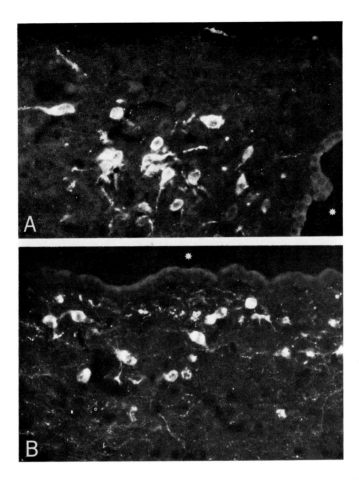

**FIG. 18.** Immunofluorescence micrographs of the dorsal part of the dorsomedial hypothalamic nucleus **(A)** and the anterior periventricular hypothalamic area **(B)** after incubation with antiserum to thyrotropin releasing hormone **(A)** and somatostatin **(B)**. Numerous TRH-positive nerve cell bodies are seen in the most dorsomedial part of the dorsomedial nucleus **(A)**. SOM-positive cell bodies are located close to the ventricle in the anterior hypothalamus **(B)**. Asterisks indicate third ventricle. ×260.

chemically characterized this factor as a tetradecapeptide and termed it somatostatin. With bioassay (327) and radioimmunoassay methods, the distribution of SOM has been described in the brain (45,91,251). Numerous immunohistochemical studies, including some ultrastructural studies (260,263), deal with the localization of SOM in the nervous system (4,14,17, 73,74,76,77,140,141,144,145,157,161,185,255,263,264,298). SOM is also present in circumventricular organs (215,263), in the pineal gland (264), in endocrine-like cells in different tissues including pancreas, gastrointestinal tract, and thyroid gland, as demonstrated with radioimmunoassay (11) and light (78,79,86,89,92,140,209,220,243–246,262,274,288) and electron

(118,136,209,287) microscopic immunocytochemistry. The distribution of SOM-positive cell bodies and terminals in the hypothalamus and the amygdaloid complex is schematically indicated in Fig. 19.

*Cell Bodies*

Within the hypothalamus a large SOM cell group is present in the anterior periventricular hypothalamic area (Figs. 8A, 18B, 30A, D). Cell bodies are also found in the entopeduncular nucleus–zona incerta region, in many extra-hypothalamic areas such as the amygdala, hippocampus, cortex, and other areas. In the periphery SOM (or SOM-like immunofluorescent material) is present in primary sensory neurons of the spinal ganglia and in several endocrine-like cells in the pancreas (D-cells or $A_1$-cells), thyroid gland (parafollicular cells), and the gut (enterochromaffin cells), as reported in references already cited.

*Nerve Terminals*

The highest concentrations of SOM nerve terminals are found in the medial internal and external layers and in the lateral external layers of the median eminence (Figs. 14D, 15D) and in the stalk (Fig. 16D). A high density is found in the arcuate (Figs. 14D, 15D, 16D, 20B), ventromedial (Fig. 20A, C), and suprachiasmatic nuclei (Fig. 21A, B) of the hypothalamus and in the OVLT. SOM-positive nerve terminals are also found in the amygdaloid complex, the caudate nucleus (medial part), the nucleus accumbens, the olfactory tubercle, in cortical areas, and in lower parts of the brainstem, e.g., in the substantia gelatinosa trigemini. In the spinal cord a high density is found in the substantia gelatinosa of the dorsal horn, representing the nerve terminals of the central branch of primary sensory neurons. Outside the central nervous system SOM-positive fibers are found in the intestinal wall.

*Pathways*

After lesions of the anterior periventricular area, all SOM-positive fibers in the median eminence disappear (86), whereas other fibers, e.g., in the ventromedial nucleus, remain. Thus, there is a projection from the periventricular SOM cell bodies that supplies the median eminence. The origin of the other hypothalamic fibers is unknown.

**Oxytocin, Vasopressin, and Neurophysin**

Oxytocin and vasopressin are octapeptides (81). The localization of these substances and of their carrier substance neurophysin also is discussed by Zimmerman (347) in this book. The main projections of fibers containing

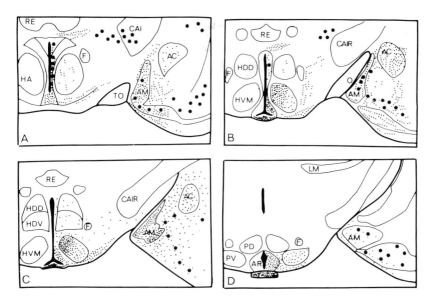

**FIG. 19.** Schematic drawing of the distribution of SOM-positive cell bodies (*asterisks*) and nerve terminals (*dots*) at four different levels of the rat hypothalamus. The frontal sections are drawn from König and Klippel (195) at the following levels: A, A 5340 μm; B, A 4620 μm; C, A 4110 μm; D, 3430 μm. For explanations of abbreviations, see legend to Fig. 5. SOM-positive cell bodies are present in the anterior periventricular region, in the zona incerta and entopeduncular nucleus, and in various parts of the amygdaloid complex including the medial and cortical nuclei. SOM-positive nerve terminals are present in high concentrations in the median eminence, the ventromedial nucleus, the arcuate nucleus, and in different parts of the amygdaloid complex.

these peptides from the supraoptic and paraventricular nuclei to the posterior pituitary have been mapped out immunohistochemically both at the light (71,84,219,305,319,330,331,333–335,345–347,349) and electron (12,265,307,309,310) microscopic level. Recently interest has been focused on the occurrence of vasopressin-containing fibers in the external layer of the median eminence (71,331,333). VP and OXY and neurophysin (NF) immunoreactivity can in addition be observed in neurons in many other hypothalamic and extrahypothalamic brain areas and in the spinal cord (see 143).

### Substance P

Substance P was originally discovered by von Euler and Gaddum (94) and was structurally characterized as an undecapeptide by Leeman and co-workers (64,211). A marked regional distribution of SP in the brain with especially high concentrations in the hypothalamus was reported early on the basis of bioassays (5,64,213,248,267,344) and more recently with radioimmunoassays (275). A more detailed description of the distribution of SP in the hypothalamus as well as in other brain regions has been reported

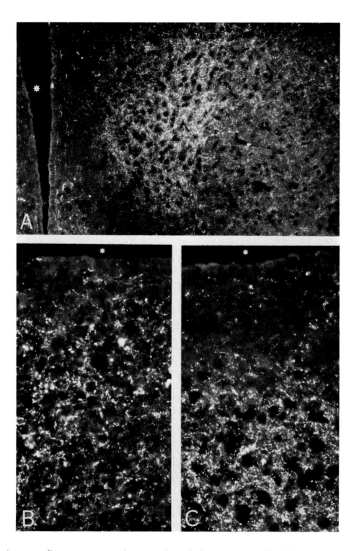

**FIG. 20.** Immunofluorescence micrographs of the ventromedial hypothalamic nucleus **(A, B)** and the arcuate nucleus **(C)** after incubation with antiserum to somatostatin. A dense plexus of SOM-positive nerve terminals is found both in the ventromedial nucleus, especially its medial part **(A)**, and in the arcuate nucleus **(C)**. Asterisks indicate third ventricle. Magnification: 105× **(A)** and 260× **(B, C)**.

(47,80,93,183), and these biochemical findings agree well with recent immunohistochemical observations (142,162,163,165,168,238,239,258).

## Cell Bodies

Numerous groups of cell bodies containing SP immunoreactive components have been found. At the present time at least 25 groups can be

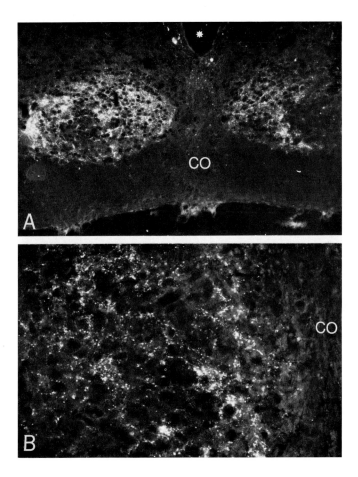

**FIG. 21.** Immunofluorescence micrographs of the suprachiasmatic nucleus after incubation with antiserum to somatostatin. A dense plexus of SOM-positive nerve terminals is seen in this nucleus. CO, chiasma opticum. Asterisk indicates third ventricle. Magnification 105× **(A)** and 260× **(B)**.

distinguished in the brain. In the hypothalamus SP-positive cell bodies are found notably, but not exclusively, in the nucleus premammillaris and nuclei dorsomedialis and ventromedialis (Fig. 22A). Some fluorescent cell bodies are also found in the lateral preoptic nucleus.

### Nerve Terminals

Extensive networks of SP-positive nerve terminals are found within the hypothalamus and the preoptic area, and almost all hypothalamic nuclei receive some SP fibers. The highest concentrations are found in the medial preoptic nucleus, with somewhat less in the periventricular area, the dorsomedial nucleus, the lateral hypothalamus, the arcuate nucleus, the ventro-

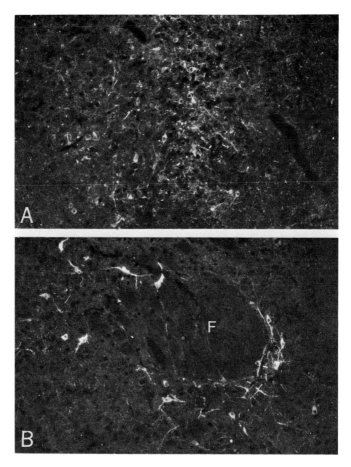

**FIG. 22.** Immunofluorescence micrographs of the ventromedial hypothalamic nucleus **(A)** and the perifornical area **(B)** after incubation with antiserum to substance P **(A)** and met-enkephalin **(B)**. Numerous small SP-positive cell bodies are present in the ventromedial nucleus and several met-ENK-positive cell bodies are found in the perifornical area **(B)**. F, fornix. ×105.

lateral part of the ventromedial nucleus, and the anterior hypothalamic nucleus. Relatively few fibers are present in the anterior and dorsomedial parts of the ventromedial nucleus, the suprachiasmatic and supraoptic nuclei, and almost no fibers are seen in the mammillary nuclei. In the median eminence only a few fibers are seen in the internal layer but almost none in the external layer.

Outside the hypothalamus, extensive networks of SP-positive fibers are present in most areas of the spinal cord and brain, except for the neocortical regions. In the periphery, SP is present in some primary sensory neurons (164,165,212,247,269,322,323), and in nerve fibers it is found in most tissues often surrounding blood vessels or related to glandular elements and in the skin.

## Pathways

Comparatively little so far is known about SP pathways in the central nervous system. Apart from the projections of SP neurons from cell bodies of spinal ganglia into the dorsal horn of the spinal cord (164,323) and a descending supraspinal projection (162), there is good evidence for a habenulo-interpeduncular SP tract (162,169,229) and a descending striatonigral tract (48,182). The hypothalamic connections are not known, but it may be pointed out that numerous axons are observed in the capsula interna, the medial forebrain bundle, and the stria terminalis. The latter finding may indicate that SP connections between hypothalamus and the amygdaloid complex exist.

## Enkephalins

Recently Hughes and collaborators (170) isolated from the brain and structurally characterized two similar pentapeptides, called enkephalins, characterized by leucine or methionine as terminal amino acids (leu- and met-ENK). These peptides were suggested to represent natural ligands for the opiate receptors. Guillemin et al. (129) described three larger peptides, termed $\alpha$-, $\beta$-, and $\gamma$-endorphin, which also possess morphinomimetic activity. It has been recognized that all these peptides may represent portions of a larger molecule, the $\beta$-lipotropin (342). ENK has been reported to have a marked regional distribution in the brain (311,314), with a generally higher content of met-ENK than leu-ENK. The distribution of these peptides agrees fairly well with that of the opiate receptor as determined by biochemical (137,205,206) and autoradiographic (13,14,268) techniques, as well as by immunohistochemical methods (87,88,142).

In our preliminary work (87) the distribution of immunoreactive ENK reported was based on an antibody raised against leu-ENK. Our subsequent analysis of this antiserum revealed that this antibody cross-reacts by about 20% with met-ENK but not at all with any of the three endorphins (146). The distribution patterns now described are based on a new antibody raised to met-ENK, which cross-reacts very little (10%) with leu-ENK and less than 0.1% with $\alpha$-, $\beta$-, and $\gamma$-endorphins. Since the distribution patterns are very similar with both antisera and since there is more met-ENK than leu-ENK in the rat brain, it is possible that we are dealing with met-ENK systems mainly. Immunohistochemical localization of $\alpha$- and $\beta$-endorphins in the intermediate lobe and in the hypothalamus has been studied with antisera to these larger polypeptides (43,128).

### Cell Bodies

Cell bodies containing immunoreactive ENK have been observed in many areas of the central nervous system. Of the approximately 25 to 30 cell groups observed so far, some are found in the hypothalamus. The arcuate, ventromedial, premammillary, and paraventricular nuclei and the perifornical region (Fig. 22B) contain ENK-positive cells.

### Nerve Terminals

Almost all hypothalamic nuclei contain moderate or high concentrations of met-ENK-positive nerve terminals. Also numerous fluorescent nerve terminals are present in the external layer of the median eminence (Fig. 23A). ENK (or ENK-like immunofluorescence) is found throughout the brainstem and spinal cord; in the periphery (e.g., in the gut and in sympathetic ganglia) we find similar ENK fluorescence in nerve terminals (87,143).

### Pathways

Little is known about ENK pathways. The existence of ENK-positive cell bodies within the hypothalamus clearly indicates that at least some of the nerve terminals may belong to intrahypothalamic neurons. In view of the recent demonstration of paraventricular neurons projecting to the external layer of the median eminence (8), ENK neurons might participate in such a projection.

## Angiotensin II

Three different forms of angiotensin (ANG) have been characterized (ANG I, a decapeptide, ANG II, an octapeptide, and ANG III, a heptapeptide) (90,312). Evidence has been presented for the occurrence of an iso-renin-angiotensin system in the brain (97,112,113). ANG has a regional distribution within the brain with the highest concentrations in the hypothalamus (97). This view is supported by our recent immunohistochemical findings of ANG II-like immunoreactivity in various parts of the nervous system (102). This immunoreactivity is not influenced by nephrectomy. So far cell bodies containing immunoreactive ANG II have been observed in the paraventricular nucleus and in the perifornical area. The ANG II-positive terminal networks are less dense than some of those described for other peptides. The highest concentrations are found in the dorsomedial nucleus and in the ventral-basal hypothalamus, with scattered terminals in most hypothalamic nuclei. An especial high density is found in the external layer of the median eminence (Fig. 24A, B), mainly in the medial part, but extending also into the lateral aspects. Outside the hypothalamus the highest concentrations are found in the sympathetic mediolateral column and sub-

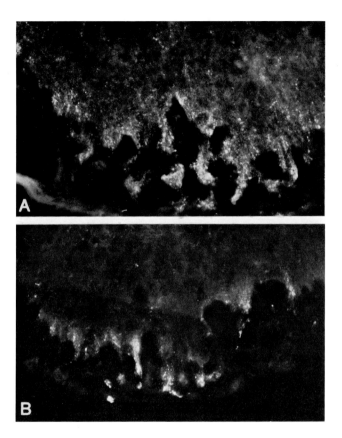

**FIG. 23.** Immunofluorescence micrographs of the medial part of the external layer of the anterior median eminence after incubation with antiserum to met-enkephalin **(A)** and gastrin **(B)**. The immunoreactive material is in both cases present in nerve endings in the most superficial layer. ×260.

stantia gelatinosa of the spinal cord. Other scattered ANG-positive fibers are found in many areas of the brainstem.

### Vasoactive Intestinal Polypeptide

A vasoactive intestinal polypeptide has recently been isolated and structurally characterized by Mutt and Said (230) as a peptide containing 28 amino acids. Although originally thought to be a primarily gut hormone, recent studies with radioimmunoassay (292) have demonstrated immunoreactive VIP in nervous tissues (293) including synaptosomes (115), with a marked regional distribution in the brain. High concentrations were found in the hypothalamus and in forebrain areas including neocortex. Recent immunohistochemical studies have demonstrated VIP in nerve fibers and

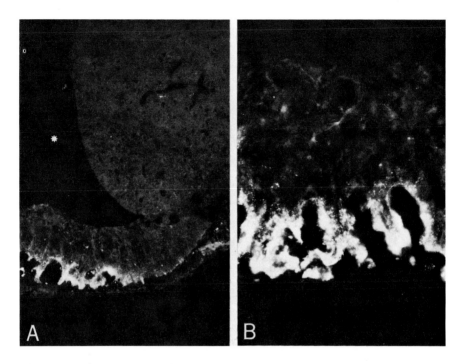

**FIG. 24.** Immunofluorescence micrographs of the median eminence after incubation with antiserum to angiotensin II. ANG-positive nerve terminals are present mainly in the medial parts of the external layer. **B**: A higher magnification (260×) of **A** (105×).

cell bodies in the intestinal wall (55,208), in cerebrovascular nerves (207), and in nerve cell bodies and fibers in the hypothalamus of the mouse (208). Immunohistochemical studies (108,110) with an antiserum prepared by Said and Faloona (292) essentially confirmed and extended the radioimmunological findings of Said and Rosenberg (293) and partly those of Larsson et al. (207,208) and Bryant et al. (55). VIP-positive cell bodies have so far not been found in the hypothalamus of the rat but are found mainly in cortical areas, in apparent contrast to the findings of Larsson et al. (208). Species differences or the different antisera used may explain this discrepancy. Moderate densities of VIP-positive nerve terminals were observed in the suprachiasmatic, medial preoptic, and anterior hypothalamic nuclei. In other hypothalamic nuclei including the median eminence, only scattered fibers were observed. Other brain areas containing considerable amounts of VIP-positive nerve terminals are the amygdaloid complex, especially the central nucleus, the neostriatum, and old and new cerebral cortex. VIP-positive cell bodies are also found in the intestinal wall. Extensive plexuses of VIP-positive fibers are found in the intestinal wall and around blood vessels in certain tissues.

## Gastrin

The structure of "gastrin" recovered from stomach and gut tumor has been found by Gregory and Tracy (125,126) to be heterogenous including at least two peptides, one of 34 amino acids and the other, a 17 amino acid fragment of the 34 amino acid sequence. Vanderhaegen et al. (329) have reported the occurrence of a brain gastrin immunoassayable peptide with an uneven regional distribution in brain, and with the highest concentrations in cortical areas. We used an antiserum raised to the 17 amino acid GAS by Rehfeld and collaborators (278) to study the histochemical distribution of immunoreactive GAS. In agreement with Vanderhaegen et al. (329), GAS-positive cell bodies and nerve terminals were observed in cortical areas, especially in the hippocampus. In the hypothalamus moderate densities of GAS-positive nerve terminals were observed in the ventromedial nucleus, the periventricular area, and the medial external layer of the median eminence (Fig. 23B), with scattered fibers in the lateral and basal parts. Recently evidence has been obtained that a brain peptide reacting with the GAS antiserum is a cholecystokinin-like peptide with a molecular size corresponding to that of an octapeptide (72, J. F. Rehfeld, *personal communication*).

## Prolactin

There is some immunohistochemical evidence that at least one of the anterior pituitary hormones, prolactin (PROL), may be present in neurons in the brain (107). Using antibodies against rat prolactin (rat prolactin 5-A, supplied by the U.S. National Institutes of Health-NIAMDD), we could observe fluorescent nerve terminals in many hypothalamic nuclei including the internal layer of the median eminence. These fibers did not disappear after hypophysectomy. The significance of PROL-like immunoreactivity is at present unclear, but it may be pointed out that prolactin appears to be present in the cerebrospinal fluid (65).

## ELECTRON MICROSCOPIC STUDIES

Electron microscopic identification of central neurons on the basis of their transmitter substance has been possible only to a limited extent. CA neurons have been identified with a permanganate fixation technique (284) in some cases (see 2 and 139) (Fig. 25), with the use of exogenous markers such as 5-hydroxy-DA (see 224 and 325), and with autoradiography (see 42, 70, and 316). The last technique has also been used to trace neurons that accumulate putative amino acid neurotransmitters (167,171,316). With the immunocytochemical techniques, new possibilities of characterizing neurons at the ultrastructural level have opened up. As indicated in the citations

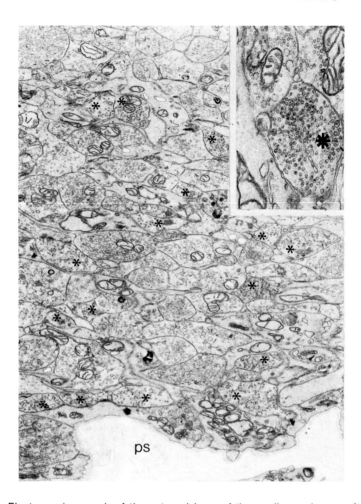

**FIG. 25.** Electron micrograph of the external layer of the median eminence after incubation with 5-hydroxy-dopamine and fixation with potassium permanganate. Numerous nerve endings (*asterisks*) contain small granular vesicles, i.e., vesicles with a diameter of about 500 Å with an electron-dense core. These nerve endings probably belong to monoamine neurons, mainly representing dopamine nerve endings. **Inset:** A higher magnification of one nerve ending containing small granular vesicles (*asterisk*) and some boutons with agranular vesicles. ps, pericapillary space. From Ajika and Hökfelt (2). 10,500×; 22,500× **(inset).**

above, several groups have identified CA and peptide neurons in the electron microscope by ultrastructural immunocytochemistry, almost exclusively with the peroxidase (PAP) technique (317). CA, 5-HT (180,269–272), and GABA neurons (see 286,294, and 341) have been identified in various brain areas, but so far to only a limited extent in the hypothalamus (Fig. 26A). On the other hand, studies on peptides have mainly been carried out in this region. LHRH- (Fig. 26B, 27) and SOM-containing nerve endings

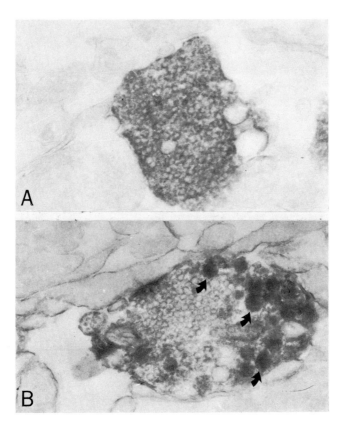

**FIG. 26.** Immunocytochemical electron micrographs of the external layer of the median eminence after incubation with antiserum to tyrosine hydroxylase **(A)** and LHRH **(B)**. The tissue was processed according to the antiperoxidase technique of Sternberger et al. (see 317), and the incubations were performed on 40–60 μm Vibratome sections. After incubation with TH antiserum, the reaction product is mainly confined to the cytoplasm, whereas after incubation with LHRH antiserum, the large granular vesicles (*arrows*) are electron-dense. This is in agreement with the view that TH is a cytoplasmic enzyme, whereas hormones in general appear to be localized to storage vesicles. ×39,000.

have been traced in the median eminence (61,117,232,260,261,263) and in the OVLT (263), and NF, VP, and OXY have been observed in the posterior pituitary and median eminence (265,307,309,310). The reaction product appears to be confined mainly to the so-called large granular vesicles, indicating that these organelles are probable peptide storage sites. On the other hand, some extravesicular NF and OXY have also been observed (307). Interestingly, Silverman and Desnoyers (308, see also 61) (Fig. 27) have convincingly demonstrated two types of LHRH-positive nerve endings distinguished by large granular vesicles of different sizes. This finding suggests that a classification and identification of peptidergic nerve endings on the basis of the size of presumptive storage vesicles may not

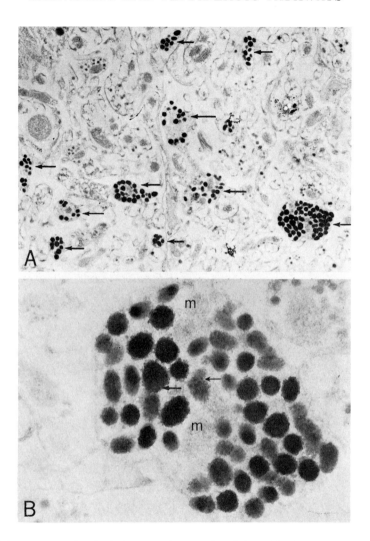

**FIG. 27.** Immunocytochemical electron micrographs of the palisade zone of the guinea pig median eminence of freeze-substituted tissue after incubation with antiserum to LHRH. The peroxidase antiperoxidase (PAP) technique of Sternberger and colleagues (see 317) was used. Numerous axon profiles (*arrows* in **A**) contain vesicles covered by PAP molecules (*arrows* in **B**). Note that mitochondria (m) and cytoplasm are negative. **B:** A higher magnification (65,250×) of part of A (12,000×). These figures were kindly supplied by Drs. Silverman and Desnoyers and are taken from ref. 308 with permission of the publisher.

be possible. Pelletier et al. (263) have, on the other hand, shown that the granular vesicles in the SOM nerve endings have a slightly larger diameter than the LHRH storage vesicles. It is obvious that studies on the ultrastructural features of immunohistochemically identified neurons will give further important information with regard to synaptology and storage sites of peptides and transmitters.

## INTERACTIONS AMONG SYSTEMS

By considering the distribution pattern of the various systems described above, it is clear that interactions among different systems are likely to occur in many hypothalamic and extrahypothalamic areas. Full morphologic evidence of such interactions can be obtained at the ultrastructural level only by identifying two types of neurons histochemically and by demonstrating synaptic connections between them. However, at the light microscopic level indications of interactions can already be obtained. We have been particularly interested in possible relations between the CA systems and the peptide-containing neurons. Principally, direct connections between two systems may occur via axodendritic (or axosomatic) or via axoaxonic contacts. In the former case cell bodies of one system should be surrounded by nerve endings of the other, whereas in the latter case nerve endings from both systems should intermingle. We would like to discuss briefly some examples of each of these alternatives.

An obvious site for axoaxonic interactions is in the median eminence. Particularly the DA and LHRH axons appear to distribute very similarly, but also DA and SOM nerve endings are both found in the lateral part of the median eminence. With the aid of a recently developed objective method based on microspectrofluorometry (1), the spatial correlation between various nerve endings can be analyzed. Interestingly, in the lateral external layer of the median eminence a good correlation was obtained for DA and LHRH nerve endings, whereas no such correlation could be found for DA and SOM fibers. These findings support our hypothesis that DA may inhibit LH release via an axoaxonic interaction with LHRH nerve endings in the median eminence (106,147). It may be pointed out that in the OVLT, which contains both LHRH and SOM nerve endings (32,49,197,251,263,337,347), such an interaction is not suggested by our immunohistochemical findings (see Fig. 17). Here the CA nerve endings do not approach the vascular plexus and thus do not intermingle with the LHRH nerve endings. Thus, the LHRH stores in the OVLT do not appear to be under a dopaminergic influence as may be the case in the median eminence.

Another example of a hypothetical interaction involves the DA and SOM neurons. The SOM cell bodies in the anterior periventricular hypothalamic area are surrounded by CA nerve terminals indicating a possible adrenergic control of SOM neurons (see Fig. 10). The reverse situation may exist in the arcuate nucleus where DA cell bodies are surrounded by SOM, SP, and ENK-positive nerve endings as can be shown by incubating consecutive sections with antisera to these peptides and to the CA-synthesizing enzyme TH. As shown in Fig. 28, a close contact between SOM nerve endings and

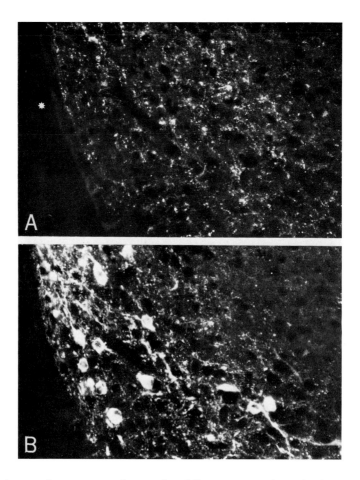

FIG. 28. Immunofluorescence micrographs of the arcuate nucleus after incubation with antiserum to somatostatin **(A)**. This section is the same as that shown in Fig. 14D and has been reincubated with antiserum to tyrosine hydroxylase **(B)**. The dopamine cell bodies of the A12 group are surrounded by numerous SOM-positive cell bodies. Asterisk indicates third ventricle. ×262.

DA cell bodies can be directly demonstrated by reincubating "SOM-stained" section after photography with the TH antiserum. These morphological findings suggest that the DA neurons of the A 12 cell group, which project to the median eminence, are controlled by SOM and other peptidergic systems.

These fragmentary examples of hypothetical interactions between different systems merely serve to focus attention on the fact that a neuron should be looked upon not as an isolated entity but as a part of a complex network. It will be an important task for immunohistochemistry to establish the connections between the great number of systems and thus to elucidate the components of the neuronal loops in the brain.

## COMMENTS

The results briefly summarized in this chapter give evidence for the existence of a large number of neuronal systems characterized on the basis of immunohistochemical criteria or, more precisely, their content of specific enzymes or peptides. In almost all cases these systems are not confined to the hypothalamus but distribute to many parts of the central nervous system. Some are mainly present in the brain system, others are located in both brainstem and cortical areas, whereas still others seem to be present mainly in cortical areas. Some substances are also present in the peripheral nervous system and even in nonneuronal cell systems. All of these substances are, however, present in the hypothalamus, and often the highest concentrations are found in this brain area.

### Specificity

It should be emphasized that certain problems of specificity are often associated with histochemical techniques. Whereas rather good agreement exists with regard to the specificity of the histochemical reactions for visualizing monoamine and possibly GABA neurons, the possibility of misleading results with the acetylcholinesterase reaction has been discussed (304). Furthermore, in immunohistochemical studies with antisera to small peptides it is not possible to exclude cross-reactivity with analogues, fragments, or sequences or larger peptides and proteins. Therefore, it is preferable to refer more cautiously, for example, to "SP-like-immunoreactivity." An example of such a cross-reaction is presented by GAS. Antibodies raised to GAS coupled to bovine serum albumin appear to cross-react with a smaller cholecystokinin-like octapeptide in the brain (72). On the other hand, although partly present in the same brain areas, each of all peptides so far studied has its own characteristic and unique distribution pattern. Thus, whatever the exact amino acid sequence of the peptide in a neuron turns out to be, it appears that we are dealing with specific subsystems in the brain.

### Median Eminence

A large number of neurohumoral substances appear to be present in the median eminence (see also 191). As might be expected, the systems containing LHRH, TRH, and SOM, the three hypothalamic releasing (inhibitory) hormones so far isolated and structurally characterized, and to which antisera are available, are all present in this area. However, a number of other peptides and smaller molecules of a probable neurotransmitter nature are also found in this area, possibly in nerve endings located in a "secretion position" close to the portal vessels. This raises the question of the functional role of all these substances in this region. In principle, substances located in the external layer of the median eminence may influence the pituitary gland in at least three ways as indicated in the schematic drawing in Fig. 29. First, they may, according to Harris' (131) basic concept, be secreted into the portal vessels and transported to the anterior pituitary to influence the hormone-producing cells, i.e., they may act as true hypothalamic hormones. Second, they may be released and influence adjacent nerve endings via an axoaxonic mechanism, i.e., they may act as neurotransmitters. Third, they may be released and exert an influence on glial elements (tanycytes), which in turn may participate in the control of secretion of substances from the median eminence (340).

It seems reasonable to assume that peptides such as LHRH, TRH, and SOM are mainly released into the portal vessels. For many other sub-

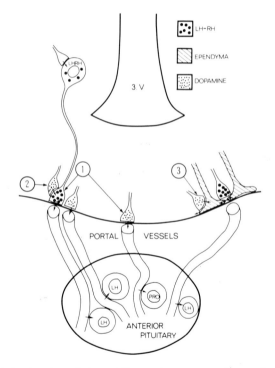

**FIG. 29.** Schematic drawing of the median eminence, portal vessels, and the pituitary gland. A substance located in the external layer of the median eminence close to the portal vessel may exert an effect on the hormone secretion in the pituitary gland principally in at least three ways as illustrated and exemplified in this figure. (1) A substance, e.g., LHRH or dopamine (perhaps representing prolactin inhibitory factor (PIF) (220), may be secreted into the portal vessels and transported to the pituitary gland. (2) It may act as a neurotransmitter and be released at an axoaxonic synapse. For example, it has been postulated that dopamine inhibits LHRH release via an axoaxonic influence (106,147). (3) It may act on the ependymal elements, the tanycytes, which in turn may control the position of other hormone- (e.g., LHRH-) containing nerve endings in their relation to the portal vessels.

stances, on the other hand, the situation is less clear. DA has been postulated to act on LHRH-containing nerve endings via axoaxonic contacts to inhibit the release of LHRH and in this way exert an inhibitory influence on the secretion of LH (106,147). DA exerts also a powerful effect on pituitary secretion of prolactin (221), and it has, in fact, been postulated that DA is identical with the prolactin inhibiting factor. DA receptors have been identified in both pituitary and hypothalamus (44a). DA may thus exert a dual effect via two different mechanisms on pituitary hormone secretion. There is evidence that the DA nerve terminals exerting these two effects have a different localization in the median eminence: the effect on prolactin secretion is mediated via DA nerve endings in the medial part, whereas DA nerve endings in the lateral part control LHRH secretion (109). In agreement with this view our present immunohistochemical results demonstrate a marked superposition of LHRH and DA (TH-positive) nerve endings in the lateral part of the external layer (compare Figs. 14 and 15). As discussed above, this impression has been confirmed by objective analysis (1).

The demonstration of VP in the external layer of the median eminence is suggestive of a role for this peptide in the regulation of anterior pituitary function (see 346), and there is evidence that VP releases corticotropin (ACTH) (201,343). Recently, ENK has been demonstrated to enhance PROL release (216). Although this effect could be mediated at a higher level in the hypothalamus, the occurrence of ENK in the external layer is suggestive either of an effect at the median eminence level or of a release into the portal vessels. The latter idea is supported by the findings of Lien et al. (216) that ENK exerts a powerful PROL releasing effect on pituitary cells.

The roles of many of the other peptides and transmitters in the median eminence at present are not clear, but it may be mentioned that GABA has been shown to influence prolactin release (226), ACTH release (223), and FSH but not LH release (242).

As shown above, the number of different substances present in the median eminence is large, and some of these systems are quantitatively very significant. Thus, it has been calculated that the monoamine, mainly DA, boutons constitute about one-third of all nerve terminals in the lateral external layer (Fig. 30; see 2) and SOM-positive boutons may make up another 30% (260). This apparent wealth of different systems may evoke the question whether each of these substances is present in separate neurons or whether a concomitant storage of two or even more substances in one neuron may occur. This is a particularly relevant question since in peripheral tissues, such as the gut, cells are known which simultaneously store a monoamine and a peptide, probably even in the same storage vesicles (256).

The differential distribution of most of the peptides within the median eminence makes a concomitant storage in the same cell in many cases unlikely. However, the high density of the nerve terminals and their small size preclude a definitive answer to this question merely by comparisons made at

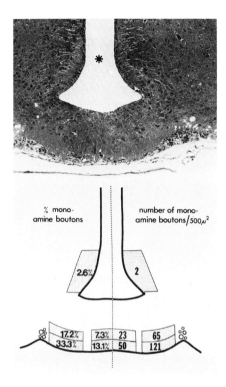

FIG. 30. Schematic presentation of the quantitative estimation of monoamine (preferentially dopamine) nerve endings in the median eminence. The highest concentrations of monoamine boutons are found in the superficial lateral part of the external layer (33% of all boutons) with the lowest concentrations in the deep medial part of the external layer. The calculations are based on 5-hydroxy-dopamine-incubated slices fixed with potassium permanganate (see Fig. 26). The micrograph is from a toluidine blue-stained semithin section of the basal hypothalamus, approximately at the level at which the counting was performed. From Ajika and Hökfelt (2).

the light microscopic level of consecutive immunohistochemical sections incubated with different antibodies. Therefore, this problem has been approached in different ways. Kizer et al. (189) have administered the destructive neurotoxin 6-OH-DA intraventricularly and obtained a marked decrease in DA but not in LHRH, indicating that these two substances are present in different neurons. Barnea et al. (22) have found a differential subcellular compartmentalization of TRH and LHRH. We have compared the histologic distribution of cell bodies containing different peptides with that of the hypothalamic DA neurons. No obvious correlation is found except for DA and SOM cell bodies which overlap *partially* (Fig. 8). However, after a comparison of sections consecutively stained with the two antisera, it appears unlikely that DA and SOM are stored in the same cells (Fig. 31). The ideal technique to resolve this question is to identify and characterize the different types of nerve endings on consecutive sections in the electron microscope. As reported above, several promising ultrastructural studies have been carried out. For example, Pelletier et al. (263) have pointed out that the storage granules in the LHRH-positive nerve endings are slightly smaller than those containing SOM, indicating that these two peptides are probably stored in different neurons. Similar studies on the other peptides and neurotransmitters in the median eminence are, however, clearly needed

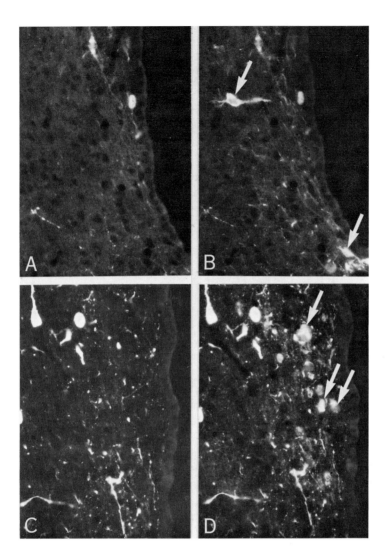

**FIG. 31.** Immunofluorescence micrographs of the anterior periventricular region (approximately at the same level as Fig. 8) after incubation with antiserum to somatostatin **(A)**, tyrosine hydroxylase **(C)**, SOM plus TH **(B)**, and TH plus SOM **(D)**. **A** and **B** show the same section, and **C** and **D** are from another section. Some SOM-positive cell bodies are shown in **A**. After photography the same section was incubated with TH antiserum to visualize dopamine cell bodies. As can be seen in **B**, newly stained, additional cell bodies can be observed (*arrows*). Similar results are obtained with the reversed procedure, i.e., consecutive staining of the same section first with TH antiserum and photography **(C)**, then restaining with SOM antiserum **(D)**. The second incubation reveals newly stained cells suggesting that SOM and dopamine are present in different neurons. ×245.

to clarify the problem of possible simultaneous storage of several substances in one neuron.

Most of the substances dealt with in this chapter occur not only in the median eminence but also in many other hypothalamic nuclei. In fact, with the probable exception of LHRH, most of the peptides are distributed over large areas of the central nervous system, and in some cases even in peripheral sites. For example, it has been pointed out that two-thirds of the brain TRH is located outside the hypothalamus (173,339). The findings support the view derived from a number of electrophysiological studies indicating widespread pharmacologic effects of peptides throughout the nervous system. It has been suggested that peptides such as LHRH, TRH, and SOM may be released at synapses and act as a neurotransmitter or modulator of neuroelectrical activity (21,82,196,228,237,282).

## CONCLUSIONS AND SUMMARY

It is apparent from the work discussed in this chapter that the recent brilliant developments in protein and peptide biochemistry have made possible an exciting new immunohistochemical morphological exploration of the hypothalamus (and the nervous system in general) by which numerous subpopulations of neurons can be characterized on the basis of their content of specific substances (including hormones and enzymes).

With antisera to all four catecholamine-synthesizing enzymes, DA, NE, and epinephrine neuron systems have been identified. GABA neurons can be mapped with antibodies to glutamic acid decarboxylase. Antisera have been raised to a number of peptides, and there appear to be neuron systems in the hypothalamus and other parts of the brain containing luteinizing hormone releasing hormone, somatostatin (growth hormone release-inhibiting hormone), thyrotropin releasing hormone, oxytocin, vasopressin, neurophysin, substance P, enkephalin, angiotensin II, vasoactive intestinal polypeptide, and possibly prolactin and a cholecystokinin-like or gastrin-like peptide.

Much of the work presented here is still in its early stages, and extensive research will be necessary to verify and elucidate the exact nature of the peptides, enzymes, and neuronal systems described. It is possible that hitherto unknown peptides will be shown to cross-react with the antisera now being used and that some of our results will have to be revised. One such example is that the brain fluorescence found with a gastrin antiserum probably represents a cholecystokinin-like octapeptide.

Nevertheless, the number and the extensive distribution of peptide-containing neurons in the brain, spinal cord, and peripheral nervous system appear to surpass our expectations. The morphological analysis of these peptide-containing neuronal systems will hopefully be useful for neuroendocrinologists, neurochemists, pharmacologists, and neurophysiologists, who at present are carrying out work aimed at revealing if and how these substances are involved in nervous system function.

## ACKNOWLEDGMENTS

This work was supported by grants from the Swedish Medical Research Council (04X-2887, 04X-2886, 04X-3521, 04X-04495, 04X-715, 25X-5065, 19X-00034), Magnus Bergwalls Stiftelse, Knut and Alice Wallenbergs Stiftelse, by a grant from the U.S. Department of Health, Education and Welfare (1R01 HL 18994-01, 555), through the National Heart and Lung Institute, by United States Public Health Service Grants MH-02717 and MH-25504-3, and by National Science Foundation Grant CB-8465. The skillful technical assistance of Mrs. W. Hiort and Miss A. Nygårds is gratefully acknowledged. For excellent secretarial help we thank Mrs. U. B. Nilsson.

## REFERENCES

1. Agnati, L., Fuxe, K., Hökfelt, T., Goldstein, M., Jeffcoate, S. L., and Elde, P. (1977): A method to measure the distribution pattern of specific nerve terminals in sampled regions. Studies on tyrosine hydroxylase. LH-RH and GHR-IH immunofluorescence. *J. Histochem. Cytochem.* (*in press*).
2. Ajika, K., and Hökfelt, T. (1973): Ultrastructural identification of catecholamine neurones in the hypothalamic periventricular-arcuate nucleus-median eminence complex with special reference to quantitative aspects. *Brain Res.*, 57:97–117.
3. Alpert, L. C., Brawer, J. R., Jackson, I. M. D., and Reichlin, S. (1976): Localization of LHRH in neurons in frog brain. *Endocrinology*, 98:910–921.
4. Alpert, L. C., Brawer, J. R., Patel, Y. C., and Reichlin, S. (1976): Somatostatinergic neurons in anterior hypothalamus: Immunohistochemical localization. *Endocrinology*, 98:255–258.
5. Amin, A. H., Crawford, T. B. B., and Gaddum, J. H. (1954): The distribution of substance P and 5-hydroxytryptamine in the central nervous system of the dog. *J. Physiol.*, 126:596–618.
6. Amoss, M., Burgus, R., Blackwell, R., Vale, W., Fellows, R., and Guillemin, R. (1971): Purification, amino acid composition and N-terminus of the hypothalamic luteinizing hormone releasing factor (LRF) of ovine origin. *Biochem. Biophys. Res. Commun.*, 44:205–210.
7. Andén, N. E., Dahlström, A., Fuxe, K., Larsson, K., Olson, L., and Ungerstedt, U. (1966): Ascending monoamine neurons to the telencephalon and diencephalon. *Acta Physiol. Scand.*, 67:313–326.
8. Antunes, J. L., Carmel, P. W., and Zimmerman, E. A. (1977): Projections from the paraventricular nucleus to the zona externa of the median eminence of the rhesus monkey: An immunohistochemical study. *Brain Res.* (*in press*).
9. Araki, S., Toran-Allerand, C. D., Ferin, M., and Vande Wiele, R. L. (1975): Immunoreactive gonadotropin-releasing hormone (Gn-RH) during maturation in the rat: Ontogeny of regional hypothalamic differences. *Endocrinology*, 97:693–697.
10. Arimura, A., Sato, H., Coy, D. H., Schally, A. V. (1975): Radioimmunoassay for GH-release inhibiting hormone. *Proc. Soc. Exp. Biol. Med.*, 148:784–789.
11. Arimura, A., Sato, H., Dupont, A. Nishi, N., and Schally, V. (1975): Somatostatin: Abundance of immunoreactive hormones in rat stomach and pancreas. *Science*, 189:1007–1009.
12. Aspelagh, M.-R., Vandesande, F., and Dierickx, K. (1976): Electron microscopic immu-

nocytochemical demonstration of separate neurophysin-vasopressinergic and neurophysin-oxytocinergic nerve fibres in the neural lobe of the rat hypophysis. *Cell Tissue Res.,* 171:31–37.
13. Atweh, S. F., and Kuhar, M. J. (1977): Autoradiographic localization of opiate receptors in rat brain. I. Spinal cord and lower medulla. *Brain Res. (in press).*
14. Atweh, S. F., and Kuhar, M. J. (1977): Autoradiographic localization of opiate receptors in rat brain. II. The brainstem. *Brain Res. (in press).*
14a. Baker, B. L., and Dermody, W. C. (1976): Effect of hypophysectomy on immunocytochemically demonstrated gonadotropin-releasing hormone in the rat brain. *Endocrinology,* 98:1116–1122.
15. Baker, B. L., Dermody, W. C., and Reel, J. R. (1974): Localization of luteinizing hormone-releasing hormone in the mammalian hypothalamus. *Am. J. Anat.,* 139:129–134.
16. Baker, B. L., Dermody, W. C., and Reel, J. R. (1975): Distribution of gonadotropin-releasing hormone in the rat brain as observed with immunocytochemistry. *Endocrinology,* 97:125–135.
17. Baker, B. L., and Yen, Y.-Y. (1976): The influence of hypophysectomy on the stores of somatostatin in the hypothalamus and pituitary stem. *Proc. Soc. Exp. Biol. Med.,* 151:599–602.
18. Bargmann, W. (1966): Neurosecretion. *Int. Rev. Cytol.,* 19:183–201.
19. Bargmann, W., and Scharrer, B. (1951): The site of origin of the hormones of the posterior pituitary. *Am. Sci.,* 39:255–259.
20. Bargmann, W., and Scharrer, B. (Eds.) (1970): *Aspects of Neuroendocrinology.* Springer-Verlag, Berlin, Heidelberg, New York.
21. Barker, J. L. (1976): Peptides: Roles in neuronal excitability. *Physiol. Rev.,* 56:435–452.
22. Barnea, A., Ben-Jonathan, N., Colston, C., Johnston, J. M., and Porter, J. C. (1975): Differential sub-cellular compartmentalization of thyrotropin releasing hormone (TRH) and gonadotropin releasing hormone (LRH) in hypothalamic tissue. *Proc. Natl. Acad. Sci. USA,* 72:3153–3157.
23. Barry, J. (1970): Recherches sur le role des monoamines infundibulaires dans le controle de la sécrétion gonadotrope chez le cobaye et la souris. In: *Aspects of Neuroendocrinology,* edited by W. Bargmann and B. Scharrer, pp. 243–252. Springer-Verlag: Berlin, Heidelberg, New York.
24. Barry, J. (1976): Characterization and topography of LH-RH neurons in the rabbit. *Neurosci. Lett.,* 2:201–205.
25. Barry, J., and Carette, B. (1975): Etude en immunofluorescence des neurones élaborateurs de LRF chez les Cébidés. *C. R. Acad. Sci. [D] (Paris),* 281:735–738.
26. Barry, J., and Carette, B. (1975): Immunofluorescence study of LRF neurons in primates. *Cell Tissue Res.,* 164:163–178.
27. Barry, J., and Dubois, M.-P. (1973): Etude en immunofluorescence des structures hypothalamiques à competénce gonadotrope. *Ann. Endocrinol. (Paris),* 34:735–742.
28. Barry, J., and Dubois, M.-P. (1974): Etude en immunofluorescence de la différenciation prénatale des cellules hypothalamiques élaboratrices de LH-RF et de la maturation de la voie neurosécrétrice préoptico-infundibulaire chez le cobaye. *Brain Res.,* 67:103–113.
29. Barry, J., and Dubois, M.-P. (1974): Immunofluorescence study of the preoptico-infundibular LH-RH neurosecretory pathway of the guinea pig during the estrous cycle. *Neuroendocrinology,* 15:200–208.
30. Barry, J., and Dubois, M.-P. (1975): Immunofluorescence study of LRF-producing neurons in the cat and the dog. *Neuroendocrinology,* 18:290–298.
31. Barry, J., and Dubois, M.-P. (1976): Immunoreactive LRF neurosecretory pathways in mammals. *Acta Anat. (Basel),* 94:497–503.
32. Barry, J., Dubois, M.-P., and Carette, B. (1974): Immunofluorescence study of the preoptic-infundibular LRF neurosecretory pathway in the normal, castrated or testosterone-treated male guinea pig. *Endocrinology,* 95:1416–1423.
33. Barry, J., Dubois, M. P., and Poulain, P. (1973): LRF producing cells of the mammalian hypothalamus. A fluorescent antibody study. *Z. Zellforsch.,* 146:351–366.
34. Bassiri, R. M., and Utiger, R. D. (1972): The preparation and specificity of antibody to thyrotropin releasing hormone. *Endocrinology,* 90:722–727.
35. Battenberg, E. L. F., and Bloom, F. E. (1975): A rapid, simple and more sensitive method for the demonstration of central catecholamine-containing neurons and axons by gly-

oxylic acid induced fluorescence: 1. Specificity. *Psychopharmacol. Comm.*, 1:3–13.
36. Bern, H. A., and Knowles, F. G. W. (1966): Neurosecretion. In: *Neuroendocrinology*, edited by L. Martini and W. F. Ganong, pp. 139–186. Academic Press, New York.
37. Bird, E. D., Chiappa, S. A., and Fink, G. (1976): Brain immunoreactive gonadotropin-releasing hormone in Huntington's chorea and in non-choreic subjects. *Nature*, 260:536–538.
38. Björklund, A., Falck, B., and Owman, Ch. (1972): Fluorescence microscopic and microspectrofluorametric techniques for the cellular localization and characterization of biogenic amines. In: *Methods in Investigative and Diagnostic Endocrinology*, edited by S. A. Berson, *Vol. 1: The Thyroid and Biogenic Amines*, edited by J. E. Rall and I. J. Kopin, pp. 318–368. North Holland Publishing Co., Amsterdam.
39. Björklund, A., Lindvall, O., and Nobin, A. (1975): Evidence of an incertohypothalamic dopamine neurone system in the rat. *Brain Res.*, 89:29–42.
40. Björklund, A., Moore, R. Y., Nobin, A., and Stenevi, U. (1973): The organization of tubero-hypophyseal and reticulo-infundibular catecholamine neuron systems in the rat brain. *Brain Res.*, 51:171–191.
41. Björklund, A., and Nobin, A. (1973): Fluorescence and microspectrofluorimetric mapping of dopamine and noradrenaline cell groups in the rat diencephalon. *Brain Res.*, 51:193–205.
42. Bloom, F. E.: The fine structural localization of biogenic monoamines in nervous tissue. (1970): *Int. Rev. Neurobiol.*, 13:27–66.
43. Bloom, F., Battenberg, E., Rossier, J., Ling, N., Leppaluoto, J., Vargo, T. M., and Guillemin, R. (1977): Endorphines are located in the intermediate and anterior lobes of the pituitary gland, not in the neurohypophysis. *Life Sci.*, 20:40–48.
44. Brazeau, P., Vale, W., Burgus, R., Ling, N., Butcher, M., Rivier, J., and Guillemin, R. (1973): Hypothalamic polypeptide that inhibits the secretion of immunoreactive pituitary growth hormone. *Science*, 179:77–79.
44a. Brown, G. M., Seeman, P., and Lee, T. (1976): Dopamine neuroleptic receptors in basal hypothalamus and pituitary. *Endocrinology*, 99:1407.
45. Brownstein, M., Arimura, A., Sato, H., Schally, A. V., and Kizer, J. S. (1975): The regional distribution of somatostatin in the rat brain. *Endocrinology*, 96:1456–1461.
45a. Brownstein, M. J., Arimura, A., Schally, A. V., Palkovits, M., and Kizer, J. S. (1976): The effect of surgical isolation of the hypothalamus on its luteinizing hormone-releasing hormone content. *Endocrinology*, 98:662–665.
46. Brownstein, M., Kobayashi, R., Palkovits, M., and Saavedra, J. M. (1975): Choline acetyltransferase levels in diencephalic nuclei of the rat. *J. Neurochem.*, 24:35–38.
47. Brownstein, M. J., Mroz, E. A., Kizer, J. S., Palkovits, M., and Leeman, S. E. (1976): Regional distribution of substance P in the brain of the rat. *Brain Res.*, 116:299–305.
48. Brownstein, M. J., Mroz, E. A., Tappaz, M. L., and Leeman, S. E. (1977): On the origin of substance P and glutamic acid decarboxylase (GAD) in the substantia nigra. *Brain Res.* (in press).
49. Brownstein, M., Palkovits, M., and Kizer, J. S. (1975): On the origin of luteinizing hormone-releasing hormone (LH-RH) in the supraoptic crest. *Life Sci.*, 17:679–682.
50. Brownstein, M., Palkovits, M., Saavedra, J. M., Bassiri, R., and Utiger, R. D. (1974): Thyrotropin-releasing hormone in specific nuclei of rat brain. *Science*, 185:267–269.
51. Brownstein, M. J., Palkovits, M., Saavedra, J. M., and Kizer, J. S. (1976): Distribution of hypothalamic hormones and neurotransmitters within the diencephalon. In: *Frontiers in Neuroendocrinology, Vol. 4: Distribution of Hypothalamic Hormones and Neurotransmitters Within the Diencephalon*, edited by L. Martini and W. F. Ganong, pp. 1–23. Raven Press, New York.
52. Brownstein, M., Palkovits, M., Tappaz, M., Saavedra, J., and Kizer, S. (1976): Effect of surgical isolation of the hypothalamus on its neurotransmitter content. *Brain Res.*, 117:287–295.
53. Brownstein, M., Saavedra, J. M., Palkovits, M., and Axelrod, J. (1974): Histamine content of hypothalamic nuclei of the rat. *Brain Res.*, 77:151–156.
54. Brownstein, M. J., Utiger, R. D., Palkovits, M., and Kizer, J. S. (1975): Effect of hypothalamic deafferentation on thyrotropin-releasing hormone levels in rat brain. *Proc. Natl. Acad. Sci. USA*, 72:4177–4179.
55. Bryant, M. G., Polak, J. M., Modlin, I., Bloom, S. R., Albuquerque, R. H., and Pearse,

A. G. E. (1976): Possible dual role for vasoactive intestinal peptide as gastrointestinal hormone and neurotransmitter substance. *Lancet,* 1:991–993.
56. Bugnon, C., Bloch, B., and Fellman, D. (1976): Mise en évidence cytoimmunologique de neurones à LH-RH chez le foetus humain. *C.R. Acad. Sci. [D] (Paris),* 282:1625–1628.
57. Bugnon, C., Bloch, B., Fellman, D., and Gouget, A. (1976): Etude comparative chez le cobaye, le chat et le chien, des fibres nerveuses hypothalamo-infundibulaires immunoréactives à un immun-sérum antiLH-RH. *C.R. Soc. Biol.,* 170:83–87.
58. Burgus, R., Dunn, T. F., Desiderio, D., Ward, D. N., Vale, W., and Guillemin, R. (1970): Characterization of ovine hypothalamic hypophysiotropic TSH-releasing factor. *Nature,* 226:321–325.
59. Burt, A. M., and Silver, A. (1973): Histochemistry of choline acetyltransferase: A critical analysis. *Brain Res.,* 62:509–516.
60. Calas, A., Kerdelhué, B., Assenmacher, I., and Jutisz, M. (1973): Les axones à LH-RH de l'éminence médiane. Mise en évidence chez le canard par une technique immunocytochimique. *C.R. Acad. Sci. [D] (Paris),* 277:2765–2768.
61. Calas, A., Kerdelhué, B., Assenmacher, I., and Jutisz, M. (1974): Les axons à LH-RH de l'éminence médiane. Etude ultrastructurale chez le canard par une technique immunocytochimique. *C.R. Acad. Sci. [D] (Paris),* 278:2557–2559.
62. Calcutt, C. R. (1976): The role of histamine in the brain. *Gen. Pharmacol.,* 7:15–25.
63. Carlsson, A., Falck, B., and Hillarp, N. Å. (1962): Cellular localization of brain monoamines. *Acta Physiol. Scand. [Suppl.* 196], 56:1–28.
64. Chang, M. M., Leeman, S. E., and Niall, H. D. (1971): Amino acid sequence of substance P. *Nature [New Biol.],* 232:86–87.
64a. Cheney, D. L., LeFevre, H. F., and Racagni, G. (1975): Choline acetyltransferase activity and mass fragmentographic measurement of acetylcholine in specific nuclei and tracts of rat brain. *Neuropharmacology,* 14:801–809.
65. Clemens, J. A., and Sawyer, B. D. (1974): Identification of prolactin in cerebrospinal fluid. *Exp. Brain Res.,* 21:399–402.
66. Coons, A. H. (1958): Fluorescent antibody methods. In: *General Cytochemical Methods,* edited by J. F. Danielli, pp. 399–422. Academic Press, New York.
67. Cottle, M. K. W., and Silver, A. (1970): Histochemical demonstration of acetylcholinesterase in the hypothalamus of the female guinea-pig. *Z. Zellforsch.,* 103:570–588.
68. Curtis, D. R., and Jonston, G. A. R. (1974): Amino acid transmitters in the mammalian central nervous system. *Rev. Physiol.,* 69:98–188.
69. Dahlström, A., and Fuxe, K. (1964): Evidence for the existence of monoamine containing neurons in the central nervous system. I. Demonstration of monoamines in the cell bodies of brain stem neuron. *Acta Physiol. Scand. [Suppl.* 232] 62:1–55.
70. Descarries, L., and Droz, B. (1970): Intraneural distribution of exogenous norepinephrine in the central nervous system of the rat. *J. Cell Biol.,* 44:385–399.
71. Dierickx, K., Vandesande, F., and DeMey, J. (1976): Identification in the external region of the rat median eminence, of separate neurophysin-vasopressin and neurophysin-oxytocin containing nerve fibers. *Cell Tissue Res.,* 168:141–151.
72. Dockray, G. J. (1976): Immunohistochemical evidence of cholecystokinin-like peptides in brain. *Nature,* 264:568–570.
72a. Doerr-Schott, J., and Dubois, M. P. (1976): LHRH-like system in the brain of *xenopus laevis* daud. *Cell Tiss. Res.,* 172:477–486.
72b. Dube, D., Leclerc, R., and Pelletier, G. (1976): Electron microscopic immunohistochemical localization of vasopressin and neurophysin in the median eminence of normal and adrenalectomized rats. *Am. J. Anat.,* 147:103–108.
73. Dubé, D., Leclerc, R., Pelletier, G., Arimura, A., and Schally, A. V. (1975): Immunohistochemical detection of growth hormone-release inhibiting hormone (somatostatin) in the guinea-pig brain. *Cell Tissue Res.,* 161:385–392.
74. Dubois, M. P. (1975): Immunoreactive somatostatin is present in discrete cells of the endocrine pancreas. *Proc. Natl. Acad. Sci. USA,* 72:1340–1343.
75. Dubois, M. P. (1976): Immunocytological evidence of LH-RF in hypothalamus and median eminence: A review. *Ann. Biol. Anim. Biochim. Biophys.,* 16:177–194.
76. Dubois, M. P., Barry, J., and Leonardelli, J. (1974): Mise en évidence par immunofluorescence et répartition de la somatostatine (SRIF) dans l'éminence médiale des Verté-

brés (Mammifères, Oiseaux, Amphibiens, Poissons). *C.R. Acad. Sci. [D] (Paris),* 279:1899–1902.
77. Dubois, M. P., and Kolodziejczyk, E. (1975): Centres hypothalamiques du rat secretant la somatostatine: repartition des pericaryons en 2 systèmes magno et parvocellulaires (édute immunocytologique). *C.R. Acad. Sci. [D] (Paris),* 281:1737–1740.
78. Dubois, P. M., and Paulin, C. (1976): Gastrointestinal somatostatin cells in the human fetus. *Cell Tissue Res.,* 166:179–184.
79. Dubois, P. M., Paulin. C., Assan, R., and Dubois, M. P. (1975): Evidence for immunoreactive somatostatin in the endocrine cells of human foetal pancreas. *Nature,* 26:731–732.
80. Duffy, M. J., Mulhall, D., and Powell, D. (1975): Subcellular distribution of substance P in bovine hypothalamus and substantia nigra. *J. Neurochem.,* 25:305–307.
81. Du Vigneaud, V. (1956): Hormones of the posterior pituitary gland: Oxytocin and vasopressin. In: *The Harvey Lectures 1954–55,* pp. 1–26. Academic Press, New York.
82. Dyer, R. G., and Dyball, R. E. J. (1974): Evidence for a direct effect of LRF and TRF on single unit activity in the rostral hypothalamus. *Nature,* 252:486–488.
83. Ehinger, B., Håkansson, R., Owman, Ch., and Sporrong, B. (1968): Histochemical demonstration of histamine in paraffin sections by a fluorescence method. *Biochem. Pharmacol.,* 17:1997–1998.
84. Elde, R. P. (1973): Methods for the localization of vasopressin-containing neurons in the hypothalamus and pars nervosa by immunoenzyme histochemistry. *Anat. Rec.,* 175:255–518.
85. Elde, R., Efendić, S., Hökfelt, T., Johansson, O., Luft, R., Parsons, J. A., Roovete, A., and Sorenson, R. L. (1977): Production of antibodies to somatostatin for radioimmunoassay and immunohistochemistry. *Acta Endocrinol. (Kbh.) (in press).*
86. Elde, R. P., Hökfelt, T., Johansson, O., Efendić, S., and Luft, R. (1976): Somatostatin containing pathways in the nervous system. *Neurosci. Abstr.,* 11:759.
87. Elde, R., Hökfelt, T., Johansson, O., and Terenius, L. (1976): Immunohistochemical studies using antibodies to leucine-enkephalin: Initial observations on the nervous system of the rat. *Neuroscience,* 1:349–351.
88. Elde, R. P., Hökfelt, T., Johansson, O., and Terenius, L. (1977): Distribution of enkephalin neurons in the central nervous system of the rat. In Preparation.
89. Elde, R. P., and Parsons, J. A. (1975): Immunocytochemical localization of somatostatin in cell bodies of the rat hypothalamus. *Am. J. Anat.,* 144:541–548.
90. Elliot, D. F., and Peart, W. S. (1956): Amino acid sequence in a hypertensin. *Nature,* 177:527–528.
91. Epelbaum, J., Brazeau, P., Tsang, D., Brawer, J., and Martin, J. (1977): Subcellular distribution of radioimmunoassayable somatostatin in rat brain. *Brain Res. (in press).*
92. Erlandsen, S. L., Hegre, O. D., Parsons, J. A., McEvoy, R. C., and Elde, R. P. (1976): Pancreatic islet cell hormones. Distribution of cell types in the islet and evidence for the presence of somatostatin and gastrin within the D cell. *J. Histochem. Cytochem.,* 24:883–897.
93. Euler, U. S. von (1963): Substance P in subcellular particles in peripheral nerves. *Ann. N.Y. Acad. Sci.,* 104:449–461.
94. Euler, U. S. von, and Gaddum, J. H. (1931): An unidentified depressor substance in certain tissue extracts. *J. Physiol.,* 72:74–87.
95. Falck, B. (1962): Observations on the possibilities of the cellular localization of monoamines by a fluorescence method. *Acta Physiol. Scand. [Suppl. 197],* 56:1–25.
96. Falck, B., Hillarp, N. Å., Thieme, G., and Torp, A. (1962): Fluorescence of catecholamines and related compounds with formaldehyde. *J. Histochem. Cytochem.,* 10:348–354.
97. Fischer-Ferraro, C., Nahmod, V. E., Goldstein, D. J., and Finkielman, S. (1971): Angiotensin and renin in rat and dog brain. *J. Exp. Med.,* 133:353–361.
98. Flament-Durand, J., and Desclin, L. (1968): A topographical study of a hypothalamic region with a thyrotrophic action. *J. Endocrinol.,* 41:531–539.
99. Fuxe, K. (1964): Cellular localization of monoamines in the median eminence and the infundibular stem of some mammals. *Z. Zellforsch.,* 61:710–724.
100. Fuxe, K. (1965): Evidence for the existence of monoamine neurons in the central nervous system. III. The monoamine nerve terminal. *Z. Zellforsch.,* 65:573–596.

101. Fuxe, K. (1965): Evidence for the existence of monoamine neurons in the central nervous sytem. IV. The distribution of monoamine nerve terminals in the central nervous system. *Acta Physiol. Scand.* [*Suppl.* 247], 64:39-85.
102. Fuxe, K., Ganten, D., Hökfelt, T., and Bolme, P. (1976): Immunohistochemical evidence for the existence of angiotensin II-containing nerve terminals in the brain and spinal cord in the rat. *Neurosci. Lett.,* 2:229-234.
103. Fuxe, K., Goldstein, M., Hökfelt, T., and Joh, T. H. (1970): Immunohistochemical localization of dopamine-β-hydroxylase in the peripheral and central nervous system. *Res. Commun. Chem. Pathol. Pharmacol.,* 1:627-636.
104. Fuxe, K., Goldstein, M., Hökfelt, T., and Joh, T. H. (1971): Cellular localization of dopamine-β-hydroxylase and phenylethanolamine-N-methyl transferase as revealed by immunohistochemistry. *Prog. Brain Res.,* 34:127-138.
105. Fuxe, K., and Hökfelt, T. (1966): Further evidence for the existence of tubero-infundibular dopamine neurons. *Acta Physiol. Scand.,* 66:243-244.
106. Fuxe, K., and Hökfelt, T. (1969): Catecholamines in the hypothalamus and the pituitary gland. In: *Frontiers in Neuroendocrinology, Vol. 1,* edited by W. F. Ganong and L. Martini, pp. 47-96. Oxford University Press, New York.
107. Fuxe, K., Hökfelt, T., Eneroth, P., Gustafsson, J.-Å., and Skett, P. (1977): Prolactin: Localization in nerve terminals of the rat hypothalamus. *Science,* 196:899-900.
108. Fuxe, K., Hökfelt, T., Johansson, O., Ganten, D., Goldstein, M., Perez de la Mora, M., Possani, L., Tapia, R., Teran, L., Palacios, R., Said, S., and Mutt, V. (1977): Monoamine neuron systems in the hypothalamus and their relation to the GABA and peptide containing neurons. In: *Colloque de Synthese des Actions Thématiques 22 et 35. Neuromédiateurs et Polypeptides Hypothalamiques a Action Relàchante ou Inhibitrice,* edited by R. Mornex and J. Barry. Institut National de la Sante et de la Recherche Medicale, Paris (*in press*).
109. Fuxe, K., Hökfelt, T., Löfstrom, A., Johansson, O., Agnati, L., Everitt, B., Goldstein, M., Jeffcoate, S., White, N., Eneroth, P., Gustafsson, J.-Å., and Skett, P. (1976): On the role of neurotransmitters and hypothalamic hormones and their interactions in hypothalamic and extrahypothalamic control of pituitary function and sexual behavior. In: *Subcellular Mechanisms in Reproductive Neuroendocrinology,* edited by F. Naftolin, K. J. Ryan, and J. Davies, pp. 193-246. Elsevier, Amsterdam.
110. Fuxe, K., Hökfelt, T., Said, S., and Mutt, V. (1977): Evidence for the existence of VIP containing nerve terminals in the rat brain. *Neurosci. Lett.,* 5:241-246.
111. Fuxe, K., and Jonsson, G. (1974): Further mapping of central 5-hydroxytryptamine neurons: Studies with the neurotoxic dihydroxytryptamines. *Adv. Biochem. Psychopharmacol.,* 10:1-12.
112. Ganten, D., Hutchinson, J. S., Schelling, P., Ganten, U., and Fischer, H. (1975): The isorenin angiotensin systems in extrarenal tissue. *Clin. Exp. Pharmacol. Physiol.,* 2:127-151.
113. Ganten, D., Minnich, J. L., Granger, P., Hayduk, K., Brecht, H. M., Barbeau, A., Boucher, R., and Genest, J. (1971): Angiotensin-forming enzyme in brain tissue. *Science,* 173:64-65.
114. Geffen, L. B., Livett, B. G., and Rush, R. A. (1969): Immunohistochemical localization of protein component of catecholamine storage vesicles. *J. Physiol. (Lond.),* 204:593-605.
115. Giachetti, A., Rosenberg, R. N., and Said, S. I. (1976): Vasoactive intestinal polypeptide in brain synaptosomes. *Lancet,* 2:741-742.
116. Gibb, J. W., Spector, S., and Udenfriend, S. (1967): Production of antibodies to dopamine-β-hydroxylase of bovine adrenal medulla. *Mol. Pharmacol.,* 3:473-478.
117. Goldsmith, P. C., and Ganong, W. F. (1975): Ultrastructural localization of luteinizing hormone-releasing hormone in the median eminence of the rat. *Brain Res.,* 97:181-193.
118. Goldsmith, P. C., Rose, J. C., Arimura, A., and Ganong, W. F. (1975): Ultrastructural localization of somatostatin in pancreatic islets of the rat. *Endocrinology,* 97:1061-1064.
119. Goldstein, M. (1972): Enzymes involved in the catalysis of catecholamine biosynthesis. In: *Methods in Neurochemistry,* edited by R. N. Ubell, pp. 317-340. Plenum Publishing Corp., New York.
120. Goldstein, M., Anagnoste, B., Freedman, L. S., Roffman, M., Ebstein, R. P., Park, D. H., Fuxe, K., and Hökfelt, T. (1974): Characterization, localization and regulation of catecholamine synthesizing enzymes. In: *Frontiers in Catecholamine Research,* edited by E. Usdin and S. Snyder, pp. 69-78. Pergamon Press, New York.

121. Goldstein, M., Fuxe, K., and Hökfelt, T. (1972): Characterization and tissue localization of catecholamine synthesizing enzymes. *Pharmacol. Rev.*, 24:293–309.
122. Goos, H. J. Th., Ligtenberg, P. J. M., and van Oordt, P. G. W. J. (1976): Immunofluorescence studies on gonadotropin releasing hormone (GRH) in the fore-brain and the neurohypophysis of the green frog, Rana esculenta L. *Cell Tissue Res.*, 168:325–333.
123. Green, J. P. (1970): Histamine. In: *Handbook of Neurochemistry*, edited by A. Lajtha, pp. 221–250. Plenum Press, New York.
124. Greer, M. A. (1951): Evidence of hypothalamic control of pituitary release of thyrotrophin. *Proc. Soc. Exp. Biol. Med.*, 77:603–608.
125. Gregory, R. A., and Tracy, H. J. (1964): The constitution and properties of two gastrins extracted from hog antral mucosa. *Gut*, 5:103–114.
126. Gregory, R. A., and Tracy, H. J. (1972): Isolation of two "big gastrins" from Zollinger-Ellison tumour tissue. *Lancet*, 2:797–799.
127. Gross, D. S. (1976): Distribution of gonadotropin-releasing hormone in the mouse brain as revealed by immunohistochemistry. *Endocrinology*, 98:1408–1417.
128. Guillemin, R. (1977): Chemistry of brain peptides. In: *The Hypothalamus, ARNMD monograph series 56*, edited by S. Reichlin, R. Baldesarrini, and J. B. Martin. Raven Press, New York (*in press*).
129. Guillemin, R., Ling, N., and Burgus, R. 1976): Endorphines, peptides, d'origine hypothalamique et neurohypophysaire a activité morphinomimetique. Isolement et structure moléculaire de l'α-endorphine. *C.R. Acad. Sci. [D] (Paris)*, 282:783–785.
130. Halász, B., Florsheim, W. H., Coreorran, N. L., and Gorski, R. A. (1967): Thyrotrophic hormone secretion in rats after partial or total interruption of neural afferents to the medial basal hypothalamus. *Endocrinology*, 80:1075–1082.
131. Harris, G. H. (1955): *Neural Control of the Pituitary Gland*. Edward Arnold, London.
132. Hartman, B. K. (1973): Immunofluorescence of dopamine-β-hydroxylase. Application of improved methodology to the localization of the peripheral and central noradrenergic nervous system. *J. Histochem. Cytochem.*, 21:312–332.
133. Hartman, B., and Udenfriend, S. (1970): Immunofluorescent localization of dopamine-β-hydroxylase in tissues. *Mol. Pharmacol.*, 6:85–94.
134. Hartman, B. K., Zide, D., and Udenfriend, S. (1972): The use of dopamine-β-hydroxylase as a marker for the noradrenergic pathways of the central nervous system in the rat. *Proc. Natl. Acad. Sci. USA*, 69:2722–2726.
135. Hebb, C. O., and Silver, A. (1956): Choline acetylase in the central nervous system of man and some other mammals. *J. Physiol. (Lond.)*, 134:718–728.
136. L'Hermite, A., Lefranc, G., Pradal, G., André, M.-J., and Dubois, M. P. (1976): Identification ultrastructurale et étude immunocytochimique des cellules á somatostatine de la muqueuse antrale du lapin et de de la souris. *Histochemistry*, 47:31–41.
137. Hiller, J. M., Pearson, J., and Simon, E. J. (1973): Distribution of stereospecific binding of the potent narcotic analgesic etorphine in the human brain: Predominance in the limbic system. *Res. Commun. Chem. Pathol. Pharmacol.*, 6:1052–1062.
138. Hoffman, G. E., Moynihan, J. A., and Knigge, K. M. (1976): Immunocytochemical localization of luteinizing hormone-releasing hormone (LH-RH). Differences with different antisera. *Neurosci. Abstr.*, II:673.
139. Hökfelt, T. (1968): *In vitro* studies on central and peripheral monoamine neurons at the ultrastructural level. *Z. Zellforsch.*, 91:1–74.
140. Hökfelt, T., Efendić, S., Hellerström, C., Johansson, O., Luft, R., and Arimura, A. (1975): Cellular localization of somatostatin in endocrine-like cells and neurons of the rat with special references to the $A_1$ cells of the pancreatic islets and to the hypothalamus. *Acta Endocrinol. (Kbh.) [Suppl.]*, 200:5–41.
141. Hökfelt, T., Efendić, S., Johansson, O., Luft, R., and Arimura, A. (1974): Immunohistochemical localization of somatostatin (growth hormone release-inhibiting factor) in the guinea pig brain. *Brain Res.*, 80:165–169.
142. Hökfelt, T., Elde, R. P., Johansson, O., Kellerth, J.-O., Ljungdahl, A., Nilsson, G., Pernow, B., and Terenius, L. (1977): Substance P and enkephalin: Distribution in the nervous system as revealed with immunohistochemistry. In: *Proc. CINP*, edited by Radouco-Thomas (*in press*).
143. Hökfelt, T., Elde, R. P., Johansson, O., Ljungdahl, Å., Schultzberg, M., Fuxe, K., Goldstein, M., Nilsson, G., Pernow, B., Terenius, L., Ganten, D., Jeffcoate, S. L.,

Rehfeld, J., and Said, S. (1978): The distribution of peptide containing neurons in the nervous system. In: *Psychopharmacology: A Generation of Progress,* edited by M. A. Lipton, A. DiMascio, and K. F. Killam. Raven Press, New York, (*in press*).

144. Hökfelt, T., Elde, R. P., Johansson, O., Luft, R., and Arimura, A. (1975): Immunohistochemical evidence for the presence of somatostatin, a powerful inhibitory peptide in some primary sensory neurons. *Neurosci. Lett.,* 1:231–235.
145. Hökfelt, T., Elde, R., Johansson, O., Luft, R., Nilsson, G., and Arimura, A. (1976): Immunohistochemical evidence for separate populations of somatostatin-containing and substance P-containing primary afferent neurons. *Neuroscience,* 1:131–136.
146. Hökfelt, T., Elde, B., Johansson, O., Terenius, L., and Stein, L. (1977): The distribution of enkephalin immunoreactive cell bodies in the rat central nervous system. *Neurosci. Lett.,* 5:25–31.
147. Hökfelt, T., and Fuxe, K. (1972): On the morphology and the neuroendocrine role of the hypothalamic catecholamine neurons. In: *Brain-Endocrine Interaction. Median Eminence: Structure and Function,* edited by K. M. Knigge, D. E. Scott, and A. Weindl, pp. 181–223. S. Karger, Basel.
148. Hökfelt, T., Fuxe, K., and Goldstein, M. (1973): Immunohistochemical localization of aromatic L-aminoacid decarboxylase (DOPA decarboxylase) in central dopamine and 5-hydroxytryptamine nerve cell bodies of the rat brain. *Brain Res.,* 53:175–180.
149. Hökfelt, T., Fuxe, K., and Goldstein, M. (1973): Immunohistochemical studies on monoamine-containing cell systems. *Brain Res.,* 62:461–469.
150. Hökfelt, T., Fuxe, K., and Goldstein, M. (1975): Applications of immunohistochemistry to studies on monoamine cell systems with special references to nervous tissues. *Ann. N. Y. Acad. Sci.,* 254:407–432.
151. Hökfelt, T., Fuxe, K., Goldstein, M., and Joh, T. H. (1973): Immunohistochemical studies of three catecholamine synthesizing enzymes: Aspects on methodology. *Histochemistry,* 33:231–254.
152. Hökfelt, T., Fuxe, K., Goldstein, M., and Johansson, O. (1973): Evidence for adrenaline neurons in the rat brain. *Acta Physiol. Scand.,* 89:286–288.
153. Hökfelt, T., Fuxe, K., Goldstein, M., and Johansson, O. (1974): Immunohistochemical evidence for the existence of adrenaline neurons in the rat brain. *Brain Res.,* 66:235–251.
154. Hökfelt, T., Fuxe, K., Goldstein, M., Johansson, O., Fraser, H., and Jeffcoate, S. (1975): Immunofluorescence mapping of central monoamine and releasing hormone (LRH) systems. In: *Anatomical Neuroendocrinology,* edited by W. E. Stumpf and L. D. Grant, pp. 381–392. Karger, Basel.
155. Hökfelt, T., Fuxe, K., Johansson, O., Jeffcoate, S. L., and White, N. (1975): Distribution of thyrotropin-releasing hormone (TRH) in the central nervous system as revealed with immunohistochemistry. *Eur. J. Pharmacol.,* 34:389–392.
156. Hökfelt, T., Fuxe, K., Johansson, O., Jeffcoate, S., and White, N. (1975): Thyrotropin releasing hormone (TRH)-containing nerve terminals in certain brain stem nuclei and in the spinal cord. *Neurosci. Lett.,* 1:133–139.
157. Hökfelt, T., Johansson, O., Efendić, S., Luft, R., and Arimura, A. (1975): Are there somatostatin-containing nerves in the rat gut? Immunohistochemical evidence for a new type of peripheral nerves. *Experientia,* 31:852–854.
158. Hökfelt, T., Johansson, O., Fuxe, K., Goldstein, M., and Park, D. (1976): Immunohistochemical studies on the localization and distribution of monoamine neuron systems in the rat brain. I. Tyrosine hydroxylase in the mes- and diencephalon. *Med. Biol.,* 54:427–453.
159. Hökfelt, T., Johansson, O., Fuxe, K., Goldstein, M., and Park, D. (1977): Immunohistochemical studies on the localization and distribution of monoamine neuron systems in the rat brain. II. Tyrosine hydroxylase in the telencephalon. *Med. Biol.,* 55:21–40.
160. Hökfelt, T., Johansson, O., Fuxe, K., Goldstein, M., and Park, D. (1977): Immunohistochemical studies on the localization and distribution of monoamine neuron systems in the rat brain. III. Three catecholamine synthesizing enzymes in the rhinencephalon. In: *Fourth Bel Air Symposium, Geneva* (*in press*).
161. Hökfelt, T., Johansson, O., Fuxe, K., Löfström, A., Goldstein, M., Park, D., Ebstein, R., Fraser, H., Jeffcoate, S., Efendic, S., Luft, R., and Arimura, A. (1975): Mapping and relationship of hypothalamic neurotransmitters and hypothalamic hormones. In: *CNS and Behavioural Pharmacology. Proceedings of the Sixth International Congress of*

*Pharmacology, Vol. 3,* edited by J. Tuomisto and M. K. Paasonen, pp. 93-110. Forssan Kirjapaino Oy, Forssa.
162. Hökfelt, T., Johansson, O., Kellerth, J.-O., Ljungdahl, Å., Nilsson, G., Nygårds, A., and Pernow, B. (1977): Immunohistochemical distribution of substance P. In: *Substance P Nobel Symposium, Vol. 37,* edited by U. S. von Euler and B. Pernow, pp. 117-143. Raven Press, New York.
163. Hökfelt, T., Kellerth, J.-O., Ljungdahl, A., Nilsson, G., Nygårds, A., and Pernow, B. (1977): Immunohistochemical localization of substance P in the central and peripheral nervous system. In: *Neuroregulators and Hypotheses of Psychiatric Disorders,* edited by J. Barchas, E. Costa, and E. Usdin, pp. 299-310. Oxford University Press, New York.
164. Hökfelt, T., Kellerth, J.-O., Nilsson, G., and Pernow, B. (1975): Experimental immunohistochemical studies on the localization and distribution of substance P in cat primary sensory neurons. *Brain Res.,* 100:235-252.
165. Hökfelt, T., Kellerth, J.-O., Nilsson, G., and Pernow, B. (1975): Substance P: Localization in the central nervous system and in some primary sensory neurons. *Science,* 190: 889-890.
166. Hökfelt, T., and Ljungdahl, A. (1972): Modification of the Falck-Hillarp formaldehyde fluorescence method using the Vibratome: Simple, rapid and sensitive localization of catecholamines in sections of unfixed or formalin fixed brain tissue. *Histochemistry,* 29:325-339.
167. Hökfelt, T., and Ljungdahl, A. (1975): Uptake mechanisms as a basis for the histochemical identification and tracing of transmitter-specific neuron populations. In: *The Use of Axonal Transport for Studies of Neuronal Connectivity,* edited by W. M. Cowan and M. Cuénod, pp. 249-305. Elsevier, Amsterdam.
168. Hökfelt, T., Meyerson, B., Nilsson, G., Pernow, B., and Sachs, Ch. (1976): Immunohistochemical evidence for substance P-containing nerve endings in the human cortex. *Brain Res.,* 104:181-186.
169. Hong, J. S., Costa, E., and Yang, H.-Y. T. (1976): Effects of habenular lesions on the substance P content of various brain regions. *Brain Res.,* 118:523-525.
170. Hughes, I., Smith, T. W., Kosterlitz, H. W., Fothergill, L. H., Morgan, B. A., and Morris, H. R. (1975): Identification of two related pentapeptides from the brain with potent opiate agonist activity. *Nature,* 258:577-579.
171. Iversen, L. L., and Schon, F. (1973): The use of autoradiographic techniques for the identification and mapping of transmitter-specific neurones in the CNS. In: *New Concepts in Transmitter Regulation,* edited by A. Mandel and D. Segal, pp. 153-193. Plenum Press, New York.
172. Iversen, L. L., and Storm-Mathisen, J. (1976): Uptake of ($^3$H) glutamate in excitatory nerve endings in the hippocampal formation of the rat. *Acta Physiol. Scand.,* 96:22.
173. Jackson, I. M. D., and Reichlin, S. (1974): Thyrotropin-releasing hormone (TRH): Distribution in hypothalamic and extrahypothalamic brain tissue of mammalian and submammalian chordates. *Endocrinology,* 95:854-862.
174. Jacobowitz, D. M. (1975): Fluorescence microscopic mapping of CNS norepinephrine systems in the rat forebrain. In: *Anatomical Neuroendocrinology,* edited by W. E. Stumpt and L. D. Grant, pp. 368-380. S. Karger, Basel.
175. Jacobowitz, D., and Palkovits, M. (1974): Topographic atlas of catecholamine and acetylcholinesterase-containing neurons in the rat brain. I. Forebrain (telencephalon, diencephalon). *J. Comp. Neurol.,* 157:13-28.
176. Jeffcoate, S. L., Fraser, H. M., Gunn, A., and Holland, D. T. (1973): Radioimmunoassay of luteinizing hormone releasing factor. *J. Endocrinol.,* 57:89-90.
177. Jeffcoate, S. L., Fraser, H. M., Gunn, A., and White, N. (1973): Radioimmunoassay of thyrotropin-releasing hormone. *J. Endocrinol.,* 59:191-192.
178. Jeffcoate, S. L., Holland, D. T., Fraser, H. M., and Gunn, A. (1974): Preparation and specificity of antibodies to the decapeptide, luteinizing hormone-releasing hormone (LH-RH). *Immunochemistry,* 11:75-77.
179. Jeffcoate, S. L., and White, N. (1975): Studies on the nature of mammalian hypothalamic thyrotrophin releasing hormone using immunochemical, chromatographic and enzymatic techniques. *J. Endocrinol.,* 65:83-90.
180. Joh, T. T., Shikimi, T., Pickel, V. M., and Reis, D. J. (1975): Brain tryptophan hydroxylase: Purification of, production of antibodies to, and cellular and ultrastructural localiza-

tion in serotonergic neurons of rat mid-brain. *Proc. Natl. Acad. Sci. USA*, 72:3575–3579.
181. Juhlin, L., and Shelley, B. (1966): Detection of histamine by a new fluorescent o-phthaldialdehyde stain. *J. Histochem. Cytochem.*, 14:525.
182. Kanazawa, I., Emson, P. C., and Cuello, A. C. (1977): Evidence for the existence of substance P-containing fibres in striato-nigral and pallido-nigral pathways in rat brain. *Brain Res.*, 119:447–453.
183. Kanazawa, I., and Jessell, T. (1976): Post mortem changes and regional distribution of substance P in the rat and mouse nervous system. *Brain Res.*, 117:362–367.
184. Kasa, P., Mann, S. P., and Hebb, C. (1970): Localization of choline acetyltransferase. *Nature*, 226:812–816.
185. King, J. C., Arimura, A., Gerall, A. G., Fishback, J. B., and Elkind, K. E. (1975): Growth hormone-release inhibiting hormone (GH-RIH) pathway of the rat hypothalamus revealed by the unlabeled antibody peroxidase-antiperoxidase method. *Cell Tissue Res.*, 160:423–430.
186. King, J. C., and Gerall, A. A. (1976): Localization of luteinizing hormone-releasing hormone. *J. Histochem. Cytochem.*, 24:829–845.
187. King, J. C., Parsons, J. A., Erlandsen, S. L., and Williams, T. H. (1974): Luteinizing hormone-releasing hormone (LH-RH) pathway of the rat hypothalamus revealed by the unlabeled peroxidase-anti-peroxidase method. *Cell Tissue Res.*, 153:211–217.
188. King, J. C., Williams, T. H., and Arimura, A. (1975): Localization of luteinizing hormone-releasing hormone in rat hypothalamus using radioimmunoassay. *J. Anat.*, 120:275–288.
189. Kizer, J. S., Arimura, A., Schally, A. V., and Brownstein, M. J. (1975): Absence of luteinizing hormone-releasing hormone (LH-RH) from catecholaminergic neurons. *Endocrinology*, 96:523–525.
190. Kizer, J. S., Palkovits, M., and Brownstein, M. J. (1976): The projections of the A8, A9 and A10 dopaminergic cell bodies: Evidence for a nigral-hypothalamic-median eminence dopaminergic pathway. *Brain Res.*, 108:363–370.
191. Kizer, J. S., Palkovits, M., Tappaz, M., Kebabian, J., and Brownstein, M. J. (1976): Distribution of releasing factors, biogenic amines, and related enzymes in the bovine median eminence. *Endocrinology*, 98:685–695.
192. Knigge, K. M. (1977): Anatomy of the endocrine hypothalamus. In: *ARNMD: The Hypothalamus*, edited by S. Reichlin, R. Baldessarini, and J. B. Martin. Raven Press, New York (*in press*).
193. Knowles, F. G. W., and Bern, H. A. (1966): The function of neurosecretion in endocrine regulation. *Nature*, 10:271–272.
194. Knowles, F., and Vollrath, L. (Eds.) (1974): *Neurosecretion—The Final Neuroendocrine Pathway*. Springer-Verlag, Berlin, Heidelberg, New York.
195. König, J. F. R., and Klippel, R. A. (1963): *The Rat Brain. A Stereotaxic Atlas of the Forebrain and Lower Parts of the Brain Stem*. Williams & Wilkins Co., Baltimore.
196. Konishi, S., and Otsuka, M. (1974): The effects of substance P and other peptides on spinal neurons of the frog. *Brain Res.*, 65:397–410.
197. Kordon, C., Kerdelhué, B., Pattou, E., and Jutisz, M. (1974): Immunocytochemical localization of LH-RH in axons and nerve terminals of the rat median eminence. *Proc. Soc. Exp. Biol.*, 147:122–127.
198. Koslow, S. H., Racagni, G., and Costa, E. (1974): Mass fragmentographic measurement of norepinephrine, dopamine, serotonin, and acetylcholine in seven discrete nuclei of the rat tel-diencephalon. *Neuropharmacology*, 13:1123–1130.
199. Koslow, S. H., and Schlumpf, M. (1974): Quantitation of adrenaline in rat brain nuclei and areas by mass fragmentography. *Nature*, 251:530–531.
200. Köves, K., and Magyar, A. (1975): On the location of TRH producing neurons: Thyroid response to PTU treatment after stereotaxic interventions in hypothalamus. *Endocrinol. Exp.*, 9:247–257.
201. Krieger, D. T., and Zimmerman, E. A. (1977): The nature of CRF and its relationship to vasopressin CRF. In: *Clinical Neuroendocrinology*, edited by M. Besser and L. Martini. Academic Press, New York (*in press*).
202. Krnjević, K. (1974): Chemical nature of synaptic transmission in vertebrates. *Physiol. Rev.*, 54:418–540.

203. Krulich, L., Dhariwal, A. P. S., and McCann, S. M. (1968): Stimulatory and inhibitory effects of purified hypothalamic extracts on growth hormone release from rat pituitary in vitro. *Endocrinology*, 83:783-790.
204. Krulich, L., Quijada, M., Hefco, E., and Sundberg, D. K. (1974): Localization of thyrotropin-releasing factor (TRF) in the hypothalamus of the rat. *Endocrinology*, 95:9-17.
205. Kuhar, M. J., Pert, C. B., and Snyder, S. H. (1973): Regional distribution of opiate receptor binding in monkey and human brain. *Nature*, 245:447-450.
206. Lamotte, C., Pert, C. B., and Snyder, S. H. (1976): Opiate receptor binding in primate spinal cord: Distribution and changes after dorsal root section. *Brain Res.*, 112:407-412.
207. Larsson, L.-I., Edvinsson, L., Fahrenkrug, J., Håkanson, R., Owman, Ch., Schaffalitzky de Muckadell, O., and Sundler, F. (1976): Immunohistochemical localization of a vasodilatory polypeptide (VIP) in cerebrovascular nerves. *Brain Res.*, 113:400-404.
208. Larsson, L.-I., Fahrenkrug, J., Schaffalitzky de Muckadell, O., Sundler, F., Håkanson, R., and Rehfeld, J. F. (1976): Localization of vasoactive intestinal polypeptide (VIP) to central and peripheral neurons. *Proc. Natl. Acad. Sci. USA*, 73:3197-3200.
209. Leclerc, R., Pelletier, G., Puviani, R., Arimura, A., and Schally, A. V. (1976): Immunohistochemical localization of somatostatin in endocrine cells of the rat stomach. *Mol. Cell. Endocrinol.*, 4:257-261.
210. Lederis, K., and Livingston, A. (1969): Acetylcholine and related enzymes in the neural lobe and anterior hypothalamus of the rabbit. *J. Physiol. (Lond.)*, 201:695-709.
211. Leeman, S. E., and Mroz, E. A. (1975): Substance P. *Life Sci.*, 15:2033-2044.
212. Lembeck, F. (1953): Zur Frage der zentralen Übertragung afferenter Impulse. III. Mitteilung. Das Vorkommen und die Bedeutung der Substanz P in den dorsalen Wurzeln des Rückenmarks. *Naunyn-Schmiedeberg's Arch. Pharmacol.*, 219:197-213.
213. Lembeck, F., and Zetler, G. (1962): Substance P: A polypeptide of possible physiological significance, especially within the nervous system. *Int. Rev. Neurobiol.*, 4:159-215.
214. Leonardelli, J., Barry, J., and Dubois, M. P. (1973): Mise en évidence par immunofluorescence d'un constituant immunologiquement apparenté au LH-RH dans l'hypothalamus et l'eminence médiane chez les Mammifères. *C.R. Acad. Sci. [D] (Paris)*, 276:2043-2046.
215. Lidbrink, P., Jonsson, G., and Fuxe, K. (1974): Selective reserpine resistant accumulation of catecholamines in central dopamine neurons after dopa administration. *Brain Res.*, 67:439-456.
216. Lien, E. L., Fenichel, R. L., Garsky, V., Sarantakis, D., and Grant, N. H. (1976): Enkephalin-stimulated prolactin release. *Life Sci.*, 19:837-840.
217. Lindvall, O., and Björklund, A. (1974): The organization of the ascending catecholamine neuron systems in the rat brain as revealed by the glyoxylic acid fluorescence method. *Acta Physiol. Scand. [Suppl. 412]*, 92:1-48.
218. Lindvall, O., and Björklund, A. (1974): The glyoxylic acid fluorescence histochemical method: A detailed account of the methodology for the visualization of central catecholamine neurons. *Histochemistry*, 39:97-127.
219. Livett, B. G., Uttenthal, L. O., and Hope, D. B. (1971): Localization of neurophysin II in the hypothalamo-neurohypophysial system of the pig by immunofluorescence histology. *Proc. R. Soc. Phil. Trans. B.*, 261:371-378.
220. Luft, R., Efendic, S., Hökfelt, T., Johansson, O., and Arimura, A. (1974): Immunohistochemical evidence for the localization of somatostatin-like immunoreactivity in a cell population of the pancreatic islets. *Med. Biol.*, 52:428-430.
221. Macleod, R. M., and Lehmeyer, J. E. (1974): Studies on the mechanism of the dopamine-mediated inhibition of prolactin secretion. *Endocrinology*, 94:1077-1085.
222. Makara, G. B., Rappay, G., and Stark, E. (1975): Autoradiographic localization of $^3$H-gamma-aminobutyric acid in the medial hypothalamus. *Exp. Brain Res.*, 22:449-455.
223. Makara, G. B., and Stark, E. (1974): Effect of gamma-aminobutyric acid (GABA) and GABA antagonist drugs on ACTH release. *Neuroendocrinology*, 16:178-190.
224. Mazzucca, M., and Poulain, P. (1971): Mise en évidence en microscopie électronique des terminaisons nerveuses monoaminergiques dans l'éminence médiane du cobaye à l'aide de la 5-hydroxydopamine. *C.R. Acad. Sci. [D] (Paris)*, 273:1044-1047.
224a. McCann, S. M. (1962): A hypothalamic luteinizing hormone-releasing factor. *Am. J. Physiol.*, 202:395-400.

225. McGeer, P. L., McGeer, E. G., Singh, V. K., and Chase, W. H. (1974): Choline acetyltransferase localization in the central nervous system by immunohistochemistry. *Brain Res.*, 81:373–379.
226. Mioduszewski, R., Grandison, L., and Meites, J. (1976): Stimulation of prolactin release in rats by GABA (39139). *Proc. Soc. Exp. Biol. Med.*, 151:44–46.
227. Moriarty, G. C. (1973): Adenohypophysis. Ultrastructural cytochemistry. A review. *J. Histochem. Cytochem.*, 21:855–894.
227a. Morris, M., Tandy, B., Sundberg, D. K., and Knigge, K. M. (1975): Modification of brain and CSF LH-RH following deafferentation. *Neuroendocrinology*, 18:131–135.
228. Moss, R. L. (1977): Role of hypophysiotrophic neurohormones in mediating neural and behavioral events. Localization of hypophysiotrophic neurohormones. *Fed. Proc.* (in press).
229. Mroz, E. A., Brownstein, M. J., and Leeman, S. E. (1976): Evidence for substance P in the habenulo-interpeduncular tract. *Brain Res.*, 113:597–599.
230. Mutt, V., and Said, S. I. (1974): Structure of the porcine vasoactive intestinal octacosapeptide. The amino-acid sequence. Use of kallikrein in its determination. *Eur. J. Biochem.*, 42:581–589.
231. Naik, D. V. (1975): Immunoreactive LH-RH neurons in the hypothalamus identified by light and fluorescent microscopy. *Cell Tissue Res.*, 157:423–436.
232. Naik, D. V. (1975): Immuno-electron microscopic localization of luteinizing hormone-releasing hormone in the arcuate nuclei and median eminence of the rat. *Cell Tissue Res.*, 157:437–455.
232a. Naik, D. V. (1976): Immuno-histochemical localization of LH-RH during different phases of estrus cycle of rat, with reference to the preoptic and arcuate neurons, and the ependymal cells. *Cell Tiss. Res.*, 173:143–166.
233. Nair, R. M. G., Barrett, J. F., Bowers, C. Y., and Schally, A. V. (1970): Structure of porcine thyrotropin releasing hormone. *Biochemistry*, 9:1103–1106.
234. Nairn, R. C. (1969): *Fluorescent Protein Tracing.* E.&S. Livingstone, Ltd., Edinburgh and London.
235. Nakane, P. K. (1971): Application of peroxidase-labelled antibodies to the intracellular localization of hormones. In: *Karolinska Symposia on Research Methods in Reproductive Endocrinology, Vol. 3: In Vitro Methods in Reproductive Cell Biology,* edited by A. Diczfalusy, pp. 190–204. Forum, Copenhagen.
236. Nauta, W. J. H., and Haymaker, W. (1969): Hypothalamic nuclei and fiber connections. In: *The Hypothalamus,* edited by W. Haymaker, E. Andersson, and W. H. J. Nauta, pp. 136–209. Charles C Thomas, Springfield, Ill.
237. Nicoll, R. A. (1975): Promising peptides. In: *Neuroscience Symposia, Vol. I: Neurotransmitters, Hormones and Receptors: Novel Approaches,* edited by J. A. Ferrendelli, B. S. McEwen, and S. H. Snyder, pp. 99–122. Society for Neuroscience, Bethesda, Md.
238. Nilsson, G., Hökfelt, T., and Pernow, B. (1974): Distribution of substance P-like immunoreactivity in the rat central nervous system as revealed by immunohistochemistry. *Med. Biol.*, 52:424–427.
239. Nilsson, G., Larsson, L. I., Håkansson, R., Brodin, E., Sundler, F., and Pernow, B. (1975): Localization of substance P-like immunoreactivity in mouse gut. *Histochemistry*, 43:97–99.
240. Nilsson, G., Pernow, B., Fischer, G. H., and Folkers, K. (1975): Presence of substance P-like immunoreactivity in plasma from man and dog. *Acta Physiol. Scand.*, 94:542–544.
240a. Okon, E., and Koch, Y. (1976): Localisation of gonadotropin-releasing and thyrotropin-releasing hormones in human brain by radioimmunoassay. *Nature*, 263:345–347.
241. Oliver, C., Eskay, R. L., Ben-Jonathan, N., and Porter, J. C. (1974): Distribution and concentration of TRH in the rat brain. *Endocrinology*, 96:540–546.
242. Ondo, J. G. (1974): Gamma-aminobutyric acid effects on pituitary gonadotropin secretion. *Science*, 186:738–739.
243. Orci, L., Baetens, D., Dubois, M. P., and Rufener, C. (1975): Evidence for the D-cell of the pancreas secreting somatostatin. *Horm. Metab. Res.*, 7:400–402.
244. Orci, L., Baetens, D., Ravazzola, M., Malaisse-Lagae, F., Amherdt, M., and Rufener, C. (1976): Somatostatin in the pancreas and the gastrointestinal tract. In: *Endocrine Gut and Pancreas,* edited by T. Fujita, pp. 73–88. Elsevier, Amsterdam.
245. Orci, M. L., Baetens, B., Rufener, C., Amherdt, M., Ravazzola, M., Studer, P., Malaisse-

Lagae, F., and Unger, R. H. (1975): Réactivité de la cellule à somatostatine de l'ilot de Langerhans dans le diabète expérimental. *C.R. Acad. Sci. [D]* (*Paris*), 281:1883-1885.
246. Orci, L., Baetens, D., Rufener, C., Amherdt, M., Ravazzola, M., Studer, P., Malaisse-Lagae, F., and Unger, R. (1976): Hypertrophy and hyperplasia of somatostatin-containing D-cells in diabetes. *Proc. Natl. Acad. Sci. USA*, 73:1338-1342.
247. Otsuka, M. (1977): Electrophysiological and neurochemical evidence for substance P as a transmitter of primary sensory neurons. In: *Nobel Symposium, Vol. 37: Substance P*, edited by B. Pernow and U.S. von Euler, pp. 207-214. Raven Press, New York (*in press*).
248. Paasonen, M. K., and Vogt, M. (1956): The effect of drugs on the amounts of substance P and 5-hydroxytryptamine in mammalian brain. *J. Physiol.*, 131:617-626.
249. Palkovits, M. (1973): Isolated removal of hypothalamic or other brain nuclei of the rat. *Brain Res.*, 59:449-450.
250. Palkovits, M., Arimura, A., Brownstein, M., Schally, A. V., and Saavedra, J. M. (1971): Luteinizing hormone-releasing hormone (LH-RH) content of the hypothalamic nuclei in rat. *Endocrinology*, 96:554-558.
251. Palkovits, M., Brownstein, M. J., Arimura, A., Sato, H., Schally, A. V., and Kizer, J. S. (1976): Somatostatin content of the hypothalamic ventro-medial and arcuate nuclei and the circumventricular organs in the rat. *Brain Res.*, 109:430-434.
252. Palkovits, M., Brownstein, M., and Kizer, S. J. (1976): Effect of total hypothalamic deafferentation on releasing hormone and neurotransmitter concentrations of the mediobasal hypothalamus in rat. *International Symposium on Cellular and Molecular Bases of Neuroendocrine Processes*, edited by E. Endroczy, pp. 575-599. Akadémiai Kiadó, Budapest.
253. Palkovits, M., Brownstein, M., Saavedra, J. M., and Axelrod, J. (1974): Norepinephrine and dopamine content of hypothalamic nuclei. *Brain Res.*, 77:137-149.
254. Park, D. H., and Goldstein, M. (1976): Purification of tyrosine hydroxylase from pheochromocytoma tumors. *Life Sci.*, 18:55-60.
255. Parsons, J., Erlandsen, S., Hegre, O., McEvoy, R., and Elde, R. P. (1976): Central and peripheral localization of somatostatin. Immunoenzyme immunocytochemical studies. *J. Histochem. Cytochem.*, 24:872-882.
256. Pearse, A. G. E. (1975): Neurocristopathy, neuroendocrine pathology and the APUD concept. *Z. Krebsforsch.*, 84:1-18.
257. Pearse, A. G. E. (1976): Peptides in brain and intestine. *Nature*, 262:92-94.
258. Pearse, A. G. E., and Polak, J. (1975): Immunocytochemical localization of substance P in mammalian intestine. *Histochemistry*, 41:373-375.
259. Pease, D. C. (1962): Buffered formaldehyde as a killing agent and primary fixative for electron microscopy. *Anat. Rec.*, 142:342.
260. Pelletier, G., Labrie, F., Arimura, A., and Schally, A. V. (1974): Electron microscopic immunohistochemical localization of growth hormone release inhibiting hormone (somatostatin) in the rat median eminence. *Am. J. Anat.*, 140:445-450.
261. Pelletier, G., Labrie, F., Puviani, R., Arimura, A., and Schally, A. V. (1974): Electron microscope immunohistochemical localization of luteinizing hormone-releasing hormone in the rat median eminence. *Endocrinology*, 95:314-317.
262. Pelletier, G., Leclerc, R., Arimura, A., and Schally, A. V. (1975): Immunohistochemical localization of somatostatin in the rat pancreas. *J. Histochem. Cytochem.*, 23:699-701.
263. Pelletier, G., Leclerc, R., and Dubé, D. (1976): Immunohistochemical localization of hypothalamic hormones. *J. Histochem. Cytochem.*, 24:864-871.
264. Pelletier, G., Leclerc, R., Dubé, D., Labrie, F., Puviani, R., Arimura, A., and Schally, A. V. (1975): Localization of growth hormone-release inhibiting hormone (somatostatin) in the rat brain. *Am. J. Anat.*, 142:397-401.
265. Pelletier, G., Leclerc, R., Labrie, F., and Puviani, R. (1974): Electron microscopic immunohistochemical localization of neurophysin in the rat hypothalamus and pituitary. *Mol. Cell. Endocrinol.*, 1:157-166.
266. Pérez de la Mora, M., Possani, L., Tapia, R., Teran, L., Palacios, R., Fuxe, K., Hökfelt, T., and Ljungdahl, Å. (1977): Immunohistochemical studies on the distribution of glutamic acid decarboxylate in the rat brain. In preparation.
267. Pernow, B. (1953): Studies on substance P. Purification, occurrence and biological actions. *Acta Physiol. Scand.* [*Suppl.* 105], 29:1-90.
268. Pert, C. B., Kuhar, M. J., and Snyder, S. H. (1976): Opiate receptor: Autoradiographic localization in rat brain. *Proc. Natl. Acad. Sci. USA*, 73:3729-3733.
269. Pickel, V. M., Joh, T. H., Field, P. M., Becker, C. G., and Reis, D. J. (1975): Cellular

localization of tyrosine hydroxylase by immunohistochemistry. *J. Histochem. Cytochem.,* 23:1-12.
270. Pickel, V. M., Joh, T. H., and Reis, D. J. (1975): Ultrastructural localization of tyrosine hydroxylase in noradrenergic neurons of brain. *Proc. Natl. Acad. Sci. USA,* 72:659-663.
271. Pickel, V. M., Joh, T. H., and Reis, D. J. (1975): Immunohistochemical localization of tyrosine hydroxylase in brain by light and electron microscopy. *Brain Res.,* 85:295-300.
272. Pickel, V. M., Joh, T. H., and Reis, D. J. (1976): Monoamine-synthesizing enzymes in central dopaminergic, noradrenergic and serotonergic neurons. Immunocytochemical localization by light and electron microscopy. *J. Histochem. Cytochem.,* 24:792-806.
273. Plotnikoff, N. P., and Kastin, A. J. (1976): Commentary: Neuropharmacology of hypothalamic releasing factors. *Biochem. Pharmacol.,* 25:363-365.
274. Polak, J. M., Grimelius, L., Pearse, A. G. E., Bloom, S. R., and Arimura, A. (1975): Growth-hormone release-inhibiting hormone in gastrointestinal and pancreatic D-cells. *Lancet,* 1:1220-1222.
275. Powell, D., Leeman, S. E., Tregear, G. W., Niall, H. D., and Potts, J. T. (1973): Radio-immunoassay for substance P. *Nature [New Biol.],* 241:252-254.
276. Prange, A. J., Jr., Nemeroff, C. B., Lipton, M. A., Breese, G. B., and Wilson, I. C. (1976): Peptides and the central nervous system. In: *Handbook of Psychopharmacology,* edited by L. L. Iversen, S. D. Iversen, and S. H. Snyder. Plenum Press, New York (*in press*).
277. Ramirez, V. C., Gautron, J. P., Epelbaum, J., Pattou, E., Zamora, A., and Kordon, C. (1975): Distribution of LH-RH in subcellular fractions of the basomedial hypothalamus. *Mol. Cell. Endocrinol.,* 3:339-350.
278. Rehfeld, J. F., Stadil, F., and Rubin, B. (1972): Production and evaluation of antibodies for the radioimmunoassay of gastrin. *Scand. J. Clin. Lab. Invest.,* 30:221-232.
279. Reichlin, S. (1960): Thyroid function, body temperature regulation and growth in rats with hypothalamic lesions. *Endocrinology,* 66:340-354.
280. Reichlin, S., Saperstein, R., Jackson, I. M. D., Boyd, A. E., III, and Patel, Y. (1976): Hypothalamic hormones. *Ann. Rev. Physiol.,* 38:389-424.
281. Renaud, L. P. (1977): TRH, LHRH and somatostatin: Distribution and physiological action in neural tissue. In: *Neuroscience Symposia,* Society for Neuroscience; Bethesda, Maryland (*in press*).
282. Renaud, L. P., Martin, J. B., and Brazeau, P. (1975): Depressant action of TRH, LH-RH and somatostatin on activity of central neurones. *Nature,* 255:233-235.
283. Réthelyi, M., and Halász, B. (1970): Origin of the nerve endings in the surface zone of the median eminence of the rat hypothalamus. *Exp. Brain Res.,* 11:145-158.
284. Richardson, K. C. (1966): Electron microscopic identification of autonomic nerve endings. *Nature,* 210:756.
285. Roberts, E. (1975): Immunocytochemistry of the GABA system—A novel approach to an old transmitter. In: *Neuroscience Symposia, Vol. I: Neurotransmitters, Hormones and Receptors: Novel Approaches,* edited by J. A. Ferrendelli, B. S. McEwen, and S. H. Snyder, pp. 123-138. Society for Neuroscience, Bethesda, Md.
286. Roberts, E., Chase, T. N., and Tower, D. B. (Eds.) (1975): *GABA in Nervous System Function.* Raven Press, New York.
287. Rufener, C., Amherdt, M., Dubois, M. P., and Orci, L. (1975): Ultrastructural immunocytochemical localization of somatostatin in rat pancreatic monolayer culture. *J. Histochem. Cytochem.,* 23:866-869.
288. Rufener, C., Dubois, M. P., Malaisse-Lagae, F., and Orci, L. (1975): Immuno-fluorescent reactivity to anti-somatostatin in the gastrointestinal mucosa of the dog. *Diabetologia,* 11:321-324.
289. Saavedra, J. M., Brownstein, M., Palkovits, M., Kizer, J. S., and Axelrod, J. (1974): Tyrosine hydroxylase and dopamine-$\beta$-hydroxylase. Distribution in individual rat hypothalamic nuclei. *J. Neurochem.,* 23:869-871.
290. Saavedra, J. M., Palkovits, M., Brownstein, M., and Axelrod, J. (1974): Serotonin distribution in the nuclei of the rat hypothalamus and preoptic region. *Brain Res.,* 77:157-165.
291. Saavedra, J. M., Palkovits, M., Brownstein, M., and Axelrod, J. (1974): Localization of phenylethanolamine N-methyl-transferase in rat brain nuclei. *Nature,* 248:695-696.
292. Said, S. I., and Faloona, G. R. (1975): A radioimmunoassay of vasoactive intestinal polypeptide. *N. Engl. J. Med.,* 293:155.

293. Said, S. I., and Rosenberg, R. N. (1976): Vasoactive intestinal polypeptide: Abundant immunoreactivity in neural cell lines and normal nervous tissue. *Science,* 192:907–908.
294. Saito, K., Barber, R., Wu, J.-Y., Matsuda, T., Roberts, E., and Vaughn, J. E. (1974): Immunohistochemical localization of glutamate decarboxylase in rat cerebellum. *Proc. Natl. Acad. Sci. USA,* 71:269–273.
295. Schally, A. V., Arimura, A., Baba, Y., Nair, R. M. G., Matsuo, J., Redding, T. W., Debeljuk, L., and White, W. F. (1971): Isolation and properties of the FSH- and LH-releasing hormone. *Biochem. Biophys. Res. Commun.,* 43:393–399.
296. Scharrer, E., and Scharrer, B. (1954): Hormones produced by neurosecretory cells. *Recent Prog. Horm. Res.,* 10:183–240.
297. Schwartz, J. C., Julien, C., Feger, J., and Garbarg, M. (1974): Histaminergic pathway in rat brain evidenced by hypothalamic lesions. *Fed. Proc. Fed. Am. Soc. Exp. Biol.,* 33:285 (Abst.).
298. Sétáló, G., Vigh, S., Schally, A. V., Arimura, A., and Flerkó, B. (1975): GH-RIH containing neural elements in the hypothalamus. *Brain Res.,* 90:352–356.
299. Sétáló, G., Vigh, S., Schally, A. V., Arimura, A., and Flerkó, B. (1975): LH-RH-containing neural elements in the rat hypothalamus. *Endocrinology,* 96:135–142.
300. Sétáló, G., Vigh, S., Schally, A. V., Arimura, A., and Flerkó, B. (1976): Immunohistological study of the origin of LH-RH-containing nerve fibers of the rat hypothalamus. *Brain Res.,* 103:597–602.
301. Sharp, P. J., Haase, E., and Fraser, H. M. (1975): Immunofluorescent localization of sites binding anti-synthetic LHRH serum in the median eminence of the greenfinch (Chloris chloris L.). *Cell Tissue Res.,* 162:83–91.
302. Shute, C. C. D., and Lewis, P. R. (1966): Cholinergic and monoaminergic pathways in the hypothalamus. *Br. Med. Bull.,* 22:221–226.
303. Shute, C. C. D., and Lewis, P. R. (1967): The ascending cholinergic reticular system: Neocortical, olfactory and subcortical projections. *Brain,* 90:497–520.
304. Silver, A. (1974): The biology of cholinesterases. In: *Frontiers of Biology, Vol. 36,* edited by A. Neuberger and E. L. Tatum, pp. 1–596. North Holland, Amsterdam.
305. Silverman, A. J. (1975): The hypothalamic magnocellular neurosecretory system of the guinea pig. I. Immunohistochemical localization of neurophysin in the adult. *Am. J. Anat.,* 144:433–444.
306. Silverman, A. J. (1976): Distribution of luteinizing hormone-releasing hormone (LH-RH) in the guinea pig brain. *Endocrinology,* 99:30–41.
307. Silverman, A. J. (1976): Ultrastructural studies on the localization of neurohypophysial hormones and their carrier proteins. *J. Histochem. Cytochem.,* 24:816–827.
308. Silverman, A. J., and Desnoyers, P. (1976): Ultrastructural immunocytochemical localization of luteinizing hormone-releasing hormone (LH-RH) in the median eminence of the guinea pig. *Cell Tissue Res.,* 169:157–166.
309. Silverman, A. J., Knigge, K. M., and Zimmerman, E. A. (1975): Ultrastructural immunocytochemical localization of neurophysin in freeze-substituted neurohypophysis *Am. J. Anat.,* 142:265–271.
310. Silverman, A. J., and Zimmerman, E. A. (1975): Ultrastructural immunocytochemical localization of neurophysin and vasopressin in the median eminence and posterior pituitary of the guinea pig. *Cell Tissue Res.,* 159:291–301.
311. Simantov, R., Kuhar, M. J., Pasternak, G. W., and Snyder, S. H. (1976): The regional distribution of a morphine-like factor enkephalin in monkey brain. *Brain Res.,* 106:189–197.
312. Skeggs, T. L., Jr., Marsh, W. H., Kahn, J. R., and Shumway, N. P. (1955): Amino acid composition and electrophoretic properties of hypertensin I. *J. Exp. Med.,* 102:435–440.
313. Smith, A. R., and Kappers, J. A. (1975): Effect of pinealectomy, gonadectomy, pCPA and pineal extracts on the rat parvocellular neurosecretory hypothalamic system; a fluorescence histochemical investigation. *Brain Res.,* 86:353–371.
314. Smith, T. W., Hughes, J., Kosterlitz, H. W., and Sosa, R. P. (1976): Enkephalins: Isolation, distribution and function. In: *Opiates and Endogenous Opioid Peptides,* pp. 57–62. North Holland, Amsterdam.
315. Snyder, S. H., and Taylor, K. M. (1972): Histamine in the brain: A neurotransmitter? In: *Perspectives in Neuropharmacology—A Tribute to Julius Axelrod,* edited by S. H. Snyder, pp. 43–73. Oxford University Press, New York.

316. Sotelo, C. (1975): Radioautography as a tool for the study of putative neurotransmitters in the nervous system. *J. Neural Transm.* [*Suppl.*], XII:75–95.
317. Sternberger, L. A. (1974): *Immunocytochemistry.* Prentice-Hall, Englewood Cliffs, N.J.
318. Stutinsky, F. (Ed.) (1967): *Neurosecretion.* Springer-Verlag, Berlin, Heidelberg, New York.
319. Swaab, D. G., Pool, C. W., and Nijveldt, F. (1975): Immunofluorescence of vasopressin and oxytocin in the rat hypothalamo-neurohypophyseal system. *J. Neural Transm.*, 36:195–215.
320. Swanson, L. W., and Hartman, B. K. (1975): The central adrenergic system. An immunofluorescence study of the location of cell bodies and their efferent connections in the rat utilizing dopamine-$\beta$-hydroxylase as a marker. *J. Comp. Neurol.*, 163:467–505.
321. Szentágothai, J., Flerko, B., Mess, B., and Halász, B. (1962): *Hypothalamic Control of the Anterior Pituitary.* Akadémiai Kiadó, Budapest.
322. Takahashi, T., Konishi, S., Powell, D., Leeman, S. E., and Otsuka, M. (1974): Identification of the motoneuron-depolarizing peptide in bovine dorsal root as hypothalamic substance P. *Brain Res.*, 73:59–69.
323. Takahashi, T., and Otsuka, M. (1975): Regional distribution of substance P in the spinal cord and nerve roots of the cat and the effect of dorsal root section. *Brain Res.*, 87:1–11.
324. Tappaz, M. L., Brownstein, M. J., and Palkovits, M. (1976): Distribution of glutamate decarboxylase in discrete brain nuclei. *Brain Res.*, 108:371–379.
325. Tranzer, J. P., Thoenen, H., Snipes, R. L., and Richards, J. G. (1969): Recent developments on the ultrastructural aspect of adrenergic nerve endings in various experimental conditions. *Prog. Brain Res.*, 31:33–46.
326. Ungerstedt, U. (1971): Stereotaxic mapping of the monoamine pathways in the rat brain. *Acta Physiol. Scand.* [*Suppl.*], 367:1–48.
327. Vale, W., Brazeau, P., Rivier, C., Brown, M., Boss, B., Rivier, J., Burgus, R., Ling, N., and Guillemin, R. (1975): Somatostatin. *Recent Prog. Horm. Res.*, 31:365.
328. Van der Gugten, J., Palkovits, M., Wijnen, H. L. J. M., and Versteeg, D. H. G. (1976): Regional distribution of adrenaline in rat brain. *Brain Res.*, 107:171–175.
329. Vanderhaegen, J. J., Signeau, J. C., and Gepts, W. (1975): New peptide in the vertebrate CNS reacting with antigastrin antibodies. *Nature*, 257:604–605.
330. Vandesande, F., and Dierickx, K. (1975): Identification of the vasopressin producing and of the oxytocin producing neurons in the hypothalamic magnocellular neurosecretory system of the rat. *Cell Tissue Res.*, 164:153–162.
331. Vandesande, F., Dierickx, K., and DeMey, J. (1975): Identification of separate vasopressin-neurophysin II and oxytocin-neurophysin I containing nerve fibers in the external region of the bovine median eminence. *Cell Tissue Res.*, 158:509–516.
332. Vogt, M. (1954): The concentration of sympathin in different parts of the central nervous system under normal conditions and after the administration of drugs. *J. Physiol. (Lond.)*, 123:451–481.
333. Watkins, W. B. (1974): Neurosecretion in the external and internal zone of the median eminence of the cat and dog. *Cell Tissue Res.*, 155:201–210.
334. Watkins, W. B. (1975): Neurosecretory neurons in the hypothalamus and median eminence of the dog and sheep as revealed by immunohistochemical methods. *Ann. N.Y. Acad. Sci.*, 248:134–152.
335. Watkins, W. B. (1975): Immunohistochemical demonstration of neurophysin in the hypothalamo-neurohypophysial system. *Int. Rev. Cytol.*, 41:241–284.
336. Watson, S. J., Jr. (1977): Cryostat technique for central nervous system histofluorescence. *J. Histochem. Cytochem.* (in press).
337. Weiner, R. I., Pattou, E., Kerdelhue, B., and Kordon, C. (1975): Differential effects of hypothalamic deafferentation upon luteinizing hormone-releasing hormone in the median eminence and organum vasculosum of the lamina terminalis. *Endocrinology*, 97:1597–1600.
338. Wheaton, J. E., Krulich, L., and McCann, S. M. (1975): Localization of luteinizing hormone-releasing hormone in the preoptic area and hypothalamus of the rat using radioimmunoassay. *Endocrinology*, 97:30–38.
339. Winokur, A., Utiger, R. D. (1974): Thyrotropin-releasing hormone: Regional distribution in rat brain. *Science*, 185:265–267.

340. Wittkowski, W., and Scheuer, A. (1974): Functional changes of the neuronal and glial elements at the surface of the external layer of the median eminence. *Z. Anat. Entwicklungsgesch*, 143:255-262.
341. Wood, J. G., McLaughlin, B. J., and Vaughn, J. E. (1975): Immunocytochemical localization of GAD in electron microscopic preparations of rodent CNS. In: *GABA in Nervous System Function*, edited by E. Roberts, T. N. Chase, and D. B. Tower, pp. 133-148. Raven Press, New York.
342. Yamashiro, D., and Li, C. H. (1974): Synthesis of a pentekontapeptide with high lipolytic activity corresponding to the carboxyl-terminal fifty amino acids of ovine β-lipotropin. *Proc. Natl. Acad. Sci. USA*, 71:4945-4949.
343. Yates, F. E., Russel, S. M., Dallman, M. F., Hedge, G. A., McCann, S. M., and Dharival, A. P. S. (1971): Potentiation by vasopressin of corticotrophin-releasing factor. *Endocrinology*, 88:3-15.
344. Zetler, G. (1970): Distribution of peptidergic neurons in mammalian brain. In: *Aspects of Neuroendocrinology*, edited by W. Bargmann and B. Scharrer, pp. 287-295. Springer-Verlag, Berlin, Heidelberg, New York.
345. Zimmerman, E. A. (1976): Localization of hypothalamic hormones by immunocytochemical techniques. In: *Frontiers in Neuroendocrinology*, edited by L. Martini and W. F. Ganong, pp. 25-62. Raven Press, New York.
346. Zimmerman, E. A., and Antunes, J. L. (1976): Organization of the hypothalamic-pituitary system: Current concepts from immunohistochemical studies. *J. Histochem. Cytochem.*, 24:807-815.
347. Zimmerman, E. A., and Defendini, R. (1977): Anatomy of neurohypophysial pathways. In: *The Hypothalamus, ARNMD, Vol. 56*, edited by S. Reichlin. R. Baldessarini, and J. B. Martin. Raven Press, New York (*in press*).
348. Zimmerman, E. A., Hsu, K. G., Ferin, M., and Kozlowski, G. P. (1974): Localization of gonadotropin-releasing hormone (Gn-RH) in the hypothalamus of mouse by immunoperoxidase technique. *Endocrinology*, 95:1-8.
349. Zimmerman, E. A., Hsu, K. G., Robinson, A. G., Carmel, P. W., Frank, A. G., and Tannenbaum, M. (1973): Studies on neurophysin secreting neurons with immunoperoxidase techniques employing antibody to bovine neurophysin. I. Light microscopic findings in monkey and bovine tissues. *Endocrinology*, 92:931-940.

*The Hypothalamus*, edited by S. Reichlin,
R. J. Baldessarini, and J. B. Martin,
Raven Press, New York, © 1978.

# The Magnocellular Neurosecretory System of the Mammalian Hypothalamus

*Richard Defendini and Earl A. Zimmerman

*The authors dedicate this presentation to
Dr. Berta Scharrer
on the occasion of her seventieth birthday*

The use of immunospecific histology, first fluorescent (12) and more recently enzymatic (42), has produced ample substantiation of the revolutionary concepts of neurosecretion presented by the Scharrers (25) at the Association for Research in Nervous and Mental Disease's milestone meeting on the hypothalamus in 1939. They proposed then that the nerve cells on the mammalian supraoptic (SON) and paraventricular (PVN) nuclei (or analogous nuclei in lower species) produced a substance which was detectable in the cytoplasm in granules and other subcellular structures, and that the material was discharged from the cell in granules or as protein "colloid" for the purpose of creating a physiologic effect at a distance—in the same manner that glandular systems were known to function. It must be emphasized that at that time neither the nature of the substance, nor its physiologic effect, nor its target organ was known. The Scharrers' theory was based on morphologic insights alone. They marshaled embryologic and histoanatomic observations in order to argue that neuroepithelium and secretory epithelium were not as far removed from each other as their many critics conventionally believed. The conceptual leap of those days landed on firm ground after the chrome alum hematoxylin stain introduced by Gomori in 1941 to study the pancreas began to be applied to "the diencephalic gland." It was finally possible to demonstrate that the axons of these strange looking nerve cells, known since Ramón y Cajal (17) to project to the neurohypophysis, did so for the purpose of transporting and releasing their secretory material into blood vessels (26).

The discovery during the past decade of several other hypothalamic peptides with releasing and inhibitory effects on adenohypophysial secretion (2,24) has greatly enriched our understanding of the endocrine uses of neural tissue. Today, however, as we stand at the threshold of a treasure room of

---

*Present address:* Division of Neuropathology, Department of Pathology, Columbia University College of Physicians and Surgeons, 630 West 168th Street, New York, New York 10032.

brain peptides with neurotransmitter-like properties, the endocrine concept of neurosecretion may need to be expanded. Hypothalamic peptides which are being found in other parts of the brain and outside the brain are no longer illuminated by the term "hypothalamic." A particular peptide may exercise different effects at different sites by different mechanisms, doubtless in response to different stimuli. For example, luteinizing hormone releasing hormone (LHRH) behaves like an endocrine hormone when it is discharged into the portal circulation and causes gonadotropic secretion, but it also seems to affect sexual behavior in the rat (16) by mechanisms in the brain that are probably not endocrine. Even vasopressin, the protagonist of the now classic neuroendocrine system of the magnocellular hypothalamus that is the subject of our chapter, is probably more than an antidiuretic hormone, since it appears to play a role in the secretion of corticotropin (44) and perhaps an additional one in memory consolidation (36).

## ANATOMIC OVERVIEW

The mammalian magnocellular neurosecretory system (MNS) resides in the cells of SON and PVN and in the internuclear cells (INC), small groups and short rows of cells which form a discontinuous bridge between PVN and SON and are mixed with the fibers of the paraventricular-supraoptic tract. This tract, which constitutes the main outflow of PVN from its lateral aspect, arches ventrally toward SON, where it joins the outflow of SON. Together the two nuclear outflows run medially in the floor of the hypothalamus as the supraopticohypophysial tract (SOHT) into the infundibulum, commonly called the median eminence (ME). Most of the tract terminates on blood vessels of the systemic circulation of the neurohypophysis along the pituitary stalk and in the posterior lobe.

The magnocellular neurosecretory system of the rat is shown in Fig. 1. Neurosecretory material is also found in the suprachiasmatic nucleus (SCN), the dorsomedial portion of which has been shown to contain vasopressin (VP) and its specific neurophysin (NP) in mouse and rat (37). This nucleus is composed of small nerve cells and is therefore not properly a part of the magnocellular system. In the monkey, the SCN is less well developed and does not contain either VP or NP (38). In man, no nucleus can be recognized in the suprachiasmatic location; the very occasional small nerve cell encountered in this region fails to stain with specific antisera to VP and NP. Whatever the neurosecretory function of SCN may be in the rodent, it appears to have no representation in the primate.

## IMMUNOHISTOCHEMICAL OVERVIEW

The only secretory peptides that have been demonstrated immunohistologically in the mammalian MNS are the two nonapeptide hormones,

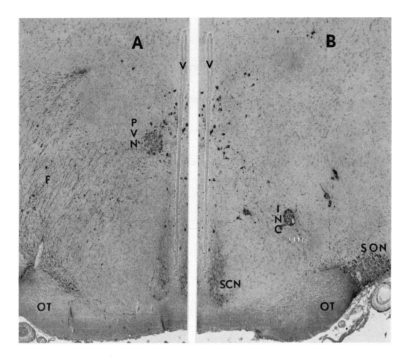

**FIG. 1.** Asymmetrical coronal section of rat hypothalamus stained by immunoperoxidase with anti-rat neurophysin (×42). **A** and **B** represent the left and the right side of the same section. OT, optic tract; V, third ventricle; SCN, suprachiasmatic nucleus; SON, supraoptic nucleus; PVN, paraventricular nucleus; INC, internuclear cells; F, fiber tract between PVN and SON.

VP and oxytocin (OT), and their associated proteins, the neurophysins. The latter substances, first isolated by Van Dyke and co-workers (35), are small proteins with a molecular weight in the range of 10,000 that are believed to be synthesized together with VP or OT from common precursor molecules (23). They are packaged in secretory granules within the magnocellular perikarya, transported by fast axonal flow to nerve terminals, and released by exocytosis (8) together with the nonapeptides into blood vessels of the systemic circulation in the neurohypophysis. Their function, other than as presumed "carriers" of the hormones, is not known. It is generally accepted that there are two chemically different NPs in the mammalian hypothalamus (21). Robinson has demonstrated that in man (19) and in monkey (20) one of them is released into the bloodstream in response to a dose of estrogen ("estrogen-stimulated neurophysin," ESN); the other one is released together with VP in response to nicotine (19) or to hemorrhage (20) ("nicotine-stimulated neurophysin," NSN).

In order to demonstrate on tissue sections that each of the two NPs is associated with only one of the hormones in a given species, it has been necessary to develop antibodies that are specific for each of the two NPs of that particular species, a task that is complicated by the structural homologies of the two NPs within species and among different species (3). It has been accomplished in the ox using purified bovine bovine NP extracts as the immunogen, followed by adsorption of each antiserum with the heterologous NP (13,34), and in monkey because of preserved specificity in cross species reactivity with the antihuman NPs (38). In the ox and monkey, one NP is associated with VP and the other one with OT. This

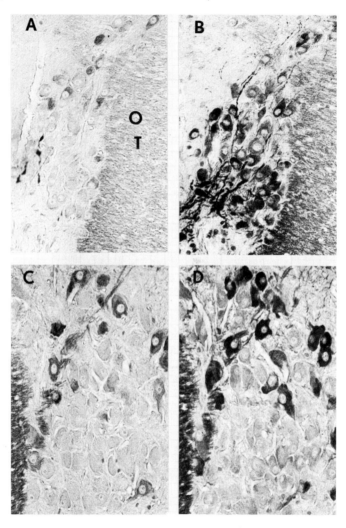

**FIG. 2.** Coronal sections of the supraoptic nucleus of the normal rat (**A** and **B**) and the Brattleboro rat with diabetes insipidus (**C** and **D**) ×225. **A:** Anti-oxytocin in the normal rat. **B.** Anti-vasopressin in a section adjacent to **A. C.** Anti-oxytocin in the Brattleboro rat. **D.** Anti-rat neurophysin in a section adjacent to **C.** OT, optic tract.

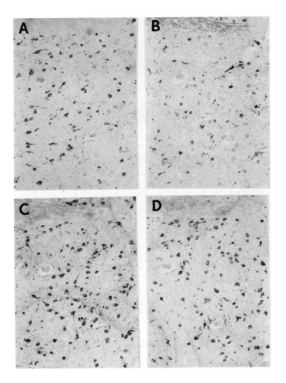

**FIG. 3.** Consecutive 5-μm sections of human PVN in the sagittal plane stained by immunoperoxidase with **(A)** anti-human ESN, **(B)** anti-OT, **(C)** anti-human NSN, and **(D)** anti-AVP. ESN is associated with OT, and NSN is associated with VP. ×22.

specific association can be inferred with rat by taking advantage of the Brattleboro strain. In the homozygote with congenital diabetes insipidus (DI rat), which by tissue analysis has no VP and is missing one NP (1), half of the MNS cells are immunoreactive to anti-NP and the same cells react to anti-OT (29,33). The cells which fail to react to NP in the DI rat are in the same area as those which in the normal rat contain VP (Fig. 2). In man, using specific antisera against the human NPs (NSN and ESN) and specific antisera against VP and OT on adjacent 5 μm sections, we have demonstrated by field patterns of staining (41) (Fig. 3) and by tracing of single cells through adjacent sections, that NSN is associated with OT in the perikarya of the MNS.

## THE HYPOTHALAMIC NUCLEI

A three-dimensional schematic reconstruction of SON and PVN in man, based on measured serial sections of autopsy specimens in the coronal, horizontal, and sagittal planes, stained conventionally and immunoenzymatically with antisera against the four human secretory peptides is shown in Fig. 4 (5). The PVN, a long, thin column of cells, is oriented dorsoventrally in the wall of the third ventricle. It begins at the level of the dorsal margin of the anterior commissure just behind the fornix and ends in the floor of the hypothalamus in front of the infundibulum. The outline of the column is altered in its midportion as the fornix moves from an anterior to a posterior position and comes to lie directly lateral to PVN. At this point, the fornix compresses PVN medially, causing a slight bulge into the third ventricle, and the column of cells becomes elongated in its anteroposterior dimension. The usual coronal section of the cerebral hemispheres cuts across the PVN column tangentially. In coronal sections, therefore, PVN appears as a somewhat oblong plate of cells oriented dorsoventrally, situated higher or lower along the wall of the ventricle depending on whether the section passes through the dorsal or the ventral segment of the column. With the exception of a somewhat broader lateral dimension in some species, the appearance of PVN in the coronal plane is similar in lower mammals and in man.

The human SON is a solid nucleus which has been divided into anterior and posterior parts by the growth of the optic tract into the floor of the hypothalamus. SON appears to be draped like saddlebags over the proximal half of the tract. Few if any cells link the crests of the two bags across the dorsum of the tract. The optic tract arises from the chiasm and runs in a

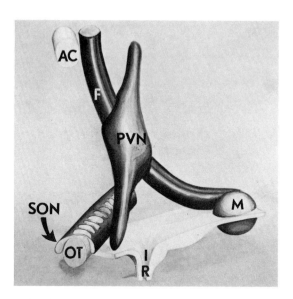

**FIG. 4.** Schematic three-dimensional reconstruction of the anatomy of the two major nuclei of the magnocellular neurosecretory system of the human hypothalamus. The plane of the foreground is a midsagittal section through the floor of the third ventricle. AC, anterior commissure; F, fornix; IR, infundibular recess; M, mammillary body; OT, optic tract; PVN, paraventricular nucleus; SON, supraoptic nucleus.

lateral, posterior, and slightly dorsal direction to the lateral geniculate body; it also rotates and changes its shape to some extent along the way. The usual coronal section through the cerebral hemispheres, therefore, is tangential to SON in all three anatomical planes, which explains the considerable variation displayed by coronal sections in the shape of the nucleus and in the apparent segregation of the cells into groups. The most common segregation of SON cells seen in mammals—one group medial and another dorsal and/or lateral to the tract—corresponds to tangential sections through the posterior and the anterior saddlebag of the nucleus, respectively. The dorsolateral group is usually larger than the medial group because the anterior component of the nucleus is usually larger than the posterior one.

The typical nerve cell of the MNS is large, polygonal, and multipolar. The nucleus is displaced to the cell membrane, the nissl substance is distributed peripherally, and the immunoreactive secretory material tends to be concentrated about the nucleus. Secretory material extends into axons and arborizing dendrites (Fig. 5) without clearly defined transition between the perikaryon and the cell process; in immunospecific preparations it is difficult to distinguish the axon from the dendrites. The nerve cells are seen at times in somatic contact with each other—an unusual feature in neural tissue. Also unusual is the fact that in these highly vascularized nuclei, nerve cells are often in direct contact with tiny capillaries. In INC, however, the nerve cells cluster about blood vessels of larger caliber. The concentration of secretory material around the nucleus and the direct contact of nerve cell bodies with each other and with nonfenestrated capillaries have been re-

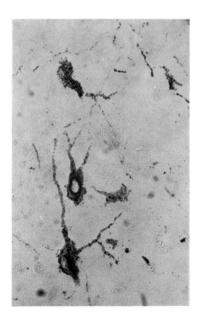

**FIG. 5.** Multipolar cells with branching processes in the human paraventricular nucleus stained with anti-human ESN. ×375.

ported at the ultrastructural level by Sloper and Bateson (28) in dog and rat.

The nerve cells of SON are rather uniform in size. Some variation occurs in man, but with rare exception they fall within the magnocellular range of 25 to 30 $\mu$m in the paraffin-embedded material. In view of the fact that all the cells contain immunoreactive material and that no clear pattern of distribution of larger and less large cells is evident, the degree of variation in cell size observed in the human SON may have no major functional significance.

The nerve cell population of PVN in man is much less homogeneous than SON's. Less than half the neurons can be considered magnocellular (25 to 30 $\mu$m). An equal number range from 15 to 20 $\mu$m in diameter. With rare exception, these large and medium-sized cells are immunoreactive. The smallest of all, however, which measure 10 $\mu$m in diameter and comprise 15% of the total nerve cell population of PVN, do not react with any of the antisera and therefore do not appear to be secreting members of the system.

## THE DISTRIBUTION OF NEUROHYPOPHYSIAL PEPTIDES

In all mammalian species studied immunohistochemically, both SON and PVN have been found to contain VP and OT and the two NPs. Functional organization in the MNS takes place at a cellular rather than a nuclear level. Nevertheless, within the nuclei, regional differences in the prevalence of the two hormones have been found in several species.

> The most constant and conspicuous pattern is a concentration of OT-containing cells in the dorsal portion of SON and of VP-containing cells in the ventral portion. This distribution has been reported in ox (34), rat (29,33), and monkey (38), and is found in man. Most of the SON cells that contain OT (as well as ESN) in the human hypothalamus are concentrated in the less densely populated dorsal cap of the anterior and posterior divisions of the nucleus. In the rat, Swaab et al. (32) have reported that the rostral levels of SON contain more anti-OT-positive cells than the caudal. Cell counts on our serial coronal sections of the DI rat confirm this second distribution pattern, since we find the same gradual reversal of the ratio of anti-NP-positive to anti-NP-negative cells as one proceeds from the rostral (medial) to the caudal (lateral) end of SON.
> 
> Regional distribution patterns of the two hormone-carrier protein pairs in PVN are less clearly established, less obvious, and probably less constant from species to species than in SON. De Mey and co-workers reported marked differences in the bovine PVN between the "apical," "middle," and "dorsocaudal" parts of the nucleus in "transverse" (13) (coronal or frontal) and sagittal (34) sections. Despite the problems of localization on sections of the long column of PVN that do not cover all three anatomical planes (especially the horizontal), the coronal sections demonstrate in the rat a prevalence of OT-containing cells in the lateral "wings" and along the periphery of the nucleus and an obverse predominance of VP cells in the more central and medial por-

tions of the nucleus (29,33). Like Swaab et al. (32), we also find in our coronal series of the rat that there are relatively more OT than VP cells in the "rostral" (dorsal ?) than in the "caudal" (ventral ?) part of PVN. In the human PVN, our observations again suggest a more complex evolution of functional organization in this nucleus than in SON. In the coronal series in man there is a slight tendency for OT-containing cells to be concentrated about the periphery of the nucleus, especially "dorsally." This distribution, however, is not borne out by the horizontal series. We conclude, therefore, that in the human PVN, in contrast to probable patterns in lower mammals that may need further definition, there is no regional distribution of the two MNS hormones.

Quantitation in immunocytology at present is relative rather than absolute, but it provides the anatomic localization that chemical assay cannot provide. The histologic approach is colorimetric, and many technical factors can affect the results. If the staining procedures are standardized, however, quantitative immunohistochemistry provides reliable values that are useful for comparative purposes.

By such methods, the relative amounts of the two hormones in the MNS have been shown to vary among mammalian species. In the bovine MNS, it has been reported that VP-NP-II cells are more numerous in SON than OT-NP-I cells, whereas the opposite was found in PVN (34). In the rat, equal proportions of VP- and OT-containing cells have been reported by Vandesande and Dierickx (33) and Sokol et al. (29). Swaab et al. (32), using immunofluorescent methods, found somewhat larger numbers of VP- than OT-containing cells in the two nuclei. They also found, however, that there are three times as many magnocellular neurons in the rat SON as in PVN and pointed out that the total number of OT cells is therefore greater in SON than in PVN.

In man, the amount of VP in the MNS is considerably greater than the amount of OT. In SON, which by cell counts Morton (14) reports is 36% larger than PVN, virtually all cells are immunoreactive. Scarcely 15% of them react with anti-OT, whereas 20% react strongly with anti-ESN. In PVN, the smallest nerve cells (15%) and an occasional large cell fail to react with either of the two pairs of antisera. Among the reactive cells, 60% are anti-VP-NSN positive and approximately 40% react with anti-OT-ESN. As in SON, the number of anti-ESN-positive cells is approximately one-third larger than the number of anti-OT-positive ones. We attribute this discrepancy to the weakness of our anti-OT antiserum (1:400 for 30 min) rather than to cross-reactivity of our anti-ESN (1:10,000 for 30 min) with NSN because the same difference obtains when we apply these antisera to the DI rat: half of the SON cells are anti-ESN positive, whereas in the adjacent section only one-third are anti-OT positive. The discrepancy represents a false-negative result with our anti-OT antiserum, which fails to recognize cells that contain lesser amounts of OT, and we therefore accept as valid the higher figures from the highly potent anti-ESN antiserum. Taking the average of Morton's values for the normal number of cells in the human SON and PVN, the ratio of the total number of cells containing VP and NSN to those containing OT and ESN in the human MNS is 2.5:1.

## ANATOMIC PATHWAYS

Little has been added to the knowledge derived from the Gomori stains about the anatomy of the principal outflow of the MNS along the length of the neurohypophysis. Immunohistologic methods, however, have demonstrated the presence of secretory products of the MNS in the zona externa (ZE) of the ME in several species, beginning with Parry and Livett's (15) report of a large amount of NP in the sheep ZE [see Zimmerman (37) for a review]. Zimmerman et al. (40) found very high levels of NP and VP on direct canulation of long portal veins in the anesthetized monkey subjected to the stress of intracranial surgery. By immunoelectron microscopy, NP and VP have now been shown in granules within axon endings near portal capillaries in rat (10), guinea pig (31), and mouse (4); Silverman (30) has also shown OT-containing granules in the guinea pig ZE. In both rat and guinea pig, the anti-VP-reacting granules in axons in the ZE are considerably smaller than those found in the zona interna (ZI) (9,30,44). The MNS fibers that reach the ZE emanate in radial fashion from the descending SOHT tract in the ZI (Fig. 6). Unlike fibers carrying other hypothalamic secretions to the ZE, some of which have been shown by Hökfelt and co-workers (see preceding chapter) to occupy distinct segments of the infundibular perimeter, the MNS fibers appear randomly distributed in the rat ME. In rat (7) and monkey (38), the amount of VP-NP in ZE is much greater than the amount of OT-NP.

The functional significance of MNS secretion in the territory of the tuberoinfundibular or parvocellular neurosecretory system remains to be elucidated. Recent studies of selective lesions in the hypothalamus of the

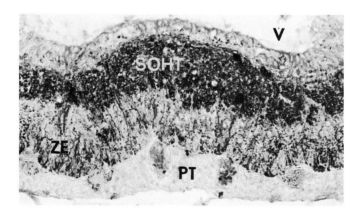

**FIG. 6.** Coronal section of the median eminence of a rat 3 weeks after bilateral adrenalectomy, stained with anti-vasopressin. Many neurosecretory fibers emanate from the densely stained supraopticohypophysial tract (SOHT) in the zona interna to end on blood vessels of the adenohypophysial portal system in the zona externa (ZE). PT, pars tuberalis; V, floor of third ventricle. ×225. From ref. 11 with permission.

monkey (39) indicate that the MNS fibers in the ZE proceed entirely from PVN; SON appears to project its fibers exclusively to the classic neurohypophysial pathway. This distribution would explain the old observation (18) that stalk section in the monkey results in total degeneration of SON but loss of only 20% of cells in PVN. These findings make it unlikely that the fibers that fan out from SOHT are collaterals of axons belonging to the classic system. The smaller size of the VP granules in the ZE already mentioned also supports the alternative supposition that the MNS fibers of the ZE proceed from a specialized population of PVN cells that are governed by a separate input.

From a functional standpoint, the long debate over the relation between VP and the elusive corticotropin releasing hormone (CRF) may slowly be approaching a consensus (11). VP is probably not CRF (alternatively, VP is not the *only* CRF), but VP undoubtedly stimulates the secretion of ACTH, possibly via the portal circulation. Moreover, the fact that bilateral adrenalectomy results in no apparent change in the amount of the MNS secretory products in the ZI but brings about a marked selective increase in the amount of VP in the ZE, unrelated to dehydration and reversible by glucocorticoid replacement therapy (44), suggests that adrenal steroid is the principal inhibitory factor in the feedback loop of this aspect of VP physiology. What remains quite unclear is the special role of VP in the overall regulation of ACTH secretion (the diurnal rhythm of ACTH secretion is preserved in the DI rat, whose ACTH response to stress is only

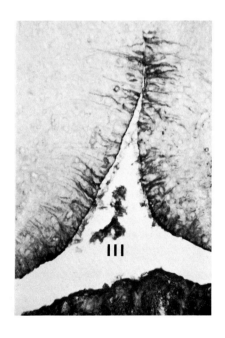

**FIG. 7.** Coronal section of the floor of the third ventricle (III) of a rat dehydrated for 5 days, stained with anti-rat neurophysin. Neurosecretory reaction products are present in the perikarya and processes of tanycytes. ×210. Courtesy of M. A. Stillman, *unpublished data.*

slightly below normal) and whether its effect on ACTH is directly mediated by or dependent on interaction with [another] CRF.

It is likely that the infundibulum does not contain all the fiber projections of the VP and OT secretory system. In the rat, the VP-NP fibers of the dorsomedial cells of SCN appear to run dorsally, although they have not been traced beyond the wall of the ventricle. The direction of these fibers suggests that the SCN VP system does not project to the infundibulum. More substantial evidence for extrainfundibular pathways comes from the recent demonstration of VP and NP in the organum vasculosum of the lamina terminalis (OVLT) of the monkey (38). The presence of neurophysin in the cytoplasm and processes of tanycytes (22,43) (Fig. 7) has prompted much speculation about the role of an intraventricular pathway for VP.

The inconsistency of tanycyte staining, like the long history of difficulties demonstrating LHRH in hypothalamic cells (37), points up the fact that at this stage in the immunohistologic pursuit of peptides false negatives may be a more serious problem than false positives. In our human autopsy material, for example, despite relatively complete definition of the MNS perikarya, we have succeeded in staining scarcely 5% of the fibers of the large SOHT in the anterior wall of the infundibulum. Under these cir-

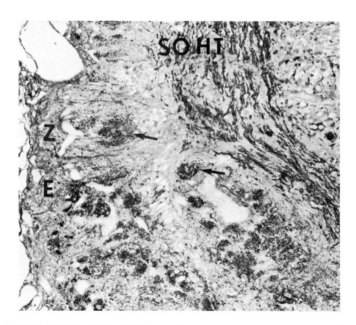

**FIG. 8.** Coronal section of the median eminence of monkey immunoreactive products stained for vasopressin, demonstrating occasional long fibers running from the supraopticohypophysial tract (SOHT) to the zona externa (ZE) and neurovascular cores (arrows) sharply outlined by immunoreactive products. ×225.

cumstances, failure to demonstrate any MNS fibers in the ventricular wall or in OVLT is not acceptable data. We see nerve fibers that are immunoreactive with antisera to the four neurohypophysial peptides in the posterior wall of the infundibulum near the subarachnoid space, but their relation to the hypophysial portal system is difficult to establish, in part because of their relatively small number, but also because in man the well-defined ZE of lower species has become completely "internalized" into "neurovascular cores" (6,27) — small complexes of finely innervated capillary skeins (the *gomitoli* of Fumagalli) scattered among the tract fibers — and these structures are not immunoreactive in our human material, whereas they are very well outlined by the numerous reactive axon terminals in the monkey (Fig. 8).

## SUMMARY

Immunohistochemical techniques in recent years have made possible precise localization of the secretory products of the magnocellular neurosecretory system in the hypothalamus of several species. In all of them, vasopressin and oxytocin are present in both the supraoptic and the paraventricular nucleus. Vasopressin is associated intracellularly with one neurophysin "carrier" protein and oxytocin with a different one. Regional differences in the distribution of the two hormones in different species are reviewed, the most constant of which is a striking concentration of oxytocin cells in the dorsal region of the supraoptic nucleus. The proportion of the two hormones in the two nuclei and in the whole system varies from species to species. In addition to its classic projection to the systemic circulation in the neurohypophysis, the supraopticohypophysial tract projects to the zona externa of the median eminence, where vasopressin may, via the adenohypophysial portal circulation, play a role in the release of corticotropin. The projections to the zona externa probably proceed from cells in the paraventricular and not the supraoptic nucleus. Extrainfundibular pathways demonstrated so far include vasopressin fibers in the organum vasculosum of the lamina terminalis of the monkey.

This chapter presents new data on the human magnocellular system based on serial sections in the three anatomical planes, from which revisions of the accepted anatomy of both the supraoptic and the paraventricular nuclei are proposed. The suprachiasmatic nucleus, which in the rodent contains vasopressin, is not represented in man. The ratio of vasopressin-positive to oxytocin-positive cells is considerably greater in man (2.5:1) than in the rat (1:1); the proportion of oxytocin cells in the human supraoptic nucleus is singularly low (20%). Several observations about the paraventricular nucleus suggest that this nucleus, in contrast to the supraoptic, has undergone a more complex functional organization vis-à-vis lower mammals.

## ACKNOWLEDGMENTS

The authors thank Dr. Susan Rosario and Ms. Fely Inong for their excellent technical assistance. We are grateful for gifts of valuable antisera to human and rat neurophysins (Dr. Alan Robinson), to synthetic oxytocin (Dr. Paul Czernichow), to bovine arginine vasopressin (Dr. Robert Utiger), and to synthetic arginine vasopressin (Dr. Richard Weitzman). This work was supported in part by The Parkinson Research Foundation and the International Institute of Human Reproduction, Columbia University. Dr. Zimmerman is the recipient of USPHS, NIH, NINCDS Teacher-Investigator Award NS 11008.

## REFERENCES

1. Burford, G. D., Jones, C. W., and Pickering, B. T. (1971): Tentative identification of a vasopressin-neurophysin and an oxytocin-neurophysin in the rat. *Biochem. J.*, 124:809–813.
2. Burgus, R., Ling, N., Butcher, M., and Guillemin, R. (1973): Primary structure of somatostatin, a hypothalamic peptide that inhibits the secretion of pituitary growth hormone. *Proc. Natl. Acad. Sci. USA*, 70:684–688.
3. Capra, J. D., and Walter, R. (1975): Primary structure and evolution of neurophysins. *Ann. N.Y. Acad. Sci.*, 248:92–111.
4. Castel, M., and Hochman, J. (1977): Ultrastructural immunohistochemical localization of vasopressin in the hypothalamo-neurohypophysial system of three murids. *Cell Tissue Res.* (*in press*).
5. Defendini, R., and Zimmerman, E. A. (1977): Immunohistologic anatomy of the magnocellular neurosecretory system in the hypothalamus of man. *In preparation*.
6. Diepen, R. (1962): *Der Hypothalamus*. In: *Handbuch der microskopischen Anatomie des Menschen*, edited by B. von Mollendorff. Springer-Verlag, Berlin.
7. Dierickx, K., Vandesande, F., and De May, J. (1976): Identification, in the external region of the rat median eminence, of separate neurophysin-vasopressin and neurophysin-oxytocin containing nerve fibers. *Cell Tissue Res.*, 168:141–151.
8. Douglas, W. W. (1974): Mechanism of release of neurohypophysial hormones: Stimulus-secretion coupling. In: *Handbook of Physiology, Section 7*, edited by E. Knobil and W. H. Sawyer. American Physiological Society, Washington, D.C.
9. Dube, D., LeClerc, R., and Pelletier, G. (1976): Electron microscopic immunohistochemical localization of vasopressin and neurophysin in the median eminence of normal adrenalectomized rats. *Am. J. Anat.*, 147:103–108.
10. Goldsmith, P. C., and Zimmerman, E. A. (1976): Ultrastructural immunocytochemical evidence of increased neurophysin and vasopressin in the median eminence of bilaterally adrenalectomized rats. Program Abstracts of the V. International Congress of Endocrinology, Abstract 183.
11. Krieger, D. T., and Zimmerman, E. A. (1977): The nature of CRF and its relationship to vasopressin. In: *Clinical Neuroendocrinology*, edited by M. Besser and L. Martini. Academic Press, New York (*in press*).
12. Livett, B. G., Uttenthal, L. O., and Hope, D. B. (1971): Localization of neurophysin II in the hypothalamo-neurohypophysial system of the pig by immunofluorescence histochemistry. *Philos. Trans. R. Soc. Lond.* [*Biol.*], 261:371–378.
13. Mey, J. De, Vandesande, F., and Dierickx, K. (1974): Identification of neurophysin producing cells. II. Identification of the neurophysin I and the neurophysin II producing neurons in the bovine hypothalamus. *Cell Tissue Res.*, 153:531–543.
14. Morton, A. (1969): A quantitative analysis of the normal neuron population of the hypothalamic magnocellular nuclei in man and of their projections to the neurohypophysis. *J. Comp. Neurol.*, 136:143–158.

15. Parry, H. B., and Livett, B. G. (1973): A new hypothalamic pathway to the median eminence containing neurophysin and its hypertrophy in sheep with natural scrapie. *Nature*, 242:63–65.
16. Pfaff, D. W. (1973): LHRH potentiates lordosis behavior in hypophysectomized ovariectomized rats. *Science*, 182:1148–1150.
17. Ramón y Cajal, S. (1911): *Histologie du Système Nerveux de l'Homme et des Vertebrés*, Vol. 2. A. Maloine, Paris.
18. Rasmussen, A. T. (1940): Effects of hypophysectomy and hypophysial stalk section on the hypothalamic nuclei of animals and man. *Res. Publ. Assoc. Res. Nerv. Ment. Dis.*, 20:245–269.
19. Robinson, A. G. (1975): Isolation and secretion of individual human neurophysin. *J. Clin. Invest.*, 55:360–367.
20. Robinson, A. G., Ferin, M., and Zimmerman, E. A. (1976): Neurophysin in monkeys: Emphasis on the hypothalamic response to estrogen and ovarian events. *Endocrinology*, 98:468–475.
21. Robinson, A. G., and Frantz, A. G. (1973): Radioimmunoassay of posterior pituitary peptides: A review. *Metabolism*, 22:1047–1057.
22. Robinson, A. G., and Zimmerman, E. A. (1973): Cerebrospinal fluid and ependymal neurophysin. *J. Clin. Invest.*, 52:1260–1267.
23. Sachs, H., Fawcett, C. P., Portanova, R., and Takabatake, Y. (1969): Biosynthesis and release of vasopressin and neurophysin. *Recent Prog. Horm. Res.*, 25:447–491.
24. Schally, A., Arimura, A., and Kastin, A.: Hypothalamic regulatory hormones. *Science*, 179:341–350.
25. Scharrer, E., and Scharrer, B. (1940): Secretory cells within the hypothalamus. *Res. Publ. Assoc. Nerv. Ment. Dis.*, 20:170–194.
26. Scharrer, E., and Scharrer, B. (1954): Hormones produced by neurosecretory cells. *Recent Prog. Horm. Res.*, 10:183–240.
27. Sloper, J. C. (1966): The experimental and cytopathological investigation of neurosecretion in the hypothalamus and pituitary. In: *The Pituitary Gland. Pars Intermedia and Neurohypophysis, Vol. 3*, edited by G. W. Harris and B. T. Donovan. University of California Press, Berkeley.
28. Sloper, J. C., and Bateson, R. G. (1965): Ultrastructure of neurosecretory cells in the supraoptic nucleus of the dog and rat. *J. Endocrinol.*, 31:139–150.
29. Sokol, H. W., Zimmerman, E. A., Sawyer, W. H., and Robinson, A. G. (1976): The hypothalamoneurohypophysial system of the rat: Localization and quantification of neurophysin by light microscopic immunocytochemistry in normal rat and in Brattleboro rats deficient in vasopressin and a neurophysin. *Endocrinology*, 98:1176–1188.
30. Silverman, A. J. (1976): Ultrastructural studies on the localization of neurohypophysial hormones and their carrier proteins. *J. Histochem. Cytochem.*, 24:816–827.
31. Silverman, A. J., and Zimmerman, E. A. (1975): Ultrastructural immunocytochemical localization of neurophysin and vasopressin in the median eminence and posterior pituitary of the guinea pig. *Cell Tissue Res.*, 159:291–301.
32. Swaab, D. F., Pool, C. W., and Nijveldt, F. (1975): Immunofluorescence of vasopressin and oxytocin in the rat hypothalamo-neurohypophyseal system. *J. Neural Transm.*, 36:195–215.
33. Vandesande, F., and Dierickx, F. (1975): Identification of the vasopressin producing and of the oxytocin producing neurons in the hypothalamic magnocellular neurosecretory system of the rat. *Cell Tissue Res.*, 164:153–162.
34. Vandesande, F., Dierickx, K., and De Mey, J. (1975): Identification of the vasopressin-neurophysin II and the oxytocin-neurophysin I producing neurons in the bovine hypothalamus. *Cell Tissue Res.*, 156:189–200.
35. Van Dyke, H. B., Chow, B. F., Greep, R. O., and Rothen, A. (1941): The isolation of a protein from pars neuralis of the ox pituitary with constant oxytocic, pressor and diuresis-inhibiting effects. *J. Pharmacol. (Kyoto)*, 74:19–209.
36. Van Wimersma Greidanus, Tj. B., Bohus, B., and De Wied, D. (1975): The role of vasopressin in memory processes. *Prog. Brain Res.*, 42:135–141.
37. Zimmerman, E. A. (1976): Localization of hypothalamic hormones by immunocytochemical techniques. In: *Frontiers in Neuroendocrinology, Vol. 4*, edited by L. Martini and W. F. Ganong, pp. 25–62. Raven Press, New York.

38. Zimmerman, E. A., and Antunes, J. L. (1976): Organization of the hypothalamic-pituitary system: Current concepts from immunohistochemical studies. *J. Histochem. Cytochem.*, 24:807–815.
39. Zimmerman, E. A., Antunes, J. L., Carmel, P. W., Defendini, R., and Ferin, M. (1977): Magnocellular neurosecretory pathways in the monkey. Immunohistochemical studies of the normal and lesioned hypothalamus using antibodies to oxytocin, vasopressin and neurophysins. *Trans. Am. Neurol. Assoc.*, 101:1–4.
40. Zimmerman, E. A., Carmel, P. W., Husain, M. K., Tannenbaum, M., Frantz, A. G., and Robinson, A. G. (1973): Vasopressin and neurophysin: High concentrations in monkey hypophyseal portal blood. *Science*, 182:925–927.
41. Zimmerman, E. A., Defendini, R., Sokol, H. W., and Robinson, A. G. (1975): The distribution of neurophysin-secreting pathways in the mammalian brain: Light microscopic studies using the immunoperoxidase technique. *Ann. N.Y. Acad. Sci.*, 248:92–111.
42. Zimmerman, E. A., Hsu, K. C., Robinson, A. G., Carmel, P. W., Frantz, A. G., and Tannenbaum, M. (1973): Studies of neurophysin secreting neurons with immunoperoxidase technique employing antibody to bovine neurophysin. I. Light microscopic findings in monkey and bovine tissues. *Endocrinology*, 92:931–940.
43. Zimmerman, E. A., Kozlowski, G. P., and Scott, D. E. (1975): Axonal and ependymal pathways for the secretion of biologically active peptides into hypophysial portal blood. In: *Brain-Endocrine Interaction. II: The Ventricular System*, edited by K. M. Knigge, D. E. Scott, H. Kobayashi, and S. Ishii, pp. 123–134. S. Karger, Basel.
44. Zimmerman, E. A., Stillman, M. A., Recht, L. A., Antunes, J. L., Carmel, P. W., and Goldsmith, P. C. (1977): Vasopressin and corticotrophin-releasing factor (CRF): An axonal pathway to portal capillaries in the zona externa of the median eminence containing vasopressin and its interaction with adrenal corticoids. *Ann. N.Y. Acad. Sci. (in press)*.

# General Discussion

*Dr. Krieger:* Dr. Knigge, does the use of these new antibodies that have shown differences in LHRH localization help clarify recent studies of the effects of deafferentation on concentrations of LHRF within and outside the deafferented area? Also, do you believe that structurally different LHRHs are responsible for the different effects of the preoptic area and the arcuate area on gonadotropin release?

*Dr. Knigge:* I cannot answer your questions at this time. It remains to be seen whether the analysis as we presented it is correct. The data certainly look as though these two cell groups in the rat hypothalamus are different. Moreover, the idea that the material stained in Field No. 1 might be FSH-RF is a fascinating possibility.

If these inferences are true, it means that immunocytochemists must be especially concerned with the specificity of the antibody and the conformation of the material in the tissue that is stained. Those using radioimmunoassay as a single and only tool for the measurement of a hormone must observe similar precautions. Our "E" antibody appears to be recognizing more than LHRH. One could hypothesize that the basis of control of FSH and LH and of ovulation could very nicely fit into a scheme where one type of releasing hormone is in the preoptic area and another is in the basal-arcuate region.

*Dr. Guillemin:* Dr. Knigge, I appreciate the elegant studies attempting to characterize the various antibodies to LHRH. This is a very interesting approach.

To offer several suggestions: We have synthesized a large number of analogues of the decapeptide LRF, which could be used to characterize the antibodies, and of course would make them available to you. Further, it would be of interest to induce antibodies to some of these LRF analogues; e.g., it would be interesting to study the series of analogues in which we have deleted the histidine in the second position. The histidine-2-deleted analogues of LRF bind to LRF receptors but do not trigger the release of gonadotropins, so that we know that these substances are partial agonists and partial antagonists of LRF.

With respect to the idea that there may be more than one gonadotropin-releasing hormone, based on very careful bioassays, we have never so far seen anything other than the decapeptide LRF stimulating the release of LH and FSH. Burgus reported that we had possibly seen minute amounts of the free acid of the decapeptide instead of the decapeptide amide. We have never seen any hypothalamic fraction that would stimulate the secretion of FSH that did not also stimulate the secretion of LH.

*Dr. Knigge:* I agree with your comments about the use of analogues as a valuable way to identify sequences of peptides in tissue sections and tissue extracts. This method will, I believe, become one of the most powerful tools for functional analysis.

*Dr. Bloom:* I have a double question. Dr. Hökfelt showed clearly that somatostatin-positive cells and tyrosine-hydroxylase-positive cells were probably not the same. But his tables of data indicate numerous cross-matches to be pursued, particularly in the case of the arcuate nucleus. I was struck by the fact that Dr. Knigge's demonstration of LRF-positive cells and Dr. Hökfelt's demonstration of tyrosine hydroxylase-positive cells seemed to be highly similar. I wonder if Dr. Hökfelt has made similar cross-matches or whether Dr. Knigge has for catecholamines?

The second half of the question deals with the median eminence: The arcuate nucleus neurons have been said to be the major source of fibers to the pars intermedia. Are there LRF-positive fibers also in this region?

*Dr. Hökfelt:* I have not excluded the possibility that peptides and amines are stored in the same neuron. We have no evidence for this. On the other hand, there are some cells in the suprachiasmatic nucleus, similar to the ones described by Dr. Zimmerman, which appear to stain both for a peptide and for the enzyme dopa decarboxylase. We do not know which of the amines are localized in the cells, and we are not absolutely certain at this time that the same cell is involved because the cells are really very small and difficult to study by means of consecutive sections.

*Dr. Reichlin:* Dr. Knigge, what is the status of knowledge regarding the role of glia in the regulation of anterior pituitary secretion?

*Dr. Knigge:* Dr. Reichlin is referring to the hypothesis that my collaborators, Joseph, Scott, and others and I have been working on for 9 or 10 years. We postulated that in addition to the classic distribution of nerve fibers to the median eminence for release of hormone into the portal vessel, another route is possible. We suggested that hypophysiotrophic hormones could be released into the cerebrospinal fluid and then delivered to the median eminence. The mechanism for transporting the hormones across the median eminence into portal blood was proposed by us to be via cells called tanycytes which are specialized ependymal cells. We still believe that this is a valid hypothesis. If this thesis is correct, then hormone must be present at some time in the cerebrospinal fluid. There is no question that LRF, TRF, and somatostatin are demonstrable in the cerebrospinal fluid. So the hormone is there.

If it is there and is passing through the tanycyte, why don't we see it? Why don't we see it immunocytochemically? Why can't we demonstrate it?

Initially, we thought that we ought to be able to easily apply the histochemical procedure. We now know from our studies with LHRH that staining capacity greatly depends on the conformation of the antibody and on the condition of the antigen—whether it is free or bound as a prohormone or with a carrier protein. We are actively looking at the possibility that there are specific receptors for LRF in the tanycytes (as in pituitary cells), both externally on their membranes as well as internally on their granules. Dr. Sternberger, for example, working in collaboration with us, is developing rather remarkable information that the staining on the cell or in the granule depends on the form and the way in which the hormone is bound in those positions. We are therefore taking the stand that if the hormone is present in the CSF and if it is going through tanycytes, it is possible that the moment the hormone enters the cell from the ventricular surface it is bound to some carrier. It is possible that we are simply not just seeing it with our present conventional immunocytochemical techniques.

*Dr. Reichlin:* Would Dr. Zimmerman comment on his report that mouse tanycytes stain for LHRH?

*Dr. Zimmerman:* We have repeated our earlier reported studies on LHRH in mouse tanycytes, and Dr. Silverman has published a paper on this. We do have LHRH immunoactivity in mouse tanycytes. This is the only species we have seen it in, although we have not seen it in all the mice studied. Among the proofs of specificity is the finding that this reaction is prevented by treating the antiserum with excess LHRH.

In addition, we have found neurophysin immunoreactivity in rat tanycytes, a finding confirmed by Dr. Stillman in my laboratory, using absorption controls. We feel certain that we have demonstrated at least these two peptides in tanycytes on some occasions in some species.

*The Hypothalamus,* edited by S. Reichlin,
R. J. Baldessarini, and J. B. Martin,
Raven Press, New York, © 1978.

# Biochemical and Physiological Correlates of Hypothalamic Peptides. The New Endocrinology of the Neuron

## Roger Guillemin

### HISTORICAL MICROSURVEY BY WAY OF AN INTRODUCTION

When the Association for Research in Nervous and Mental Diseases called its meeting on the hypothalamus in December, 1939, the milestone volume that appeared a few months later contained what I will call incipient evidence for a hypothalamic participation in the regulation of pituitary functions both neurohypophysial and adenohypophysial—not much more than that. If there were a control by the hypothalamus on the secretion of adenohypophysial hormones, Uotila's paper (80) proposed that it would be a partial one. There was no evidence of what its nature would be, whether neuronal in nature or neurohumoral. There was, however, in that volume the remarkable chapter by Ernst and Berta Scharrer (71): based on their morphological observations, these two investigators were unquestionably proposing that some neurons showed signs of secretory activity of the type observed in classic endocrine cells known to secrete peptides or proteins. In view of what has happened in the intervening years between that meeting of 1939 and the one of today, it is fitting indeed that Berta Scharrer should be recognized today for the remarkable observations made by her and her late husband. Much of the material underlying the subtitle of this lecture, "The New Endocrinology of the Neuron," was already in its formative stages in the lecture the Scharrers presented at the 1939 meeting.

Today we know that the ventral hypothalamus secretes several peptides endowed with hypophysiotropic activity. They are thus involved in the immediate control of the secretory functions of the adenohypophysis. We also know that the hormones of the posterior pituitary, involved in the control of water metabolism and producing contractions of the uterus [the antidiuretic hormone (ADH), or vasopressin and oxytocin], are actually two small, closely related nonapeptides secreted by specialized cells of the anterior hypothalamus. The hypothalamic peptides with hypophysiotropic

---

*Present address:* Laboratories for Neuroendocrinology, Salk Institute for Biological Studies, La Jolla, California 92037

activity which have been isolated, characterized, and reproduced by synthesis are the thyrotropin releasing factor (ZHP),[1] the gonadotropin releasing factor or LRF (ZHWSYGLRPG), and somatostatin (AGCKNFFW-KTFTSC), which inhibits the secretion of growth hormone and thyrotropin. The long and arduous road from that early meeting in 1939 to the isolation of TRF in 1968 and its characterization and synthesis in 1969 by my laboratory then at the Baylor College of Medicine in Houston involved collecting millions of fragments of the brain and dissecting the hypothalamus—50 tons of it—from which we eventually isolated 1 mg of TRF. Characterization of TRF (13–15) was the turning point which separated doubt—and often confusion—from unquestionable knowledge in that field; it contributed most importantly to the establishment of neuroendocrinology as a true science.

Then followed determination of the structure of porcine TRF (60), the isolation and characterization of porcine LRF in 1971 by Schally's group (56), then of ovine LRF by our group (12). A year later, we described (82) the first analogue of LRF with partial agonist and antagonist activities, as well as the first analogue of TRF with increased potency (67). One more year, and we isolated, characterized, and synthesized somatostatin (9). Analogues of LRF are now available with increased potency and extended duration of activity as well as powerful analogues of LRF with antagonist properties. Unexpectedly, TRF was shown to stimulate the secretion of prolactin (76); also to be present in parts of the brain other than the hypothalamus and to exert profound effects in the central nervous system both on electrophysiological parameters and on behavioral patterns. Somatostatin was shown to inhibit not only the secretion of growth hormone and thyrotropin by the pituitary but also that of glucagon, insulin, gastrin, and acetylcholine by direct peripheral actions. Analogues of somatostatin have recently been described with increased potency at the level of all its target organs; others have been prepared with remarkably dissociated activities at the level of several of the target tissues. Somatostatin was found, not only in the extrahypothalamic central nervous system, like TRF, but also in specialized cells of the endocrine pancreas, the gastric mucosa, the intestinal mucosa, and in neurons of the myenteric plexus (22,40). The still unresolved question of the existence and nature of a growth hormone releasing factor led me about a year ago to look into the nature of the endogenous ligand for the opiate receptors known to exist on brain synaptosomes: an endogenous

---

[1] Single letter code for amino acids, as in *Atlas of Protein Sequence and Structure, Vol. 5,* edited by M. O. Dayhoff, Georgetown University Press, 1972. A, Ala; C, Cys; D, Asp; E, Glu; F, Phe; G, Gly; H, His; I, Ile; K, Lys; L, Leu; M, Met; N, Asn; P, Pro; Q, Gln; R, Arg; S, Ser; T, Thr; V, Val; W, Trp; X, undetermined; Y, Tyr; Z, Pyroglutamic ac.

Other abbreviations throughout the text are: TRF, thyrotropin releasing factor; LRF, luteinizing hormone releasing factor; TSH, thyroid stimulating hormone; LH, luteinizing hormone; FSH, follicle stimulating hormone; GH, growth hormone; LPH, lipotropin; MSH, melanocyte stimulating hormone; ACTH, adrenocorticotropin; cyclic AMP, cyclic 3′,5′-adenosine monophosphate; CMC, carboxymethyl cellulose; CLIP, corticotrophin-like intermediate lobe peptide; VIP, vasoactive intestinal peptide; GIP, gastric inhibitory peptide.

morphine-like peptide might be involved in the release of GH since morphine releases GH. Less than a year ago the two enkephalins were isolated in Scotland (42), while at the Salk Institute I isolated several endorphins which with Ling and Burgus (36,51) we have characterized to be identical to several fragments of the β-lipotropin as are the enkephalins.

The clinical significance of these several discoveries has also become apparent, both in ways that were obvious from the start, and in others that were not so obvious. For instance, the use of TRF for the study of pituitary-thyroid functions; the use of LRF for the treatment of some types of infertility, including induction of ovulation; the use of somatostatin to show the role of glucagon in juvenile diabetes with the possibility of a role of somatostatin or one of its analogues in the treatment of that disease; recent observations have led to speculations that the newly characterized endorphins along with their likely precursor the β-lipotropin may play a role in the pathogenesis of mental illnesses.

I will review here some of the major achievements in the biochemistry and physiology of the characterized hypothalamic peptides, essentially to draw from these many new facts, the new concepts and working hypotheses that as a physiologist I see in them. There will be ample reference to the specific publications in which procedures or other technical details are described.

## THYROTROPIN RELEASING FACTOR AND LUTEINIZING HORMONE RELEASING FACTOR

Of the many analogues of TRF which have been synthesized and studied biologically, the only one which has an increased specific activity over that of the native compound is one described by our group and synthesized by Rivier et al. (84) a few years ago. It is the analogue [3N-methyl-his]-TRF. Its specific activity is approximately 10 times that of the native molecule, on the secretion of TSH as well as of prolactin. Of the several hundreds of TRF analogues synthesized, none has been found so far to be even a partial antagonist. They are all agonists with full intrinsic activity but variable specific activity; no true antagonist of TRF has been reported.

In contradistinction to TRF, antagonist as well as extremely potent agonist analogues of LRF have been prepared by a number of laboratories. We now have available preparations of a series of what we may accurately call "super-LRFs," analogues which have as much as 150 times the specific activity of the native compound. In fact, in certain assays such as ovulation, they may have 1,000 times the specific activity of the native peptide. All the agonist-analogues of super-LRFs possess structural variations around two major modifications of the amino acid sequence of native LRF: They all have a modification of the C-terminal glycine, as originally reported for a series of analogues by Fujino et al. (28). The Fujino modification consists of deletion of $Gly^{10}$-$NH_2$ and replacement by primary or secondary amide on the (now C-terminal) $Pro^9$. In addition to the Fujino modification, they have an additional modification at the Gly$^6$ position by substitution of one of several D-amino acids as originally discovered by Monahan and co-workers (57) in our laboratories. The most potent of the LRF-analogue agonists are [D-Trp$^6$]-LRF; des-Gly$^{10}$-[D-Trp$^6$-Pro$^9$-N-Et]-LRF (83); and [D-Leu$^6$, Pro$^9$-N-Et]-LRF (85).

In an *in vitro* assay in which the peptides stimulate release of LH and FSH by surviving adenohypophysial cells in monolayer cultures, these analogues of LRF have a specific activity 50 to 100 times greater than that of the synthetic replicate of native LRF. There is no evidence of dissociation of the specific activity for the release of LH from that of FSH. All agonist analogues release LH and FSH in the same ratio (in that particular assay system) as does native LRF. Probably because of their much greater specific activity, when given in doses identical in weight to the reference doses of LRF, the super-LRFs are remarkably long acting. Whereas the elevated secretion of LH (or FSH) induced by LRF is returned to normal in 60 min, identical amounts in weight of [D-Trp$^6$-des-Gly$^{10}$]-N-Et-LRF lead to statistically elevated levels of LH up to 24 hr in several *in vivo* preparations, including man. These analogues are ideal agents to stimulate ovulation (85). Marks and Stern (55) have reported that these analogues are considerably more resistant than the native structures to degradation by tissue enzymes.

Rivier and Vale in our laboratories have recently observed that the injection of 1 to 10 $\mu$g of the analogues [D-Trp$^6$-des-Gly$^{10}$]-N-ethylamide-LRF in pregnant rats, either once or on consecutive days, over the first 7 days of gestation causes resorption of the fetuses and prevents normal pregnancy 83a. Johnson et al. (43) have reported similar observations. The mechanism involved in these effects is not clear at the moment.

All of the antagonist LRF-analogues as originally found by our group (82) or as later reported by others have deletion or a D-amino acid substitution of His$^2$. For reasons not clearly understood, addition of the Fujino modification on the C-terminal (28) does not increase the specific activity (as antagonists) of the antagonist analogues. Administered simultaneously with LRF, the antagonist analogues inhibit LRF in weight ratios ranging from 5:1 to 15:1. The most potent of these antagonists inhibit activity of LRF not only *in vitro*, but also in various tests *in vivo*. They inhibit the release of LH and FSH induced by an acute dose of LRF; they also inhibit endogenous release of LH-FSH and thus prevent ovulation in laboratory animals. The clinical testing of some of these LRF antagonists prepared in our laboratory has recently started in collaboration with Yen at the University of California in San Diego.

## SOMATOSTATIN

It is now well recognized that somatostatin has many biological effects other than the one on the basis of which we isolated it in extracts of the hypothalamus, i.e., as an inhibitor of the secretion of growth hormone (9). Somatostatin inhibits the secretion of thyrotropin, but not prolactin, normally stimulated by TRF; it also inhibits the secretion of glucagon, insulin, gastrin, and secretin by acting directly on the secretory cells of these peptides. I have recently shown that somatostatin also inhibits the secretion of acetylcholine from the (electrically stimulated) myenteric plexus of the guinea pig ileum. This may explain in part the inhibitory effects of somatostatin on gut contraction *in vivo* and *in vitro*.

It is also now well recognized that somatostatin occurs in many locations other than the hypothalamus, from which we originally isolated it. Somatostatin has been found in neuronal elements and axonal fibers in multiple locations in the central nervous system, including the spinal cord. It has been found also in discrete secretory cells of classic epithelial appearance in all the parts of the stomach, gut, and pancreas in which it had been first recognized to have an inhibitory effect (34,81).

Somatostatin does not inhibit indiscriminately the secretion of everything or anything. For instance, somatostatin does not inhibit the secretion of prolactin concomitant to that of thyrotropin when stimulated by a dose of TRF; this is true *in vivo* with normal animals or *in vitro* with normal pituitary tissue. Somatostatin does not inhibit the secretion of the gonadotropins (LH or FSH), calcitonin, ACTH in normal animals or from normal pituitary tissues *in vitro;* it does not inhibit the secretion of steroids from adrenal cortex or gonads under any known circumstances. Regarding the secretion of polypeptides or proteins from abnormal tissues of experimental or clinical sources, such as pituitary adenomas, gastrinomas, insulinomas, or other secreting tumors, somatostatin has been shown to be inhibitory according to its normal pattern of activity or in some cases indiscriminantly. The latter must reflect one of the differences between normal and neoplastic tissue. I have long thought of acromegaly as a disease of the plasma membrane receptors of the somatotrophs, in which any and all stimuli attaching somehow to these abnormal receptors can induce the secretion of growth hormone. This is in keeping with the observation that TRF, or LRF, can stimulate release of growth hormone from the pituitaries of acromegalics, although that does not happen with normal tissues.

Clinical studies have confirmed, in man, all observations obtained in the laboratory. The powerful inhibitory effects of somatostatin on the secretion not only of growth hormone but also of insulin and glucagon have led to extensive studies over the last 3 years of a possible role of somatostatin in the management or treatment of juvenile diabetes. First of all, the ability of somatostatin to inhibit insulin and glucagon secretion has provided a useful tool for studying the physiological and pathological effects of these hormones on human metabolism. Infusion of somatostatin lowers plasma glucose levels in normal man despite concomitant lowering of both plasma insulin and glucagon levels (1,30,59). These observations provided the first clear-cut evidence that glucagon has an important physiological role in human carbohydrate homeostasis. Somatostatin itself has no direct effect on either hepatic glucose production or peripheral glucose utilization, since the fall in plasma glucose levels could be prevented by exogenous glucagon (31).

In juvenile-type diabetics, somatostatin diminishes fasting hyperglycemia by as much as 50% in the complete absence of circulating insulin (30,31). Although somatostatin impairs carbohydrate tolerance after oral or intravenous glucose challenges in normal man by inhibiting insulin secretion, carbohydrate tolerance after ingestion of balanced meals is improved in patients with insulin-dependent diabetes mellitus through the suppression of excessive glucagon responses (30,31). The combination of somatostatin and a suboptimal amount of exogenous insulin (which by itself had prevented neither excessive hyperglycemia nor hyperglucagonemia in response to meals) completely prevents plasma glucose levels from rising after meal ingestion in insulin-dependent diabetics (30). Through its suppression of glucagon and growth hormone secretion, somatostatin has also been shown to moderate or prevent completely the development of diabetic ketoacidosis after the acute withdrawal of insulin from patients with insulin-dependent diabetes mellitus (30).

At the moment, clinical studies with somatostatin as provided by our group at the Salk Institute are proceeding in several clinical centers in the United States with the concurrence of the FDA. Ascertaining the high purity of the peptides for such clinical studies is of primary importance. It implies not only absence of peptides other than somatostatin as can be generated during the process of synthesis but also absence of any contaminant, organic or inorganic, possibly remaining from the process of synthesis. Such precautions in characterizing the peptides for clinical studies are even more important when dealing with a molecule such as somatostatin with multiple biological activities: "side effects" attributed to the preparation of somatostatin because they resemble one or another of the multiple effects of the peptide could be due to some contaminant unless carefully excluded. In that regard, our laboratory has always not only been demanding in terms of the criteria of purity requested for such clinical grade peptides; we also have devised the methodology, such as high-pressure liquid chromatography (15a), routinely to ascertain such purity.

From the foregoing description of the ability of somatostatin to inhibit the secretion of various hormones, it would appear that somatostatin may be of therapeutic use in certain clinical conditions such as acromegaly, pancreatic islet cell tumors, and diabetes mellitus. With regard to endocrine tumors, it must be emphasized that although somatostatin will inhibit hormone secretion by these tissues, it would not be expected to diminish tumor growth. Thus, in these conditions it is unlikely that somatostatin will find use other than as a symptomatic or temporizing measure.

In diabetes mellitus, however, somatostatin might be of considerable clinical value. First, it has already been demonstrated that it can acutely improve fasting as well as postprandial hyperglycemia in insulin-requiring diabetics by inhibiting glucagon secretion. Second, since growth hormone has been implicated in the development of diabetic retinopathy, the inhibition of growth hormone secretion by somatostatin may lessen this complication of diabetes. Finally, through suppression of both growth hormone and glucagon secretion, somatostatin may prevent or diminish the severity of diabetic ketoacidosis and find application in "brittle diabetes." These optimistic expectations must be considered with the facts that the multiple effects of somatostatin on hormone secretions and its short duration of action make its clinical use impractical at the present time and that its long-term effectiveness and safety have not been established as yet. Regarding the clinical use of somatostatin, see the recent review by Guillemin and Gerich (34).

With the considerable interest in somatostatin as a part of the treatment of diabetics, "improved" analogues of somatostatin would be highly desirable. Analogues of somatostatin have been prepared in attempts to obtain substances of longer duration of activity than the native form of somatostatin; this has not been very successful so far. Other analogues have been sought that would have dissociated biological activities on one or more of the multiple targets of somatostatin. Remarkable results have recently been obtained. The first such analogue was reported by the Wyeth Research Laboratories: [des-Asn$^5$]-somatostatin has approximately 4%, 10%, and 1% the activity of somatostatin to inhibit secretion of growth hormone, insulin, and glucagon, respectively (70). Although such an analogue is not of clinical interest, it showed that dissociation of the biological activities of the native somatostatin on three of its receptors could be achieved. The most interesting analogues with dissociated activities reported so far, all prepared and studied by Rivier, Brown, and Vale in our laboratories, are [D-Ser$^{13}$]-somatostatin, [D-Cys$^{14}$]-somatostatin, and [D-Trp$^8$, D-Cys$^{14}$]-somatostatin. When compared to somatostatin, the latter compound has ratios of activity of 300%, 10%, and 100% to inhibit the secretions of growth hormone, insulin, and glucagon, respectively (10). These and other analogues are obviously of clinical interest.

## THE ENDORPHINS

The recent demonstration of the existence in the brain of vertebrates of (synaptosomal) opiate receptors has led to the search for what has been termed the endogenous-ligand(s) of these opiate receptors. The generic name *endorphins* (from endogenous and morphine) was proposed for these (then hypothetical) substances by Eric Simon and will be used here. Some

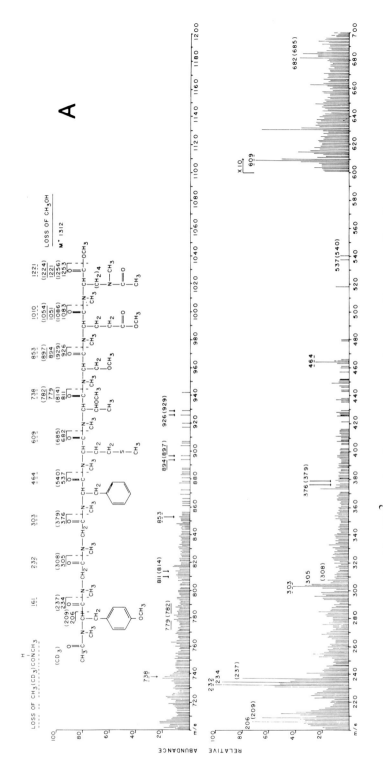

α-endorphin. NH₂-terminal fragment H - Tyr - Gly - Gly - Phe - Met - Thr - Ser

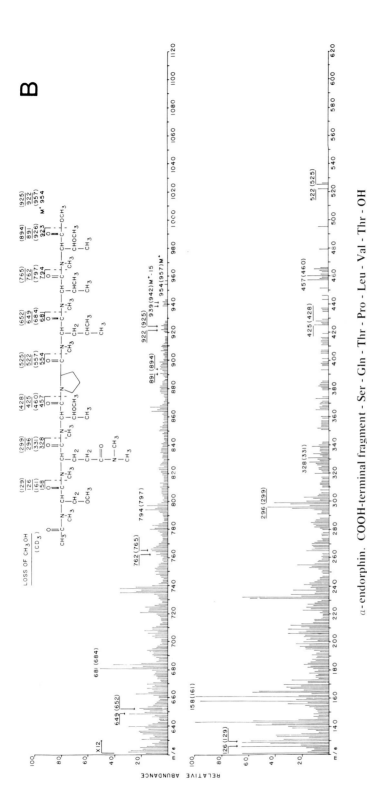

**FIG. 1.** Mass spectra of α-endorphin after trypsin digestion, acetic and deuterioacetic anhydride acetylation, and permethylation. The sequences are: **A**: NH²-terminal fragment H-Tyr-Gly-Gly-Phe-Met-Thr-Ser-; **B**: COOH-terminal fragment H-Ser-Gln-Thr-Pro-Leu-Val-Thr-OH. (From ref. 51.)

time last year, we became interested in these early observations. Besides the challenge of characterizing an endogenous substance as the ligand of the brain opiate receptors, we could not ignore the possibility that endorphins, like morphine, might stimulate the secretion of growth hormone. The nature of the growth hormone releasing factor remains unknown; endorphins might have been either releasing factors for growth hormone or involved in the physiological mechanisms controlling the secretion of growth hormone and possibly prolactin. I thus decided to engage in the isolation and characterization of the endogenous ligand(s) for the opiate receptors. As will be seen in this short review, the isolation of these endogenous ligands of the opiate receptors turned out to be a relatively simple problem to which a solution was indeed provided in less than a couple of months of effort.

### Isolation of the Endorphins: Chemical Characterization

Dilute acetic acid-methanol extracts of whole brain (ox, pig, rat) were confirmed to contain substances presumably peptidic in nature, with naloxone-reversible, morphine-like activity in the bioassay using the myenteric plexus longitudinal muscle of the guinea pig ileum. Evidence of such biological activity in our laboratory was in agreement with earlier results of Hughes (41), Terenius and Wahlstrom (78), Teschemacher et al. (79), and Pasternak et al. (62). Searching for an enriched source of endorphins in available concentrates from our earlier efforts toward the isolation of CRF, TRF, LRF, and somatostatin, it was recognized that acetic acid-methanol extracts of porcine hypothalamus-neurohypophysis contained much greater concentrations of the morphine-like activity than extracts of whole brain. From such an extract of approximately 250,000 fragments of pig hypothalamus-neurohypophysis we have isolated several oligopeptides (endorphins) with opioid activity (36,48,51). The isolation procedure involved successively gel filtration, ion-exchange chromatography, liquid partition chromatography, and high-pressure liquid chromatography (36, 48,51). Met$^5$-enkephalin and Leu$^5$-enkephalin recently isolated by Hughes and co-workers (42) have not been observed in these extracts. The primary structure of α-*endorphin* was established by mass spectrometry and classic Edman degradation of the enzymatically cleaved peptide and is H-Tyr-Gly-Gly-Phe-Met-Thr-Ser-Glu-Lys-Ser-Gln-Thr-Pro-Leu-Val-Thr-OH (Fig. 1). The primary structure of γ-*endorphin* was similarly established by mass spectrometry and by Edman degradation: γ-endorphin has the same primary structure as α-endorphin with one additional Leu as the COOH-terminal residue in position 17 (Fig. 2). Thus Met-enkephalin is the N-terminal pentapeptide of the endorphins, which have the same amino acid sequence as β-lipotropin [61–76] and [61–77]. β-LPH[61–91],

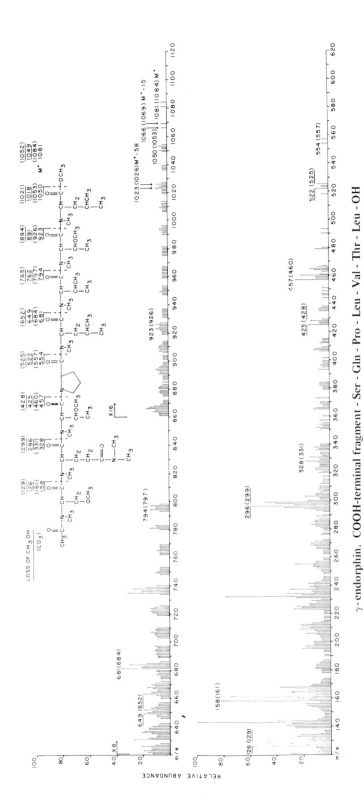

**FIG. 2.** Mass spectra of γ-endorphin after trypsin digestion, acetic and deuterioacetic anhydride acetylation, and permethylation. Only the COOH-terminal fragment resulting from trypsin cleavage is shown. The sequence is: H-Ser-Gln-Pro-Leu-Val-Thr-Leu-OH. (From ref. 51.)

a fragment of β-LPH isolated earlier on the basis of its chemical characteristics (7,50), was shown later to have opiate-like activity (6,19) and has been named β-endorphin (50). Recently we have isolated from the same starting material of hypothalamus-neurohypophysis origin from which we originally isolated α- and γ-endorphin two peptides characterized by amino acid composition as *β-endorphin* (β-LPH[61–91]) and *δ-endorphin* (β-LPH[61–87]). No effort was made to obtain the amino acid sequences of these two samples. The synthetic replicates of these two polypeptides have the same chromatographic behavior as the native materials in several systems.

Recently Goldstein and his collaborators (33,68) have stated that extracts of the pituitary gland contain peptides with morphinomimetic activity that would be different from that of the characterized endorphins and which reportedly have greater specific activity. On the other hand, Chrétien et al. (18) have stated that the morphinomimetic activity of extracts of sheep whole pituitary glands can be accounted for by the exclusive presence of only β-endorphin, which they have indeed isolated and characterized in such extracts. With such conflicting statements and in view of the fact that the starting material from which we isolated α, β, γ, and δ-endorphins is known to contain tissues of the posterior pituitary gland, I decided to investigate the question of the nature of pituitary endorphins with similar starting material as used by Goldstein et al. and Chrétien et al., i.e., frozen whole pituitary glands.

The starting material was 1.2 kg of frozen sheep whole pituitary glands. The whole sample was extracted exactly as described by Li (49) for the isolation of β-LPH; the same methodology was carefully followed through the stage of ion exchange on carboxymethyl cellulose (CMC) using the same gradients described by Li (Table 1). Morphinomimetic activity was ascertained by bioassay with the guinea pig myenteric plexus-longitudinal ideal muscle preparation, as well as by a recently developed radioimmunoassay for α-endorphin (see below) with a well-characterized antiserum. At the CMC stage, an elution pattern identical to that reported earlier by Li was obtained (Fig. 3). Fraction 190–220 corresponds to expected β-lipotropin. Biological assays for morphinomimetic activity of the whole effluent of the CMC chromatography revealed only two zones of activity (see Fig. 3): The very sharp fraction 328–333 (see Fig. 3) was further resolved by high-pressure liquid chromatography (HPLC), showing essentially one peak of biological activity. Amino acid composition of this material, now homogeneous, was identical to that of ovine β-endorphin (β-LPH[61–91]); its retention time on HPLC was identical to that of pure synthetic ovine β-endorphin. It is thus concluded that this material in the extract of whole pituitary gland is β-endorphin. Fraction 348–355 (Fig. 3) was similarly further purified by HPLC. At that stage it contained only one zone of biological activity. The active component is not fully characterized as yet; partial results are consistent with the statement that the material is closely

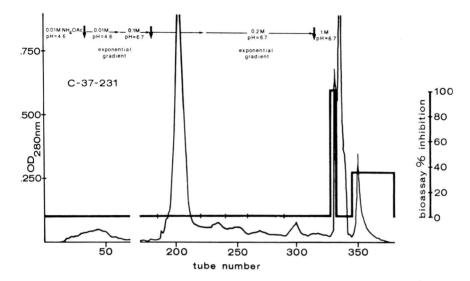

**FIG. 3.** Chromatography on carboxymethyl cellulose of material from stage 5 (Table 1) remaining in the dialysis bag. Whole (ovine) pituitary glands. Fraction 190–220 corresponds to β-lipotropin. Morphinomimetic activity is seen in two fractions of the effluent: fraction 328–333 after HPLC has a single biologically active component characterized as β-endorphin (β-LPH[61–91]); fraction 348–355 contains also a single biologically active component, not fully characterized at the moment but appearing to be related to β-endorphin and of lower specific activity.

related to but not identical with β-endorphin. Its specific activity in the bioassay is of the order of that of β-endorphin. In addition, α-endorphin was located, using the radioimmunoassay, in the external dialysis bath of step 4 as in the Li's procedure (see Table 1). α-Endorphin would indeed be ex-

TABLE 1. *Steps in the purification of LPH and endorphins from whole sheep pituitary glands*

1. HCl-acetone extraction
2. Acetone precipitation
3. NaCl precipitation
4. Dialysis
   Out: α-endorphin
   Retained: next step
5. Ion-exchange chromatography on carboxymethyl cellulose
   A: β-LPH fraction
   B: β-Endorphin fraction
6. Fraction A, filtration on Sephadex-G 50: homogenous β-LPH
   Fraction B, high-pressure liquid chromatography: homogenous β-endorphin

Purification scheme for β-LPH as described by Li (49), showing also localization of α-endorphin and use of high-pressure liquid chromatography for isolation of β-endorphin.

pected to dialyze through the Spectrapor 1 membrane (molecular weight cutoff = 6,000 to 8,000 daltons) used here.

Thus, with an extraction method and starting material very similar to that used by Goldstein et al. (14) and in contradistinction to the statements by these investigators, we have no evidence, in the pituitary gland, of peptides endowed of morphinomimetic activity other than those related to the active fragments of β-lipotropin. Moreover, our results showing the presence of α-endorphin in the external dialysis bath at stage 4 of the same extraction as used by Chrétien et al. (18) explain the statement of these authors (of not observing material other than β-endorphin) since they had apparently not tested the dialysate in the purification scheme. Moreover, the small amounts of α-endorphin as measured in that fraction by the radioimmunoassay would probably have made it difficult to detect by the much less sensitive bioassay, the only detection method used by Chrétien et al.

### Immunocytochemical Localization of Endorphins

Cells of pars intermedia of the rat pituitary gland stain brightly and uniformly with antisera to α-endorphin or β-endorphin (Fig. 4). Cells of the pars intermedia show reactivity within small granules evenly distributed throughout the cytoplasm. In the pars distalis (adenohypophysis), discrete cells give bright fluorescence with either antiserum; the fluorescence observed is associated with granules somewhat larger than those of pars intermedia cells and located only at the periphery of the cells. The endorphin-reactive adenohypophysial cells appear often to be adjacent to blood vessels. The pars nervosa (neurohypophysis, posterior lobe) and the interlobular stroma of pars intermedia are completely unstained. The results are consistent in all animals (rats) so prepared, also using one cat and with pituitary glands from several cows. The fluorescence of all cells disappears entirely when the antisera are incubated with the respective peptide antigens. Discrete nerve fibers in the hypothalamus stain also by this technique with antiserum to either α-endorphin or β-endorphin. The immunofluorescence staining of nerve fibers in the hypothalamus and other rat brain areas such as the hippocampus has been observed as late as 7 weeks after total hypophysectomy with no obvious difference from what was seen in normal animals (5). These results strongly support the suggestion that some cellular elements in the brain have the ability to synthesize β-lipotropin as the precursor of the endorphins; alternatively, they may possess enzymes that would cleave β-lipotropin as a prohormone of the endorphins, available either from its pituitary origin or from some other site in the brain. The only source of β-lipotropin reported so far has been the pituitary gland.

The results presented here raise the possibility that α- and β-endorphins may exist simultaneously with β-lipotropin in the same pituitary cells of either intermediate or anterior lobe. They could exist in the same or in differ-

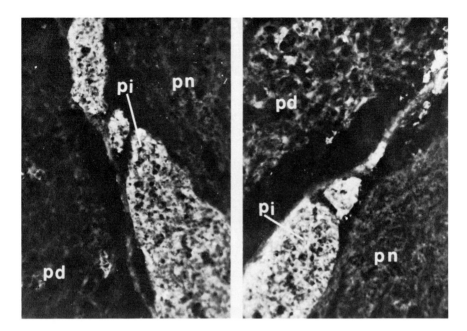

**FIG. 4. Left:** Immunocytochemical localization of α-endorphin in rat pituitary. Staining is confined to cells of the intermediate lobe (pi, pars intermedia) in which staining is present throughout the cytoplasm of every cell, but not in the cell nuclei or in the basement membrane between the lobules. Discrete cells in the anterior lobe (pd, pars distalis) are stained as well; these cells are located around capillaries, have a polygonal shape, and exhibit staining within large cytoplasmic granules. Note that the posterior pituitary (pn, pars nervosa) is completely unreacted. ×70. (See also ref. 5a.) **Right:** Immunocytochemical staining of the same pituitary as shown in an adjacent section reacted with immunoglobulin to β-endorphin. The staining is identical in distribution to that for anti-α-endorphin. Magnification 100×. (See also ref. 5a.)

ent compartments. Moreover, these same cells are known also to contain ACTH (1–39) as well as several fragments of ACTH such as CLIP (i.e., ACTH 18–39).

In view of the novel information relating the presence of α- and β-endorphin to cells of the pars intermedia and not of the neurohypophysis, presence of the endorphins in the crude extract "pitressin intermediate" used earlier to isolate α-, β-, and γ-endorphins (36,51) is explained by realizing that tissues of the intermediate lobe always remain attached to the posterior lobe as it is dissected away from the whole pituitary gland. In this respect Ling et al. (51) had reported that the extract used to isolate the endorphins had indeed melanophoretic activity—approximately 1/10 of the specific activity of pure α-MSH, which in retrospect should be realized to be non-negligible; it is further evidence for the presence of pars intermedia tissues in that starting material, originally considered to be composed of neurohypophysis, and attached pituitary stalk and hypothalamus.

## Relation of Endorphins to β-Lipotropin

So far, all morphinomimetic peptides isolated from natural sources on the basis of a bioassay or displacement assay for $^3$H-opiates on synaptosomal preparations, and chemically characterized, have all been related to a fragment of the C-terminal of the molecule of β-lipotropin, starting at Tyr$^{61}$. In the case of Leu$^5$-enkephalin, the relationship still holds for the sequence Tyr-Gly-Gly-Phe; no β-lipotropin with a Leu residue in position 65 has been observed.

β-LPH has no opioid activity by bioassay or radioligand assay (see Fig. 3, also ref. 48). Incubation of β-LPH at 37°C with the $10^5$ g supernatant of a neutral sucrose extract of rat brain generates opioid activity suggesting the presence of peptidase activity capable of generating one or several opioid peptides. Thus, β-LPH may be a prohormone for the opiate-like peptides (48). This would imply that the biogenesis of endorphins may be similar to that of angiotensin with cleaving enzymes available in the central nervous system. β-LPH[61–63] has no opioid activity at $10^{-4}$ M; β-LPH[61–64], β-LPH[61–65]-NH$_2$, (Met (0)$^{65}$)-β-LPH[61–65], β-LPH[61–69], β-LPH-[61–76], and β-LPH[61–91] all have opioid activity. β-LPH[61–65]-NH$_2$, β-LPH[61–65]-NEt, and all peptides larger than β-LPH[61–65] have longer duration of biological activity than Met-enkephalin in the guinea pig ileum-myenteric plexus bioassay. All these peptides were prepared by solid phase synthesis (see 52). β-Endorphin is by far the longest acting peptide when compared at equimolar ratios with all other fragments of the 61–91 COOH-fragment of β-LPH. In quantitative assays using the guinea pig ileum-myenteric plexus, β-endorphin is approximately five times more potent than Met$^5$-enkephalin; the two analogues of Met$^5$-enkephalin amidated on the C-terminal residue have also two to three times greater specific activity than the free acid form of the peptide, with 95% fiducial limits of the assays overlapping those of β-endorphin.

When tested in the myenteric plexus-ileum bioassay, β-LPH[61–64], Phe$^1$-Met$^5$-enkephalin, and (O-Me)Tyr$^1$-enkephalin, although of low specific activity when compared to Met$^5$-enkephalin, all have full intrinsic activity. [Arg-Tyr]$^1$-Met$^5$-enkephalin, i.e., β-LPH[60–65], is equipotent to Met$^5$-enkephalin. Thus an intact Tyr NH$_2$-terminal is not a requisite for full intrinsic activity. Acetyl-Tyr$^1$-Met$^5$-enkephalin has very low specific activity, which, however, cannot be further quantitated as the log-dose response function is totally divergent from that of α-endorphin or Met$^5$-enkephalin as reference standard (52).

A series of analogues of the endorphins was synthesized by Ling and further purified to high purity. Table 2 shows a list of some of these analogues with their biological activity in the myenteric plexus bioassay; it shows also the potency of the same analogues in an opiate-displacement assay from rat brain synaptosomes (48), normorphine being considered as

TABLE 2. *Biological potency in the guinea pig ileum-myenteric plexus assay in vitro (48) of synthetic replicates of isolated endorphins and of several analogues[a]*

| Analogue | Guinea pig ileum potency ($Met^5$ − Enk = 100) | $^3$H-etorphine displacement relative affinity (Normorphine = 100) |
|---|---|---|
| β-LPH[61–68] | 66 | |
| β-LPH[61–69] | 60 | |
| [b]β-LPH[61–76] ≡ α-endorphin | 36 | 33 |
| [b]β-LPH[61–77] ≡ γ-endorphin | 23 | 400 |
| β-LPH[61–79] | 37 | |
| [b]β-LPH[61–87] | 48 | 200 |
| [b]$β_p$-LPH[61–91] ≡ $β_p$-endorphin | 450 | 350 |
| [b]$β_o$-LPH[61–91] ≡ $β_o$-endorphin | 450 | |
| α-endorphin amide | 72 | |
| [$Gln^8$]-α-endorphin | 72 | |
| [$Leu^5$]-α-endorphin | 17 | 1 |
| γ-endorphin amide | 50 | |
| [$Leu^5$]-γ-endorphin | 9 | 540 |
| [$Leu^5$]-β-endorphin | 75 | 950 |
| $β_p$-LPH[66–91] | 0 | |
| $β_p$-LPH[62–91] | 0 | <1 |

[a] Also, relative affinity (normorphine = 100) of several of the same materials when tested in an opiate-displacement assay using rat brain synaptosomes (48).

[b] Molecules isolated from natural sources on the basis of the same bioassay for morphinomimetic activity as used here (N. Ling, L. H. Lazarus, S. Minick, and R. Guillemin, *unpublished observations*).

a reference standard. All these peptides have parallel competition curves when studied at 5 to 6 dose levels with the exception of β-LPH[62–91], which is definitely divergent from the other curves. Comparing the values obtained in the bioassay (myenteric plexus-longitudinal muscle) and the synaptosomal displacement assay, it is obvious that the two assay systems may give different potencies for the same peptide. Similar dissociations between potency results obtained in bioassays versus receptor assays have been observed in entirely different systems, for analogues of insulin, somatostatin, LRF, and TRF (82a). These results have been interpreted as reflecting differences between ability of the peptide to bind to receptors and ability to initiate the various steps involved in expressing the ultimate biological activity, after the ligand has attached to the receptor. At the moment, the values obtained for the endorphin analogues in the synaptosome displacement assays and shown in Table 2 are still preliminary. Should they be confirmed, they would lead to interesting speculations.

Of interest are the results observed with the analogues of α-, β-, and γ-endorphins in which a residue of leucine has been substituted for methionine in position 5 from the $NH^2$-terminus. [$Leu^5$]-β-endorphin and [$Leu^5$]-γ-endorphin appear to be considerably more potent than their native congeners in the synaptosomal displacement assay. One could speculate that the brain variety of endorphins might contain a residue of leucine in position 5. To this date, no [Leu$^{65}$]-β-lipotropin has been recognized and characterized. On the other hand, Hughes and collaborators (42) and later Simantov and Snyder (75) have isolated from brain extracts not only Met$^5$-enkephalin but also Leu$^5$-enkephalin. Leu$^5$-enkephalin might come from an allele of β-lipotropin of brain origin. The observed greater potency of [Leu$^5$]-β-endorphin over that of [Met$^5$]-β-endorphin is of interest as such a molecule could explain some of the results of Goldstein et al. (33) — such as resistance of the biological activity to CnBr treatment of the extract (which would oxydize methionine residues and destroy biological activity). The presence of such a molecule as [Leu$^5$]-γ-endorphin in extracts of the brain or any other tissue will not be ascertained until it is isolated from these tissues and properly characterized. This is no simple endeavor. Specific antisera to [Leu$^5$]-β-endorphin (should they be possible) might help approach the question in a manner somewhat easier although not definitive.

### Release of Pituitary Hormones by Endorphins

One of our original interests in engaging in the isolation and characterization of the endorphins was that the opiate-like peptides might be involved in the secretion of pituitary hormones, particularly growth hormone and prolactin, long known to be acutely released following injection of morphine.

We have recently shown with Rivier et al. (66) that β-endorphin is a potent releaser of growth hormone and prolactin when administered to rats by intracisternal injection. Plasma levels of growth hormone or prolactin were measured by radioimmunoassays. These effects were prevented by prior administration of naloxone. The endorphins are not active directly at the level of the pituitary cells: they show no effect, even in large doses, when added directly to monolayer cultures of (rat) pituitary cells. Thus the hypophysiotropic effects of the endorphins, like those of the opiate alkaloids, are mediated by some structure in the central nervous system and are not directly at the level of the adenohypophysis.

### Neuronal Actions of Endorphins and Enkephalins Among Brain Regions: A Comparative Microiontophoretic Study

The existence of endogenous peptides with opiate-like actions suggests that these substances may function as neuromodulators or neurotransmitters in the CNS. Indeed, recent iontophoretic studies have shown that the enkephalins can modify the excitability of a variety of neurons in the CNS. Most neurons tested were inhibited by these peptides (21,26,39,58,90), although Renshaw cells responded with an excitation (20). Studies have recently appeared exploring systematically the sensitivity of neurons to the endorphins or reporting a systematic regional survey of neurons responsive to the peptides.

## Characterization of Responses

In most of the brain regions the majority of neurons encountered in the study of Nicol et al. (61) were inhibited in a dose-dependent and reversible fashion by endorphins or normorphine (Table 3, Figs. 5–8). An exception to

TABLE 3. Neuronal effects of opioid peptides

| Region (cell type) | Met-enkephalin % Exc. % Inh. (N) | | $\beta$-Endorphin % Exc. % Inh. (N) | | Normorphine % Exc. % Inh. (N) | |
|---|---|---|---|---|---|---|
| Cerebellum (Purkinje) | 18 | 21 (34) | 23 | 23 (13) | 20 | 60 (5) |
| Cerebral cortex (unidentified) | 1 | 79 (58) | 25 | 48 (44) | 26 | 52 (27) |
| Brainstem (Lateral ret. nuc. +) | 3 | 47 (113) | 23 | 45 (35) | 10 | 75 (20) |
| Caudate nucleus (unidentified) | 0 | 83 (18) | 10 | 86 (35) | 9 | 73 (20) |
| Thalamus (unidentified) | 0 | 100 (15) | 0 | 100 (5) | 0 | 100 (4) |
| Hippocampus (pyramidal) | 90 | 5 (19) | 86 | 7 (14) | 92 | 0 (12) |

In each category the total number of cells tested and the percentage of this total that were inhibited or excited is shown [from Nicol et al. (61)].

this generalization was that almost all hippocampal pyramidal cells were excited by these agents (see below). The inhibitions to the pentapeptides were typically rapid in onset (10 to 30 sec), and neuronal activity recovered within 1 to 2 min after terminating the ejection current (Fig. 5). With spontaneously firing units, the responses to the endorphins and normorphine generally had a slower time course than the responses to the enkephalins (Fig. 5D). The inhibitory action of the peptides could also be demonstrated on glutamate-evoked activity (Fig. 6). The inhibition of glutamate-evoked responses followed a time course similar to that of the inhibition of spontaneous activity recorded in the same cell (Fig. 6A). This did not appear to be a specific antiglutamate effect, since the spontaneous firing was also depressed along with responses to glutamate. In all neurons studied sequentially, $\beta$-endorphin appeared to be more potent than $\alpha$-endorphin or Met$^5$-enkephalin (on a molar basis) and also to be considerably longer acting.

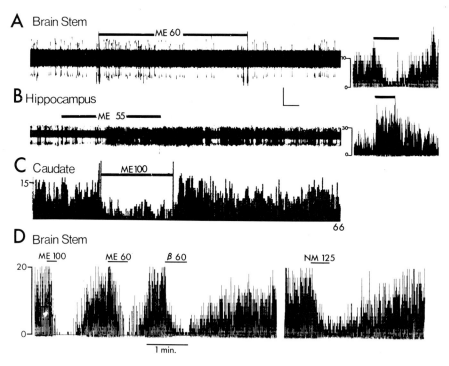

**FIG. 5.** Effect of opioid substances on the spontaneous activity of CNS neurons. **(A):** Oscilloscope record of the discharge of a brainstem neuron which is inhibited by 60 nA met-enkephalin (ME). The rate meter record (1 sec rate integration) of the same response is shown to the right **(B):** An excitatory response from a hippocampal neuron to met-enkephalin. The calibration in A is 159 μV and 4.7 sec and in B is 500 μV and 5.7 sec. **(C):** Computer-generated drug histogram of the inhibition of a caudate neuron to met-enkephalin. This histogram sums the spikes of 6 sweeps each with a duration of 66 sec. The number on the ordinate refers to counts per bin. **(D):** Comparison of the action of met-enkephalin, β-endorphin (B), and normorphine (NM) on the activity of a brainstem neuron. (From ref. 61.)

## Distribution of Responsive Cells

The highest proportion of tested cells inhibited by the peptides and normorphine was found in the caudate nucleus and the thalamus, whereas the cerebellum had the lowest percentage of inhibitions. Since the same electrodes were used to record Purkinje cells in the cerebellum and neurons in the brainstem, the greater sensitivity of brainstem neurons is particularly striking and not attributable to artifacts of the equipment. Furthermore, in the cerebellum approximately as many cells were excited as were depressed. In the brainstem, sensitive cells often occurred in clusters and a higher percentage of cells responded in the lateral reticular nucleus than in the medial reticular formation. In the cerebral cortex and brainstem a somewhat higher proportion of cells were excited by β-endorphin than by met-enkephalin.

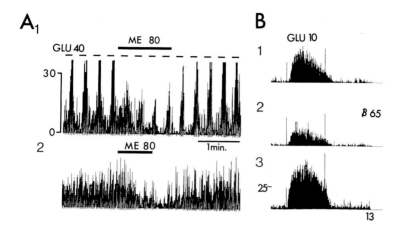

**FIG. 6.** Cerebral cortex. Opioid peptides decrease glutamate excitation. **(A):** Comparison of the effect of met-enkephalin (ME) on glutamate-evoked activity ($A_1$) and spontaneous activity ($A_2$) in the same cell. Note that the time course of inhibition is the same in both. During each of the horizontal bars 40 nA of glutamate was applied. The vertical gain and time scale are the same in both records. **(B):** The excitatory action of glutamate as recorded by serial computer-generated drug histograms ($B_1$) is reduced by $\beta$-endorphin ($B_2$). $B_3$ shows the recovery of the glutamate excitation. Each histogram is composed of 10 sweeps, each of 13 sec duration. The ordinate calibration (25 counts per bin) in $B_3$ also applies to the other responses in B.

Destruction of the dopaminergic nigral-striatal pathway with 6-hydroxydopamine did not significantly alter the predominantly inhibitory action of the peptides on caudate neurons. A surprising finding (61) was the potent excitatory effects of the peptides and normorphine on hippocampal pyramidal cells (Fig. 6B). The regional specificity of this excitatory action could be clearly demonstrated with the same electrode by recording from cells in the overlying cerebral cortex and the underlying thalamus during a single penetration. Thus, as the electrode was advanced through the cortex, cells responded with inhibition to the peptides. As soon as the electrode entered the pyramidal cell layer, only marked excitatory responses were observed. Further advancement of the electrode into the thalamus again revealed exclusively inhibitory responses.

No tachyphylaxis was observed to either the excitatory or the inhibitory action of the peptides in any of the regions examined, even though the peptides were often applied repeatedly to the same cell for periods in excess of 1 hr.

## Naloxone Antagonism of Peptide Effects

To determine whether the responses observed with the peptides were related to the activation of opiate receptors, we administered the specific opiate antagonist naloxone both by iontophoresis from an adjacent barrel of the microelectrode and by subcutaneous injections. Administered by either route, naloxone antagonized both the excitations and the inhibitions (Figs. 7 and 8).

Both the percentage of cells affected by the opiate-like peptides and the type of response encountered (excitation and inhibition) depended on the region of the CNS examined. Of all the regions examined the cerebellum had the fewest responsive cells, and both excitations and inhibitions were observed. This finding is consistent with the low enkephalin levels (74) and the low density of stereospecific opiate binding sites in the cerebellum (46). However, the fact that naloxone antagonized the responses indicates that opiate receptor mechanisms can be detected in the cerebellum with the iontophoretic technique. In cerebral cortex, brainstem, caudate nucleus, and thalamus, all areas reported to have a high density of stereospecific opiate binding sites (4), a higher proportion of cells responded to the peptides with naloxone-sensitive inhibition. These results agree with previous results on the enkephalins (8,26,29,39,88), but other investigators have had difficulty in blocking the peptide responses with naloxone (29,39).

Hippocampal pyramidal cells, unlike all other neurons examined in this study, were strongly excited by the peptides and these excitations were blocked by naloxone. A similar excitatory action, although of faster time course (20), has been found with enkephalin on Renshaw cells. Interestingly, both types of cells are known to be strongly excited by acetylcholine (4,20), and the opiate and peptide excitations of Renshaw cells have been closely linked to nicotinic receptors (20,23). The acetylcholine responses in the hippocampus have been both nicotinic and muscarinic properties (4). Further studies will be required to determine whether the excitation of hippocampal pyramidal cells by opiate-like peptides is related to endogenous cholinergic mechanisms in these regions.

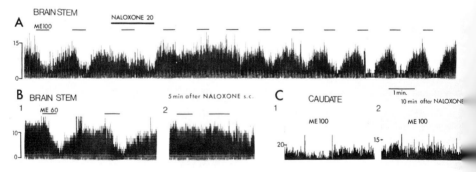

**FIG. 7.** Antagonism of met$^5$-enkephalin inhibition by naloxone. **(A):** A short iontophoretic application of naloxone reversibly blocks the inhibitory response of a brainstem. **(B):** A subcutaneous injection of 8 mg/kg of naloxone blocks the inhibitory response of another brainstem neuron. **(C):** A similar action of naloxone (8 mg/kg subcutaneously) in a caudate neuron, as documented by computer-generated histograms immediately before and 3 min after naloxone. A histogram is composed of 6 sweeps of 66 sec duration. (From ref. 61.)

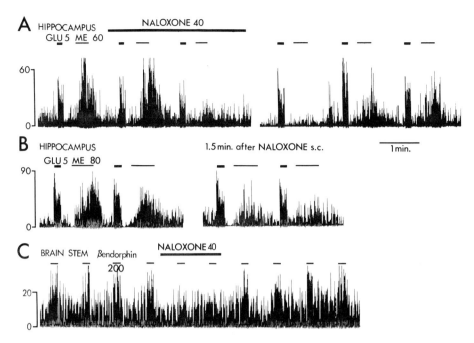

FIG. 8. Naloxone antagonism of excitatory responses to the opioid. **(A):** Iontophoresis of naloxone reversibly and selectively blocks the excitatory action of met-enkephalin on a hippocampal neuron. The break in the record is 8.5 min. **(B):** A subcutaneous injection of naloxone (10 mg/kg) also selectively antagonizes the met-enkephalin excitation of another hippocampal neuron. **(C):** The excitation of a brainstem neuron by $\beta$-endorphin is also reversibly blocked by iontophoretically applied naloxone. (From ref. 61.)

It has been suggested that opiate receptors may be present on presynaptic terminals (47), which raises the possibility that the neuronal responses observed with the iontophoretic technique may result from a presynaptic action altering the tonic synaptic input onto the neuron under investigation. Although such an action cannot be excluded, the fact that the postsynaptic excitatory action of glutamate is depressed by the peptides in addition to spontaneous activity indicates that the inhibitory action of the opiate-like peptides observed in this study is to a large extent a postsynaptic action. The mechanism underlying this postsynaptic effect is not entirely clear, although a block in the sodium conductance increase elicited by excitatory transmitters has been suggested (88).

All these effects of opiate-like peptides on neuronal activity taken with biochemical and histochemical evidence for their existence in brain are consistent with the hypothesis that these peptides are neurotransmitters in the CNS. When the cells of origin of these peptide-containing fibers have been determined, it may then be possible to proceed with studies into the effects on cellular activity and the secretion of the peptides in order to satisfy more completely the criteria for a neurotransmitter. Crucial points in such future analyses will be the questions of whether the endorphin- and enkephalin-containing fibers are mutually inclusive systems, and whether the size of the peptide released by neuronal activity is subject to physiological modulation. Although the endorphins and $\beta$-LPH may be prohormones for Met-enkephalin, there are at present no such candidates for Leu-enkephalin. The results presented here indicate to us that the cellular roles of endorphin and enkephalin peptides cannot now be generalized across all brain regions where they are found, and that no simple cellular action of any peptide will yield an integrative picture of the way in which opiate alkaloids produce complex analgesic, euphoric, and addictive responses.

### Behavioral Effects of Endorphins

The pharmacological properties of endorphins have so far been screened through application of tests *in vitro* or *in vivo* previously used to characterize opiate agonists and antagonists.

When injected into the cerebrospinal fluid, endorphins affect several behavioral and physiological measures in addition to responses to noxious agents, and each of the peptides exhibits different dose-effect profiles on these measures: $\beta$-endorphin induces a marked catatonic state (27) lasting for hours (5a) at molar doses 1/100 those at which $Met^5$-enkephalin transiently inhibits responses to noxious agents (3,16,53). This potent effect of a naturally occurring substance suggests its regulation could have etiological significance in mental illness.

Peptides were injected into the cerebrospinal fluid of rats either into the cisterna magna or through a permanently implanted lateral ventricular cannula. Other animals were injected similarly with saline vehicle, morphine sulfate, or normorphine chloride. After injections, rats were observed for changes of behavior; responses to noxious stimuli were tested before injections and at 5-min intervals after injections by evocation of corneal and eyelid reflexes, and by tail pinch and pin-prick evoked responses. Rectal temperature was monitored at 5- to 10-min intervals for 1 to 4 hr. In general, the overall effects of the individual peptides were similar, regardless of the route of injection into the cerebrospinal fluid, but no major behavioral effects or responses to noxious stimulation could be seen after intravenous injection in doses up to 1 mg/kg.

In terms of molar dose effectiveness, $\beta$-endorphin is clearly the most potent substance tested (Table 4). Within 5 to 10 min after injection, corneal reflexes disappeared and general locomotor activity became depressed; transient episodes of nystagmus could be seen in this period. Within 15 min, at doses as low as $3 \times 10^{-10}$ moles (administered intracisternally), animals showed a total lack of responsiveness to pin-prick or tail pinch stimuli. After 15 to 30 min, animals injected with $7.4 \times 10^{-9}$ moles of $\beta$-endorphin

TABLE 4. Dose-effect profiles of endorphin peptides on rats after injection into a lateral cerebral ventricle

| Substance | Rats (N) | Rigidity | | Loss of corneal reflex | | Loss of tail-pinch reflex | | Wet-dog shakes | |
|---|---|---|---|---|---|---|---|---|---|
| | | Dose (nmole) | Response | Dose (nmole) | Response | Dose (nmole) | Response | Dose (nmole) | Response |
| Met$^5$-enkephalin | 8 | 1,030 | None | 340–1,030 | + | 1,030 | None | 340–1,030 | + |
| α-Endorphin | 24 | 1,210 | None | 120–1,210 | + | 1,210 | None | 76–1,210 | + |
| γ-Endorphin | 9 | 281 | None | 281 | + | 281 | None | 110–281 | + |
| β-Endorphin | 9 | 7.4–14.9 | + | 3.0–14.9 | + | 3.0–14.9 | + | 3.0–14.9 | + |
| Morphine | 16 | 132 | None | 8–132 | + | 13.2–132 | + | 132 | " |
| Saline | 8 | | None | | None | | None | | None |

For each of the four test criteria (generalized rigidity, loss of corneal reflex, loss of tail pinch response, and elicitation of wet-dog shakes), the number of animals tested with each substance is indicated, as are the dose ranges. [See also Bloom, et al. (5a).]
"One of three animals tested with 5.4 nmole of morphine showed episodic shakes, but this effect was not obtained on repeated testing at this dose, or at doses above or below this dose.

began to exhibit a profound catatonic state (27) characterized by extreme generalized muscular rigidity, loss of the righting reflex, and total absence of spontaneous movement (5a). As a result of these effects, animals could be placed in and would retain abnormal body positions (Fig. 10) for indefinite periods. Respiratory movements shifted to the abdomen, and rectal temperature decreased. While in this state, the animals' eyes remained widely open (Fig. 9), with no spontaneous blinking, and showed loss of corneal and lid reflexes and often showed exophthalmos. With doses of $7.4 \times 10^{-9}$ moles, rats remained in this state for approximately $2\frac{1}{2}$ hr. Full spontaneous recovery then occurred rapidly, with no detectable aftereffects. All these actions of $\beta$-endorphin were reversed within seconds after intravenous injection of naloxone (1.0 mg/kg); after naloxone-induced recovery, rats frequently showed several episodes of "wet-dog" shakes even though they had no prior exposure to endorphins, to exogenous opiates, or to opiate antagonists. Rats given seven daily intracisternal injections of $14.9 \times 10^{-9}$ moles of $\beta$-endorphin continued to show the full set of responses and duration of action. However, 8 to 24 hr after as few as five daily injections, such animals also showed spontaneous wet-dog shakes.

The catatonic state induced by $\beta$-endorphin was not observed with the other endorphin peptides, even at considerably higher doses (Table 4). Doses of morphine or normorphine ($8 \times 10^{-9}$ to $132 \times 10^{-9}$ moles) which suppressed responses to noxious stimuli, did produce marked sedation with fixed open eyes and loss of corneal reflexes; but such animals retained considerable spontaneous locomotion (until doses of $1.3 \times 10^{-7}$ moles), and showed no muscular rigidity and only a moderate decrease in rectal temperature (0.8 to 1.2°C). At very high doses of $\alpha$-endorphin, $\gamma$-endorphin, or Met⁵-enkephalin, transient losses of corneal reflexes were also observed, and $\alpha$-endorphin seemed more potent in this regard than either $\gamma$-endorphin or Met⁵-enkephalin. No significant depressions of responsiveness to tail

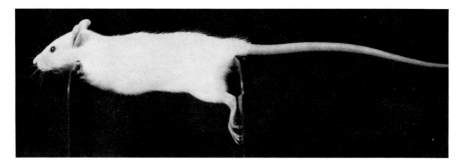

**FIG. 9.** Thirty minutes after the intracisternal injection of $\beta$-endorphin ($14.9 \times 10^{-9}$ moles), this rat exhibited sufficient rigid immobility to remain totally self-supporting when placed across metal bookends which were in contact only at the upper neck and base of the tail. Such postures were maintained for prolonged periods. Note the erect ears and tail, widely opened eyelids, and extended lower limbs. (From ref. 5a.)

pinch or pin-prick stimuli were observed with Met$^5$-enkephalin, $\alpha$-endorphin, or $\gamma$-endorphin, but such effects (7,16,68) could have been missed by the 5-min interval after injection and before testing began. In contrast to rats with the syndrome induced by $\beta$-endorphin, rats given $\gamma$-endorphin showed consistent elevations in rectal temperature (about 2.0°C ± 0.2°C at 30 min after 281 × $10^{-9}$ moles) and sometimes exhibited some degree of hyperresponsivity to sensory testing and handling, although there were individual variations in this response.

Thus, $\beta$-endorphin in relatively small amounts induced in rats a naloxone-reversible, catatonic-like (27) state reminiscent of some aspects of schizophrenia. Depending on the dose level, $\alpha$- and $\gamma$-endorphins and Met$^5$-enkephalin also exhibited some of the other behavioral and physiological effects of $\beta$-endorphin, in which morphine-like effects (e.g., loss of response to noxious stimuli) appeared to be only a portion of a larger behavioral and pharmacological picture. As has long been known with the separate nicotinic and muscarinic actions of acetylcholine, all effects produced by endorphins may not necessarily be explicable in terms of the alkaloid agonist morphine. Extremely puzzling in this regard is the finding that when injected in the lateral ventricle all three endorphin peptides and Met$^5$-enkephalin can elicit in drug-naive rats the wet-dog shaking behavior ordinarily attributable to opiate withdrawal; this effect is counteracted by naloxone (44,87–89).

All of our observations suggest that normal variations—either qualitative or quantitative—in the homeostatic mechanisms regulating the postulated (48) conversion of $\beta$-LPH as a prohormone to its several endorphin cleavage products could constitute a system fundamentally involved in maintaining "normal" behavior; alterations of the mechanisms normally regulating $\beta$-lipotropin-endorphin homeostasis could lead to signs and symptoms of mental illness. Such a potential psychophysiological role of endorphins could logically be testable through the therapeutic administration of available opiate antagonists. In fact, at a recent presentation of these results and concepts (37), Terenius and co-workers (38,77) reported the observation that administration of naloxone to several chronic schizophrenics halted their auditory hallucinations within minutes. Recently in collaboration with Janowsky at the Veterans Administration Hospital in San Diego, investigators have administered naloxone in a double-blind design to 10 patients with recognized active schizophrenia. So far, results observed show no improvement of the clinical picture of these patients. The ultimate identification of endorphin-sensitive behavioral events and specific treatment of their dysfunctional states may require the development of more specific "anti-endorphins" than those now available, and other naturally occurring brain peptides (68) have already been reported to be endorphin antagonists. Moreover, the matter of doses of the opiate antagonist remains to be more fully investigated. Although their endogenous levels in physiological circumstances have not been measured as yet, there are reasons to believe that the endorphins are physiologically meaningful substances: they have already been observed by immunocytochemical methods in neurons of the CNS, both in cell bodies and in axons; they may well act as neurotransmitters, or neuromodulators, at the level of multiple CNS loci. Moreover, the striking behavioral effects observed upon their injection in the CNS suggest that endorphins and their likely precursor (prohormone) $\beta$-lipotropin may be of significance in the etiology of some neurologic and psychiatric illnesses in man.

## LOOKING TO THE FUTURE: TWO HYPOTHESES

After this rapid review of some of the newer developments in the expanding field of neuroendocrinology, I would like to conclude with a look to the future. This will take the form of two rather novel concepts. One of these is the concept of low-voltage processing of information by brain cells, as already named by Schmitt and his collaborators (72). The other I will call the neural origin of endocrine glands, or is endocrinology a branch of neuroendocrinology? Both concepts have remarkable implications and point to areas of research which I think will be of importance in the next few years. What we will see also is that these two concepts lead to recognition of a remarkable unity of the mechanisms involved in physiological phenomena as widely separate as the stimulation of the secretion of ACTH or growth hormone by pituitary cells and the inhibition by $\beta$-endorphin of the firing pattern of a neuron in the cerebral cortex.

### Low-Voltage Processing of Information by Brain Cells

Until recently the neuron has been seen primarily as a one-way communication system with a central processor for proximally received inputs and a one-way cable for output, the axon. The axon is characterized by its self-regenerating ability to conduct waves of high-voltage depolarization for rapid transmission of an essentially binary type of information expressed at the axon terminal. The axon is usually of considerable length, many times the average diameter of the cell body. Although the dendritic surface has long been recognized morphologically and its vastness well observed, it was not granted much of an active role in the performance of the neuron, principally because experimental evidence of such activity was simply lacking. Any integrative ability or capability of the system was located at the axon hillock. One of the major characteristics of this view of the projection neuron, well studied for over 50 years, is the high-voltage action potential, from a few millivolts to as much as 100 mV. Classically, such a neuron will deliver its ultimate message at the limited address of its axon terminal(s) in the form of the packets of a discrete neurotransmitter. For most projection neurons we still do not know whether norepinephrine, acetylcholine, dopamine, or in a few instances serotonin is involved. No consensus exists as to the ultimate significance of neuropharmacologically active substances such as certain amino acids, γ-aminobutyric acid, substance P, and, lately, the small peptides—e.g., somatostatin, the hypothalamic releasing factors, and the endorphins—traced by immunoassays or immunocytochemistry to increasing numbers of neuronal fibers and neuronal bodies far removed from the ventral hypothalamus and the pituitary.

Although this simplified picture of the projection neuron is still accepted, the majority appear to welcome this multiplicity of effectors and may in the next few years permit satisfactory integration of these multiple effectors, with specific functional processes of individual neurons as members of a neuronal network. This new view of the neuron is based on new morphology as seen by the electron microscope. Much of the discussion that follows is based on a recent review by Schmitt, Dev, and Smith (72) and on the text of the proceedings of a Meeting of the Neurosciences Research Program devoted to local circuits (65). The dendrite is seen no longer as a "passive receptor surface," but rather as a locus for transmitting as well as receiving information in traffic with dendrites of other neurons or with extracellular compartments including capillaries. The means of such communications are seen in the release or uptake of diverse small or large molecules. The electric phenomena involved are those of pinpoint depolarizations and are measured in a few microvolts. Such electrotonic currents spread only over distances measured in microns, not millimeters or centimeters. This type of extremely low-voltage communication constitutes the so-called local circuits. The local circuit neuron can also modify the ultimate behavior of one or more projection neurons which can send responses to a remote contact by a high-voltage long axon pathway. Such systems have been well studied in the retina, the olfactory bulb, and the lateral geniculate body. There is increasing evidence that such local circuitry is actually present in all parts of the central nervous system. It actually represents the structure of the greatest mass or volume of the CNS, with the projection neurons in their classic anatomical arrangements, probably a minority in number as well as in occupied space.

Such multiple dendritic connections have been observed in the electron microscope. Figure 10 is a diagram of possible connections [redrawn from the recent review by Schmitt et al. (72) mentioned above] between one axon terminal and two dendrites, one of which also is in contact with three other dendrites at about ten other locations. The diagram conveys the observation that dendrites may be both presynaptic and postsynaptic to each other as in reciprocal synapses. There is also electron microscope evidence of gap junctions between dendrites. Such electrotonic coupling has been demonstrated in the central nervous system. Several neurons so coupled will respond synchronously with extremely low activating voltages required. Oscillatory behavior has been observed in populations of neurons in some invertebrates (32) when electrically coupled. Such electronic junctions are

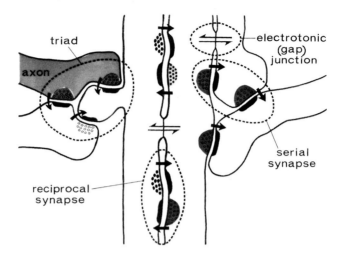

**FIG. 10.** Diagram illustrating the proposal that dendrites in their reciprocal interactions can be pre- and post-synaptic to one another. One axon only is shown in the diagram; all other connections are between dendrites. The hypothesis is that several neurochemicals may be involved at the several junction points (peptides, enzymes, catecholamines, etc.), with either inhibitory or excitatory activity. [Redrawn from Schmitt et al. (72), Fig. 1.]

frequently observed in immediate proximity to chemical synapses (72). None of these phenomena require and none of these structures produce high-voltage spikes. Information transfer by such mechanisms is relatively slow, in seconds or longer, not milliseconds.

### Peptidergic Neurons and Peptidergic Endocrine Cells

What are the relationships between this new view of neurons and the peptides of neuroendocrinology? They become apparent when one asks about the nature of the chemicals involved in these dendro-dendritic contacts. The fragmentary emerging picture is most interesting. It has long been considered that dendrites are involved in uptake of necessary metabolites such as sugars, free amino acids, and adenosine. Such uptake can proceed from extracellular fluid and also from capillary vessels, through endothelial cells. Molecules of much larger size appear similarly to be taken up by dendrites. For instance, an enzyme such as acetylcholine esterase, after being released in extracellular compartments, has been reported (45) to be first bound to the outer surface of the dendritic membrane and later to be taken up by dendrites.

Thus an interesting hypothesis would be that at dendritic points are released still to be characterized enzymes of neuronal origin that would specifically cleave a biologically active peptide, such as one of the endorphins, from a biologically inactive precursor, such as β-lipotropin available in extracellular compartments or perhaps to be found in the axoplasm.

Classic transmitters appear to be released and taken up by the dendrites. Similar release and reciprocal uptake of the small peptides such as substance P, neurotensin, somatostatin, TRF, and endorphins have not been demonstrated as yet. That this is possible and actually happens is a working hypothesis worth investigating. It would go a long way to explain the multiplicity of effects of the polypeptides on the CNS. See for instance (Table 5) the multiplicity of effects of the tripeptide TRF on biological events which have nothing to do with the release of pituitary thyrotropin, the well-known hypophysiotropic activity of TRF, for which it was originally named and recognized. The possibility of an enormous number of such dendritic transfer sites might also explain the psychotropic effects of some of these peptides. Cajal as early as 1899 had made the comment that local circuit neurons may well play an important role as the substrate of complex behavior because of their "prodigious abundance and unaccustomed wealth of forms." Considering such enormous dendritic trees, with each dendrite ending possibly involved in multiple synaptic junctions (Fig. 10), the number of contacts and control points for a single neuron defies the imagination. Such cellular anatomy when considered with the hypotheses mentioned above

TABLE 5. *CNS-mediated actions of TRF[a]*

Increases spontaneous motor activity
  Alters sleep patterns
  Produces anorexia
  Inhibits condition avoidance behavior
  Causes head to tail rotation
Opposes actions of barbiturates on sleeping time, hypothermia, and lethality
  Opposes actions of ethanol, chloral hydrate, chlorpromazine, and diazepam on sleeping time and hypothermia
  Enhances convulsion time and lethality of strychnine
  Increases motor activity in morphine-treated animals
Potentiates DOPA-pargyline effects
Ameliorates human behavioral disorders?
Causes central inhibition of morphine-mediated secretion of GH and PRL
Alters brain cell membrane electrical activity
Increases norepinephrine turnover
  Releases norepinephrine and dopamine from synaptosomal preparations
  Enhances disappearance of norepinephrine from nerve terminals
Potentiates excitatory actions of acetylcholine on cerebral cortical neurons

[a] Information obtained from data used in review by W. Vale, M. Brown, and C. Rivier in (82a).

for chemical inputs and outputs shows the considerable possible functional significance of an expanded dendritic connection network.

The hypothesis of networks with multiple terminals capable of releasing and/or utilizing biologically active peptides may also be involved in another way in explaining some current data. It has been recognized that the total amounts as well as the concentrations of TRF, or LRF or somatostatin in the extrahypothalamic CNS as measured by bioassays or radioimmunoassays, are considerably higher than can be accounted for by the number of cell bodies shown to contain such peptides by immunocytochemistry. A hypothesis to consider is that there would be relatively few neurons manufacturing, say, TRF, LRF (primarily) located in the hypophysiotropic area of the hypothalamus with perhaps a few more cells in the amygdala, and that these neurons have long axons with multiple axon collaterals, all with peptide containing and secreting bouton terminals.

There is already evidence that the dendritic traffic of chemicals works both ways with release and uptake. Thus, in a reciprocally functioning system, if the endorphins and enkephalins are enzymatically cleaved extracellularly from $\beta$-lipotropin as a circulating extracellular precursor in a manner reminiscent of the biogenesis of angiotensin, they could then be picked up by the multiple dendritic endings and carried by retrograde axoplasmic flow whatever distance is necessary for their physiological function.

In summary, we see the small peptides as substances released locally and perhaps produced locally at innumerable possible source points. The specific functions each would subserve would thus be dependent on the effector cell each would be affecting (homonymic information). The same peptide could have vastly different ultimate effects depending on the target cell it would be acting on. This is now well recognized, for instance, for somatostatin. In other words, the same message could have multiple meanings depending on which receptor-effector receives it. ("Go," received-effected by the conductor of a train has an entirely different result than when received-effected by a swimmer ready to dive.) They could also be either transmitters or modulators, depending on their locus of release. If the effect of the small peptides is then the activation of the adenylcyclase-cyclicAMP system of their effector neurons as they may well do in other target tissues such as the adenohypophysis, their effects in neuronal networks could be amplified, long-lasting as well as possibly expanding from their exact source point. Such a system involving neuronal cyclic AMP has already been demonstrated by Siggins et al. (73) for neurons of the locus coeruleus.

### Endocrinology Versus Paracrinology

This concept of a local release and local immediate effect of the peptides from multiple sources in the CNS is a point to remember for future discussion. It does not belong to substances classically defined as hormones. This is true even if each local event may lead to ultimately widespread effects, which is the usual result of hormonal actions. Needless to say, the technology involved in exploring such secretory functions of the dendrites or of the boutons of axon collaterals will be particularly challenging.

TRF activity, LRF activity, and somatostatin activity were demonstrated by bioassays and radioimmunoassays in the extrahypothalamic central nervous system. Later reports appeared showing effects of somatostatin in inhibiting secretion of glucagon and insulin by direct action at the level of the endocrine pancreas. Because somatostatin has a short biological half-life, upon injection in peripheral blood it was unlikely that any physiological effect of endogenous somatostatin on the endocrine pancreas would be due to somatostatin of hypothalamic origin. Looking originally for somatostatin in pancreatic nerve endings, investigators found somatostatin by immunofluorescence in discrete endocrine cells of the pancreas, now well characterized as the D cells (22,54). The same studies showed somatostatin in discrete cells in the jejunum, the colon, the duodenum, and the gastric mucosa. Other observations indicated also that somatostatin can inhibit the secretion of gastrin, secretin, gastric HCl, and acetylcholine from the myenteric plexus. TRF and LRF, although found in extrahypothalamic CNS, have not been found as yet, to my knowledge, in extra-CNS tissues. This is certainly worth investigating. As early as 1957 I had observed CRF activity in extracts of gut tissues also containing substance P (35). Brodish has also described extrahypothalamic CRF. Besides somatostatin, other peptides are now known to be present, and most likely synthesized by cellular elements, in both the central and peripheral nervous system and also in glandular elements of the gastrointestinal tract. The first peptide so observed was substance P in the remarkable experiments of Ulf von Euler and Gaddum as early as 1931 (86). There is now evidence that neurotensin, gastrin, VIP, GIP, the endorphins, and enkephalin(s) are found in the brain as well as in the gastrointestinal tract and the pancreas. This is also true for several of the small peptides such as bombesin, caerulein, and physalaemin, isolated years ago from extracts of the skin of frogs. Furthermore, there are remarkable analogies and homologies among the amino acid sequences of several of these peptides of CNS origin and gastrointestinal origin as well as those isolated from the frog skin. These peptides have been found by immunocytochemistry essentially in two types of cells: (a) they are seen in cell bodies and nerve fibers, i.e., neural and dendritic processes of neurons in brain, spinal cord, spinal ganglia, and the myenteric plexus; (b) they are seen also in typical endocrine cells, for instance, the pancreatic islets of Langerhans, the enterochromaffin cells of the gut, and the adrenal medulla. Neuroblastomas have been reported to contain high levels of vasointestinal peptide (69). An undifferentiated mediastinoma has been found to contain somatostatin, calcitonin, ACTH, and prolactin (I. MacIntyre, *personal communication*). All these results are based on radioimmunoassays and immunocytochemistry, in some instances also bioassays, in most cases with evidence of parallelism of the responses to the known peptide-reference standard and the crude tissue extracts. In cases involving immunological methods there is modest evidence for specificity of the antibodies used. More significant, reports are beginning to appear that demonstrate identity of the primary structure of the gastrointestinal variety of a peptide when compared to its central nervous system variety, as for instance neurotensin (17) and substance P. Our laboratory has already reported the complete sequencing of hypothalamic $\alpha$-MSH and found it to be identical to that of pituitary $\alpha$-MSH (11). Thus there is every reason to believe that we are dealing with the same peptides regardless of their tissue origin.

What is the message to be read in these observations of startling commonalities between the central nervous system and endocrine tissues, and what does it imply for future research?

### The Neural Origin of Peptidergic Endocrine Cells

There is already an interesting unifying concept: much credit must go to A. G. E. Pearse (63) for his visionary concept of the APUD cells formulated some 10 years ago. Pearse observed that neurons and some endocrine cells producing polypeptide hormones shared a set of common cytochemical features and ultrastructural characteristics. APUD is an acronym referring to amine content and/or amine precursor uptake and decarboxylation, as common qualities of these cells (63). The APUD concept postulated that these endocrine cells were derived from a common neuroectodermal ancestor, the transient neural crest. Pearse postulated further that a still larger number of endocrine cells would be eventually found sharing these common properties if one were to explore further in the adult, endocrine tissues derived from the neural crest. Recent observations with refined techniques, particularly the work of Le Douarin on topical chimeras with chromosomal markers, have led Pearse to modify the original APUD concept, but, as we will see, in a remarkable manner. The new evidence regarding the multiple sources of the several peptides mentioned above showed that tissues were involved that were not of neural crest origin; this is particularly true for the peptide-secreting cells of the gut. All these cells have been shown to arise from specialized neuroectoderm (64), that is, not only the neural crest but also the neural tube, the neural ridges, and the placodes.

The expanded concept now postulates that all peptide hormone-producing cells are derived from the neural ectoderm, as are all neurons. For instance, recently Takor and Pearse (64) have reexamined the early stages of development of both the hypophysis and the hypothalamus. They confirmed and expanded earlier conclusions of Ballard who as early as 1964 (2) had recognized that Rathke's pouch does not come from the stomodeum (the pharyngeal origin) as is classically written, but that it originates from the ventral neural ridge (from studies in the chick embryo).

Thus, the hypophysis would share with the hypothalamus the same ventral neural ridge of the neuroectoderm for its origin. In recent work Ferrand and Hraoui (24) using the chromosomal markers in topical chimeras have also concluded an exclusively neuroectodermic origin of the adenohypophysis. Thus, Takor and Pearse (64) have recently concluded: "it is therefore necessary to postulate a neuroectodermal derivation for *all* the endocrine cells of the adenohypophysis and to regard the whole hypothalamohypophysial complex as a neuroendocrine derivative of the ventral neural ridge."

I mentioned earlier that we have recently observed with well-characterized antisera to $\alpha$-endorphin and $\beta$-endorphin that these peptides can be seen by immunocytochemistry in discrete nerve fibers in the hypothalamus and in all cells of the intermediate lobe plus some cells of the adenohypophysis (5). These same cells in the pars intermedia have long been known from the work of several groups of investigators also to contain ACTH[1-39], CLIP, i.e., ACTH[18-39], $\alpha$-MSH, i.e., ACTH[1-13], $\beta$-LPH[1-91], and $\gamma$-LPH[1-48]. Cells of the pars intermedia have also been considered to belong to the APUD series.

The conclusion from all this is that the peptide-secreting cells and tissues appear to be as much part of the nervous system as is the adrenal medulla or what has been called traditionally the neurohypophysis. The word *neuroendocrinology* is now taking a fuller meaning than ever. Pearse has gone so far as proposing that the nervous system should be recognized as composed of three divisions: somatic, autonomic, and endocrine.

Perhaps time has come to redefine the word *hormone*. The hypothalamic hypophysiotropic peptides, TRF, LRF, and somatostatin, are really not hormones according to the current definition, which is still that proposed by Starling in 1905. The definition implies a single source of secretion and, fundamentally, the direct ingress of the secretory product into blood vessels for distribution to a distant receptor tissue, itself thus triggered to respond by its own secretion, change in metabolism, etc. At the level of the dendritic network discussed above, between median eminence and adenohypophysial cells, in the endocrine pancreas or in the gastric mucosa, TRF, LRF, and somatostatin appear to have remarkably localized ranges of extracellular movements ranging from angstroms to microns, at most a few millimeters through the hypothalamohypophysial portal capillaries. There is no incontrovertible evidence so far that these peptides circulate in peripheral blood in physiologically significant concentrations. They have multiple sources, possibly, as hypothesized above, innumerable sources. They would thus not be hormones in the classic sense. The concept of *paracrine secretion* first proposed by Feyrter in 1938 (25) seems much more appropriate to describe products from cells which act on immediate neighbors. The distinction with neurotransmitters is not obvious. I proposed earlier the name *cybernin* for these substances, the etymology of the word implying "local information" or "local control." I have not pushed that new terminology too boldly as it is still another word, and an entirely new root (in biology). As we discussed above, there is good evidence that the small peptides can act as modulators of the function of neuronal system. They are not necessarily neurotransmitters in the classic sense and so far have not been demonstrated as such. We may want to redefine a hormone to be any substance released by a cell and which acts on another cell near or far, regardless of the singularity or ubiquity of the source and regardless of the means of conveyance; bloodstream, axoplasmic flow, or immediate extracellular space. If we do not do so, because of the old definition and to be consistent, hormones would be the steroids, the products of the adenohypophysis, of the thyroid, insulin, glucagon, etc., i.e., those messengers which really circulate wide and far in peripheral blood. The neuroendocrine peptides would have to be something else. I think that it is becoming of heuristic significance to reconsider that terminology. The choice of words or of definitions that will be proposed should take into considerations the remarkable developments I have summarized briefly here.

## CONCLUSIONS

The remarkable picture that emerges is not only that more and more neurophysiologists and endocrinologists are dealing with similar concepts, but that they are and have been all along talking about various forms or embodiments of a single anatomical structure fundamentally devoted to the centrifugal dispatch of information. This structure comes as the classic neuron or as the classic endocrine cell and as several overlapping forms.

From the increasing number of studies using specific radioimmunoassays as well as methods of immunocytochemistry, more and more neurons are being recognized as containing specific peptides. Gastrin (immunoreactive gastrin) is found in cells of the cerebral cortex; substance P, angiotensin, somatostatin, vasoactive intestinal peptide, and peptide sequences corresponding to fragments of ACTH, to fragments of $\beta$-LPH are found in neurons, their axons and dendrites. When one realizes that all our current classic neurophysiology as well as all current classic thinking in neurology and neuropsychiatry simply ignore this increasing overwhelming morphological and biochemical information, one wonders at what will be the significance of these peptides in the physiology of the brain. This has to be the beginning of a new era in our knowledge and prospective of the central nervous system, both normal and diseased.

## ACKNOWLEDGMENTS

Research of the Laboratories for Neuroendocrinology at the Salk Institute is currently supported by research grants from the NIH (HD-09690-02 and AM-18811-02), the National Foundation (1–411), and the William Randolph Hearst Foundation.

## REFERENCES

1. Alford, A. P. (1974): Glucagon control of fasting glucose in man. *Lancet,* 2:974–977.
2. Ballard, W. W. (1964): *Comparative Anatomy and Embryology.* Ronald Press, New York.
3. Belluzzi, J. D., Grant, N., Garsky, V., Sarantakis, D., Wise, C. C., and Stein, L. (1976): Analgesia induced *in vivo* by central administration of enkephalin in rat. *Nature,* 260: 625–627.
4. Bird, S. J., and Aghajanian, G. K. (1976): Impulse generation in type 1 receptors. *Neuropharmacology,* 15:273–282.
5. Bloom, F., Battenberg, E., Rossier, J., Ling, N., Leppaluoto, J., Vargo, T. M., and Guillemin, R. (1977): Endorphins are located in the intermediate and anterior lobes of the pituitary gland, not in the neurohypophysis. *Life Sci.,* 20:43–48.
5a. Bloom, F., Segal, D., Ling, N., and Guillemin, R. (1976): Endorphins: Profound behavioural effects in rats suggest new etiological factors in mental illness. *Science,* 194:630–632.
6. Bradbury, A. F., Smyth, D. G., and Snell, C. R. (1975): Biosynthesis of $\beta$-MSH and ACTH. In: *Peptides, Chemistry, Structure and Biology,* edited by R. Walter and G. Meienhofer, pp. 609–615. Ann Arbor Science, Michigan.
7. Bradbury, A. F., Smyth, D. G., Snell, C. R., Birdsall, N. J. M., and Hulme, E. C. (1976): Structural and conformational relationships between the enkephalins and the opiates. *Nature,* 260:793–795.
8. Bradley, P. B., Briggs, I., Gayton, R. J., and Lambert, L. A. (1976): Effects of microiontophoretically applied methionine-enkephalin on single neurons in rat brainstem. *Nature,* 261:425–426.
9. Brazeau, P., Vale, W., Burgus, R., Ling, N., Butcher, M., Rivier, J., and Guillemin, R. (1973): Hypothalamic polypeptides that inhibit the secretion of immunoreactive pituitary growth hormone, *Science,* 179:423–425.
10. Brown, M., Rivier, J., and Vale, W. (1976): Somatostatin analogs with selected biologic activities. *Metabolism* (Suppl. 1), 25:1501–1503.
11. Burgus, R., Amoss, M., Brazeau, P., Brown, M., Ling, N., Rivier, C., Rivier, J., Vale, W., and Villarreal, J. (1976): Isolation and characterization of hypothalamic peptide hormones. In: *Hypothalamus and Endocrine Functions,* edited by F. Labrie, J. Meites, and G. Pelletier, pp. 355–372. Plenum Press, New York.
12. Burgus, R., Butcher, M., Ling, N., Monahan, M., Rivier, J., Fellows, R., Amoss, M., Blackwell, R., Vale, W., and Guillemin, R. (1971): Structure moléculaire du facteur hypothalamique (LRF) d'origine ovine controlant la sécrétion de l'hormone gonadotrope hypophysaire de lutéinisation (LH). *C.R. Acad. Sci. [D] (Paris),* 273:1611–1613.
13. Burgus, R., Dunn, T. F., Desiderio, D., and Guillemin, R. (1969): Structure moléculaire du facteur hypothalamique hypophysiotrope TRF d'origine ovine: Evidence par spectrométrie de masse de la séquence PCA-His-Pro-NH$_2$. *C.R. Acad. Sci. [D] (Paris),* 269: 1870–1873.
14. Burgus, R., Dunn, T. F., Desiderio, D., Vale, W., and Guillemin, R. (1969): Dérivés polypeptidiques de synthése dous d'activité hypophysiotrope TRF. Nouvelles observations. *C.R. Acad. Sci. [D] (Paris),* 269:226–228.
15. Burgus, R., Dunn T. F., Ward. D. N., Vale, W., Amoss, M., and Guillemin, R. (1969): Dérivés polypeptidiques de synthèse doués d'activité hypophysiotrope TRF. *C.R. Acad. Sci. [D] (Paris),* 269:2116–2118.
15a. Burgus, R., and Rivier, J. (1976): Use of high pressure liquid chromatography in the purification of peptides. In: *Peptides 1976: Proceedings of the Fourteenth European Peptide Symposium, Wepion, Belgium,* pp. 85–94. Editions de l'Universite de Bruxelles, Brussels, Belgium.
16. Buscher, H. H., Hill, R. C., Romer, D., Cardinaux, F., Cloose, A., Hauser, D., and Pless, J. (1976): Evidence for analgesic activity of enkephalin in the mouse. *Nature,* 261:423–425.
17. Carraway, R. E., Kitabgi, P., and Leeman, S. (1976): Hyperglycemic effects of neurotensin, a hypothalamic peptide. VI International Congress on Endocrinology, Hamburg, pp. 1452–1462.
18. Chrétien, M., Benjannet, S., Dragon, N., Seidah, N., and Lis, M. (1976): Isolation of peptides with opiate activity from sheep and human pituitaries: Relationship to $\beta$-lipotropin. *Biochem. Biophys. Res. Commun.,* 72:427–428.

19. Cox, B. M., Goldstein, A., and Li, C. H. (1976): Opioid activity of a peptide (β-LPH-(61-91)) derived from β-lipotropin. *Proc. Natl. Acad. Sci U.S.A.*, 73:1821-1823.
20. Davies, J., and Dray, A. (1976): Effects of enkephalin and morphine on Renshaw cells in feline spinal cords. *Nature*, 262:603-604.
21. Dubois, M. (1972): Nouvelles données sur la localisation au niveau de l'adenohypophyse des hormones polypeptiques: ACTH, MSH, LPH. (Etude en immunofluorescence de l'hypophyse "in situ" de bovins, ovins et porcins et d'homogreffes hypophysaires dans l'hypothalamus du rat.) *Lille Med.*, 17:1391-1394.
22. Dubois, M. (1975): Immunoreactive somatostatin is present in discrete cells of the endocrine pancreas. *Proc. Natl. Acad. Sci. U.S.A.*, 72:1340-1343.
23. Duggan, A. W., Davies, J., and Hall, J. G. (1976): Effects of opiate agonists and antagonists on central neurons of the cat. *J. Pharmacol. Exp. Ther.*, 196:107-120.
24. Ferrand, R., and Hraoui, S. (1973): Origine exclusivement ectodermique de l'adènohypophyse chez la caille et le poulet; demonstrations par la mèthode des associations tissulaires interspecifiques. *C.R. Seanc. Soc. Biol.*, 167:740-743.
25. Feyrter, F. (1938): *Uber Diffuse Endokrine Epitheliale Organ, Vol. 1*. Barth, Leipzig.
26. Frederickson, R. C. A., and Norris, F. H. (1976): Enkephalin-induced depression of single neurons in brain areas with opiate receptors–antagonism by naloxone. *Science*, 194:440-442.
27. Freedman, A. M., Kaplan, H. I., and Saddock, B. J. (Eds.) (1975): *Comprehensive Textbook of Psychiatry*. William & Wilkins, Baltimore.
28. Fujino, M., Yamazaki, I., Kobayashi, S., Fukuda, T., Shinagawa, S., and Nakayama, R. (1974): Some analogs of luteinizing hormone releasing hormone (LH-RH) having intense ovulation-inducing activity. *Biochem. Biophys. Res. Commun.*, 57:1248-1256.
29. Gent, T. P., and Wolstencroft, J. H. (1976): Effects of methionine-enkephalin and leucine-enkephalin compared with thoremorphine on brainstem neurons in cats. *Nature*, 261:426-427.
30. Gerich, J. E., Lorenzi, M., Gustafson, G., Guillemin, R., and Forsham, P. (1975): Evidence for a physiological role of pancreatic glucagon in human glucose homeostasis: Studies with somatostatin. *Metabolism*, 24:175-182.
31. Gerich, J. E., Lorenzi, M., Schneider, V., Karam, J., Rivier, J., and Guillemin, R. (1974): Effects of somatostatin on plasma glucose and glucagon levels in human diabetes mellitus: Pathophysiologic and therapeutic implications. *N. Engl. J. Med.*, 291:544-547.
32. Gettings, P. A., and Willows, A. O. D. (1974): Responses of primate spinothalamic tract neurons to natural stimulation of hind limb. *J. Neurophysiol.*, 37:358-361.
33. Goldstein, A. (1976): Opioid peptides (endorphins) in pituitary and brain. *Science*, 193:1081-1086.
34. Guillemin, R., and Gerich, J. E. (1976): Somatostatin: Physiological and clinical significance. *Annu. Rev. Med.*, 27:379-388.
35. Guillemin, R., Hearn, W. R., Cheek, W. R., and Housholder, D. E. (1957): Control of corticotrophin release: Further studies with *in vitro* methods. *Endocrinology*, 60:488-506.
36. Guillemin, R., Ling, N., and Burgus, R. (1976): Endorphins, peptides d'origine hypothalamique et neurohypophysaire a activité morphinomètique. Isolement et structure moléculaire d'α-endorphin. *C.R. Acad. Sci. [D] (Paris)*, 282:783-785.
37. Guillemin, R., Ling, N., Burgus, R., Bloom, F., and Segal, D. (1977): Characterization of the endorphins, novel hypothalamic and neurohypophysial peptides with opiate-like activity. Evidence that they induce profound behavioral changes. *Psychoneuroendocrinology*, 2:59-62.
38. Gunne, L. M., Lindstrom, L., and Terenius, L. (1977): Naloxone-induced reversal of schizophrenic hallucinations. *Neural Transm.*, 40:13-19.
39. Hill, R. G., Pepper, C. M., and Mitchell, J. F. (1976): Depression of nociceptive and other neurons in the brain by iontophoretically applied met-enkephalin. *Nature*, 262:604-606.
40. Hökfelt, T., Efendic, S., Hellerstrom, C., Johansson, O., Luft, R., and Arimura, A. (1975): Cellular localization of somatostatin in endocrine-like cells and neurons of the rat with special inferences to the $A_1$ cell of the pancreatic islets and to the hypothalamus. *Acta Endocrinol. (Kbh.) [Suppl. 200]*, 80:1-41.
41. Hughes, J. (1975): Isolation of an endogenous compound from the brain with pharmacological properties similar to morphine. *Brain Res.*, 88:295-308.
42. Hughes, J., Smith, T. W., Kosterlitz, H. W., Fothergil, L. A., Morgan, B. A., and Morris,

H. R. (1975): Identification of two related pentapeptides from the brain with potent opiate agonist activity. *Nature*, 258:577–579.
43. Johnson, E. S., Gendrich, R. L., and White, W. F. (1976): Delay of puberty and inhibition of reproductive processes in the rat by a gonadotropin-releasing hormone agonist analog. *Fertil. Steril.*, 27:853–860.
44. Kaymakcalan, S., and Wood, L. A. (1956): Nalorphine-induced "abstinence syndrome" in morphine-tolerant albino rats. *J. Pharmacol. Exp. Ther.*, 117:112–116.
45. Kreutzberg, G. W., Schubert, P., and Lux, H. D. (1975): Neuroplasmic transport in axons and dendrites. In: *Golgi Centennial Symposium*, pp. 161–166, edited by M. Santini. Raven Press, New York.
46. Kuhar, M., Pert, C., and Snyder, F. H. (1973): Regional distribution of opiate receptors binding in monkey and human brains. *Nature*, 245:447–450.
47. La Motte, C., Pert, C. B., and Snyder, S. H. (1976): Opiate receptors binding in primate spinal cord: Distribution and changes after dorsal root section. *Brain Res.*, 112:407–412.
48. Lazarus, L. H., Ling, N., and Guillemin, R. (1976): $\beta$-Lipotropin as a prohormone for the morphinomimetic peptides endorphin and enkephalin. *Proc. Natl. Acad. Sci. U.S.A.*, 73:2156–2159.
49. Li, C. H. (1964): Lipotropin a new active peptide from pituitary glands. *Nature*, 201:924.
50. Li, C. H., and Chung, D. (1976): Isolation and structure of an untriakontapeptide with opiate activity from camel pituitary glands. *Proc. Natl. Acad. Sci. U.S.A.*, 73:1145–1148.
51. Ling, N., Burgus, R., and Guillemin, R. (1976): Isolation, primary structure and synthesis of $\alpha$-endorphin and $\gamma$-endorphin, two peptides of hypothalamic-hypophysial origin with morphinomimetic activity. *Proc. Natl. Acad. Sci. U.S.A.*, 73:3942–3946.
52. Ling, N., and Guillemin, R. (1976): Morphinomimetic activity of synthetic fragments of $\beta$-lipotropin and analogs. *Proc. Natl. Acad. Sci. U.S.A.*, 73:3308–3310.
53. Loh, H. H., Tseng, L. F., Wei, E., and Li, C. H. (1976): $\beta$-Endorphin is a potent analgesic agent. *Proc. Natl. Acad. Sci. U.S.A.*, 73:2895–2896.
54. Luft, R., Efendic, S., Hökfelt, T., Johansson, O., and Arimura, A. (1974): Immunohistochemical evidence for the localization of somatostatin-like immunoreactivity in a cell population of the pancreatic islets. *Med. Biol.*, 52:428–430.
55. Marks, N., and Stern, F. (1975): Inactivation of somatostatin (GH-RIH) and its analogs by crude and partially purified rat brain extracts. *FEBS Lett.*, 55:220–224.
56. Matsuo, H., Baba, Y., Nair, R. M., Arimura, A., and Schally, A. (1971): Structure of the porcine LH- and FSH-releasing hormone. I. The proposed amino acid sequence. *Biochem. Biophys. Res. Commun.*, 43:1334–1339.
57. Monahan, M., Amoss, M., Anderson, H., and Vale, W. (1973): Synthetic analogs of the hypothalamic luteinizing hormone releasing factor with increased agonist or antagonist properties. *Biochemistry*, 12:4616–4620.
58. Moon, H. D., Li, C. H., and Jennings, B. M. (1973): Immunohistochemical and histochemical studies of pituitary $\beta$-lipotrophs in sheep. *Anat. Rec.*, 175:529–538.
59. Mortimer, C. H. (1974): Effects of growth-hormone releasing-inhibiting hormone on circulating glucagon, insulin, and growth hormone in normal diabetic, acromegalic, and hypopituitary patients. *Lancet*, 1:697–701.
60. Nair, R. M. G., Barrett, J. F., Bowers, C. Y., and Schally, A. W. (1970): Structure of porcine thyrotropin releasing hormone. *Biochemistry*, 9:1103–1106.
61. Nicol, R., Siggins, G., Ling, N., Bloom, F., and Guillemin, R. (1977): Neuronal actions of endorphins and enkephalins among brain regions: A comparative microiontophoretic study. *Proc. Natl. Acad. Sci.*, 74:2584–2588.
62. Pasternak, G., Goodman, R., and Snyder, S. H. (1975): An endogenous morphine-like factor in mammalian brain. *Life Sci.*, 16:1765–1770.
63. Pearse, A. G. E. (1968): Common cytochemical and ultra-structural characteristics of cells producing polypeptide hormones (The APUD Series) and their relevance to thyroid and ultimobranchial C cells and calcitonin. *Proc. R. Soc. Lond. [Biol.]*, 170:71–80.
64. Pearse, A. G. E., and Takor, T. (1976): Neuroendocrine embryology and the apud concept. *Clin. Endocrinol.*, 5:229s–244s.
65. Rakic, P. (1975): Definition of the term "local circuit neuron" and the concept of local neuronal circuits. *Neurosci. Res. Prog. Bull.*, 13:291–319.
66. Rivier, C., Vale, W., Ling, N., Brown, M., and Guillemin, R. (1977): Stimulation *in vivo* of the secretion of prolactin and growth hormone by $\beta$-endorphin. *Endocrinology*, 100:238–241.

67. Rivier, J., Burgus, R., and Vale, W. (1971): 3-Methyl-TRF, a synthetic analogue with specific activity greater than that of TRF. *Endocrinology,* 88:A86.
68. Ross, M., Su, T.-P., Cox, B. M., and Goldstein, A. (1976): Brain endorphins. In: *Opiates and Endogenous Opioid Peptides,* pp. 35–40. Elsevier, Amsterdam.
69. Said, S. I., and Rosenberg, R. N. (1976): Vasoactive intestinal polypeptide: Abundant immunoreactivity in neural cell lines and normal nervous tissue. *Science,* 192:907–908.
70. Sarantakis, D., McKinley, W. A., Jaunakais, I., Clark, D., and Grant, N. (1976): Structure activity studies on somatostatin. *Clin. Endocrinol.,* 5:275s–278s.
71. Scharrer, E., and Scharrer, B. (Eds.) (1940): Secretory cells within the hypothalamus. In: *The Hypothalamus and Central Levels of Autonomic Function,* pp. 170–194. Williams & Wilkins Co., Baltimore.
72. Schmitt, F. O., Dev, P., and Smith, B. H. (1976): Electronic processing of information by brain cells. *Science,* 193:114–120.
73. Siggins, G. R., Battenberg, E. F., Hoffer, B. J., and Bloom, F. E. (1973): Noradrenergic stimulation of cyclic adenosine monophosphate in rat Purkinje neurons: An immunocytochemical study. *Science,* 179:585–588.
74. Simantov, R., Kuhar, M. J., Pasternak, G. W., and Snyder, S. H. (1976): The regional distribution of a morphine-like factor enkephalin in monkey brain. *Brain Res.,* 106:189–197.
75. Simantov, R., and Snyder, S. H. (1976): Isolation and structure identification of a morphine-like peptide "enkephalin" in bovine brain. *Life Sci.,* 18:781–788.
76. Tashjian, A. H., Barowsky, N. J., and Jensen, D. K. (1971): Thyrotropin releasing hormone: Direct evidence for stimulation of prolactin production by pituitary cells in culture. *Biochem. Biophys. Res. Commun.,* 43:516–523.
77. Terenius, L. (1976): Discussion of reference No. 79 at Strasbourg.
78. Terenius, L., and Wahlstrom, A. (1975): Morphine-like ligand for opiate receptors in human CSF. *Life Sci.,* 16:1759–1764.
79. Teschemacher, H., Opheim, K. E., Cox, B. M., and Goldstein, A. (1975): A peptide-like substance from pituitary that acts like morphine. 1. Isolation. *Life Sci.,* 16:1771–1776.
80. Uotila, U. U. (1940): Hypothalamic control of anterior pituitary function. In: *The Hypothalamus and Central Levels of Autonomic Function,* pp. 580–588. Williams & Wilkins Co., Baltimore.
81. Vale, W., Brazeau, P., Rivier, C., Brown, M., Boss, B., Rivier, J., Burgus, R., Ling, N., and Guillemin, R. (1975): Somatostatin. *Recent Prog. Horm. Res.,* 31:365–397.
82. Vale, W., Grant, G., Rivier, J., Monahan, M., Amoss, M., Blackwell, R., Burgus, R., and Guillemin, R. (1972): Synthetic polypeptide antagonists of the hypothalamic luteinizing hormone releasing factor. *Science,* 176:933–934.
82a. Vale, W., Rivier, C., and Brown, M. (1977): Regulatory peptides of the hypothalamus. *Annu. Rev. Physiol.,* 39:473–527.
83. Vale, W., Rivier, C., Brown, M., Leppaluoto, J., Ling, N., Monahan, M., and Rivier, J. (1976): Pharmacology of hypothalamic regulatory peptides. *Clin. Endocrinol.,* 5:261s–273s.
83a. Vale, W., Rivier, C., Rivier, R., and Brown, M. (1977): Diverse roles of hypothalamic regulatory peptides. In: *Medicinal Chemistry V,* pp. 25–62. Elsevier, Amsterdam.
84. Vale, W., Rivier, J., and Burgus, R. (1971): Synthetic TRF (thyrotropin releasing factor) analogues: II. pGlu-$N^{3im}$Me-His-Pro-$NH_2$: A synthetic analogue with specific activity greater than that of TRF. *Endocrinology,* 89:1485–1488.
85. Vilchez-Martinez, J., Coy, D., Coy, E., De la Cruz, A., Nishi, N., and Schally, A. V. (1975): Prolonged inhibition of gonadotropin release and suppression of ovulation by synthetic antagonists of LH-RH. *Endocrinology,* 96:354A.
86. von Euler, U. S., and Gaddum, J. H. (1931): An unidentified depressor substance in certain tissue extracts. *J. Physiol.,* 72:74.
87. Wei, E., Loh, H. H., and Way, E. L. (1973): Neuroanatomical correlates of wet shake behaviour in the rat. *Life Sci.,* 12:489–496.
88. Wei, E., Loh, H. H., and Way, E. L. (1973): Brain sites of precipitated abstinence in morphine-dependent rats. *J. Pharmacol. Exp. Ther.,* 185:108–115.
89. Wei, E., Sigel, S., Loh, H., and Way, E. L. (1975): Thyrotropin releasing hormone and shaking behaviour in rat. *Nature,* 253:739–740.
90. Zieglgansberger, W., Fry, J. P., Herz, A., Moroder, L., and Wunsch, E. (1976): Enkephalin-induced inhibition of cortical neurones and the lack of this effect in morphine tolerant/dependent rats. *Brain Res.,* 115:160–164.

*The Hypothalamus*, edited by S. Reichlin,
R. J. Baldessarini, and J. B. Martin,
Raven Press, New York, © 1978.

# Biosynthesis, Packaging, Transport, and Release of Brain Peptides

## Jeffrey F. McKelvy and Jacques Epelbaum

In a current monograph on the hypothalamus it seems appropriate to embrace the concept of the peptidergic neuron as an ancient kind of cell which, in the vertebrate central nervous system (CNS), is represented principally in the hypothalamus, but which is involved in other integrative activities of a non-neuroendocrine nature elsewhere in the brain. Accordingly, this chapter will consider the topic of the cellular biochemistry of brain peptides generally.

The study of the biochemical processes operative in peptide-secreting neurons is of great importance to an understanding of the roles of these peptidergic neuronal subsystems in the brain. While this statement can quite justifiably be made for the low molecular weight amine utilizing neuronal pathways as well, the existing evidence, though scanty, suggests that new modes of neuronal integration may be revealed in the elucidation of the actions of peptide-secreting neurons, and that these neurons may utilize novel biochemical mechanisms, at least as compared to conventional "fast traffic" neurons. Noteworthy among novel interactions is the possible direct chemical interaction of peptides with membranes (7,23), which could result in a change in membrane structure and in the disposition of receptor or pore molecules within the membrane to influence the excitability of neurons in a manner different from that of amine synaptic transmission. Thus, knowledge of the genesis, delivery, and biotransformations of these peptides is essential to an understanding of peptide-mediated forms of neuronal communication.

This chapter will review the extant experimental information on the neurobiochemistry of vertebrate biologically active peptides of defined structure. This will include the neurohypophysial hormones oxytocin and vasopressin and their associated proteins, the neurophysins; the hypothalamic releasing factors thyrotropin releasing factor (TRH), luteinizing hormone releasing factor (LRH), and somatostatin; the endogenous opiate peptides, the endorphins, and the vasoactive peptides substance P and

---

Department of Biochemistry, University of Texas Health Science Center at Dallas, Dallas, Texas 75235

neurotensin. In addition, we will include the dipeptides carnosine and homocarnosine, since recent studies suggest that these peptides may play a role in the primary olfactory pathway.

## BIOSYNTHESIS

### Neurohypophysial Hormones and Neurophysins

The work of Sachs and co-workers, reviewed in ref. 42, laid the foundations for the study of neurohypophysial hormone and protein biosynthesis and still constitutes the most complete and rigorous body of work on this subject. These studies, employing a radioisotopic labeling approach in the dog and the guinea pig, revealed the following features of the biosynthesis of the neuropeptides: (a) due to the small endogenous pools of the neurohypophysial principles (e.g., approximately 2 $\mu$g vasopressin per guinea pig neural lobe), high specific activity amino acid precursors had to be used for isotopic studies, followed by extensive purification and rigorous demonstration of the purity of the labeled neurohypophysial hormones and neurophysins; (b) labeled vasopressin and neurophysin could be formed only in regions of neurohypophysis which contained neuronal somata; (c) constant ventricular infusion of labeled precursor amino acids, followed by subcellular fractionation of the neurohypophysis, revealed that vasopressin of highest specific radioactivity was found not in ribosome-enriched fractions, but rather in a particulate fraction sedimenting at low centrifugal fields; (d) ventricular infusion of labeled precursor amino acids for 1.5 hr did not give rise to labeled vasopressin whereas survival of the animal or *in vitro* incubation of hypothalamic fragments for an additional 4.5 hr permitted isolation of labeled neurohormone; puromycin added at the time of introduction of isotope abolished the appearance of labeled vasopressin, while, if the antibiotic were added after the 1.5-hr infusion period, labeled neurohormone could be isolated; (e) the presence of certain amino acid analogues in *in vitro* biosynthetic studies resulted in the inhibition of both neurophysin and vasopressin biosynthesis. The above observations led Sachs and co-workers to propose a precursor model for neurohypophysial hormone biosynthesis, in which a biologically inactive common precursor of vasopressin and neurophysin was synthesized on ribosomes in neuronal perikarya, moved through intracellular membranous components to a site of formation of secretory granules, and then transported to neurosecretory nerve terminals by axoplasmic flow. According to this model, precursor cleavage could take place *in granulo,* with a minimum time for all of these processes to yield detectable quantities of native vasopressin being about 1.5 hr (in the guinea pig). In subsequent studies, carried out on organ-cultured neonatal guinea pig hypothalamus (35), Sachs and his group confirmed the inhibitory effect of puromycin and cycloheximide on vasopressin biosynthesis originally observed in infusion-acute incubation experiments, with inhibition of hormone synthesis being rapidly effected. When, however, inhibitors of RNA synthesis were tested for their effects on vasopressin biosynthesis in the organ culture system, a progressive decline in, rather than an abrupt cessation of, hormone biosynthesis occurred. These authors suggested that these results are consistent with a transcriptional regulation of hormone biosynthesis involving a vasopressin message with a half-life of approximately 28 hr. As noted by these authors, the observed half-time for loss of vasopressin biosynthetic ability is actually a resultant half-time for the restitution of a number of RNA species involved directly or indirectly in vasopressin biosynthesis. These authors claim that the bromotubercidin experiments suggest a transcriptional control of neurohormone biosynthesis, but much further work is needed to substantiate this claim, since the actual half-life of the vasopressin message *per se* could be quite long.

Pickering and co-workers have studied a number of aspects of vasopressin and neurophysin biosynthesis (36). In studies on neurophysin synthesis in the diabetes insipidus Brattleboro rat (DI rat), it was found that the homozygote, in which vasopressin and vasopressin-associated neurophysin is absent from the posterior pituitary gland, exhibited $^{35}$S-cysteine labeling only of the oxytocin-associated neurophysin and that the heterozygote, whose posterior pituitary contains more oxytocin than vasopressin, showed greater labeling of the oxytocin neurophysin than of the vasopressin-associated neurophysin.

Evidence accumulated in the work of Sachs and of Pickering and their respective collaborators is certainly suggestive of a coordinate biosynthetic origin for the neurohypophysial hormones and neurophysins, with the implication that there is a precursor macromolecule, but it is indirect. Recently, reports have appeared which have directly focused on the identification of neurohypophysial hormone-protein precursors. Walter's laboratory (31,47) has provided initial evidence for the existence of higher molecular weight protein species which give rise to neurophysin- and vasopressin-like peptides in the dog and guinea pig. Gainer and co-workers have similarly observed a higher molecular weight labeled species in the rat which declines as a neurophysin-like protein increases in amount (12). These studies are a welcome addition to the body of evidence dealing with the question of the existence of a neurohypophysial hormone-neurophysin precursor molecule because they represent the preliminary identification of such a precursor(s). However, it will be of critical importance to isolate postulated prohormone molecules and establish their amino acid sequence before this question can be satisfactorily resolved.

## Hypothalamic Releasing Factors

Knowledge about the biosynthesis of releasing factors is not extensive at this time. The history of the elucidation of releasing-factor biosynthesis parallels the history of the elucidation of the chemical structures of the releasing factors, and for similar reasons. The extremely small pool of the releasing factors necessitates the use of high levels of high specific activity radioactive amino acid precursors for dynamic studies of biosynthesis. In addition, the synthetic rates for these peptides appear to be low (28). As a result of the above factors, small amounts of labeled releasing factor peptides have to be isolated from a sea of other labeled species in order to make reliable estimates of biosynthetic phenomena. The technical difficulties encountered in dynamic biosynthetic studies have not yet permitted the establishment of the mechanism of biosynthesis of the releasing factor peptides, a crucial fact on which studies on the regulation of biosynthesis need to be based. Recent resolution of many of the technical problems associated with biosynthetic studies opens the way for a greater understanding of this most important aspect of hypothalamic (and other CNS) function.

As indicated above, the question of the mechanism of releasing hormone biosynthesis is of fundamental importance, since knowledge of this must underlie rational studies on the regulation of synthesis, desired information for a complete understanding of neuroendocrine integration. *A priori,* several possible mechanisms exist for peptide bond formation for the releasing hormones. The first is ribosomal synthesis, either of the biologically active peptides themselves or of precursor polypeptides which are processed to yield the biologically active forms. This mechanism represents an attractive possibility for the synthesis of LRH and somatostatin, based on their larger size, but is attractive for any releasing hormone due to the potential advantage accruing to a neurosecretory neuron of precursor processing in nerve terminals, on functional demand, independent of axoplasmic transport over short time periods. The regulation of peptide bond formation with such a mechanism could be exerted via the interaction of regulatory molecules (cyclic nucleotide-dependent protein kinases or metabolites such as monoamines directly) with the transcription or translation processes. A potentially very important post-translational modification—the processing of a higher molecular weight precursor (possibly also involving such additional processes as pyroglutamyl cyclization and ammonolysis of peptide bonds or amidation of C-terminal residues)—could represent a key control point in such a mechanism.

A second possibility is synthesis via an RNA-independent protein template mechanism. *A priori,* this possibility is reasonably attractive for TRH biosynthesis but is much less so for the biosynthesis of larger hormonal peptides whose biological activity is for the most part critically dependent on the correct amino acid sequence which can vary for the products of this type of mechanism. A final possibility which must be considered is enzymatic synthesis by an RNA-dependent ribosome-independent mechanism in which amino acid activation occurs via aminoacyl RNA species.

At present, none of the above possibilities can be ruled out for the biosynthesis of any of the releasing factors. What is needed is the isolation and purification of the synthesizing activity from a cell-free system.

In 1974, we reported on our experience with isotope studies of TRH biosynthesis and described the multiplicity of labeled peptide species formed after incubation of guinea pig hypothalamic organ cultures with [$^3$H]L-proline (24). In the guinea pig system, for example, a multistep sequential purification scheme was required to achieve adequate radiochemical purity of labeled TRH. In the course of the purification, from 99.97 to 99.98% of the crude homogenate radioactivity had to be eliminated (yield: 40%). The studies demonstrated several important features of the investigation of TRH biosynthesis: (a) sequential purification is mandatory in labeling experiments in order to remove radioactive species (derived both from precursor [$^3$H]L-proline and from the metabolizing tissue) whose mobility matched that of standard synthetic TRH in many separation systems; (b) until more sophisticated separation methods for peptides are generally available (vide infra), the multi-step purification procedure requires the use of carrier synthetic TRH and purification to constant specific activity (cpm/unit of biological or immunological activity); (c) the putative pure biosynthetically derived TRH of constant specific activity must be further tested for purity by derivatization with evidence for coincidence of chromatographic properties of the biosynthetic derivative with those of derivatives prepared from synthetic TRH. Appropriate derivatization methods include dinitrophenylation (of histidine), butanolysis (of pyroglutamate), and enzymatic deamidation (26). This is a most important point since, in studies on TRH biosynthesis in cell-free systems we have observed that radioactive species which co-migrate with synthetic TRH after six chromatographic separations can be shown to be still impure on dinitrophenylation (J. F. McKelvy, *unpublished observations*). Each derivatization procedure selectively alters one of each of three amino acid residues, affording discrimination of TRH from other peptides, in order to more confidently identify biosynthetically derived TRH; (d) a final feature is that a separate purification scheme must be worked out for each experimental system of interest. For example, the sequence of steps which was successful in the guinea pig hypothalamic organ culture system was not adequate for the isolation of highly pure labeled TRH from newt brain tissue.

Based on our findings and a report by Reichlin (40), it is now generally felt that the simple purification methodology used in the earliest studies of TRH biosynthesis renders the subjects of these studies open to reinvestigation. Our investigations, utilizing more extensive purification of labeled TRH, have to date provided some information on aspects of TRH biosynthesis. Proceeding from our observation of differential cell survival in organ cultures of the entire guinea pig mediobasal hypothalamus, we cultured the median eminence alone for periods up to 13 days *in vitro* (30). We examined the cultures in three ways: (a) morphologically, by transmission electron microscopy and light microscopic autoradiography, (b) biochemically, by determination of the ability of the cultures to accumulate [$^3$H]L-proline, and to incorporate this isotope into [$^3$H]TRH, and (c) by bioassay in which the content of TRH in tissue and culture medium was determined by the ability of extracts of both to release radioimmunoassayable TSH from rat pituitary tissue incubated *in vitro*. We found that the cultures exhibited a high degree of preservation of ependymal cell morphology and a sustained ability to accumulate [$^3$H]L-proline over the entire 13-day time course and that neuronal elements showed a progressive degeneration with time in culture. The ability of the cultures to incorporate [$^3$H]L-proline into TRH was lost concomitantly with this loss in neuronal integrity, suggesting that neurons, and not other cell types, are the cellular site of origin of TRH, an unsettled question at the time this work was carried out and one which is still not well documented today.

Studies on TRH biosynthesis in the newt have shown that both hypothalamus and forebrain contain TRH and that fragments of both these brain areas incorporate [$^3$H]L-proline into TRH. This represents the first demonstration of the ability of an extrahypothalamic brain tissue to biosynthesize the tripeptide. This kind of evidence strengthens the contention, for this species at least, that extrahypothalamic TRH is involved in neurotransmission in the central nervous system (14).

These acute *in vitro* studies on TRH biosynthesis by fragments of newt hypothalamus have also provided evidence for a pulsatile and preferential release of newly synthesized TRH (27). To the extent that this *in vitro* phenomenon reflects the *in vivo* behavior of TRH release, this demonstration of pulsatile patterns of TRH release could represent the basis for pulsatile release of pituitary hormones.

Our studies in the newt have also provided some information on the mechanism of TRH biosynthesis. It was found that the inhibitors of cyto- and mitoribosomal protein synthesis, cycloheximide, chloramphenicol, and diphtheria toxin, do not inhibit the incorporation of [$^3$H]L-proline into TRH by fragments of hypothalamic tissue. In addition, a cell-free system for TRH biosynthesis by newt whole brain homogenates was achieved, and it was shown that [$^3$H]L-proline incorporation was completely abolished by native ribonuclease A, which was shown to be active in degrading [$^{32}$P]-labeled newt brain RNA under the conditions of incubation used to demonstrate TRH biosynthesis. Inactive ribonuclease A, prepared by carboxymethylation of the active site histidine (His-119 CM RNase A) and in which the overall chemical properties of the molecule were virtually unchanged, had no inhibitory effect on TRH biosynthesis. Partial inactivation of RNase gave rise to partial TRH biosynthetic activity (15). Although these results might seem to suggest the possibility of an enzymatic mechanism of synthesis involving aminoacyl-tRNA precursors, this suggestion is compromised by the lack of testing of the effects of ribosomal protein synthesis inhibitors, in addition to RNase, in the cell-free system since a lack of effect of the inhibitors in the fragment studies could have been due to compartmentation. Although the newt system exhibits certain desirable properties—a relatively simple spectrum of labeled peptides after [$^3$H]L-proline exposure and a relatively great stability of [$^3$H]TRH in a homogenate of whole brain—the seasonality of TRH synthesis and the small amounts of tissue have not permitted resolution of the mechanism of synthesis in this system.

In attempting to study the mechanism of biosynthesis in vertebrate hypothalamus, we found at once that a potent peptidase activity was present in incubates of guinea pig fragments or homogenates which rapidly degraded TRH, thereby preventing the detection of biosynthesis. This peptidase activity had been described for porcine and rat hypothalamic homogenates by Bauer (5). In order to pursue biosynthetic studies in a cell-free system, we set out to develop inhibitors of the peptidase(s). Using crude homogenates of guinea pig and bovine hypothalamus or whole brain, or a highly purified TRH amidase (L. Hersh, J. Rivier, and J. F. McKelvy, *in preparation*) as peptidase sources, we developed two competitive inhibitors of TRH degradation, pGlu($N^{im}$Dnp) His-pro-NH$_2$ and pGlu-His ($\beta$-Pro-NH$_2$) (McKelvy and Freer, *in preparation*). Neither of these compounds was found to undergo His-Pro bond cleavage, thus underwriting their utility as inhibitors of TRH degradation for use in biosynthetic studies, since no free proline will be liberated from them to dilute the specific activity of the labeled precursor amino acid, and since transpeptidation reactions resulting in the incorporation of labeled proline into the inhibitor, with concomitant possibility of a "false-positive" biosynthetic result, can be avoided. Another inhibitor which has been found to be useful is bacitracin. This antibiotic peptide, originally used to protect glucagon against enzyme degradation (9), can inhibit the degradation of TRH and LRH by tissue, but not by plasma peptidases (29).

Just as inhibitors of releasing factor degradation were necessary in order to pursue biosynthetic studies, so was the development of more rapid purification procedures for labeled neuropeptides.

> Strategies which we have found useful for achieving this goal are illustrated diagrammatically in Fig. 1. Crude alcoholic extracts of tissue from an isotope incorporation study of TRH biosynthesis are subjected to rapid absorption chromatography to rid the sample of most of the unincorporated radioactivity. The effluent of this column (containing TRH and other nonabsorbed peptides) is then subjected to dinitrophenylation, yielding $N^{im}$Dnp TRH, which is uncharged owing to the existence of blocked N- and C-terminals on TRH and to abolition of histidine charged by dinitrophenylation, and other peptides (the majority of tissue peptides) with Dnp N-termini and free carboxyl groups. The dinitrophenylation mixture is passed through an anion-exchange column where $N^{im}$Dnp TRH is recovered in the passthrough. At this point, advantage is taken of the specific thiolytic cleavage of the Dnp group from histidine under mild conditions (43) followed by acidification and organic extraction to

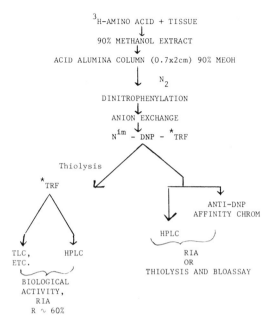

**FIG. 1.** New purification strategies for the isolation of radiolabeled thyrotropin releasing hormone in biosynthetic studies. *DNP:* dinitrophenyl; *HPLC:* high-pressure liquid chromatography; *RIA:* radioimmunoassay; *TLC:* thin-layer chromatography.

eliminate any possible nonimidazole Dnp peptides. Final purification is achieved by high-pressure liquid chromatography (HPLC) or more conventional chromatographic or electrophoretic techniques. An example of the use of HPLC for the purification of labeled TRH is shown in Fig. 2. A homogenate of guinea pig brain was incubated with [$^3$H]L-proline and bacitracin, and, after the dinitrophenylation-thiolysis shown in Fig. 1, subjected to chromatography on the cation-exchange medium Partisil SCX, using 0.1 M ammonium acetate pH 4.6–30% acetonitrile as eluant in an isocratic elution. The single UV-absorbing species, corresponding to exogenous carrier TRH, eluted with a single radioactive peak which comprised 30% of the total radioactivity applied to the column. This peak was collected and reinjected until constant specific activity (cpm/µg TRH by RIA) was reached. The total time for such a purification of biosynthetic TRH, using HPLC, is 2 days as opposed to 3 to 4 weeks by the previous methodology. An alternative route from $N^{im}$Dnp TRH is passage over an anti-Dnp-Sepharose column followed by elution with excess dinitrophenol, and RIA estimation of the Dnp peptide (Y. Grimm-Jorgensen and J. F. McKelvy, *unpublished observations*) or RIA and bioassay of its thiolysis product. Finally, $N^{im}$Dnp TRH can be purified by HPLC using either cation exchange (SCX) or reverse phase $C_{18}$ systems.

With the availability of the inhibitors, and rapid purification methodologies, we are currently investigating the mechanism of TRH and LRH biosynthesis in a cell-free system (low-speed supernatant of guinea pig and rat

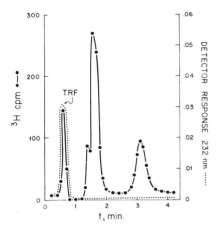

**FIG. 2.** Elution profile of biosynthetically labeled thyrotropin releasing hormone from a cation-exchange high-pressure liquid chromatography column (Partisil SCX) eluted isocratically with 70% 0.1 M ammonium acetate, pH 4.6–30% acetonitrile.

hypothalamus with additions appropriate for ribosomal protein synthesis). An experimental protocol was devised in which duplicate aliquots at a given time point were taken for estimation of TRH content by RIA, and for TRH labeling with [$^3$H]L-proline as precursor. The results from such an experiment showed that the rate of TRH production assessed by measurement of immunoreactive TRH was greater than that derived from TRH labeling with [$^3$H]L-proline (26). This suggests that two pathways may exist for TRH production in this system: (a) liberation of tripeptide from a preexisting precursor and (b) *de novo* synthesis of TRH, its precursor, or both. The implication of the existence of precursor species suggested a ribosomal mechanism, and we therefore tested the effects of inhibitors of cytoribosomal protein synthesis inhibitors on both TRH and LRH production in the cell-free system, using [$^3$H]L-proline plus [$^3$H]L-tyrosine as precursor amino acids in order to measure synthesis of both releasing factors in the same experiment. We found that inclusion of the antibiotics cycloheximide and puromycin abolished the labeling of both TRH and LRH (30a). Efforts are currently being directed toward the identification of possible higher molecular weight precursors of these peptides.

In a recent study (19) in which TRH labeling was measured by chromatographing labeled products of an *in vitro* incubation on anti-TRH affinity columns, no effect of cycloheximide on TRH labeling could be demonstrated. It should be noted that no characterization was made of the labeled species retained by the column and eluted by synthetic TRH. Another consideration in studies of this type is that the sequence -Glu-His-Pro- can be found in at least 11 polypeptides of known sequence, and two possibilities exist for a false-positive biosynthetic result using antibody binding: (a) the sequence -Glu-His-Pro- *as part of a larger peptide* could be recognized by the anti-TRH antibody; even if the cross-reactivity of such a sequence, either as part of a larger peptide, or as a free tripeptide, is very low, a greater abundance of these molecules, relative to the low concentration of TRH in tissues, could result in detection of this sequence; (b) the tripeptide Glu-His-Pro could be liberated from a non-TRH-related species and undergo pyroglutamyl formation and amidation.

A final statement with regard to immunochemical purification methods for peptides is that monospecific antibody populations should be used for the preparation of immunoabsorbents, since ion-exchange columns can result from the coupling of immune serum to stationary supports. Ideally, the Fab fragments should be prepared from monospecific antipeptide antisera, to allow for greatest specificity and recovery of peptide.

Biosynthetic studies on LRH have been reported in which the use of isotopic techniques was not accompanied by rigorous proof of the identity of the labeled product as LRH (16,18,32). In view of the many pitfalls in carrying out this type of study, consideration of these results will await substantiation of the identity of the labeled product as LRH.

At the present time, no reports exist in the literature on the biosynthesis of somatostatin, substance P, neurotensin, or endorphins by CNS tissue. The problems inherent in assessment of releasing factor synthesis exist for these peptides as well.

Recent studies from the laboratory of Margolis (17) have provided evidence that the dipeptide carnosine ($\beta$-alanyl-L-histidine) may be in some way involved in synaptic transmission in the primary olfactory pathway. The synthetic enzyme carnosine synthetase, a soluble enzymatic system, is being purified from olfactory tissue and shows interspecies differences (21). Further studies on the involvement in neural tissue of a low molecular weight peptide synthesized by a nonribosomal mechanism will be of interest in defining the breadth of participation of peptides in the function of the nervous system.

## PACKAGING, TRANSPORT, AND RELEASE OF BRAIN PEPTIDES

At the present time, of brain peptides, only the neurohypophysial hormones and neurophysins have been studied to any significant extent with regard to their transport and release. These molecules have been shown to move from the magnocellular nuclei to the pars nervosa at rates commensurate with those described for "fast axoplasmic flow," and to be interrupted in their movement by alkaloids which bind to microtubular protein (for reviews, see 20,34). Currently, however, it has not been shown by critical methods whether neurohypophysial principles, contained in neurosecretory granules or otherwise organized, are transported in the axoplasm of neurosecretory neurons in association with microtubules, or in a smooth endoplasmic reticulum compartment in the axoplasm. In this chapter, therefore, we will focus on the question of the subcellular distribution of other brain peptides, with a view toward evaluating this most basic aspect of the intracellular organization of these molecules.

Most of the information concerning the subcellular localization of brain peptides has been gathered by two different approaches: a qualitative one, immunocytochemistry (see Hökfelt et al., *this volume*), and a quantitative one, subcellular fractionation. The latter technique, which is usually performed according to methods originally described by the Whittaker (48) and DeRobertis (8) groups, permits one to isolate pinched-off nerve endings, known as synaptosomes, and to separate them from other soluble or particulate components of the nerve cells. The technique has been widely used to study the subcellular distribution of neurotransmitters and related enzymes as well as, more recently, peptides in brain. The results obtained so far with this approach are summarized below.

## TRH

Barnea et al. (1) and Winokur et al. (49) reported that most of hypothalamic (1,49) and brain (49) TRH was found in a crude mitochondrial fraction. Winokur et al. showed that TRH concentration was higher in the synaptosomal band isolated by density gradient centrifugation and characterized by electron microscopy. Arguing that standard preparations of synaptosomal fractions are not homogeneous with regard to their content of other subcellular organelles, Barnea et al. (1,2) fractionated hypothalami by means of continuous sucrose density gradient centrifugation. TRH was associated with two populations of particles separable by nonequilibrium density centrifugation; however, after equilibrium centrifugation, a single peak of TRH was observed. The authors concluded that the two subpopulations of particles differ in size but are similar in density. Both groups of workers exposed TRH-containing particles to hypoosmotic shock which resulted in solubilization of the peptide. Winokur et al. recovered the activity in a fraction corresponding to synaptic vesicles. To further characterize the two populations of TRH-containing particles, Barnea et al. (3) studied their ontogeny. In 22-day-old fetuses, TRH was almost exclusively associated with the small particles. In neonates, there was an age-dependent increase in the fractional amount of the TRH confined to the larger particles which was completed by 7 days of age. They concluded that, in hypothalamic homogenates, TRH was contained in synaptosome-like particles.

## LRH

Shin et al. (44) recovered most of the LRH from an 80,000 $g \times 11'$ pellet which was further fractionated on a discontinuous sucrose gradient, but no clear analysis of the various particulate fractions tested by these authors was provided. Using bioassay, Taber and Karavolis (46) reported that after differential centrifugation at 17,000 $g \times 60'$, only the crude mitochondrial fraction caused an increase in LH release from rat pituitaries. When this pellet was layered on a discontinuous sucrose gradient, only one fraction, containing electron-dense vesicles of varying sizes, was biologically active.

Due to some differences in preparation of the fractions, it is difficult to compare these two former reports to those of Ramirez et al. (39) and Barnea et al. (1) who found that 50 to 70% of the LRH was contained in a 17,000 g × 20' crude mitochondrial pellet containing synaptosomes which were subsequently purified and found to contain most of the bioassayable and radioimmunoassayable LRH. For the same reason as before, Barnea et al. used a continuous gradient of sucrose and found that LRH (as was TRH) was associated with two populations of particles separable by nonequilibrium density centrifugation. After equilibrium centrifugation, both sets of LRH-containing particles sedimented as a single peak which could be partially separated from the TRH. Osmotic shock only partially solubilized LRH contained in the crude mitochondrial fraction and released the peptide from the large but not the small set of particles (2). Deoxycholate had the same effect whereas Triton X-100 disrupted both types of particles. The ontogeny (3) of the subcellular compartmentalization of LRH differed markedly from that of TRH. LRH was barely detectable in hypothalami from 22-day-old fetuses but was assayable in 2-day-old neonates; at this age the peptide was confined to the large particles, and in 7-day-old rats LRH was similarly distributed. Association of LRH with small particles became evident in 14-day-old males and 21-day-old females. As the time of development of the small LRH particles coincides with the time reported for the appearance of synaptoid junctions between axons and ependymal processes, Barnea postulated that some of the small particles could derive from *synaptoid* axonal-ependymal functions while the large particles were *synaptosome* like.

### Somatostatin

Subcellular preparations of medial basal hypothalamus preoptic area and amygdala indicate (11) that over 70% of somatostatin (SRIF) immunoactivity is recovered in the synaptosomal band of a discontinuous sucrose gradient as characterized by enzyme marker and electron microscopy of the different fractions.

### Enkephalin

Simantov et al. (45) have examined in detail the distribution of enkephalin activity as determined by the opiate binding assay. Opiate receptor binding of subcellular fractions was also performed. Assay of the fractions obtained by differential centrifugation showed that both enkephalin activity and opiate receptor binding are most enriched in the crude mitochondrial pellet.

When this pellet is subfractionated on a discontinuous sucrose gradient, both enkephalin activity and opiate receptor activity are most concentrated in a fraction enriched in synaptosomes.

### Substance P

When subcellular fractions of rat hypothalamic tissue were analyzed for relative contents of immunoactive substance P (38), it was shown to be absent from the myelin layer, was found in intermediate amounts in the mitochondrial fraction, and was present in highest concentrations in the nerve ending particles. More recently, the subcellular distribution of substance P in bovine tissue has been shown to be similar in substantia nigra and hypothalamus (10). More than half of the peptide originally present in the homogenate is recovered in the synaptosomal fractions.

### α-Melanocyte Stimulating Hormone

Barnea et al. (4) found that 36% of hypothalamic α-melanocyte stimulating hormone (α-MSH) was recovered in the crude mitochondrial fraction and 31% sedimented between 11,500 and 105,000 g. When the 900 g supernatant was subjected to a continuous sucrose gradient, two populations of α-MSH-containing particles were observed under nonequilibrium conditions while under equilibrium conditions the two types of particles migrated as a single band. The sedimentation properties of the α-MSH particles were similar to those containing TRH. Hypoosmotic shock released α-MSH from the large but not the small particles. The authors suggest that the large particles containing α-MSH are synaptosomes.

In summary, it can be seen that the majority of biologically active brain peptides appear to be concentrated in fractions in which synaptosomes and their inclusions appear during subcellular fractionation. Much work remains to be done, however—careful subsynaptosomal fractionation, purification of peptide-containing organelles, electron microscopic cytochemistry of such fractions—in order to clarify the intracellular pools of these peptides.

### Peptidases

It has been known for several years (22) that brain homogenates contain enzymatic activities which degrade peptides; the subcellular localizations of the enzyme(s) responsible for the degradation of carnosine (β-alanyl-histidine) and homocarnosine (γ-amino butyryl-histidine) in the rabbit brain have been described (33). The specific activity of carnosinase was highest in cell sap where it was 12 times more potent than in synaptosomes. Homocarnosinase activity had a similar subcellular distribution (13 times more in cell sap than synaptosomes). In the rat hypothalamus, TRH and LRH degrading activities are also primarily localized in supernatant, with trace amounts in the synaptosomal band (J. F. McKelvy and J. Epelbaum, *unpublished observations*). The role of these peptidases in regulating the intra- and extraneuronal concentrations of biologically active peptides is presently unclear. It will be necessary to purify these enzymes and obtain antisera to them for immunohistochemical localization of their presence in neural tissue.

Studies on the release of biologically active peptides from neurons have been most extensively carried out on the neurohypophysial principles. As reviewed by Poisner (37), "excitation-secretion coupling" involving excytosis is the major route for release. It should be noted, however, that not all of the neurohormone and neurophysin can be recovered in the neurosecretory granule fraction. Recent immunocytochemical studies at the ultrastructural level employing antineurophysin antibodies reveal extragranular reaction product (A. Silverman, *personal communication*) in the axoplasm.

Research on the biochemical mechanism of release of the neurohypophysial peptides is in an early stage of development. We have found that neurosecretory granules (NSG) contain a membrane-associated cyclic AMP-stimulated protein kinase capable of phosphorylating NSG membrane proteins (25) and that one of these membrane proteins has the properties of myosin (J. F. McKelvy, *in preparation*). The possibility thus exists that release of peptides from neurons may be via a contractile process, as pro-

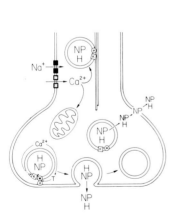

**FIG. 3.** Hypothetical model for elements participating in the mechanism of release of neurohypophysial hormones and proteins from magnocellular neurosecretory neurons. Neurosecretory granules (*NSG*) contain neurophysin (*NP*) and hormone (*H*). NSG membranes contain myosin (*M*) and cyclic nucleotide-stimulated protein kinase (*K*). Nerve terminal membrane contains actin (*A*), and a troponin-tropomyosin complex (T) is present in axoplasm. Upon synaptic activation of the neurosecretory neuron, the entry of calcium ion modulates the interaction of NSG membrane myosin and nerve terminal actin by an inhibitory action on myosin phosphorylation and a stimulatory action (for actin-myosin interaction) on the troponin-tropomyosin complex. It is postulated that granular neurophysin may, by virtue of its essentially freely dissociated state *in granulo* (13), and its lipoprotein nature, influence events in the NSG membrane which lead to exocytosis. In addition, it is postulated that extragranular neurophysin may interact with nerve terminal plasma membrane so as to result in the release of extragranular neurophysin and neurohormone.

posed by Berl, for release of amine neurotransmitters. In studies of adrenal medullary exocytosis, Poisner has isolated an actomyosin-like complex from secretory tissue (37). A model for neurohypophysial hormone release which takes into account recent findings is shown in Fig. 3 and postulates interactions among NSG membrane-associated myosin (M), protein kinases (K), and nerve terminal actin A. We would also like to take this opportunity to propose that neurophysins should be considered as possible modulators of the membrane interactions involved in exocytosis, and that they may interact directly with nerve terminal membranes to provide for their own release. Such a role may be a property of other oligo- and polypeptide products of prohormone cleavage. Recent studies have also suggested that peptidergic nerve terminals may be an important site of action of modulators of release. For example, dopamine releases LRH from synaptosomal fractions (6) and from regions of hypothalamus containing both LRH and dopamine nerve terminals (41). The role of biogenic animals and of hormones in regulation of peptidergic neuron secretion is a major area for future study.

## REFERENCES

1. Barnea, A., Ben-Jonathon, N., Colston, C., Johnston, J. M., and Porter, J. C. (1975): Differential sub-cellular compartmentalization of thyrotropin releasing hormone (TRH) and gonadotropin releasing hormone (LRH) in hypothalamic tissue. *Proc. Natl. Acad. Sci. U. S. A.,* 72:3153-3157.
2. Barnea, A., Ben-Jonathon, N., and Porter, J. C. (1976): Characterization of hypothalamic subcellular particles containing luteinizing hormone-releasing hormone and thyrotropin releasing hormone. *J. Neurochem.,* 27:471-484.
3. Barnea, A., Neaves, W. B., and Porter, J. C. (1977): Ontogeny of the subcellular compartmentalization of thyrotropin releasing hormone and luteinizing hormone releasing hormone in the rat hypothalamus. *Endocrinology,* 100:1068-1079.
4. Barnea, A., Oliver, C., and Porter, J. C. (1977): Subcellular localization of $\alpha$-MSH in the rat hypothalamus. *J. Neurochem. (in press).*
5. Bauer, K. (1974): Degradation of thyrotropin releasing hormone (TRH). Its inhibition by pGlu-His-OCH$_3$ and the effects of the inhibitor in attempts to study the biosynthesis of TRH. In: *Lipmann Symposium,* pp. 53-62. de Gruyter, Berlin.
6. Bennett, G. W., Edwardson, J. A., Holland, D. T., Jeffcoate, S. L., and White, N. (1975): Release of immunoreactive luteinizing hormone releasing hormone and thyrotropin releasing hormone from hypothalamic synaptosomes. *Nature,* 257:323-325.
7. Bleich, H. E., Cutnell, J. D., Day, A. R., Freer, R. J., Glasel, J. A., and McKelvy, J. F. (1976): NMR observations of the interaction of small oligopeptides with phospholipid vesicles. *Biochem. Biophys. Res. Commun.,* 71:168-175.
8. DeRobertis, E., and Rodriguez De Lores Arnaiz, G. (1969): Synaptosomes. In: *Handbook of Neurochemistry, Vol. II,* edited by A. Lajtha, pp. 365-392. Plenum Press, New York.
9. Desbuquois, B., and Cuatrecasas, P. (1972): Independence of glucagon receptors and glucagon inactivation in liver cell membranes. *Nature [New Biol.],* 236:202-204.
10. Duffy, M. J., Mulhall, D., and Powell, D. (1975): Subcellular distribution of substance P in bovine hypothalamus and substantia nigra. *J. Neurochem.,* 25:305-307.
11. Epelbaum, J., Brazeau, P., Tsang, D., Brawer, J., and Martin, J. B. (1977): Subcellular distribution of radioimmunoassayable somatostatin in rat brain. *Brain Res. (in press).*
12. Gainer, H., Sarne, Y., and Brownstein, M. J. (1977): Neurophysin biosynthesis: Conversion of a putative precursor during axonal transport. *Science,* 195:1354-1356.

13. Glasel, J. A., McKelvy, J. F., Hruby, V., and Spatola, A. (1976): Binding studies of polypeptide hormones to bovine neurophysins. II. Equilibrium dialysis. *J. Biol. Chem.,* 251: 2929–2937.
14. Grimm-Jorgensen, Y. and McKelvy, J. F. (1974): Biosynthesis of thyrotropin releasing factor by newt (*Triturus viridescens*) brain *in vitro*. Isolation and characterization of thyrotropin releasing factor. *J. Neurochem.,* 23:471–478.
15. Grimm-Jorgensen, Y. and McKelvy, J. F. (1976): TRF biosynthesis *in vitro,* effect of inhibitors of protein synthesis. *Brain Res. Bull.,* 1:171–175.
16. Hall, R. W., and Steinberger, E. (1976): Synthesis of LH-RH by rat hypothalamic tissue *in vitro:* I. Use of a specific antibody to LH-RH for immunoprecipitation. *Neuroendocrinology,* 21:111–119.
17. Harding, J. and Margolis, F. L. (1976): Denervation in the primary olfactory pathway of mice. III. Effect on enzymes of carnosine metabolism. *Brain Res.,* 110:351–372.
18. Johansson, N. G., Hooper, F., Sievertsson, H., Currie, B. L., and Folkers, K. (1972): Biosynthesis *in vitro* of the luteinizing releasing hormone by hypothalamic tissue. *Biochem. Biophys. Res. Commun.,* 73:507–515.
19. Kubek, M., Lorincz, M., Emanuele, N., Shambaugh, G. E., and Wilber, J. (1977): Thyrotropin releasing hormone (TRH): Biosynthesis by extrahypothalamic and hypothalamic tissues *in vitro*. Abstracts, 59th Annual Meeting of the Endocrine Society, June 8–10, 1977, Chicago, Ill., p. 125.
20. Livett, B. G. (1975): Immunochemical studies on the storage and axonal transport of neurophysins in the hypothalamo-neurohypophyseal system. *Ann. N.Y. Acad. Sci.,* 248: 112–133.
21. Margolis, F. L. (1976): A model neural system: The primary olfactory pathway. Roche Institute of Molecular Biology Annual Report, 1976, pp. 109–110.
22. Marks, N. (1976): Biodegradation of hormonally active peptides in the central nervous system. In: *Subcellular Mechanisms in Reproductive Neuroendocrinology,* edited by F. Naftolin, pp. 129–148. Elsevier, Amsterdam.
23. Mayer, M. M. (1972): Mechanism of cytolysis by complement. *Proc. Natl. Acad. Sci. U.S.A.,* 69:2954–2958.
24. McKelvy, J. F. (1974): Biochemical neuroendocrinology. I. Biosynthesis of thyrotropin releasing hormone (TRH) by organ cultures of mammalian hypothalamus. *Brain Res.,* 65:489–502.
25. McKelvy, J. F. (1975): Phosphorylation of neurosecretory granules by cyclic AMP-stimulated protein kinase and its implication for transport and release of neurophysin proteins. *Ann. N.Y. Acad. Sci.,* 248:80–91.
26. McKelvy, J. F. (1977): Biosynthesis of hypothalamic peptides. In: *Hypothalamic Peptide Hormones and Pituitary Regulation,* edited by J. C. Porter, pp. 77–98. Plenum Press, New York.
27. McKelvy, J. F. and Grimm-Jorgensen, Y. (1975): Studies on the biosynthesis of thyrotropin releasing hormone *in vitro*. In: *Hypothalamic Hormones,* edited by M. Motta, P. G. Crosignani, and L. Martini, pp. 13–26. Academic Press, New York.
28. McKelvy, J. F. and Grimm-Jorgensen, Y. (1976): Biosynthesis and degradation of hypothalamic hypophysiotropic peptides. In: *Endocrinology, Proceedings of the V International Congress of Endocrinology, Hamburg, July 18–24, 1976, Vol. 1,* edited by V. H. T. James, pp. 175–179. Excerpta Medica, Amsterdam.
29. McKelvy, J. F., Leblanc, P., Loudes, C., Perrie, S., Grimm-Jorgensen, Y., and Kordon, C. (1976): The use of bacitracin as an inhibitor of the degradation of thyrotropin releasing factor and luteinizing hormone releasing factor. *Biochem. Biophys. Res. Commun.,* 73: 507–515.
30. McKelvy, J. F., Sheridan, M., Joseph, S., Phelps, C. H., and Perrie, S. (1975): Biosynthesis of thyrotropin releasing hormone in organ cultures of the guinea pig median eminence. *Endocrinology,* 97:908–918.
30a. McKelvy, J. F., Loudes, C., Snyder, J., and Kordon, C. Inhibition of biosynthesis of TRH and LRH by inhibitors of ribosomal protein synthesis. (*Submitted to Nature*.)
31. Mendelson, I. S. and Walter, R. (1977): On the biosynthesis of putative neurophysin-vasopressin precursors in the hypothalamo-neurohypophysial gland of the guinea pig. In: *Hypothalamic Hormones, Second European Colloquium,* edited by W. Voelter and D. Gupta. Verlag Chemie, Weinheim (*in press*).

32. Moguilevsky, J. A., Enero, M. A., and Szwarcfarb, B. (1974): Luteinizing hormone releasing hormone—biosynthesis by rat hypothalamus *in vitro. Proc. Soc. Exp. Biol. Med.,* 147:434–435.
33. Ng, R. H., Marshall, F. D., Henn, F. A., and Sellstrom, A. (1977): Metabolism of carnosine and homocarnosine in subcellular fractions and neuronal and glial cell enriched fractions of rabbit brain. *J. Neurochem.,* 28:449–452.
34. Norstrom, A. (1975): Axonal transport and turnover of neurohypophyseal proteins in the rat. *Ann. N.Y. Acad. Sci.,* 248:46–63.
35. Pearson, D., Shainberg, A., Malamed, S., and Sachs, H. (1975): The hypothalamo-neurohypophysial complex in organ culture: Effects of metabolic inhibitors, biologic and pharmacologic agents. *Endocrinology,* 96:994–1003.
36. Pickering, B. T., Jones, C. W., Burford, G. D., McPherson, M., Swann, R. W., Heap, P. F., and Morriss, J. J. (1975): The role of neurophysin proteins: Suggestions from the study of their transport and turnover. *Ann. N.Y. Acad. Sci.,* 248:15–35.
37. Poisner, A. M. (1976): The role of calcium in neuroendocrine secretion. In: *Subcellular Mechanisms in Reproductive Neuroendocrinology,* edited by F. Naftolin, pp. 45–62. Elsevier, Amsterdam.
38. Powell, D., Leeman, S., Tregear, G. W., Niall, H. D., and Potts, J. T. (1973): Radioimmunoassay for substance P. *Nature [New Biol.],* 241:252–254.
39. Ramirez, V. D., Gautron, G. P., Epelbaum, J., Pattou, E., Zamora, A., and Kordon, C. (1975): Distribution of LH-RH in subcellular fractions of the mediobasal hypothalamus. *Mol. Cell. Endocrinol.,* 3:339–350.
40. Reichlin, S. (1976): Biosynthesis and degradation of hypothalamic hypophysiotropic factors. In: *Subcellular Mechanisms in Reproductive Neuroendocrinology,* edited by F. Naftolin, pp. 109–127. Elsevier, Amsterdam.
41. Rotsztejn, W. H., Charli, J. L., Pattou, E., and Kordon, C. (1977): Stimulation by dopamine of luteinizing hormone-releasing hormone (LHRH) release from mediobasal hypothalamus in male rats. *Endocrinology (in press).*
42. Sachs, H., Fawcett, C. P., Takabatake, Y., and Portanova, R. (1969): Biosynthesis and release of vasopressin and neurophysin. *Recent Prog. Hormone Res.,* 25:447–491.
43. Shaltiel, S., and Soria, M. (1969): Dinitrophenylation and thiolysis in the reversible labeling of a cysteine residue associated with the nicotinamide-adenine dinucleotide site of rabbit muscle glyceraldehyde-3-phosphate dehydrogenase. *Biochemistry,* 8:4411–4415.
44. Shin, S. N., Norris, A., Snyder, J., Hymer, W. C., and Milligan, J. V. (1974): Subcellular localization of LH-releasing hormone in the rat hypothalamus. *Neuroendocrinology,* 16:191–201.
45. Simantov, R., Snowman, A. M., and Snyder, S. H. (1976): A morphine-like factor "enkephalin" in rat brain: Subcellular distribution. *Brain Res.,* 107:650–657.
46. Taber, C. A., and Karavolis, H. J. (1975): Subcellular localization of LH releasing activity in the rat hypothalamus. *Endocrinology,* 96:446–452.
47. Walter, R., Audhya, T. K., Schlesinger, D. H., Shin, S., Saito, S., and Sachs, H. (1977): Biosynthesis of neurophysin proteins in the dog and their isolation. *Endocrinology,* 100:162–174.
48. Whittaker, V. P. (1969): The synaptosome. In: *Handbook of Neurochemistry, Vol. II,* edited by A. Lajtha, pp. 327–364. Plenum Press, New York.
49. Winokur, A., Davis, R., and Utiger, R. D. (1977): Subcellular distribution of thyrotropin releasing hormone (TRH) in rat brain and hypothalamus. *Brain Res.,* 120:423–444.

# General Discussion

*Dr. Scharrer:* Let me first say how excited I am about the rapid and beautiful progress in this area of neuroendocrinology concerned with neuropeptides.

One point to be concerned with at this juncture is terminology which must go hand-in-hand with conceptual adjustments that come from all of these new insights.

I see a slight danger, in that we may be led to abandon older but still useful terms such as neurotransmitter and neurohormone. Such terms are no longer considered adequate by some investigators because they do not reflect neurochemical classifications. We should keep in mind, however, that both invertebrates and vertebrates make multiple uses of the same few classes of peptides and catecholamines for a variety of functions. It would seem to be a mistake to be guided by the chemical nature of a neurochemical messenger rather than by its mode of operation in categorizing it as either a transmitter or a hormone.

*Dr. Guillemin:* In keeping with the remarks made by Dr. Scharrer regarding terminology, I agree that some of us should think seriously in the near future about the problems of current nomenclature and terminology.

Probably all of these peptides we are describing as hypothalamic hormones do not really meet the criteria of a hormone as given by Starling so many years ago. There is no good evidence that they circulate in the peripheral blood to affect the function of a distant organ as is true for pituitary, thyroid, and pancreatic hormones.

Perhaps one of the easiest things to do would be to redefine the word "hormone" as a substance made by certain (secretory) cells known to affect the function of other cells. Such a simple statement would not be a contradiction to the definition given by Starling in 1905; it would be somewhat more encompassing. It must be emphasized that it is time to think seriously about questions of nomenclature before they become problems of nomenclature.

*Dr. Michael:* Dr. Guillemin, can you or Dr. Bloom comment further about the behavioral effects of the endorphins? Is our knowledge of drug effects sufficiently advanced to allow a look at responses in human volunteers?

*Dr. Bloom:* The comments that Dr. Guillemin made were based on a series of experiments done with David Segal in which dose-response curves varying from 1 to 1,000 micrograms were evaluated for each of the peptides of the series described — methionine-enkephalin, $\alpha$-endorphin, $\beta$-endorphin, and $\gamma$-endorphin. The effects of $\beta$-endorphin on analgesia and the behavioral effect demonstrated in the "stiff rat" slide on body temperature and on a few other physiological indicators were 100 to 500 times more potent than any of the others.

The excitatory effects with $\gamma$-endorphin are somewhat strain-dependent. Certain species of rat, particularly the one in California called the "Simonson-Sprague-Dawley," demonstrate the $\gamma$-endorphin excitation response; the aggression is illustrated perhaps by the fact that I was bitten by several rats while handling them under ordinary conditions. We have not yet given any of these substances to primates in any significant number of experiments.

*Dr. Sachar:* In some of the papers describing these exciting effects on the animals, it has been proposed that they are a model of catatonia, or perhaps more properly "ratatonia." If you believe that you may be producing a model psychosis, have you tested the effects of neuroleptic drugs such as haloperidol?

*Dr. Bloom:* Such experiments have been done. Haloperidol does not remove the catatonic-like effects seen with β-endorphin. Only naloxone is the specific and highly sensitive material.

If one wanted to postulate a synaptic arrangement in which neuroleptics could account for this antipsychotic activity on the basis of an interaction with the endorphin peptides, it would be necessary only to point to the known dopamine innervation of the pituitary's intermediate lobe: That lobe is the major source of endorphin in the pituitary and the rat innervated by dopamine fibers. Such innervation could be a point at which antipsychotic drugs exert an antidopamine action on pituitary endorphins. The problem is that in the human being (where schizophrenia occurs), there is no intermediate lobe in the adult and therefore, the extent of true intrinsic central nervous system control of pituitary factors is unknown.

Secondly, it should be stated for the broadest audience that Dr. Bunny's lab at NIMH and Dr. Janowski's investigations at the Veteran's Hospital at the University of California, San Diego, have attempted to test the possibility that naloxone might be an antipsychotic substance in man, as was reported by a Swedish group led by Lars Terhenius. So far, we and Dr. Bunny have been unable to confirm any antipsychotic effect of naloxone in double-blind studies.

Our designation of catatonia-producing effects of β-endorphin in rats is based on the American Handbook of Psychiatry's definition of catatonia which is stated independent of the species: "A state of rigid immobility." That is precisely what is seen following administration of β-endorphin. It is a very profound and interesting kind of rigid immobility because the animal is so rigid that he will support himself when placed across bookends. Yet, if the rat is taken from its home cage and placed into a new cage, it will get up and explore the new cage for 1 to 3 min, followed by gradual diminution of spontaneous movement and a complete resumption of the β-endorphin rigid immobility. We find this form of behavioral state very interesting, even if it does not precisely duplicate schizophrenia.

*Dr. Baldessarini:* Is it technically feasible to synthesize peptides by combining neurophysin with a nonapeptide which could then be used as an antigen? The resulting antibodies might be used to identify precursors in the cell bodies where the precursor would be expected to arise and compare the staining properties of nerve terminals.

*Dr. Guillemin:* Yes, I think it can be done.

*Dr. Kety:* I would like to follow up a point that Dr. Guillemin has made regarding the need for a more precise concept of neurotransmitter. The situation may be clarified a bit if we think of the brain as sharing some of the regulatory mechanisms seen in the periphery. In the peripheral neuromuscular system, there is the requirement for highly specific circuitry such as that involved in playing a Liszt concerto. This requires a specific temporospacial pattern of activity at millions of myoneural junctions. We also recognize acetylcholine as the neurotransmitter involved. In the brain as well there is a need for precise and specific transmission at particular synapses such as those involved in perception or in the programming of a behavior pattern.

On the other hand, we recognize in the periphery neurogenic mechanisms such as the sympathetic and parasympathetic systems that the effects are considerably less specific both in time and in place; in fact, their adaptive function is to affect a large number of targets all at once. In the brain, we see what I believe to be a counterpart of the peripheral sympathetic nervous system in the locus ceruleus, now recognized as a most important neuroadrenergic center with a widespread distribution throughout the brain.

Finally, there are mechanisms in the periphery for broadcasting chemical substances throughout the entire organism, eliciting adaptive responses from widely dispersed target cells. This is the endocrine system. I think it is likely that the brain also employs that mode of distribution, using a humoral mechanism to distribute activating substances through the cerebrospinal fluid, the lymphatics of the brain, the intrinsic portal systems, as well as the transport of hormones from the periphery to neuronal regions via the cerebral circulation. It is quite possible that the same substance may be used as a neurotransmitter in one instance, a modulator in another, and a hormone in still a third mode of operation.

# Extrahypothalamic and Phylogenetic Distribution of Hypothalamic Peptides

## *Ivor M. D. Jackson

The portal vessel chemotransmitter hypothesis postulates that hormones, synthesized by neurons in the hypothalamus, are transported to nerve endings in the stalk-median eminence (SME) region where they are released into the interstitial space in contiguity with the primary portal capillary plexus, and thence transported by the portal veins to the adenohypophysis (25). Following the isolation, synthesis, and subsequent development of highly specific radioimmunoassays for three of the hypophysiotrophic hormones, thyrotropin releasing hormone (TRH), luteinizing hormone releasing hormone (LRH), and somatostatin (growth hormone release inhibiting hormone), an unexpected finding was the presence of large quantities of these peptides not only in parts of the brain outside the hypothalamus, but, in the case of somatostatin, outside the central nervous system altogether. So pronounced is the extrahypothalamic distribution of TRH, particularly in inframammalian species, that it seems reasonable to speculate that these hormones subserve an important function in neurotransmission quite apart from their role in anterior pituitary regulation. The finding of extrahypothalamic sources of hypophysiotrophic hormones provides some support for the view (ependymal tanycyte theory) that a portion of the releasing hormones reach the primary portal plexus by trans-median eminence transport, it being postulated that the releasing hormones are secreted into the ventricular system, taken up by the lumenal processes of the tanycytes of the median eminence, and then actively transported for release at the capillary end of the cell (38).

This chapter reviews recent findings concerning the extrahypothalamic and phylogenetic distribution of TRH, LRH, and somatostatin, and discusses their extrapituitary functional significance. Other hypophysiotrophic releasing factors which have not as yet been isolated but whose existence is postulated on the basis of biological activity and physiological inference are also mentioned briefly.

---

* Endocrine Division, Department of Medicine, New England Medical Center, Tufts University School of Medicine, Boston, Massachusetts 02111

## THYROTROPIN RELEASING HORMONE

### Extrahypothalamic Brain TRH

Small but significant concentrations of TRH (compared with the hypothalamus) are found in the rat extrahypothalamic brain (31) (Table 1), but quantitatively over 70% of total brain TRH is found outside the hypothalamus (45,58). In an attempt to determine the source of extrahypothalamic TRH, we have studied the effects of classic thyrotrophic area lesions which bring about a reduction of hypothalamic TRH by two-thirds. The extrahypothalamic brain TRH content was unaffected in rats so treated providing support for the intriguing hypothesis that synthesis occurs *in situ* outside the hypothalamus (32). Complementary studies to these experiments by Brownstein et al. (13), using hypothalamic deafferentation, demonstrated that such procedures not only leave the levels of TRH in the extrahypothalamic brain unaltered, but cause a marked reduction in hypothalamic content, suggesting that much of hypothalamic TRH may be synthesized by cells outside this region. Studies by Hökfelt et al. (28), using immunohistochemical staining, have localized TRH in the lower brainstem and spinal cord. Networks of TRH-positive nerve terminals were observed in several cranial nerve nuclei and around the motoneurons in the spinal cord. It is of interest that TRH-positive fibers were observed extending into the posterior pituitary, for we have previously shown that TRH concentration in the posterior pituitary is as much as 10 times that in the anterior pituitary (30,32), raising the possibility of TRH being of importance in posterior pituitary function. Unlike the situation in inframammalian species (see later), we have found only trace amounts of TRH in the rat pineal, and these are unaltered by environmental lighting.

As in the fetal rat (21), significant concentrations of TRH are present in the extrahypothalamic brain of the human fetus (59), TRH being detected in the cerebellum as early as 9 weeks.

Interestingly, the cerebellum of an anencephalic infant contained a relatively high concentration of TRH (59). It should be noted, however, that the area cerebrovasculosa—an area lacking in nerve cells—taken from an anencephalic fetus has been reported to synthesize a TSH releasing substance *in vitro* (29). Extrahypothalamic brain tissue from normal human adults (killed in traffic accidents) contains significant concentrations of immunoassayable TRH (IR-TRH) in the thalamus and cerebral cortex (44).

### Phylogenetic Distribution of TRH

The hypothalamus of all classes of vertebrates examined (rat, chicken, snake, frog, tadpole, and salmon) has high concentrations of TRH (31). Amphibia have especially high TRH levels in the hypothalamus compared

TABLE 1. Mean TRH concentration (pg/mg tissue wet weight) in various parts of the rat brain

| Brain-stem | Cerebellum | Mesencephalon | Diencephalon | Olfactory lobe | Cerebral cortex | Dorsal hypothalamus | Ventral hypothalamus | Stalk median eminence | Posterior pituitary |
|---|---|---|---|---|---|---|---|---|---|
| 5 (4–5)[a] | 2 (1–3) | 1 (1–2) | 6 (3–12) | 6 (5–8) | 1 (1–3) | 49 (41–61) | 64 (23–106) | 3,570 (920–7,600) | 155 (150–160) |

[a] Indicates range of values in individual determinations.

TABLE 2. *TRH concentration (pg/mg tissue wet weight) in the cerebral cortex (forebrain) and hypothalamus of various vertebrate species*

| Species | Rat | Chicken | Snake | Frog | Tadpole | Salmon | Lamprey (*Ammocoetes*) |
|---|---|---|---|---|---|---|---|
| Cerebral cortex (forebrain) | 1 (1–3) | 9 (8–10) | 264 | 111 (71–150) | 477 | 37 (22–52) | — |
| Hypothalamus | 280 (260–300) | 41 (34–49) | 393 | 2,270 (1,520–3,620) | 947 (568–1,225) | 235 (188–264) | (Whole head) 38 (25–60) |

For the lamprey only the head region was examined. Range of concentration is given in parentheses ($N=4$) except for snake tissue and tadpole brain where the tissue from only 1 animal was studied).

with rat (3,620 pg/mg tissue versus 300 pg/mg tissue), although TRH has no effect in stimulating pituitary-thyroid function in species lower than *aves*. In snake, frog, tadpole, and salmon, TRH is found outside the hypothalamus in high concentrations, values much higher than those in the respective brain area in the rat (Table 2). The cerebellum of the frog contains 520 pg/mg tissue, and salmon olfactory lobe 165 pg/mg tissue. The identity of these substrates with TRH is suggested by showing parallel inhibition curves by immunoassay (Fig. 1). Evidence that this material is indeed true biologically active TRH is shown by the ability of an extract of frog extrahypothalamic brain to release rat TSH *in vivo*, commensurate with its IR-TRH content. TRH is also found in the whole brain of the larval lamprey, in the head end of the amphioxus (31), and in the circumesophageal ganglia of the snail (22).

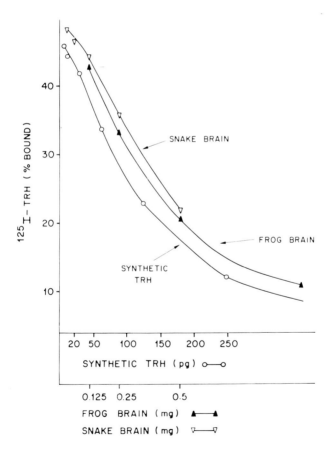

**FIG. 1.** Effect of dried methanol extracts of snake brain (cortex) and pooled extrahypothalamic frog brain on the inhibition of $^{125}$I-TRH binding with anti-TRH. Note parallelism of these inhibition curves with that of synthetic TRH. From Jackson and Reichlin (31).

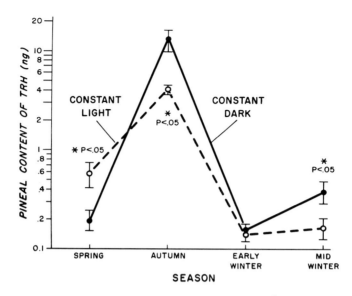

**FIG. 2.** The effect of season and illumination (constant light or darkness for 72 hr) on the pineal content of immunoreactive TRH (mean ± SEM) in the frog (*Rana pipiens*). In each group 4 to 6 animals were studied. Significant differences between groups exposed to constant light or darkness were observed in spring, autumn, and mid-winter, but not in early winter. Differences in content relative to the season were highly significant. (Note use of log scale.) From Jackson et al. (33).

As the lamprey lacks TSH and the amphioxus and snail lack a pituitary, we suggest that the TSH-regulating functions of TRH may be a late evolutionary development representing an example of an organism acquiring a new function for a preexisting chemical substance or hormone, analogous to the evolution of neurohypophysial hormones. In a sense, the pituitary has "co-opted" TRH as a regulatory hormone.

TRH is present in the frog (*Rana pipiens*) pineal in high concentrations which are influenced by the degree of photoillumination. Changing seasons are associated with swings in pineal TRH concentrations as much as 10 to 20-fold (33) (Fig. 2). The function of TRH in the frog pineal is unknown, but a role in neuronal transmission is postulated.

### TRH in Body Fluids

The presence of IR-TRH in mammalian blood, urine, and cerebrospinal fluid has been reported (see 51 for review). The significance of these findings, as well as the source and identity of the immunoreactive material reported, is controversial. In the frog (*Rana pipiens*), we have found that the IR-TRH in the circulation is chromatographically identical with native TRH and releases rat TSH *in vivo*.

## LUTEINIZING HORMONE RELEASING HORMONE

### Extrahypothalamic Brain LRH

LRH bioactivity was reported some years ago to be present in the mesencephalon of the dog and rabbit (20). By immunohistochemical techniques, cells rostral to the anterior commissure and the lamina terminalis stain with anti-LRH antibody. The organum vasculosum of the lamina terminalis (OVLT) also stains for LRH. This tissue is one of a group of specialized ependymal midline structures, known as circumventricular organs, which include the median eminence (ME), the subfornical organ (SFO), the area postrema (AP), and the subcommissural organ (SCO). Using a microdissection technique, Kizer et al. (37) have demonstrated that the circumventricular organs of the rat all contain large concentrations of LRH and somewhat lesser quantities of TRH. The origin of the releasing factors in these tissues is uncertain, but the possibilities include synthesis *de novo* and transport via ventricular CSF or axoplasmic flow to nerve terminals in the circumventricular organs (37). Gross (23), however, was unable to find LRH in the subfornical or subcommissural organs of the mouse.

Initial reports of large quantities of LRH in extracts of whole rat brain and ovine, bovine, and porcine pineals (56) have not been confirmed by others (3). Extracts of various regions of rat brain in our hands have revealed relatively low quantities of LRH in the total extrahypothalamic brain when compared with the levels of TRH. We find the concentration of LRH in the rat pineal to be at the limits of assay sensitivity, consistent with the reports from other laboratories (3,23). The claim that ovine, bovine, and porcine pineal glands contain large quantities of LRH has been retracted (57).

Using a sensitive radioimmunoassay for LRH, Cramer and Barraclough (16) were unable to detect immunoreactive LRH in third ventricle cerebrospinal fluid of the rat under a variety of experimental conditions. These results suggest that the CSF does not serve as a vehicle for transport of LRH to the median eminence under physiological conditions. However, Morris and Knigge (43) report that adult male rats show a significant rise of IR-LRH in the CSF as well as in blood after ether stress.

Immunoassayable and bioassayable LRH is present in the medial basal hypothalamus—especially in the arcuate nucleus (ARC) and in preoptic suprachiasmatic tissue (55). As with TRH, deafferentation of the rat hypothalamus causes a marked reduction in the LRH content of the medial basal hypothalamus, suggesting that such LRH arises from, or is controlled by, cells elsewhere in the brain (12,36). This view is supported by work from our group. Mice given monosodium glutamate show degeneration of over 80% of the cell bodies in the ARC without the IR-LRH content being affected (Table 3), suggesting that LRH may not be synthesized there but may be transported to the ME by axons passing through the ARC (40). The

TABLE 3. *Failure of monosodium glutamate (MSG) treatment to affect the hypothalamic content of LRH in 10-day-old albino mice*

|  | LRH | | TRH | |
|---|---|---|---|---|
|  | pg/tissue | pg/mg protein | pg/tissue | pg/mg protein |
| Hypothalamus |  |  |  |  |
| Glutamate | 1,275 ± 106 | 7,734 ± 1,389 | 2,178 ± 165 | 12,914 ± 1,637 |
| Control | 1,371 ± 343 | 7,493 ± 2,427 | 2,532 ± 159 | 13,393 ± 2,297 |
| Extrahypothalamic Brain |  |  |  |  |
| Glutamate | <66 | <18.5 | 3,601 ± 178 | 829 ± 68 |
| Control | <66 | <16.3 | 3,417 ± 144 | 840 ± 57 |

TRH was measured as a control for LRH and also shows no effect from MSG. The levels of LRH in the extrahypothalamic brain extracts from both glutamate-treated and control animals were undetectable in contrast with the large quantities of TRH present in the extrahypothalamic brain of both groups. Results show mean ± SEM. Six animals were studied in each group. From Lechan et al. (40).

data are consistent with the hypothesis of a dual central influence on pituitary LH—the LRH from the medial basal hypothalamus controlling the tonic, and the LRH from preoptic-suprachiasmatic nuclei the cyclic, secretion of LH. However, the LRH in the ARC of the rat (and mouse), involved in the tonic discharge of LH from the anterior pituitary, may in fact be synthesized in the preoptic area (36). It seems likely that species differences exist, for surgical isolation of the medial basal hypothalamus (MBH) of the guinea pig causes only a slight reduction in LRH content of the median eminence (53). These data imply an LRH-synthesizing locus *intrinsic* to the MBH, and in this regard the guinea pig resembles the monkey (39) rather than the rat. The situation with regard to the human is uncertain, but large quantities of LRH have been observed in the preoptic region of men and women (8).

### Phylogenetic Distribution of LRH

Immunoreactive LRH is reported in chicken (34) and amphibian (1,17) but not piscine (17) hypothalamus. Studies from our group have shown that approximately 16% of total frog brain LRH (*Rana pipiens* and *Rana catesbeiana*) is located within the telencephalon-septum-optic chiasm regions, whereas the remainder is distributed within the infundibular hypothalamic-pituitary complex (1).

In the frog, immunohistochemical studies using the "PAP" immunoglobulin enzyme bridge technique demonstrated LRH within neuronal perikarya in the diagonal band of Broca and in the median septal nucleus intermingled with nonimmunoreactive neurons. Fibers containing immunoreactive LRH were seen in the vicinity of these neurons and in the medial and lateral septal nuclei. In addition, LRH-containing fibers extended to the median eminence and posterior pituitary traversing a course beneath the preoptic recess, or through the medial forebrain bundle, and then through the lateral infundibular hypothalamus to enter the median eminence bilaterally. LRH within the median eminence was located in both the inner subependymal and outer palisade zones. On the basis of earlier physiological studies, it was proposed that this LRH peptidergic septoinfundibular pathway is involved in control of cyclic gonadotropin activity in the frog (1).

Exogenous LRH can activate gonadal function in amphibia and fish as well as mammals. This contrasts with TRH, which although having a wider phylogenetic and extrahypothalamic distribution than LRH, stimulates pituitary-thyroid function only in mammalian and avian species.

### LRH in Body Fluids

LRH has been detected by both bioassay and immunoassay in the peripheral circulation and CSF in some laboratories but not in others. The identity, concentration, and source of this material are controversial. This issue has recently been discussed *in extenso* (51).

## SOMATOSTATIN

### Extrahypothalamic Distribution of Somatostatin

#### Central Nervous System

Somatostatin has been found by both immunoassay (11,46) and bioassay (54) to be widely distributed in extrahypothalamic brain, including the pineal gland. Using an immunoperoxidase technique, investigators have localized somatostatin in the circumventricular organs, SCO and OVLT, in addition to the external zone of the ME (47). Somatostatinergic neurons have been detected immunohistochemically in the anterior periventricular hypothalamus and in the preoptic area, in part outside the confines of the hypothalamus (2). Physiologic studies suggest that such neurons inhibit the release of GH, and possibly TSH, from the anterior pituitary.

Hypothalamic deafferentation caudal to the optic chiasm has been shown to reduce the immunoassayable and immunohistochemical content of somatostatin in the MBH (19). These studies provide further support for the view that the somatostatin-positive perikarya of the periventricular nucleus are the source of the somatostatin-positive fibers and terminals in the MBH and stalk median eminence.

Somatostatin-immunopositive nerve fibers have also been found in the posterior pituitary (26). With the indirect immunofluorescence technique, somatostatin has been detected in some neuronal cell bodies in spinal dorsal root ganglia, as well as in fibers in the substantia gelatinosa of the spinal cord (27).

*Gastrointestinal Tract*

Somatostatin is present in the mammalian stomach and pancreas (4,46) where it is localized to the argyrophilic D ($A_1$) cells (26), and in nerves in different layers of the walls of the small and large intestine (26). The distribution of somatostatin in the gastrointestinal tract corresponds with its site of action in inhibiting glucagon, insulin, gastrin, and HCl secretion. It seems likely that somatostatin is formed *in situ* both in the extrahypothalamic brain and in the gastrointestinal tract.

Somatostatin, so named because it was initially found to inhibit growth hormone secretion, might now be termed "endocrinostatin" in recognition of the ubiquity of its site of action!

## FUNCTION OF THE EXTRAHYPOTHALAMIC DISTRIBUTION OF THE HYPOTHALAMIC RELEASING FACTORS

The widespread distribution of TRH, LRH, and somatostatin throughout the extrahypothalamic brain suggests a role for these substances in neuronal function apart from anterior pituitary regulation. The evidence supporting this view is discussed below.

### Anatomic and Phylogenetic Distribution

The widespread distribution of TRH in the nervous tissue of submammalian species is very suggestive of a role in neuronal function. Further, the location of TRH in several cranial nerve cell nuclei of the brainstem and motor nuclei of the spinal cord supports a transmitter role for this peptide, particularly in the motor system (28). The finding that a certain population of primary sensory nerves contain somatostatin suggests that this substance may act as a depressant neurotransmitter in sensory neurons (27).

The finding of TRH and LRH in synaptosomes by tissue fractionation techniques (7) is strong evidence for a role for these peptides in neuronal function.

### Neurophysiologic Studies

Using microiontophoresis, it has been shown that TRH, LRH, and somatostatin have a depressant action on the excitability of neurons in several areas of the CNS (6,52).

## Specific Brain Receptors and Enzymatic Degrading Systems

High-affinity binding sites for TRH in the synaptic membrane fraction of brain tissue have been demonstrated. These receptors have properties similar to that for pituitary membranes (14). In addition, active enzymatic degrading systems exist for TRH, LRH, and somatostatin in brain tissue (51).

## Behavioral Effects (see 51 for review)

TRH affects behavior of both mouse and man in circumstances unrelated to pituitary-thyroid activation. LRH excites sexual activity in rats in whom gonadal function is held constant. Somatostatin has been found to have behavioral effects in rats distinct from, and in many cases directly opposite to, those of TRH (10).

## Posterior Pituitary Function

The demonstration of TRH, LRH, and somatostatin by either immunoassay or immunohistochemistry in the posterior pituitary provides support for the possible existence of a third neurosecretory hypothalamo-neurohypophysial system (26).

# OTHER HYPOTHALAMIC RELEASING FACTORS

## Corticotropin Releasing Factor

The status of corticotropin releasing factor (CRF), still chemically unidentified, has been recently summarized (35). CRF activity has been reported in the CSF of the dog and peripheral circulation of the rat (41).

## Melanocyte Stimulating Hormone Inhibitory and Releasing Factors

The status of hypothalamic hormones regulating the secretion of melanocyte stimulating hormone (MSH) is controversial (50). Extracts of frog and rat extrahypothalamic brain inhibit MSH release (48). One of the peptides proposed to be the physiological MSH inhibitory factor is a tripeptide, prolyl-leucyl-glycinamide, derived by enzymatic cleavage from oxytocin (15). In man, this peptide has been reported to have antidepressant activity (18).

## Prolactin Release and Inhibiting Factors

TRH is active in releasing prolactin from the anterior pituitary, but there is evidence of another peptide with prolactin releasing factor (PRF) activity in the hypothalamus (9). There are preliminary reports of a PRF in human plasma not associated with TSH release (42).

It has been suggested that the prolactin inhibiting factor (PIF) activity in porcine hypothalami is due to dopamine (51). However, studies in our laboratory indicate the presence of an additional PIF activity, possibly peptide in nature, not due to catecholamines. PIF activity is widely distributed throughout extrahypothalamic rat brain (54), but whether such activity is due to a peptide or dopamine is uncertain at this time.

### Growth Hormone Releasing Factor

Crude extracts of hypothalamic tissue are capable of releasing immunoassayable growth hormone from the pituitary (see 49 for review). The chemical structure of growth hormone releasing factor (GRF) is unknown. In the human, a GRF has been detected in acromegalic plasma (24) and in cerebrospinal fluid (5).

The significance of the extrahypothalamic distribution of these putative hypothalamic releasing factors (? low molecular weight peptides) awaits their isolation and synthesis. It is speculated that they subserve a similar neuronal function to that postulated for TRH, LRH, and somatostatin.

## SUMMARY AND CONCLUSIONS

The anatomic distribution of TRH, LRH, and somatostatin in the brain, body fluids, and extraneural tissues of mammalian and submammalian species has been reviewed. It seems clear that these peptides, as well as other putative hypothalamic hypophysiotropic releasing factors, subserve a role in brain function quite distinct from their place in the regulation of pituitary gland secretion. Studies using microiontophoresis, immunohistochemistry, brain receptors, and behavioral psychology lend support to the hypothesis that they are central neurotransmitters.

## ACKNOWLEDGMENT

The work reported from the author's laboratory was supported in part by USPHS Grant No. AM 16684.

## REFERENCES

1. Alpert, L. C., Brawer, J. R., Jackson, I. M. D., and Reichlin, S. (1976): Localization of LHRH in neurons in frog brain (*Rana pipiens* and *Rana catesbeiana*). *Endocrinology*, 98:910–921.
2. Alpert, L. C., Brawer, J. R., Patel, Y. C., and Reichlin, S. (1976): Somatostatinergic neurones in anterior hypothalamus: Immunohistochemical localization. *Endocrinology*, 98: 255–258.

3. Araki, S., Ferin, M., Zimmerman, E. A., and Vande-Wiele, R. L. (1975): Ovarian modulation of immunoreactive gonadotropin-releasing hormone (Gn-RH) in the rat brain: Evidence for a differential effect on the anterior and mid-hypothalamus. *Endocrinology*, 96:644–650.
4. Arimura, A., Sato, H., Dupont, A., Nishi, N., and Schally, A. V. (1975): Somatostatin: Abundance of immunoreactive hormone in rat stomach and pancreas. *Science*, 189:1007–1009.
5. Barbato, T., Lawrence, A. M., and Kirsteins, L. (1974): Cerebro-spinal fluid stimulation of pituitary protein synthesis and growth hormone release in vitro. *Lancet*, 1:599–600.
6. Barker, J. L. (1976): Peptides: Roles in neuronal excitability. *Physiol. Rev.*, 56:435–452.
7. Bennett, G. W., Edwardson, J. A., Holland, D., Jeffcoate, S. L., and White, N. (1975): Release of immunoreactive luteinizing hormone-releasing hormone and thyrotrophin-releasing hormone from hypothalamic synaptosomes. *Nature*, 257:323–325.
8. Bird, E. D., Chiappa, S. A., and Fink, G. (1976): Brain immunoreactive gonadotropin-releasing hormone in Huntington's chorea and in non-choreic subjects. *Nature*, 260:536–538.
9. Boyd, A. E., Spencer, E., Jackson, I. M. D., and Reichlin, S. (1976): Prolactin releasing factor (PRF) in porcine hypothalamic extract distinct from TRH. *Endocrinology*, 99:861–871.
10. Brown, M., and Vale, W. (1975): Central nervous system effects of hypothalamic peptides. *Endocrinology*, 96:1333–1336.
11. Brownstein, M., Arimura, A., Sato, H., Schally, A. V., and Kizer, J. S. (1975): The regional distribution of somatostatin in the rat brain. *Endocrinology*, 96:1456–1461.
12. Brownstein, M. J., Arimura, A., Schally, A. V., Palkovits, M., and Kizer, J. S. (1976): The effect of surgical isolation of the hypothalamus on its luteinizing hormone-releasing hormone content. *Endocrinology*, 98:662–665.
13. Brownstein, M. J., Utiger, R. D., Palkovits, M., and Kizer, J. S. (1975): Effect of hypothalamic deafferentation on thyrotropin releasing hormone levels in rat brain. *Proc. Natl. Acad. Sci. USA*, 72:4177–4179.
14. Burt, D. R., and Snyder, S. H. (1975): Thyrotropin releasing hormone (TRH): Apparent receptor binding in rat brain membranes. *Brain Res.*, 93:309–328.
15. Celis, M. E., Taleisnik, S., and Walter, R. (1971): Regulation of formation and proposed structure of the factor inhibiting the release of melanocyte stimulating hormone. *Proc. Natl. Acad. Sci. USA*, 68:1428–1433.
16. Cramer, O. M., and Barraclough, C. A. (1975): Failure to detect luteinizing hormone-releasing hormone in third ventricular cerebrospinal fluid under a variety of experimental conditions. *Endocrinology*, 96:913–921.
17. Deery, D. J. (1974): Determination by radioimmunoassay of the luteinizing hormone-releasing hormone (LHRH) content of the hypothalamus of the rat and some lower vertebrates. *Gen. Comp. Endocrinol.*, 24:280–285.
18. Ehrensing, R. H., and Kastin, A. J. (1974): Melanocyte-stimulating hormone release inhibiting hormone as an antidepressant: A pilot study. *Arch. Gen Psychiatry*, 30:63–65.
19. Elde, R., Hokfelt, T., Johansson, O., Effendic, S., and Luft, R. (1976): Immunohistochemical and radioimmunoassay studies on hypothalamic somatostatin. Program V International Congress of Endocrinology. Hamburg, Fed. Rep. Germany, p. 271 (Abst.).
20. Endröczi, E., and Hilliard, J. (1965): Luteinizing hormone releasing activity in different parts of rabbit and dog brain. *Endocrinology*, 77:667–673.
21. Eskay, R. L., Oliver, C., Grollman, A., and Porter, J. C. (1974): Immunoreactive LRH and TRH in the fetal, neonatal and adult rat brain. Program 56th Meeting Endocrine Soc., P A-83 (Abst.).
22. Grimm-Jørgensen, Y., Mckelvy, J. F., and Jackson, I. M. D. (1975): Immunoreactive thyrotrophin releasing factor in gastropod circumoesophageal ganglia. *Nature*, 254:620.
23. Gross, D. S. (1976): Distribution of gonadotropin-releasing hormone in the mouse brain as revealed by immunohistochemistry. *Endocrinology*, 98:1408–1417.
24. Hagen, T. C., Lawrence, A. M., and Kirsteins, L. (1972): In vitro release of monkey growth hormone by acromegalic plasma. *J. Clin. Endocrinol. Metab.*, 33:448–451.
25. Harris, G. W. (1955): The function of the pituitary stalk. *Johns Hopkins Med. J.*, 97:358.
26. Hökfelt, T., Effendic, S., Hellerstrom, C., Johansson, O., Luft, R., and Arimura, A. (1975):

Cellular localization of somatostatin in endocrine-like cells and neurons of the rat with special references to the $A_1$ cells of the pancreatic islets and to the hypothalamus. *Acta Endocrinol. (Kbh.) [Suppl.]*, 5–41.
27. Hökfelt, T., Elde, R., Johansson, O., Luft, R., and Arimura, A. (1975): Immunohistochemical evidence for the presence of somatostatin, a powerful inhibitory peptide in some primary sensory neurons. *Neurosci. Lett.*, 1:231–235.
28. Hökfelt, T., Fuxe, K., Johansson, D., Jeffcoate, S., and White, N. (1975): Distribution of thyrotropin releasing hormone (TRH) in the central nervous system as revealed with immunohistochemistry. *Eur. J. Pharmacol.*, 34:389–392.
29. Ishikawa, H., Nagayama, T., Kato, C., and Niizuma, K. (1976): Establishment of a TSH-releasing-hormone-secreting cell line from the area cerebrovasculosa of an anencephalic fetus. *Am. J. Anat.*, 145:143–148.
30. Jackson, I. M. D., Gagel, R., Papapetrou, P., and Reichlin, S. (1974): Pituitary hypothalamic and urinary thyrotropin releasing hormone (TRH) concentration in altered thyroid states of rat and man. *Clin. Res.*, 22:342a.
31. Jackson, I. M. D., and Reichlin, S. (1974): Thyrotropin-releasing hormone (TRH): Distribution in hypothalamic and extrahypothalamic brain tissues of mammalian and submammalian chordates. *Endocrinology*, 95:854–862.
32. Jackson, I. M. D., and Reichlin, S. (1977): Brain thyrotrophin-releasing hormone is independent of the hypothalamus. *Nature* 267:853–854.
33. Jackson, I. M. D., Saperstein, R., and Reichlin, S. (1977): Thyrotropin releasing hormone (TRH) in pineal and hypothalamus of the frog: Effect of season and illumination. *Endocrinology*, 100:97–100.
34. Jeffcoate, S. L., Sharp, P. J., Fraser, H. M., Holland, D. T., and Gunn, A. (1974): Immunochemical and chromatographic similarity of rat, rabbit, chicken and synthetic luteinizing hormone releasing hormones. *J. Endocrinol.*, 62:85–91.
35. Jones, M. T., Hillhouse, E., and Burden, J. (1976): Secretion of corticotropin-releasing hormone in vitro. In: *Frontiers in Neuroendocrinology, Vol. 4*, edited by L. Martini and W. F. Ganong, pp. 195–226. Raven Press, New York.
36. Kalra, S. P. (1976): Tissue levels of luteinizing hormone-releasing hormone in the preoptic area and hypothalamus, and serum concentration following anterior hypothalamic deafferentation and estrogen treatment of the female rat. *Endocrinology*, 99:101–107.
37. Kizer, J. S., Palkovits, M., and Brownstein, M. J. (1976): Releasing factors in the circumventricular organs of the rat brain. *Endocrinology*, 98:311–317.
38. Knigge, K. M., Scott, D. E., and Weindl, A. (Eds.) (1972): *Brain-Endocrine Interaction. Median Eminence: Structure and Function.* S. Karger, Basel.
39. Krey, L. C., Butler, W. R., and Knobil, E. (1975): Surgical disconnection of the medial basal hypothalamus and pituitary function in the rhesus monkey. I. Gonadotropin secretion. *Endocrinology*, 96:1073–1093.
40. Lechan, R. M., Alpert, L. C., and Jackson, I. M. D. (1976): Synthesis of luteinizing hormone releasing factor and thyrotropin releasing factor in glutamate-lesioned mice. *Nature*, 264:463–465.
41. Lymangrover, J. R., and Brodish, A. (1973): Physiological regulation of tissue-CRF. *Neuroendocrinology*, 13:234–235.
42. Malarkey, W. B., and Pankratz, K. (1974): Evidence for prolactin releasing activity (PRA) in human plasma not associated with TSH release. *Clin. Res.*, 22:600A.
43. Morris, M., and Knigge, K. M. (1975): Effect of ether anesthesia on LH-releasing hormone (LH-RH) secretion. *Fed. Proc.*, 34:239.
44. Okon, E., and Koch, Y. (1976): Localization of gonadotrophin-releasing and thyrotrophin-releasing hormones in human brain by radioimmunoassay. *Nature*, 263:345–347.
45. Oliver, C., Eskay, R. L., Ben-Jonathan, N., and Porter, J. C. (1974): Distribution and concentration of TRH in the rat brain. *Endocrinology*, 95:540–546.
46. Patel, Y. C., Weir, G. C., and Reichlin, S. (1975): Anatomic distribution of somatostatin (SRIF) in brain and pancreatic islets as studied by radioimmunoassay. Program 57th Annual Meeting Endocrine Society, New York, p. 127 (Abst.).
47. Pelletier, G., Leclerc, R., Dube, D., Labrie, F., Puviani, R., Arimura, A., and Schally, A. V. (1975): Localization of growth hormone-release-inhibiting hormone (somatostatin) in the rat brain. *Am. J. Anat.*, 142:397–401.

48. Ralph, C. L., and Sampath, S. (1966): Inhibition by extracts of frog and rat brain of MSH release by frog pars intermedia. *Gen. Comp. Endocrinol.*, 7:370–374.
49. Reichlin, S. (1975): Regulation of somatotrophic hormone secretion. In: *Handbook of Physiology, Endocrinology IV, Part 2*, pp. 405–447. American Physiological Society, Washington, D.C.
50. Reichlin, S., and Mitnick, M. A. (1973): Biosynthesis of hypothalamic hypophysiotrophic hormones. In: *Frontiers in Neuroendocrinology*, edited by W. F. Ganong and L. Martini, pp. 61–68. Raven Press, New York.
51. Reichlin, S., Saperstein, R., Jackson, I. M. D., Boyd, A. E., and Patel, Y. (1976): Hypothalamic hormones. *Annu. Rev. Physiol.*, 38:389–424.
52. Renaud, L. P., Martin, J. B., and Brazeau, P. (1975): Depressant action of TRH, LH-RH and somatostatin on activity of central neurones. *Nature*, 255:233–235.
53. Silverman, A. J. (1976): Distribution of luteinizing-hormone-releasing hormone (LHRH) in the guinea pig brain. *Endocrinology*, 99:30–41.
54. Vale, W., Rivier, C., Palkovits, M., Saavedra, J. M., and Brownstein, M. (1974): Ubiquitous brain distribution of inhibition of adenohypophysial secretion. Program 56th Meeting, Endocrine Society, Atlanta, p. A-128 (Abst.).
55. Wheaton, J. E., Krulich, L., and McCann, S. M. (1975): Localization of luteinizing hormone-releasing hormone in the preoptic area and hypothalamus of the rat using radioimmunoassay. *Endocrinology*, 97:30–38.
56. White, W. F., Hedlund, M. T., Weber, G. F., Rippel, R. H., Johnson, E. S., and Wilbur, J. F. (1974): The pineal gland: A supplemental source of hypothalamic-releasing hormones. *Endocrinology*, 94:1422–1426.
57. Wilbur, J. F., Montoya, E., Plotnikoff, N. P., White, W. F., Gendrich, R., Renaud, L., and Martin, J. B. (1976): Gonadotropin-releasing hormone and thyrotropin-releasing hormone: Distribution and effects in the central nervous system. *Recent Prog. Horm. Res.*, 32:157.
58. Winokur, A., and Utiger, R. D. (1974): Thyrotropin releasing hormone. Regional distribution in rat brain. *Science*, 185:265–267.
59. Winters, A. J., Eskay, R. L., and Porter, J. C. (1974): Concentration and distribution of TRH and LRH in the human fetal brain. *J. Clin. Endocrinol. Metab.*, 39:960–963.

## DISCUSSION

*Dr. Bloom:* Do you have comparable data on the phylogenetic distribution of the peptides other than TRH?

*Dr. Jackson:* We find that LHRH is present in the frog. Others have reported that LHRH is present in the chicken. Whereas TRH has no pituitary-thyroid function in species lower than aves, LRH can activate the gonadal function in the fish. Recently, we have found LHRH in platyfish hypothalamic extracts.

*Dr. Guillemin:* An interesting use of LHRH in fish is in the promotion of spawning in fish ponds in mainland China. A delegation of scientists from Mainland China, recently visiting my laboratory, told me that they had synthesized LHRH in kilogram amounts, and that its greatest use was for fish culture where it may double or even triple the number of spawnings and hence increase productivity.

*The Hypothalamus*, edited by S. Reichlin,
R. J. Baldessarini, and J. B. Martin,
Raven Press, New York, © 1978.

# Peptide Neurotransmitter Candidates in the Brain: Focus on Enkephalin, Angiotensin II, and Neurotensin

## *Solomon H. Snyder

In recent years interest has mushroomed in the possible role of peptides as neurotransmitters in the brain. Part of this interest derived from findings that some peptide releasing factors of the hypothalamus were distributed ubiquitously throughout the brain. Other peptides were identified for a variety of seemingly accidental reasons. Enkephalins, the CNS opioid peptides, were characterized in a systematic search for an endogenous ligand of the opiate receptor. In this chapter we will focus only on enkephalins, angiotensin II, and neurotensin.

### ENKEPHALIN: THE BRAIN'S MORPHINE-LIKE PEPTIDE

Opiate-like peptides in the central nervous system were discovered in a fairly indirect fashion. The major impetus derived from the demonstration in recent years of specific opiate receptor sites in vertebrate brain which mediate pharmacological responses to opiates (17). In subcellular fractionation studies the opiate receptor appears to be localized to synaptic membranes as would be expected for a neurotransmitter receptor. Autoradiographic studies show an extremely discrete localization of opiate receptor sites to particular regions throughout the central nervous system (13).

The heterogeneous regional localization of the opiate receptor, with highest concentrations in areas related to pain perception and emotional behavior, fits with known actions of opiates. The regional and subcellular studies also suggested that the opiate receptor might interact with some naturally occurring substance. The specificity of the receptor seemed too great for an accidental membrane protein which interacts only with exogenous drugs. To identify such a hypothethical morephine-like factor, Hughes (7) took advantage of the ability of morphine and other opiates to inhibit electrically induced contractions of smooth muscle such as the guinea

---

* Departments of Pharmacology and Experimental Therapeutics, and Psychiatry and Behavioral Sciences, Johns Hopkins University School of Medicine, 725 North Wolfe Street, Baltimore, Maryland 21205

pig ileum or mouse vas deferens. He found a substance in brain extracts which, like morphine, inhibited electrically induced contractions of the mouse vas deferens and guinea pig ileum.

In our own laboratory (1-2) and that of Terenius and Wahlstrom (18), opiate receptor binding was used as an assay to identify and then purify the brain's morphine-like factor, which will hereafter be referred to as "enkephalin."[1]

Enkephalin activity of brain extracts represents their ability to inhibit the binding of radioactive opiates to the opiate receptor. Since many substances, especially ions, can interfere with opiate receptor binding, careful attention was devoted in purification studies to ensure that only a physiologically relevant substance was being studied. Our primary means of ensuring specificity was to show that the regional distribution of enkephalin activity throughout the brain paralleled that of the opiate receptor. We found negligible levels of both in the cerebellum, very high densities of receptor and enkephalin in the corpus striatum and hypothalamus, and intermediate values in other regions.

Using effects on smooth muscle as a bioassay, Hughes et al. (8) isolated enkephalin from pig brain and showed it to be a mixture of two pentapeptides, Tyr-Gly-Gly-Phe-Met-OH (methionine-enkephalin; m-enk) and Tyr-Gly-Gly-Phe-Leu (leucine-enkephalin; l-enk). Independently, by monitoring influences on opiate receptor binding, we isolated and identified the structures of the same two peptides in bovine brain (14). However, in pig brain, Hughes et al. (8) found four times more m-enk than l-enk, whereas in bovine brain we observed four times more l-enk than m-enk (14).

With synthetic enkephalin it has been possible to prepare antibodies with selectivity for m-enk and l-enk, respectively. We also succeeded in developing specific and sensitive radioimmunoassays for the two enkephalins (15). The radioimmunoassays provide discrete measurements of m-enk and l-enk, respectively. Using the ability of brain extracts to compete for opiate receptor binding, we can also measure total levels of enkephalin activity, but this radioreceptor assay of opioids cannot distinguish between the two forms of enkephalin. Relative amounts of enkephalin in different brain regions of the rat measured by the two techniques are similar, although absolute values by radioimmunoassay tend to be higher than those measured by radioreceptor assay. In most rat brain regions m-enk levels are 10 to 15 times higher than those for l-enk, and in the hippocampus l-enk is essentially undetectable.

Earlier studies in several species using a radioreceptor assay have shown

---

[1] Opioid activities of tissue extracts have been referred to with different terminology by various laboratories. The term "endorphin" is used generally to refer to peptides with opioid activities. $\alpha$-Endorphin and $\beta$-endorphin are specific opioid peptides isolated from the pituitary gland. Enkephalin is the opioid peptide which has been isolated only from the brain.

that the levels of enkephalin in different regions vary in close parallel to those of the opiate receptor (17). If enkephalin is a neurotransmitter, then the opiate receptor is likely to be its physiological synaptic receptor. Using antibodies to synthetic enkephalins, Hökfelt and collaborators (6) and ourselves (R. Simantov, M. J. Kuhar, and S. H. Snyder, *unpublished observations*) have recently visualized enkephalin neurons by immunofluorescence. Enkephalin is contained in systems of arborizing axons and terminals with numerous varicosities closely resembling the terminal networks of norepinephrine-, dopamine-, and serotonin-containing neurons. Cell bodies for enkephalin-containing neurons have not yet been identified. The microscopic mapping of enkephalin terminal systems throughout the brain closely resembles the autoradiographic mapping of opiate receptors. In the spinal cord enkephalin fluorescence is most concentrated in the dorsal gray matter in laminae I and II. In the medulla oblongata, fluorescent fibers are most dense in the lateral reticular nucleus, the nucleus ambiguus, and the nucleus tractus solitarius. In the midbrain the periaqueductal gray matter and zona compacta of the substantia nigra, both enriched in opiate receptors, contain enkephalin terminals. Most hypothalamic nuclei also display enkephalin fluorescence. The intralaminar nuclei of the thalamus contains a considerable number of enkephalin fibers, although other parts of the thalamus are not as richly endowed. The globus pallidus of the corpus striatum displays many enkephalin fibers, which accords with biochemical evidence that it possesses the highest levels of enkephalin in the brain, as determined by radioreceptor assay (16). The caudate nucleus contain a patchy distribution of enkephalin fibers, reminiscent of the patchy distribution of opiate receptors, with highest densities in the rostral and ventral portions of the nucleus as well as the lateral part of the interstitial nucleus of the stria terminalis. The amygdaloid nuclei, well known for their high concentration of opiate receptors, display substantial enkephalin fibers, especially in the central amygdaloid nucleus. Very little enkephalin fluorescence is observed in the cerebral cortex or the pituitary gland.

Since the distribution of enkephalin terminals and opiate receptors coincides so closely, one can draw some provocative inferences about the disposition of enkephalin systems in the brain by knowing the detailed localization of opiate receptors. In the spinal cord opiate receptors are highly localized to laminae I and II of the dorsal gray matter, the substantia gelatinosa, which is the first weigh station in the integration of sensory information.

Lesions of the dorsal root of the spinal cord in monkeys (9) result in a depletion of opiate receptors in the dorsal gray matter as detected by biochemical or autoradiographic techniques. The time course of this depletion is consistent with the degeneration of the sensory nerve terminals and not with postsynaptic trophic alterations. Thus it is likely that opiate receptors are contained on the nerve terminals of the sensory afferent fibers. Such a presynaptic localization is consistent with the existence of axoaxonic synapses between enkephalin terminals and afferent nerve terminals as occurs for synapses associated with presynaptic inhibition in the spinal cord. A similar presynaptic localization of opiate receptors occurs in the nucleus tractus solitarius and the nucleus ambiguus of the rat's brainstem, where opiate receptors, assayed by autoradiography, disappear after destruction of the vagus nerve in the neck (Atweh and Kuhar, *in preparation*). Similarly, opiate receptors in the terminal nuclei of the inferior accessory optic pathway, which is completely crossed in the rat, disappear after removal of the contralateral eye.

Opioid peptides are not restricted to the central nervous system. Enkephalin occurs in the gastrointestinal tract, and recently other larger opioid peptides have been detected in the pituitary gland. The amino acid sequences of two of these, $\alpha$-endorphin and $\beta$-endorphin, have been determined (5,10). Interestingly, m-enk, $\alpha$-endorphin, and $\beta$-endorphin are all contained in $\beta$-lipotropin, a 91 amino acid containing peptide isolated from pituitary, which may serve as a precursor of the small endorphin and enkephalin peptides (11) (Table 1). $\beta$-Endorphin and $\alpha$-endorphin are similar in potency to the enkephalins in competing for opiate receptor binding. Because $\beta$-endorphin, which contains the five amino acid sequence of m-enk within its structure, is less likely than enkephalin to be destroyed by proteolysis, it apparently reaches the brain after parenteral administration, and is five times as potent as morphine in producing analgesia when injected intravenously (19). Whether $\beta$-lipotropin, $\alpha$-endorphin, or $\beta$-endorphin is a natural precursor of the enkephalins is not yet clear, since these larger peptides and specific peptidase activities that would be required to liberate the smaller products have not been demonstrated in substantial amounts in the brain.

Some insight into the function of the pituitary opioid peptides has been obtained by binding studies demonstrating that the pituitary possesses opiate receptors (15). The characteristics of the pituitary opiate receptors are fairly similar to those of brain receptors except that enkephalins have less affinity for pituitary than brain receptors. Conceivably, the opioid peptides in the pituitary act directly on pituitary opiate receptors to produce effects known to be elicited by opiates, such as release of antidiuretic hormone. Such an action would accord with the observation that opiate receptor binding is much more enriched in the posterior than in the anterior pituitary.

## ANGIOTENSIN

There appear to be a large number of other peptides with possible central nervous functions. In the interest of brevity, we have focused on the enkephalins, although substance P is particularly well characterized as a

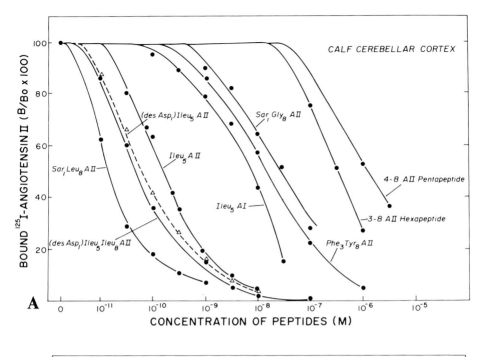
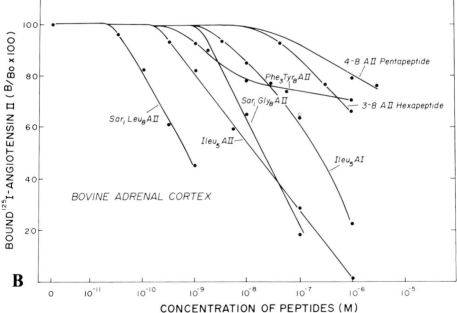

**FIG. 1.** Displacement of $^{125}$I-angiotensin II bound to calf brain **(A)** and bovine adrenal cortex **(B)** membranes by angiotensin peptides. $^{125}$I-angiotensin II (0.05 nM) was incubated with membranes, and bound radioactivity in the presence of increasing concentrations of various angiotensin peptides was assayed by filtration. $B_o$, binding in absence of unlabeled peptide; B, binding in presence of unlabeled peptides (1). More potent compounds are those which at lower concentrations reduce $^{125}$I-angiotensin II binding.

probable sensory transmitter candidate. There is also substantial evidence for a role of angiotensin II and neurotensin in the brain. Direct administration of small doses of angiotensin II into the brain can influence cardiovascular reflexes, blood pressure, and drinking behavior. Renin-like and angiotensin-converting enzymes exist in the brain. Recently we detected angiotensin II receptor binding in brain tissue similar to levels of this type of receptor binding that are found in the adrenal cortex—the peripheral tissue richest in angiotensin receptors (1). Indeed, in a survey of various tissues, only brain and adrenal cortex demonstrated detectable levels of angiotensin II receptor binding. The substrate specificities of these receptors in the brain and adrenal cortex are quite similar (Fig. 1). Taken together these data suggest a role for angiotensin in the brain.

Angiotensin receptor binding sites in the brain have about one order of magnitude greater affinity for angiotensin and its analogues than those in the adrenal cortex (1). The dissociation constant for angiotensin II at calf and rat brain receptors is about 0.2 nM. This high affinity, as well as the striking selectivity of the angiotensin receptor of brain tissue, has facilitated development of a radioreceptor assay for endogenous angiotensin II (J. P. Bennett, Jr. and S. H. Snyder, *unpublished observations*). Examination of tissue components by thin-layer chromatography or high-voltage paper electrophoresis indicates that the only substance reacting is angiotensin. In this assay, angiotensin II can be separated from angiotensin III (the hexapeptide lacking the N-terminal asparagine of angiotensin II) and endogenous levels of both measured. Using this technique brain levels of angiotensin II have been measured. The hypothalamus, with highest concentration region in rat brain, contains only about 0.7 pmoles/g wet weight. Levels in other regions are about 0.1 to 0.2 pmoles/g.

## NEUROTENSIN

Neurotensin is a tridecapeptide whose 13 amino acid sequence is <glu-leu-tyr-glu-asn-lys-pro-arg-arg-pro-tyr-ile-leu-OH (2). Neurotensin was isolated from hypothalamic extracts as an apparent by-product of the effort to isolate substance P. It was detected as a substance which in small doses could elicit hypotension, increased vascular permeability, pain sensation, increased hematocrit, cyanosis, morphine-inhibitable stimulation of ACTH secretion, increased LH secretion, increased FSH secretion, hyperglycemia, and a variety of smooth muscle effects including contraction of the rat uterus and guinea pig ileum.

Like enkephalin, somatostatin, and substance P, neurotensin is localized only to the central nervous system and gut (3,4). Radioimmunoassay reveals marked regional differences in neurotensin levels throughout the brain (3,4,20). A summary of our results appears in Table 2. In calf brain highest concentrations occur in the hypothalamus and basal ganglia with much

Table 1. Amino acid sequence of β-lipotropin and its biologically active constituent peptides

| | |
|---|---|
| β-Lipotropin: | $^{1}$NH$_2$Glu-Leu-Ala-Gly-Ala-Pro-Pro-Glu-Pro-Ala-Arg-Asp-Pro-Glu-Ala-Pro-Ala-Glu-Gly-Ala-Ala-Ala-Arg-Ala$^{20}$ |
| β-Lipotropin: (Continued) | Glu-Leu-Glu-Tyr-Gly-Leu-Val-Ala-Glu-Ala-Ala-Gln-Ala-Ala-Glu-Lys-Lys-Asp-Glu-Gly-Pro-Tyr-Lys$^{37}$ |
| β-MSH: | Asp-Glu-Gly-Pro-Tyr-Arg- |
| β-Lipotropin: (Continued) | $^{47}$Met-Glu-His-Phe-Arg-Try-Gly-Ser-Pro-Pro-Lys-Asp-Lys-Arg-Tyr-Gly-Gly-Phe-Met-Thr-Ser-Glu-Lys-Ser-$^{61}$ |
| β-MSH: (Continued) | Met-Glu-His-Phe-Arg-Try-Gly-Ser-Pro-Pro-Lys-Asp |
| ACTH$_{4-10}$: | Met-Glu-His-Phe-Arg-Try-Gly |
| α-Endorphin: | Tyr-Gly-Gly-Phe-Met-Thr-Ser-Glu-Lys-Ser |
| β-Endorphin: | Tyr-Gly-Gly-Phe-Met-Thr-Ser-Glu-Lys-Ser |
| Methionine Enkephalin: | Tyr-Gly-Gly-Phe-Met |
| β-Lipotropin: (Continued) | $^{76}$Gln-Thr-Pro-Leu-Val-Thr-Leu-Phe-Lys-Asn-Ala-Ile-Val-Lys-Asn-Ala-His-Lys-Lys-Gly-Gln-OH$^{91}$ |
| α-Endorphin: (Continued) | Gln-Thr-Pro-Leu-Val-Thr |
| β-Endorphin: (Continued) | Gln-Thr-Pro-Leu-Val-Thr-Leu-Phe-Lys-Asn-Ala-Ile-Val-Lys-Asn-Ala-His-Lys-Lys-Gly-Gln-OH |

TABLE 2. *Characteristics of neurotensin membrane binding and radioimmunoassay*

| Brain membrane binding | Radioimmunoassay |
|---|---|
| Dissociation constant ($K_D$) 3 nM | 50% displacement 1 nM (200 fmoles) |
| Rate constant, association ($k_1$) 4.1 × 10⁵/M/sec | Sensitivity less than 75 fmoles |
| Rate constant, dissociation ($k_{-1}$) 1.5 × 10⁻³/sec | Recovery ≥ 90% |
| Binding site number ($B_{max}$) 3.1 pmoles/g rat cerebral cortex | Specificity less than 0.1% cross reactivity: (2–13), (4–13), (6–13), (8–13) and (9–13) neurotensin sequence fragments, enkephalins, endorphins, GnRH, TRH, angiotensins, bradykinin, substance P, glucagon |
| Specificity ($IC_{50}$ values, nM) (2–13) 2.0 (4–13) 2.7 (6–13) 5.8 (8–13) 48 (9–13) 830 $IC_{50}$ values greater than 1μM: angiotensin, bacitracin, enkephalins, glucagon, substance P, prolactin, | Subcellular distribution (rat hypothalamus) Fraction immunoreactive neurotensin (pmoles/g) |
| Calf brain regional distribution High regions (2.5–1.2 × frontal pole): thalamus (dorsomedial, ventral, anterior) hypothalamus (anterior, medial, mamillary body) cerebral cortex (parahippocampal, cingulate, occipital) | Whole 30<br>$P_1$ (10,000 ×g, 10 min) 4<br>$P_2$ ( 7,500 ×g, 20 min) 15<br>$S_2$ 21<br>$P_2$ subfractionation:<br>A (0.32–0.8 M sucrose) 2<br>B (0.8–1.2 M sucrose) 9<br>C (1.2 M sucrose and pellet) 2 |
| Intermediate regions (1.2–0.7 × frontal pole): cerebral cortex (precentral, postcentral, frontal, insular, colliculi, pulvinar of thalamus, caudate, putamen, globus pallidus, hippocampus, amygdala | Calf brain regional distribution |
| Low regions (less than 0.7 × frontal pole): pons, cerebellar cortex, cervical spinal cord, medulla oblongata, cerebral white | High regions (≥ 8 pmoles/g wet): hypothalamus (anterior, medial, mamillary body) caudate, globus pallidus<br><br>Intermediate regions (2.5–8 pmoles/g wet): cortex (parahippocampal, cingulate, occipital) thalamus (anterior), hippocampus, colliculi<br><br>Low regions (1–2.5 pmoles/g wet): cortex (frontal, precentral, parietal), amygdala, pons, medulla oblongata, cervical spinal cord<br><br>Very low regions (less than 1 pmole/g wet): cerebral white, cerebellar cortex |

Data adapted from Uhl and Snyder (21).

lower values in cerebellum and white matter. In the cerebral cortex, there are pronounced variations with highest levels in the parahippocampal gyrus. Subcellular fractionation studies indicate a localization of neurotensin to synaptosomal fractions (20,21).

Using $^{125}$I-neurotensin, it has been possible to detect specific receptor binding (22). The dissociation constant of neurotensin is about 3 nM. Slopes of saturation curves indicate only a single population of binding sites. The density of binding sites in rat cerebral cortex is about 3 pmoles/g wet weight, similar to that of several neurotransmitter receptors. Marked regional variations exist in neurotensin binding which parallel, in part, variations in endogenous neurotensin. Highest binding is found in the dorsomedial thalamus, parahippocampal cerebral cortex (which possesses the highest endogenous levels of the cerebral gyri), and hypothalamus. Lowest levels occur in the cerebellum and brainstem. The strongest evidence that the binding sites involve physiological neurotensin receptors emerges from an examination of the relative abilities of five partial-sequence fragments of neurotensin to compete for binding. Their relative potencies correspond fairly well to their relative activities in a number of peripheral systems (S. Leeman, *personal communication*). The amino acid 2 to 13 and 4 to 13 fragments of the peptide chain have about the same potency as neurotensin in displacing $^{125}$I-neurotensin binding, whereas the 6 to 13 fragment is one-half, the 8 to 13 fragment is one-tenth, and the 9 to 13 fragment is 0.5% as potent as neurotensin itself. Numerous other peptides and nonpeptides have negligible affinity for neurotensin binding sites (Table 2).

In summary, we have reviewed evidence favoring central nervous system roles for enkephalin, angiotensin II, and neurotensin, possible as neurotransmitters. Similar evidence exists for substance P, somatostatin, TRH, vasoactive intestinal peptide, and perhaps even gastrin. Since these peptides were discovered in most cases fortuitously, it is conceivable that numerous other peptide transmitter candidates exist in the brain. This multiplicity of transmitters greatly enlarges the scope of problems and promises in elucidating neurotransmission in the central nervous system.

## REFERENCES

1. Bennett, J. P., Jr., and Snyder, S. H. (1976): Angiotensin II binding to mammalian brain membranes. *J. Biol. Chem.*, 251:7423–7430.
2. Carraway, R., and Leeman, S. E. (1975): The amino acid sequence of a hypothalamic peptide, neurotensin. *J. Biol. Chem.*, 250:1907–1911.
3. Carraway, R., and Leeman, S. (1976): Radioimmunoassay for neurotensin, a hypothalamic peptide. *J. Biol. Chem.*, 251:7035–7044.
4. Carraway, R., and Leeman, S. E. (1976): Characterization of radioimmunoassayable neurotensin in the rat. *J. Biol. Chem.*, 251:7045–7052.
5. Cox, B. W., Goldstein, A., and Li, C. H. (1976): Opioid activity of a peptide, β-lipotropin-(61–91), derived from β-lipotropin. *Proc. Natl. Acad. Sci. U.S.A.*, 73:1821–1823.
6. Elde, R., Hökfelt, T., Johansson, O., and Terenius, L. (1977): Immunohistochemical studies using antibodies to leucine-enkephalin: Initial observations of the nervous system of the rat. *Neurosciences (in press)*.
7. Hughes, J. (1975): Isolation of an endogenous compound from the brain with pharmacological properties similar to morphine. *Brain Res.*, 88:295–308.
8. Hughes, J., Smith, T. W., Kosterlitz, H. W., Foghergill, L. A., Morgan, B. A., and Morris,

H. R. (1975): Identification of two related pentapeptides from the brain with potent opiate agonist activity. *Nature,* 258:577–579.
9. Lamotte, C., Pert, C. B., and Snyder, S. H. (1976): Opiate receptor binding in primate spinal cord: Distribution and changes after dorsal root section. *Brain Res.,* 112:407–412.
10. Lazarus, L. H., Ling, N., and Guillemin, R. (1976): β-Lipotropin as a prohormone for the morphinomimetic peptides endorphins and enkephalins. *Proc. Natl. Acad. Sci. U.S.A.,* 73:2156–2159.
11. Li, C. H., Barnafi, L., Chretien, M., and Chung, D. (1965): Isolation and amino acid sequence of β-LPH from sheep pituitary glands. *Nature,* 208:1093–1094.
12. Pasternak, G. W., Goodman, R., and Snyder, S. H. (1975): An endogenous morphine-like factor in mammalian brain. *Life Sci.,* 16:1849–1854.
13. Pert, C. B., Kuhar, M. J., and Snyder, S. H. (1976): Opiate receptor: Autoradiographic localization in rat brain. *Proc. Natl. Acad. Sci. U.S.A.,* 73:3729–3733.
14. Simantov, R., and Snyder, S. H. (1976): Morphine-like peptides in mammalian brain: Isolation, structure elucidation and interactions with the opiate receptor. *Proc. Natl. Acad. Sci. U.S.A.,* 73:2515–2519.
15. Simantov, R., and Snyder, S. H. (1976): Brain pituitary opiate mechanisms: Pituitary opiate receptor binding, radioimmunoassays for methionine enkephalin and leucine enkephalin, and $^3$H-enkephalin interactions with the opiate receptor. In: *Opiates and Endogenous Opioid Peptides,* edited by H. Kosterlitz, pp. 41–48. North Holland Publishing Co., Amsterdam.
16. Simantov, R., Kuhar, M. J., Pasternak, G. W., and Snyder, S. H. (1976): The regional distribution of a morphine-like factor, enkephalin, in monkey brain. *Brain Res.,* 106:189–197.
17. Snyder, S. H. (1975): The opiate receptor in normal and drug altered brain function. *Nature,* 257:185–189.
18. Terenius, L., and Wahlstrom, A. (1975): A morphine-like ligand for opiate receptors in human CSF. *Life Sci.,* 16:1759–1764.
19. Tseng, L.-F., Loh, H. H., and Li, C. H. (1976): β-Endorphin as a potent analgesic by intravenous injection. *Nature,* 263:239–240.
20. Uhl, G. R., and Snyder, S. H. (1976): Regional and subcellular distributions of brain neurotensin. *Life Sci.* 19:1827–1832.
21. Uhl, G. R., and Snyder, S. H. (1977): Neurotensin receptor binding, regional and subcellular distributions favor transmitter role. *Eur. J. Pharmacol. (in press).*
22. Uhl, G. R., Bennett, J. P., Jr., and Snyder, S. H. (1977): Neurotensin, a central nervous system peptide: Apparent receptor binding in brain membranes. *Brain Res.* 130:299–313.

## DISCUSSION

*Dr. Bloom:* Have you observed uptake of neurotensin or other peptides by subcellular fractions?

*Dr. Snyder:* George Uhl has been looking at neurotensin uptake under conditions in which it is not degraded. There is a saturable uptake with "respectable" affinity constants but we have not done enough work to know what it means. We found no significant enkephalin uptake, a system which is hard to study because the peptide is readily degraded. The larger oligopeptides are much more stable.

It would be quite interesting to see if any of these peptide neuromodulators (neurohormones) have nerve terminal uptake systems because of the potential use in further research. The high-affinity nerve terminal uptakes of conventional neurotransmitters have proved to be a valuable tool to study the characteristics of the neurons, to label them for release study, and to study influences of drugs.

*Dr. Guillemin:* Dr. Snyder, do you have any knowledge as to whether your two antibodies against leucine- and methionine-enkephalin will react with any of the smaller endorphins such as α-endorphin or β-endorphin?

*Dr. Snyder:* There was negligible overlap. Our antibodies did not react with α- or β-endorphin, which were the only two we were able to test.

*Dr. Reichlin:* Dr. Snyder, what do you think is the significance of the finding of endorphin in the dorsal root and the dorsal root entry zone in relation to current knowledge of pain perception?

*Dr. Snyder:* We do not know exactly how endorphins are related to pain perception. One issue has to do with the site of opiate analgesia. Many studies indicate that opiate analgesia is supraspinal, but there is a convincing body of evidence to indicate that some component of the analgesia takes place at the spinal cord level.

The existence of presynaptic opiate receptors does not rule out the possibility that some opiate receptors are postsynaptic. However, the existence of presynaptic opiate receptors suggests that perhaps enkephalin terminals are making axoaxonic synapses, causing something analogous to presynaptic inhibition. One could speculate that it is similar to the GABA system. GABA is involved in presynaptic synapses in the same area, presumably on terminals of primary sensory afferents and, of course, it is also involved in postsynaptic inhibition at other sites.

Walter Zieglgansberger and Albert Herz have neurophysiologic evidence that in some parts of the brain opiates and enkephalins inhibit cell firing; these are postsynaptic, not presynaptic effects. Thus, opiates and enkephalins probably act on both postsynaptic and presynaptic opiate receptors.

# Peptide and Steroid Hormones and the Neural Mechanisms for Female Reproductive Behavior

## Donald W. Pfaff

Detailed study of neural mechanisms for female reproductive behavior has been facilitated by concentrating on a stereotyped, hormone-sensitive element of this behavior, the lordosis reflex in female rodents. First, neural circuitry for the control of the lordosis reflex in the female rat will be described. Then recent work on the control of lordosis by steroid and peptide hormones will be reviewed.

## CIRCUITRY FOR LORDOSIS BEHAVIOR

The lordosis reflex, studied with frame-by-frame film analyses and x-ray cinematography, is a standing response coupled with marked vertebral dorsiflexion (for review see 17). Initiation of lordosis occurs an average of 161 msec after first contact on skin by the male rat.

### Ascending Pathways

Since female rats which have been surgically blinded, deafened, and rendered anosmic perform lordosis behavior in a normal manner, somatosensory input above must be sufficient for lordosis. Moreover, lordosis never occurs until the female is touched by the male, so cutaneous input must be necessary. During mounting by the male, the female's skin is contacted on the flanks, posterior rump, tailbase, and perineum. None of the cutaneous stimuli applied by the male is stationary. Clearly repetitive stimuli such as palpations by the male's forepaws and thrusting by the male's pelvis have dominant frequencies between 10 and 20 per sec. Cutaneous desensitization by local anesthesia or by surgical cutaneous denervation shows that input from skin receptors on the flanks, posterior rump, and perineum is essential for triggering lordosis behavior. Conversely, manual stimulation by an experimenter, used to analyze the constellation of somatosensory stimuli provided by the male, shows that pressure on the flanks, posterior rump, and perineum is sufficient for evoking lordosis [sensory physiology

---

*Present address:* The Rockefeller University, York Avenue & 66th Street, New York, New York 10021

of lordosis is reviewed by Kow and Pfaff (8)]. From these studies, the candidates for somatosensory receptors involved in triggering lordosis can be narrowed to a specific subset of cutaneous mechanoreceptors on the skin of the ventral flanks, posterior rump, and perineum.

Cutaneous input from the perineal region of the female rat is carried by the pudendal nerve. Sensory information from the skin regions crucial for lordosis (perineum, rump, flanks) enters the spinal cord over dorsal roots $L_1$, $L_2$, $L_5$, $L_6$, and $S_1$ (reviewed in 8). Female rats with complete transections of spinal cord at low thoracic levels cannot perform lordosis, and those

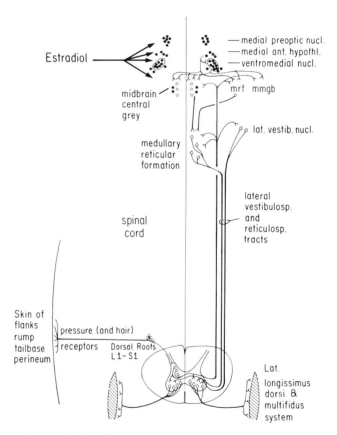

**FIG. 1.** Schematic description of neural circuitry for lordosis behavior in the female rat. In this working model, ascending spinal fibers carrying sensory information for lordosis control are shown running in the anterolateral columns of the spinal cord. The lateral reticulospinal tracts run in the anterolateral columns, while the lateral vestibulospinal tracts run in the anterolateral and ventral columns. Descending hypothalamic outputs relevant for lordosis are to the midbrain central gray and dorsolateral portions of the midbrain reticular formation. (From ref. 17.) Abbreviations: lat, lateral; ant, anterior; mmgb, medial (magnocellular) division of the medial geniculate body; mrf, mesencephalic reticular formation.

reflexes remaining have not been shown to be estrogen sensitive (7). Thus, a supraspinal control loop is required for the elaboration and hormone sensitivity of lordosis behavior. Selective transections of individual spinal cord columns showed that ascending information in the dorsal columns and dorsolateral columns was not important for lordosis, whereas information carried in the anterolateral somatosensory system was essential for the lordosis response (7). Among the fibers in this ascending system, spinoreticular, spinovestibular, and spinotectal fibers are candidates for lordosis control (Fig. 1). Anatomical work has shown that anterolateral column fibers reach the reticular formation of the medulla, the lateral vestibular nucleus, and the midbrain (including deep layers of the tectum, the intercollicular region, and the lateral portions of the central gray) (11). Indeed, single-unit recordings from neurons in the deep layers of the tectum and near the lateral borders of the central gray of the mesencephalon revealed strong responses to cutaneous stimulation on skin regions crucial for lordosis (10).

### Hypothalamic Control

Hypothalamic participation in the circuitry for the control of reproductive behavior has been well documented and reviewed [male behavior (4), female behavior (17); see brief summary in Table 1]. Male reproductive be-

TABLE 1. *Comparison of medial preoptic area (and basal forebrain) with basomedial hypothalamus: brief summary of some major differences*

|  | Medial preoptic area (and basal forebrain) | Basomedial (and posterior) hypothalamus |
|---|---|---|
| Reproductive function |  |  |
| Male mating behavior | ↑ | 0 |
| Female mating behavior | ↓ | ↑ |
| Effect on luteinizing hormone | Ovulation surge ("positive feedback") | Negative feedback |
| Effect of estradiol on unit activity | ↓ | ↑ |
| Autonomic function |  |  |
| Blood pressure | ↓ | ↑ |
| Heart rate | ↓ | ↑ |
| Diameter of pupil | ↓ | ↑ |
| Micturition, defecation, salivation | ↑ |  |
| Anatomical projections |  |  |
| Descending axon trajectories | MFB[a] | Outside MFB |
| To midbrain central gray | Weak | Strong |
| To lateral midbrain reticular formation | Weak | Strong |
| To lateral septum | Dorsal | Midlateral |

[a] MFB, medial forebrain bundle.

havior in a variety of experimental animals depends on the medial preoptic area and not on the basomedial hypothalamus. Conversely, female reproductive behavior, at least in rodents, depends on the integrity of neurons in and around the ventromedial nucleus of the hypothalamus ("basomedial hypothalamus") and not on the preoptic area. Electrophysiological effects of estrogen on neurons in these locations are consistent with the demonstrated roles of these neurons in female reproductive behavior. Estradiol, which facilitates female reproductive behavior, increases the activity of a subset of basomedial hypothalamic neurons (which in turn facilitate female reproductive behavior) and suppresses the activity of medial preoptic neurons (which do not facilitate female reproductive behavior).

Distinctions between preoptic-forebrain and basomedial posterior hypothalamic regions in autonomic function were emphasized by Hess. He (5,6) found that electrical stimulation of the preoptic area and basal forebrain in freely moving cats caused increased parasympathetic and decreased sympathetic autonomic functions. For instance, preoptic and septal stimulation decreased blood pressure, decreased heart rate, constricted the pupil, and led to micturition, defecation, and salivation. In contrast, electrical stimulation of the basomedial and posterior hypothalamus, including also stimulation of the periventricular system leading to the central gray, was followed by increased blood pressure, heart rate, and pupillary diameter, and was not followed by parasympathetic responses. These basic observations have been repeated in a large number of laboratories (references in 17).

Recent anatomic studies have provided evidence for selective pathways that may explain these different responses. Studies of axonal outputs of preoptic and hypothalamic neurons, with the use of tritiated amino acid autoradiography have for the first time demonstrated long-axon connections from these regions (2,3). Rather than cataloging all of these axonal projections, we shall concentrate on those of particular relevance for the present chapter. Projections from the ventromedial nucleus of the hypothalamus are of special interest, because those ventromedial nucleus axons descending to the midbrain are likely to be involved in the primary control of lordosis-relevant circuits. Axons from ventromedial nucleus neurons descend to terminate in the central gray of the mesencephalon (bilaterally), in the dorsal lateral mesencephalic reticular formation, and in the medial (magnocellular) division of the medial geniculate body (Fig. 1). Differences in axonal projections between preoptic area and basomedial hypothalamus may account for physiological differences in the functions of these two regions. For instance, medial preoptic axons descend in the medial portion of the medial forebrain bundle, whereas basomedial hypothalamic neurons descending to the midbrain avoid medial forebrain bundle trajectories (summary in Table 1). Preoptic neurons project weakly to the ventral portion of the midbrain central gray, whereas ventromedial nucleus neurons project strongly to both the dorsolateral and ventrolateral central gray. Basomedial hypothalamic

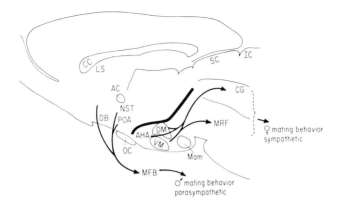

**FIG. 2.** Schematic summary of theoretical division of preoptic and hypothalamic tissue, according to its participation in the control of male or female mating behavior in rats. This division may correspond to differential participation of preoptic and basomedial (posterior) hypothalamic tissue in the control of sympathetic and parasympathetic autonomic nervous system functions. In the case of male mating behavior, there are likely causal relationships with preoptic tissue participation in parasympathetic functions, whereas for female mating behavior control and sympathetic nervous system functions no such causal relationship is implied. (From ref. 17.) Abbreviations: AC, anterior commissure; AHA, anterior hypothalamic area; CC, corpus callosum; CG, central gray; DB, diagonal bands; DM, dorsomedial nucleus of hypothalamus; IC, inferior colliculus; Mam, mammillary bodies; MFB, medial forebrain bundle; MRF, mesencephalic reticular formation; NST, bed nucleus of stria terminalis; LS, lateral septal nucleus; OC, optic chiasm; POA, preoptic area; SC, superior colliculus; VM, ventromedial nucleus of hypothalamus.

neurons project strongly to the lateral portions of the midbrain reticular formation, whereas preoptic neurons have weaker projections to the ventral aspect of this region. Finally, ventromedial and anterior hypothalamic axons project strongly to a specific, oval-shaped cell group in the midlateral septum, whereas preoptic axons avoid this cell group and project instead to the dorsal portion of the lateral septum.

Based on these data, a division, in anatomy and function, between preoptic area and basomedial hypothalamus can be roughly characterized. Medial preoptic neurons send their descending axons through the medial forebrain bundle, are responsible for facilitating male but not female reproductive behavior, and facilitate the parasympathetic division of the autonomic nervous system (Fig. 2). Neurons in the basomedial or posterior hypothalamus send their descending axons outside the medial forebrain bundle, are responsible for facilitating female but not male mating behavior, and are also related to the sympathetic division of the autonomic nervous system.

### Descending Pathways

Results in rats with selective spinal column transections show that descending fibers in the anterolateral columns are sufficient for lordosis (7) (see Fig. 1). Large transections which destroy the anterolateral columns abolish lordosis. These results are consistent with the notion that the lateral vestibulospinal and reticulospinal tracts are for the lordosis response (reviewed in 17). Among lesions in a large number of brainstem and cerebellar sites, only lesions of the lateral vestibular nucleus or the medullary gigantocellular reticular formation resulted in decrements in lordosis (reviewed in 17). Conversely, electrical stimulation of the lateral vestibular nucleus facilitates lordosis. All of these facts suggest that among the major well-recognized descending systems, only the lateral vestibulospinal and medullary reticulospinal tracts control lordosis behavior.

Since rats with complete spinal transections do not perform lordosis (7), the net influence of the obligatory descending pathways must be facilitatory for the reflex.

Film analyses and X-ray cinematography have shown that the essential element of lordosis is reflex vertebral dorsiflexion. The motoneurons comprising the final common pathway for this reflex have been identified by determining the muscles involved in this reflex vertebral dorsiflexion. Anatomical dissections of the deep back muscles in rats and direct electrical stimulation of these muscles and their nerve supply prove that the lateral longissimus dorsi and the transverso-spinalis (multifidus) muscle systems in the lumbar region are competent to execute vertebral dorsiflexion by elevating the rump and tailbase (reviewed in 17). Further, ablations of these muscles in female rats tested for lordosis showed that they are required for normal lordotic behavior. The motoneurons controlling these muscles have been located by horseradish peroxidase and electrical microstimulation techniques. Single-unit recordings from these motoneurons are expected to show how sensory input and hormone-dependent descending influences interact to account for the lordosis reflex and its hormone sensitivity.

## STEROID HORMONE CONTROL OF LORDOSIS

The dependence of reproductive behavior in the female rat on estrogen priming and progesterone facilitation has been well documented and reviewed (1,18). For the analysis of steroid hormone effects elsewhere in the body (for instance, uterus and seminal vesicles), the strategy of locating and characterizing hormone-binding cells has been highly successful. Thus, autoradiographic studies of estradiol and testosterone accumulation by cells in the brain were initially carried out in the rat (for instance, 16). These studies led to the definition of a specific limbic-hypothalamic system of cells which concentrate estradiol or testosterone. We have now extended this autoradiographic work showing binding of steroid sex hormones by specific nerve cells to include studies of all classes of vertebrates: fish, amphibia, reptiles, birds, and mammals (including rhesus monkeys). Such studies have been reviewed (15) and have led to a set of conclusions which, remarkably, seem to hold true across all vertebrates (Table 2). Thus, neuroanatomical patterns of steroid binding cells are stable and are correlated with the neural control of hormone-dependent functions. Moreover, recent studies using antiestrogens (reviewed in 9 and 17) show that if estrogen binding is blocked, estrogen effects on female reproductive behavior similarly will be blocked. All these facts suggest that the binding process, as studied autoradiographically and biochemically, is one important step in sex hormone action on the brain.

TABLE 2. *Autoradiographic studies of steroid sex hormone-binding cells in the brains of vertebrates*

Features common to all species studied:
1. All species have hormone-concentrating nerve cells in specific locations.
2. Hormone-concentrating cells are in medial preoptic area, medial (tuberal) hypothalamus, and limbic forebrain structures.
3. Nerve cell groups which bind hormone participate in control of hormone-modulated functions.

## PEPTIDE HORMONE INFLUENCE ON FEMALE REPRODUCTIVE BEHAVIOR

Systemic injections of luteinizing hormone releasing hormone (LRF) have been shown to facilitate lordosis behavior in ovariectomized, estrogen-primed female rats (12,14). Since the LRF effect can be demonstrated in hypophysectomized animals (14), the pituitary is not a necessary intermediary, so the effect is probably a direct neural one. In fact, LH and TRH injections are not effective (12). Subsequent studies have shown the dose range over which systemic LRF injections are effective in facilitating lordosis (13) and have shown that the effect can be obtained in adrenalectomized female rats, eliminating adrenal progesterone as a possible intermediary cause. We, also, have carried out further studies (D. Modianos and D. W. Pfaff, *unpublished observations*) to determine the relationship between the LRF effect and the progesterone effect on lordosis in ovariectomized, hypophysectomized, estrogen-primed female rats. Are the effects independent? Does one effect occlude the other? Does LRF "lower the threshold" for the progesterone effect? As seen in Table 3, the LRF effect on lordosis is most prominent when the base-line lordosis quotient is low, and the LRF effect neither added perfectly to nor multiplied the progesterone effect on lordosis.

From these results it appears that, as initially hypothesized (14), the LRF effect on lordosis is best considered as a synchronizing function, helping to ensure that the cycling female rat which is about to ovulate will also begin to enter a period of behavioral receptivity. The LRF effect does not appear to

TABLE 3. Lordosis quotient in hypophysectomized, ovariectomized, estrogen-primed female rats (mean ± SEM)

|          | Oil          | 50 μg Progesterone |
|----------|--------------|--------------------|
| 0 LRF    | 15.5 ± 3.6   | 30.5 ± 5.6         |
| 1 μg LRF | 31.5 ± 5.7   | 39.5 ± 6.6         |
| 4 μg LRF | 32.0 ± 6.1   | 39.0 ± 6.7         |

be identical with the estrogen effect because it has not been shown to substitute for estrogen priming and has appeared only in female rats pretreated with estrogen. Also, as studied so far (12–14), LRF does not have the magnitude of effect on lordosis that progesterone has and does not occlude the progesterone effect, and thus is unlikely to be the sole mechanism of the progesterone effect. Rather, LRF is likely to aid in causing an adaptive temporal relationship between behavioral receptivity and ovulation.

## REFERENCES

1. Beach, F. A. (1948): *Hormones and Behavior.* Hoeber, New York.
2. Conrad, L. C. A., and Pfaff, D. W. (1976): Autoradiographic study of efferents from medial basal forebrain and hypothalamus in the rat. I. Medial preoptic area. *J. Comp. Neurol.,* 169:185.
3. Conrad, L. C. A., and Pfaff, D. W. (1976): Autoradiographic study of efferents from medial basal forebrain and hypothalamus in the rat. II. Medial anterior hypothalamus. *J. Comp. Neurol.,* 169:221.
4. Hart, B. (1977): Neural mechanisms of male reproductive behavior. In: *Neurobiology of Reproduction (Handbook of Behavioral Neurobiology),* edited by R. Goy and D. W. Pfaff. Plenum Press, New York.
5. Hess, W. R. (1954): *Diencephalon, Autonomic and Extrapyramidal Functions.* Grune & Stratton, New York.
6. Hess, W. R. (1957): *The Functional Organization of the Diencephalon.* Grune & Stratton, New York.
7. Kow, L. M., Montgomery, M. O., and Pfaff, D. W. (1977): Effects of spinal cord transections on lordosis reflex in female rats. *Brain Res.,* 123:75.
8. Kow, L. M., and Pfaff, D. W. (1977): Sensory control of reproductive behavior in female rodents. In: Tonic functions of sensory systems. *Ann. N.Y. Acad. Sci.*
9. Luine, V., and McEwen, B. S. (1977): In: *Neurobiology of Reproduction (Handbook of Behavioral Neurobiology),* edited by R. Goy and D. W. Pfaff. Plenum Press, New York.
10. Malsbury, C., Kelley, D. B., and Pfaff, D. W. (1972): Responses of single units in the dorsal midbrain to somatosensory stimulation in female rats. In: *Progress in Endocrinology, Proceedings of the IV International Congress of Endocrinology,* edited by C. Gaul, p. 205. Excerpta Medica, Amsterdam.
11. Mehler, W. R. (1969): Some neurological species differences—a posteriori. *Ann. N.Y. Acad. Sci.,* 167:424.
12. Moss, R. L., and McCann, S. M. (1973): Induction of mating behavior in rats by luteinizing hormone-releasing factor. *Science,* 181:177.
13. Moss, R. L., and McCann, S. M. (1975): Action of luteinizing hormone-releasing factor (LRF) in the initiation of lordosis behavior in the estrone-primed ovariectomized female rat. *Neuroendocrinology,* 17:309.
14. Pfaff, D. W. (1973): Luteinizing hormone releasing factor (LRF) potentiates lordosis behavior in hypophysectomized ovariectomized female rats. *Science,* 182:1148.
15. Pfaff, D. W. (1976): The neuroanatomy of sex hormone receptors in the vertebrate brain.

In: *Neuroendocrine Regulation of Fertility,* edited by T. C. A. Kumar, p. 30. S. Karger, Basel.
16. Pfaff, D. W., and Keiner, M. (1973): Atlas of estradiol-concentrating cells in the central nervous system of the female rat. *J. Comp. Neurol.,* 151:121.
17. Pfaff, D. W., and Modianos, D. (1977): Neural mechanisms of female reproductive behavior. In: *Neurobiology of Reproduction (Handbook of Behavioral Neurobiology),* edited by R. Goy and D. W. Pfaff. Plenum Press, New York.
18. Young, W. C. (1961): The hormones and mating behavior. In: *Sex and Internal Secretions, Vol. 2 (3rd ed.),* edited by W. C. Young. Williams & Wilkins Co., Baltimore.

## DISCUSSION

*Dr. Martin:* Would you comment, Dr. Pfaff, on the existence of a blood-brain barrier for peptides? Most workers give large systemic doses of peptides to assess effects on brain. This question is also relevant to the possible effects of peptides that are found in the gastrointestinal tract and which could reach the brain via the circulation. How much peptide do you think enters the brain after systemic administration?

*Dr. Pfaff:* It must be a very small amount. We assume that we have to use high doses to elicit a behavioral effect with systemic LRF for at least three reasons: The short time LRF remains in the blood, the blood-brain barrier, and the wealth of opportunities for the peptide to be inactivated before it ever gets to the brain.

We do not especially like using the systemic route. What one normally wants to do for this kind of study, of course, as one can do with steroids, is to place the hormone into its normal route of transport. In the case of LRF that means that we have to get the LRF into an intracellular position in a population of LRF-producing cells.

*The Hypothalamus*, edited by S. Reichlin,
R. J. Baldessarini, and J. B. Martin,
Raven Press, New York, © 1978.

# Steroid Hormone Action in the Neuroendocrine System: When is the Genome Involved?

*Bruce S. McEwen, Lewis C. Krey, and Victoria N. Luine

Some actions of steroid hormones at the cellular level are now recognized to occur via an activation of the target cell genome. This is elegantly illustrated by recent demonstrations of steroid-dependent increases in production of messenger RNAs for ovalbumin and avidin in oviduct (75) and tryptophan oxygenase in liver (66). In such activation of the genome, the first step is believed to be the binding of the steroid to intracellular receptor sites, and the second step a translocation of the hormone-receptor complex to the cell nucleus.

Like oviduct and liver, brain and pituitary also contain receptors for steroid hormones of the type which mediate genomic activation. Some of these

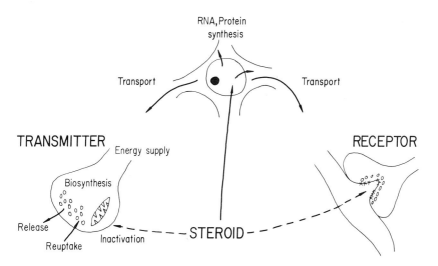

**FIG. 1.** Genomic and nongenomic effects of steroid hormones on pre- and postsynaptic events. Nongenomic effects (*dashed line*) may involve the action of the hormone on the pre- or postsynaptic membrane to alter permeability to neurotransmitters or their precursors and/or functioning of neurotransmitter receptors. Genomic action of the steroid (*unbroken line*) leads to altered synthesis of proteins, which, after axonal or dendritic transport, may participate in pre- or postsynaptic events.

---

* The Rockefeller University, 1230 York Avenue, New York, New York 10021

receptors participate in stimulating the formation of gene products which are involved in growth and differentiation of the brain. These are referred to as "organizational" effects and are believed to underlie the sexual differentiation of the brain. "Organizational" effects are irreversible and occur only during a critical period of early brain development. Steroid receptors also participate in "activational" effects on the brain. "Activational" effects are reversible and appear to involve the temporary modification of neuronal efficiency within neural circuits already fixed during brain maturation.

The purpose of this essay is to examine the characteristics of genomic involvement in steroid hormone effects on the brain. Special emphasis will be placed on the activational aspects of steroid hormone action because it is important to distinguish effects of this category which are mediated by the genome from those which are not (Fig. 1).

## INTERACTIONS OF STEROID HORMONES WITH NEUROENDOCRINE TISSUES

All five major subclasses of steroid hormones influence neural events underlying behavior and neuroendocrine function in adult vertebrates. Intracellular receptors for estradiol, androgens, and glucocorticoids of the type found to be associated with genomic effects in peripheral tissues have also been identified in neuroendocrine tissues (47,48). Less extensive evidence supports the existence of intracellular receptors for progestins (22,29,30) and for mineralocorticoids (1,35). Steroid receptor systems are not uniformly distributed in neuroendocrine tissues but are concentrated in certain brain regions and largely absent from others. For example, estradiol receptor levels are high in pituitary, hypothalamus, preoptic area, and amygdala and absent from cortex of adult mammals, whereas glucocorticoid receptors are especially numerous in hippocampus and septum. This is illustrated, in part, in Fig. 2 for one species, the rhesus monkey, using autoradiographs which show neuronal localization of radioactivity from a $^3$H steroid injection. In the few cases where brain hormone implants have been made and specific effects have been assessed (e.g., activation of sex behavior by estradiol and testosterone), it appears that the most responsive brain regions are those corresponding to the above-mentioned intracellular receptors (26,37,38).

At least two steroid hormones, testosterone and progesterone, are subject to metabolic transformations by neuroendocrine tissues; these metabolites interact with the hormone receptors to activate behavior and alter pituitary function. The A ring of each of the steroids is enzymatically reduced by $\Delta$ 5, 3 keto steroid reductase (Fig. 3) and the 3 keto group further reduced to form $5\alpha$, $3\alpha$, and $5\alpha$, $3\beta$ metabolites (not shown). These transformations occur both in pituitary and in brain tissue. In addition, the A ring of testosterone is also aromatized to form estradiol by an enzyme system found in hypothala-

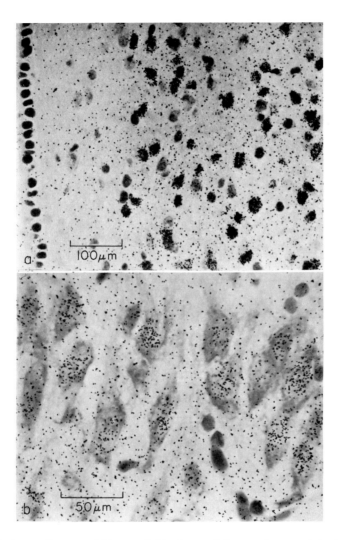

**FIG. 2. (a):** Autoradiogram of ³H-estradiol in the medial preoptic area of the brain of an ovariectomized adult rhesus monkey. An unfixed, unembedded frozen section was exposed 387 days and stained with cresyl violet acetate. Neurons in the medial preoptic area which concentrated radioactivity are intensely labeled with silver grains and lie near a row of darkly stained ependymal cells of the third ventricle. The neuronal labeling illustrated in this autoradiogram corresponds to the high degree of cell nuclear retention of estradiol radioactivity in this area demonstrated by nuclear isolation. **(b):** Autoradiogram of ³H-corticosterone in the hippocampus of the brain of an adrenalectomized adult rhesus monkey. An unfixed, unembedded frozen section was exposed 470 days and stained with methyl green-pyronin Y Shown are pyramidal neurons in the cornu ammonis, with radioactivity concentrated primarily in cell nuclei. The neuronal labeling illustrated in this autoradiogram corresponds to the high degree of cell nuclear retention of corticosterone radioactivity in this area demonstrated by nuclear isolation. Reprinted from ref. 15 with permission.

mus and limbic structures but not in pituitary nor in cerebral cortex (Fig. 3). One of the important reduction products of testosterone, 5α-dehydrotestosterone (5α-DHT), has been implicated in the feedback regulation of gonadotropin secretion (26), and testosterone-derived estradiol has been implicated as a mediator of the androgen-facilitation of male sexual behavior in the rat (8). Progesterone effects on ovulation and on sexual behavior may also be mediated by metabolites (29). Estradiol and estrone are hydroxylated at the 2 position to form catechol estrogen derivatives (Fig. 3), a transformation limited to the hypothalamus (13). Although the structure of catechol estrogens suggests that they might have a direct effect on noradrenergic or dopaminergic receptors (see Fig. 1), their neuroendocrine function is not well established.

The metabolic transformations of testosterone are known to occur in developing brain (9,36). In neonatal rats and hamsters, estrogens mimic the differentiative effects of testosterone; 5α-reduced androgens are far less effective (61,74). Estrogen receptors have been identified in neonatal rat and mouse brain tissue during the "critical" period for sexual differentiation. Because testosterone-induced masculine differentiation of hypothalamic function during the critical period is blocked by antagonists of estrogen

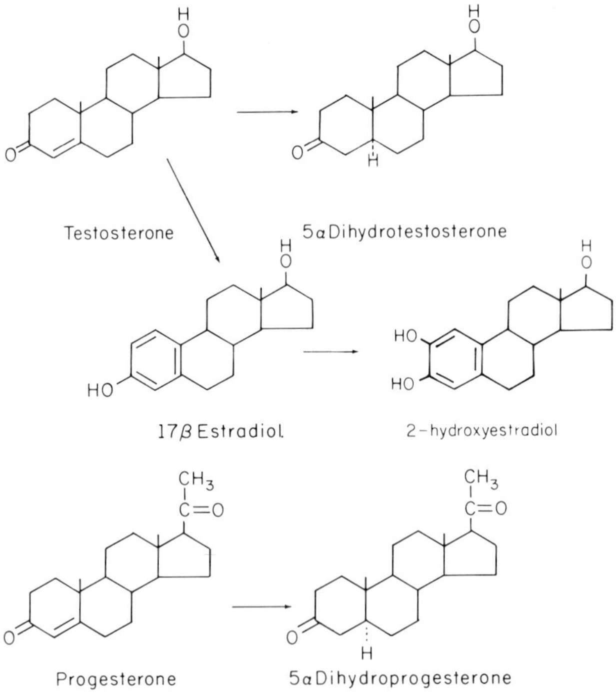

**FIG. 3.** Some of the major steroid transformations which occur in brain and/or pituitary tissue of the rat.

action (49,61), it is likely that these estrogen receptors mediate testosterone effects. Moreover, the fact that the antiestrogens MER 25 and CI 628 appear to act by blocking estrogen-receptor interactions suggests that the estrogen receptors may play some major role in sexual differentiation via a genomic mechanism.

## IDENTIFICATION OF GENOMIC INVOLVEMENT IN NEUROENDOCRINE EFFECTS OF STEROID HORMONES

Aspects of steroid action mediated by the genome that have been considered include latency and duration of effects, sensitivity to inhibitors of macromolecular synthesis, and induction of long-lasting cellular changes. On the basis of extensive studies of genomic effects in peripheral tissues, it is to be expected that they would be delayed in onset and prolonged in duration (Fig. 4), and in these respects not directly correlated with the levels of hormone in blood and target tissue. For example, there is a 24-hr time delay for the induction of lordosis behavior in the female rat by single 2- to 10-$\mu$g doses of estradiol 17$\beta$; this delay is longer than the occupation of the receptor sites by the hormone (17,50). A long latency is not, however, a necessary condition for a genomic effect; genomic actions of cortisol in thymus lymphocytes result in an onset of suppression of glucose transport within as little as 15 min (53). Thus, it is conceivable that some relatively rapid actions of glucocorticoids on hippocampal neural activity such as effects on single-unit activity (51) and evoked potential size (58) may be mediated by a genomic mechanism. With respect to duration, "activational" effects of steroid hormones on behavior such as lordosis are measured in hours or days. In contrast, the duration of "developmental" effects of steroids is the lifetime of the individual.

Since steroid actions at the genomic level affect cell function by stimulat-

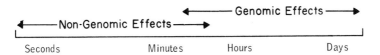

Non-Genomic Effects
    RAPID IN ONSET - Seconds, Minutes
    SHORT IN DURATION - following disappearance of steroid from tissue

Genomic Effects
    SLOWER IN ONSET - Minutes, Hours
    LONGER IN DURATION - persist after steroid disappears from tissue

**FIG. 4.** Time consideration for genomic and nongenomic effects.

ing the synthesis of RNA and proteins (Fig. 1), these actions should be inhibited by agents which block RNA and protein synthesis. A number of steroid hormone-induced behavioral and neuroendocrine effects are in fact inhibited by these agents. This is true for estrogen induction of lordosis behavior in the female rat [inhibited by actinomycin D (Act D) and by cycloheximide (63,70,73)] and for estrogen induction of the preovulatory LH surge [blocked by Act D (23,24,27)]. Estradiol effects on locomotor activity are also inhibited by cycloheximide (67). Negative feedback effects of dexamethasone on ACTH secretion are inhibited by Act D (3,14,62). Even in the case of progesterone action, where evidence for the existence of intracellular receptors is presently weak, Act D does appear to block progesterone facilitation of LH release in the rat (25), and cycloheximide has been reported to block progesterone inhibition of estrous behavior in the female guinea pig (72).

Genomic effects of steroid hormones should result in long-lasting changes of specific gene products. Indeed, the developmental actions of steroids on brain sexual differentiation appear to result in changes in neuronal growth and connectivity which involve many gene products. The observed changes in neuronal morphology include hormone-dependent outgrowth of neurites in cultures of hypothalamic explants from animals sacrificed during the critical period for sexual differentiation (71). There are also observations of sex differences in patterns of dendritic spread and synaptic connections in later life which may reflect the consequences of the early differences in neurite outgrowth (18,64).

Activational effects of steroids, on the other hand, are reflected in reversible but still relatively long-lasting changes in gene products. For example, tissue levels of both LH and GnRH are influenced by the presence or absence of gonadal steroids, and this may be indicative of changes in both synthesis and release of these hormones (2,28,54,60). A number of enzymes are influenced in pituitary and brain tissue by estrogen treatment in such a way to suggest that enzyme induction may be involved. These will be discussed in more detail below.

## ACTIVATIONAL STEROID EFFECTS IN BRAIN AND PITUITARY WHICH MAY BE MEDIATED BY THE GENOME

Attempts to directly assess genomic activity by measuring the hormonal stimulation of the formation of gene products has met with mixed success (44,48). Changes in RNA content of pituitary have been observed as a consequence of estrogen treatment, but changes in RNA and protein formation as a result of such treatment have been difficult to demonstrate in brain tissue (44). The most effective means of assessing changes in gene products has been to measure enzyme activities in pituitary and brain tissue. The enzymes which have been studied and the directions of the effects of estrogen treatment on ovariectomized rats are indicated in Table 1. In general, these estro-

TABLE 1. *Estrogen effects on enzyme activities in neuroendocrine tissues*

| Enzyme | Structure[a] | $E_2$ effect (%) | Reference |
|---|---|---|---|
| Glucose-6-phosphate dehydrogenase | Pit, MBH | 30–150 ↑ | 39,40 |
| 1 Cystine aminopeptidase | Hyp, Amygd | 50–100 ↑ | 20 |
| LH-RH peptidase | Hyp | 30 ↑ | 19 |
| Choline acetyltransferase | MPOA, CMA | 30–40 ↑ | 41 |
| Tyrosine hydroxylase | MBH | 30–40 ↓ | 45 |
| Monoamine oxidase | MBH, CMA | 30 ↓ | 41,42 |

[a] Abbreviations: Pit, pituitary; Hyp, hypothalamus; MBH, medial basal hypothalamus, MPOA, medial preoptic area; Amygd, amygdala; CMA, corticomedial amygdala.

gen effects are confined to brain areas which contain substantial numbers of estrogen receptors, and they require at least 24 hr to be manifested after *in vivo* treatment with estrogen. The effects cannot be reproduced by adding estradiol to the enzyme incubation mixture.

With respect to the direction of the effects shown in Table 1, hormone-induced increases may well be due to increased synthesis of enzyme or to decreased enzyme degradation. However, information is lacking on this point. Additional information is also necessary to show that increases in catalytic activity are due to increases in the actual amount of enzyme protein rather than catalytic efficiency of existing enzyme. Perhaps this latter point can be studied successfully by means of immunological techniques. Hormone-induced decreases in enzyme activity (Table 1) may be explained by either decreased synthesis or increased degradation. There is some information on this point for monoamine oxidase (MAO), and it has been concluded that estrogen action on hypothalamus and amygdala results in increased rates of MAO degradation without a change in synthesis (42).

One characteristic of estrogen-genome interactions in the uterus is that nonsteroidal antiestrogens such as nafoxidine and CI628 inhibit estrogen effects while showing weak or moderate estrogenic action of their own. Antiestrogens also block estrogen actions on brain and pituitary enzyme levels, and these results strengthen the case for genomic involvement in neuroendocrine effects of estrogens (48). However, a comparison of CI628 effects in uterus, pituitary, and brain reveals a relatively greater estrogenicity than antiestrogenicity in brain compared to pituitary and uterus. This has been correlated with differences in dynamics of CI628 interaction with receptors in these three tissues (43).

## ACTIVATIONAL STEROID EFFECTS THAT DO NOT APPEAR TO INVOLVE THE GENOME

Steroid hormone effects on neural function which do not appear to involve the genome have been recognized (Fig. 1). Some of these effects occur so rapidly and have such short durations as to preclude any genomic intermediation (Fig. 4). This is true of rapid (<5 min) electrical responses of neurons to iontophoretically or systemically applied steroids such as estradiol hemisuccinate, dexamethasone phosphate, or cortisol (12,31,52,59,65). This appears also to be the case for inhibition of pulsatile LH release by estradiol, an effect which appears within minutes of intravenous steroid infusion and which is mimicked almost exactly by an α-adrenergic blocking agent, phenoxybenzamine (6,7,34).

Other effects, which also have short latencies and durations, can be reproduced *in vitro* in isolated nerve ending particles. This is true of fast negative feedback effects of glucocorticoids on the ability of hypothalamic synaptosomes to release corticotropin releasing factor (CRF) (11) as well as of the glucocorticoid stimulation of the uptake of tryptophan by synaptosomes (56).

An interesting example of a possible nongenomic steroid effect comes from the observation that stressful stimuli and adrenocortical hormones both appear to be required to increase the activity of the serotonin biosynthetic pathway. For example, a number of stimuli that increase tryptophan hydroxylase activity such as ethanol intoxication, cold exposure, electric footshock, and reserpine treatment are effective only in the presence of adrenocortical secretions (4,5,56,68,69). Such effects appear simultaneously in forebrain and in midbrain and thus cannot easily be explained by a genomic steroid action, i.e., by the *de novo* synthesis and axoplasmic transport of the enzyme (4). It may be possible to explain these effects by the local action of corticosterone, noted above, to increase tryptophan uptake in synaptic endings (56), whereupon this amino acid may influence the catalytic activity of tryptophan hydroxylase (10).

## STEROID EFFECTS AT SOME DISTANCE FROM RECEPTOR SITES

Because hormone-sensitive neurons project from one brain region to another, it is likely that chemical changes resulting from hormone action on neurons in one brain region may be found in another. This appears to have been the case for hippocampal choline acetyltransferase (CAT) activity, which is increased by estrogen treatment in ovariectomized rats. Most, if not all, hippocampal CAT arises from the septum, a neural area with some estrogen receptor sites (41). It is also possible that an action in a brain region lacking estrogen receptor sites, or even in one which contains them, might be mediated by other hormonal factors. Thus hypophysectomy might be expected to block some steroid effects if indeed pituitary hormones mediate them. In one such experiment hypophysectomy was shown to block estradiol effects on hypothalamic MAO activity, but not its actions to reduce amygdala MAO activity (41). In this experiment, however, it is not clear whether hypophysectomy may have also resulted in some degenerative change within the hypothalamic cells in which MAO is capable of changing with estrogen treatment. Finally, it must be borne in mind that effects at some distance from hormone target areas might be mediated by trans-synaptic influences of neurons, electrical activities of which are altered by hormonal treatment.

One special case in which to consider "action at a distance" is the highly branched noradrenergic and serotonergic innervation of the forebrain. Steroid effects on the cell body region of the locus coeruleus or midbrain raphe might be expected to spread widely through the forebrain. With the exception of the serotonin response discussed above, there is little indication that such widespread effects actually occur. With respect to steroid receptors in these areas, there is no indication that midbrain raphe neurons contain intracellular steroid hormone receptor sites. Locus coeruleus, on the other hand, has been shown by autoradiographic study to have a small amount of estrogen-binding activity associated with scattered catecholamine-fluorescing cells (21). It is likely that there are estrogen receptors in the tuberoinfundibular (cell group A12) and periventricular (cell group A14) dopaminergic systems (16,57), and there is some evidence as well for gonadal steroid influences on tyrosine hydroxylase activity in these systems but not elsewhere in the nervous system (32,33,45,55). These findings suggest that localized change of tyrosine hydroxylase activity or monoamine turnover may be traced to hormone effects on discrete dopaminergic systems. Effects on the locus coeruleus as stated above would be expected to manifest themselves in many cerebral areas innervated by this noradrenergic system; effects in only one branch of the system are not anticipated. The localized changes in hypothalamic tyrosine hydroxylase (TH) activity, as reported by Kizer et al. (33), do not appear to be exceptions to this view. Rather, they appear to represent alteration of a more localized hypothalamic dopaminergic system because dopamine hydroxylase activity did not change at the same time as TH (33), and disruption of the ventral noradrenergic bundle did not abolish TH increase occurring after castration (32).

## CRITERIA FOR IDENTIFYING STEROID-DEPENDENT NEUROCHEMICAL CHANGES UNDERLYING NEUROENDOCRINE EVENTS

We have learned enough about the distribution and properties of steroid receptor sites in neuroendocrine tissues and about some of the neurochemical changes resulting from steroid action that we can now begin to specify criteria for recognizing a chemical change induced by a steroid-genome interaction and crucial for a neuroendocrine event.

1. If the steroid activates the genome, the steroid-induced change should be confined to tissues (e.g., brain regions) that contain the intracellular receptors, although it may also be found, with a time delay representing axoplasmic transport, in a brain region receiving projections from the hormone-sensitive area.

2. It should occur within the time period for hormonal activation of the neuroendocrine event.

3. If the steroid facilitation of the neuroendocrine event is blocked by antagonists (e.g., antiestrogens in the case of estradiol), the steroid-induced chemical change should also be blocked.

4. It should be a change which is translatable in terms of some aspect of neuronal function (e.g., neurotransmitter or releasing factor biosynthesis, release, inactivation, or receptor mechanism; oxidative metabolism) and which is ultimately testable by pharmacological experiments. For example, inhibitors of the dopaminergic system should modify ovulation and/or lordosis if indeed estradiol influence on the dopaminergic system is a critical neurochemical event.

Of the chemical changes summarized in Table 1, none so far appears to satisfy all of these criteria. Estrogen-dependent decreases in activities of two enzymes, MAO and TH, appear on the basis of present knowledge to occur too slowly (41,42,45). CAT, on the other hand, changes rapidly enough (within 24 hr), but the change is not blocked by an antiestrogen, CI628, which appears to block estrogen effects on lordosis and ovulation (43). The other estrogen effects summarized in Table 1 have not yet been adequately investigated.

Whereas the above-mentioned estrogen-dependent changes in MAO, TH, and CAT may not be events crucial to the triggering of ovulation or lordosis behavior, they may nevertheless be involved in the maintenance of the neuroendocrine system in a state of readiness for the crucial triggering events, whatever they may be. This statement is based on the finding that long-term ovariectomy (OVX) results in a decreased responsiveness to lordosis-facilitating effects of estrogen replacement (8a). Moreover, the estrogen receptor mechanism does not appear to decrease after long-term OVX (50), whereas estrogen-sensitive enzymes such as MAO and glucose-6-phosphate dehydrogenase do change in the expected direction with long-term OVX (46).

## ACKNOWLEDGMENTS

This work was supported by Research Grant NS 07080 from the United States Department of Health, Education and Welfare, and by Institutional Grant RF 70095 from The Rockefeller Foundation.

## REFERENCES

1. Anderson, N. S., III, and Fanestil, D. D. (1976): Corticoid receptors in rat brain: Evidence for an aldosterone receptor. *Endocrinology*, 98:676–684.
2. Araki, S., Ferin, M., Zimmerman, E. Z., and Van de Wiele, R. L. (1975): Ovarian modulation of immunoreactive gonadotropin-releasing hormone (Gn-RH) in the rat brain: Evidence for a differential effect on the anterior and mid-hypothalamus. *Endocrinology*, 96:644–650.
3. Arimura, A., Bowers, C. Y., Schally, A. V., Saito, M., and Miller, M. C. (1969): Effect of

corticotropin-releasing factor, dexamethasone and actinomycin D on the release of ACTH from rat pituitaries *in vivo* and *in vitro*. *Endocrinology,* 85:300–311.
4. Azmitia, E. C., Jr. (1973): Rapid changes in brain tryptophan hydroxylase activity; effects of glucocorticoids, stressors and drugs. Doctoral dissertation, The Rockefeller University, New York City.
5. Azmitia, E. C., Jr., and McEwen, B. S. (1974): Adrenocortical influence on rat brain tryptophan hydroxylase activity. *Brain Res.,* 78:291–302.
6. Blake, C. A. (1974): Localization of the inhibitory actions of ovulation blocking drugs on the release of luteinizing hormone in ovariectomized rats. *Endocrinology,* 95:999–1004.
7. Blake, C. A., Norman, R. L., and Sawyer, C. H. (1974): Localization of the inhibitory action of estrogen and nicotine on release of luteinizing hormone in rats. *Neuroendocrinology,* 16:22–35.
8. Christensen, L. W., and Clemens, L. G. (1975): Blockade of testosterone-induced mounting behavior in the male rat with intracranial application of the aromatization inhibitor, androst-1,4,6-triene-3,17-dione. *Endocrinology,* 97:1545–1551.
8a. Damassa, D., and Davidson, J. M. (1973): Effects of ovariectomy and constant light on responsiveness to estrogen in the rat. *Horm. Behav.,* 4:269–279.
9. Denef, C., Magnus, C., and McEwen, B. S. (1974): Sex-dependent changes in pituitary $5\alpha$-dihydrotestosterone and $3\beta$-androstanediol formation during postnatal development and puberty in the rat. *Endocrinology,* 94:1265–1274.
10. Diez, J. A., Sze, P. Y., and Ginsburg, B. E. (1976): Tryptophan regulation of brain tryptophan hydroxylase. *Brain Res.,* 104:396–400.
11. Edwardson, J. A., and Bennett, G. W. (1974): Modulation of corticotrophin-releasing factor release from hypothalamic synaptosomes. *Nature,* 251:425–427.
12. Feldman, S., and Sarne, Y. (1970): Effect of cortisol on single cell activity in hypothalamic islands. *Brain Res.,* 23:67–75.
13. Fishman, J., and Norton, B. (1975): Catechol estrogen formation in the central nervous system of the rat. *Endocrinology,* 96:1054–1059.
14. Fleischer, N., and Battarbee, H. (1967): Inhibition of dexamethasone suppression of ACTH secretion *in vivo* by actinomycin D. *Proc. Soc. Exp. Biol. Med.,* 126:922–925.
15. Gerlach, J. L., McEwen, B. S., Pfaff, D. W., Moskovitz, S., Ferin, M., Carmel, P. W., and Zimmerman, E. A. (1976): Cells in regions of rhesus monkey brain and pituitary retain radioactive estradiol, corticosterone and cortisol differentially. *Brain Res.,* 103:603–612.
16. Grant, L. D., and Stumpf, W. E. (1973): Localization of [3]H-estradiol and catecholamines in identical neurons in the hypothalamus. *J. Histochem. Cytochem.,* 404 (Abst. #6, Histochemical Society).
17. Green, R., Luttge, W. G., and Whalen, R. E. (1970): Induction of receptivity in ovariectomized rats by a single intravenous injection of estradiol-17 beta. *Physiol. Behav.,* 5:137–141.
18. Greenough, W. T., and Carter, S. C. (1975): Androgen-dependent sexual dimorphism in dendritic branching pattern (clustering) of preoptic area neurons in the hamster. Neuroscience Abstracts, Society for Neuroscience, 5th Annual Meeting, New York, Abstract #1210, p. 789.
19. Griffiths, E. C., Hooper, K. C., Jeffcoate, S. L., and Holland, D. T. (1975): The effects of gonadectomy and gonadal steroids on the activity of hypothalamic peptidases inactivating luteinizing hormone-releasing hormone (LH-RH). *Brain Res.,* 88:384–388.
20. Heil, H., Meltzer, V., Kuhl, H., Abraham, R., and Taubert, H. D. (1971): Stimulation of L-cystine-aminopeptidase activity by hormonal steroids and steroid analogs in the hypothalamus and other tissues of the female rat. *Fertil. Steril.,* 22:181–187.
21. Heritage, A. S., Grant, L. D., and Stumpf, W. E. (1976): Autoradiographic localization of estradiol in the catecholamine neurons of the rat brain stem. Neuroscience Abstracts, Society for Neuroscience, 6th Annual Meeting, p. 672.
22. Iramain, C. A., Danzo, B. M., Strott, C. A., and Toft, D. O. (1973): Progesterone binding in the hypothalamus and hypophysis of female guinea pigs and rabbits. 4th International Congress of International Society of Psychoneuroendocrinology, Berkeley, p. 171.
23. Jackson, G. L. (1972): Effect of actinomycin D on estrogen-induced release of luteinizing hormone in ovariectomized rats. *Endocrinology,* 91:1284–1287.
24. Jackson, G. L. (1973): Time interval between injection of estradiol benzoate and LH release in the rat and effect of actinomycin D or cycloheximide. *Endocrinology,* 93:887–891.

25. Jackson, G. L. (1975): Blockage of progesterone-induced release of LH by intrabrain implants of actinomycin D. *Neuroendocrinology,* 17:236-244.
26. Johnston, P., and Davidson, J. M. (1972): Intracerebral androgens and sexual behavior in the male rat. *Horm. Behav.,* 3:345-357.
27. Kalra, S. P. (1975): Studies on the site(s) of blockade by actinomycin-D of estrogen-induced LH release. *Neuroendocrinology,* 18:333-344.
28. Kalra, S. P., Krulich, L., and McCann, S. M. (1973): Changes in gonadotropin-releasing factor content in the rat hypothalamus following electrochemical stimulation of anterior hypothalamic area and during the estrous cycle. *Neuroendocrinology,* 12:321-333.
29. Karavolas, H. J., and Nuti, K. M. (1976): Progesterone metabolism by neuroendocrine tissues. In: *Subcellular Mechanisms in Reproductive Neuroendocrinology,* edited by F. Naftolin, K. J. Ryan, and J. Davies, pp. 305-326. Elsevier, Amsterdam.
30. Kato, J. (1977): Characterization and function of steroids receptors in the hypothalamus and hypophysis. Proceedings of the Fifth International Congress on Endocrinology Symposium, "Steroid Receptors in the Brain," Hamburg, July, 1976. Excerpta Medica, Amsterdam (*in press*).
31. Kelly, M. J., Moss, R. L., and Dudley, C. A. (1976): Steroid specific changes in preoptic-septal (POA-S) unit activity in normal cyclic female rats. Endocrine Society Program and Abstracts, 58th Annual Meeting, p. 360.
32. Kizer, J. S., Muth, E., and Jacobowitz, D. M. (1976): The effect of bilateral lesions of the ventral noradrenergic bundle on endocrine-induced changes of tyrosine hydroxylase in the rat median eminence. *Endocrinology,* 98:886-893.
33. Kizer, J. S., Palkovits, M., Zivin, J., Brownstein, M., Saavedra, J. M., and Kopin, I. J. (1974): The effect of endocrinological manipulations on tyrosine hydroxylase and dopamine hydroxylase activities in individual hypothalamic nuclei of the adult male rat. *Endocrinology,* 95:799-812.
34. Knobil, E. (1974): On the control of gonadotropin secretion in the rhesus monkey. *Recent Prog. Horm. Res.,* 30:1-36.
35. Lassman, M. N., and Mulrow, P. M. (1974): Deficiency of deoxycorticosterone-binding protein in the hypothalamus of rats resistant to deoxycorticosterone-induced hypertension. *Endocrinology,* 94:1541-1546.
36. Lieberburg, I., and McEwen, B. S. (1975): Estradiol-17$\beta$: A metabolite of testosterone recovered in cell nuclei from limbic areas of neonatal rat brains. *Brain Res.,* 85:165-170.
37. Lisk, R. S. (1967): Sexual behavior: Hormonal control. In: *Neuroendocrinology, Vol. 2,* edited by L. Martini and W. F. Ganong, pp. 197-239. Academic Press, New York.
38. Lisk, R. D., and Barfield, M. A. (1975): Progesterone facilitation of sexual receptivity in rats with neural implantation of estrogen. *Neuroendocrinology,*19:28-35.
39. Luine, V. N., Khylchevskaya, R. I., and McEwen, B. S. (1974): Oestrogen effects on brain and pituitary enzyme activities. *J. Neurochem.,* 23:925-934.
40. Luine, V. N., Khylchevskaya, R. I., and McEwen, B. S. (1975): Effect of gonadal hormones on enzyme activities in brain and pituitary of male and female rats. *Brain Res.,* 86:283-292.
41. Luine, V. N., Khylchevskaya, R. I., and McEwen, B. S. (1975): Effect of gonadal steroids on activities of monoamine oxidase and choline acetylase in rat brain. *Brain Res.,* 86: 293-306.
42. Luine, V. N., and McEwen, B. S. (1977): Effect of oestradiol on turnover of Type A monoamine oxidase in brain. *J. Neurochem.,* 28:1221-1227.
43. Luine, V. N., and McEwen, B. S. (1977): Effects of an estrogen antagonist on enzyme activities and $^3$H estradiol nuclear binding in uterus, pituitary and brain. *Endocrinology,* 100:903-910.
44. Luine, V. N., and McEwen, B. S. (1977): Steroid hormone receptors in brain and pituitary: Topography and possible function. In: *Sexual Behavior,* edited by R. W. Goy and D. W. Pfaff. Plenum Press, New York (*in press*).
45. Luine, V. N., McEwen, B. S., and Black, I. B. (1977): Effect of 17$\beta$-estradiol on hypothalamic tyrosine hydroxylase activity. *Brain Res.,* 120:188-192.
46. Luine, V., Wallach, G., and McEwen, B. (1975): Neurochemical correlates of estrogen receptor function. Society for Neuroscience 5th Annual Meeting, New York, 1975. Abst. 683, p. 439.
47. McEwen, B. S. (1976): Steroid receptors in neuroendocrine tissues: Topography, sub-

cellular distribution and functional implications. In: *International Symposium on Subcellular Mechanisms in Reproductive Neuroendocrinology,* edited by F. Naftolin, K. J. Ryan, and J. Davies, pp. 277–304. Elsevier, Amsterdam.
48. McEwen, B. S. (1977): Gonadal steroid receptors in neuroendocrine tissues. In: *Hormone Receptors, Vol. I, Steroid Hormones,* edited by B. O'Malley and L. Birnbaumer. Academic Press, New York *(in press).*
49. McEwen, B. S., Lieberburg, I., Maclusky, N., and Plapinger, L. (1977): Do estrogen receptors play a role in the sexual differentiation of the rat brain? *J. Steroid Biochem. (in press).*
50. McEwen, B. S., Pfaff, D. W., Chaptal, C., and Luine, V. (1975): Brain cell nuclear retention of $^3H$ estradiol doses able to promote lordosis: Temporal and regional aspects. *Brain Res.,* 86:155–161.
51. McGowan-Sass, B. K., and Timiras, P. S. (1975): The hippocampus and hormonal cyclicity. In: *The Hippocampus, Vol. I,* edited by R. L. Isaacson and K. H. Pribram, pp. 355–374. Plenum Press, New York.
52. Mandelbrod, I., Feldman, S., and Werman, R. (1974): Inhibition of firing is the primary effect of microelectrophoresis of cortisol to units in the rat tuberal hypothalamus. *Brain Res.,* 80:303–315.
53. Mosher, K. M., Young, D. A., and Munck, A. (1971): Evidence for irreversible, actinomycin D-sensitive, and temperature-sensitive steps following the binding of cortisol to glucocorticoid receptors and preceding effects on glucose metabolism in rat thymus cells. *J. Biol. Chem.,* 246:654–659.
54. Moszkowska, A., and Kordon, C. (1965): Controle hypothalamique de la fonction gonadotrope et variation du taux des GRF chez le Rat. *Gen. Comp. Endocrinol.,* 5:596–613.
55. Nakahara, T., Uchimura, H., Hirano, M., Saito, M., and Ito, M. (1976): Effects of gonadectomy and thyroidectomy on the tyrosine hydroxylase activity in individual hypothalamic nuclei and lower brain stem catecholaminergic cell groups of the rat. *Brain Res.,* 117:351–356.
56. Neckers, L., and Sze, P. Y. (1975): Regulation of 5-hydroxytryptamine metabolism in mouse brain by adrenal glucocorticoids. *Brain Res.,* 93:123–132.
57. Pfaff, D. W., and Keiner, M. (1973): Atlas of estradiol-concentrating cells in the central nervous system of the female rat. *J. Comp. Neurol.,* 151:121–158.
58. Pfaff, D. W., Silva, M. T. A., and Weiss, J. M. (1971): Telemetered recording of hormone effects on hippocampal neurons. *Science,* 172:394–395.
59. Phillips, M. E., and Dafny, N. (1971): Effect of cortisol on unit activity in freely moving rats. *Brain Res.,* 25:651–655.
60. Piacsek, B. E., and Meites, J. (1966): Effects of castration and gonadal hormones on hypothalamic content of luteinizing hormone releasing factor (LRF). *Endocrinology,* 79:432–439.
61. Plapinger, L., and McEwen, B. S. (1977): Gonadal steroid-brain interactions in sexual differentiation. In: *Biological Determinants of Sexual Behavior,* edited by J. Hutchison. John Wiley & Sons, New York *(in press).*
62. Portanova, R., and Sayers, G. (1974): Corticosterone suppression of ACTH secretion: Actinomycin D sensitive and insensitive components of the response. *Biochem. Biophys. Res. Commun.,* 56:928–933.
63. Quadagno, D. M., and Ho, G. K. W. (1975): The reversible inhibition of steroid-induced sexual behavior by intracranial cycloheximide. *Horm. Behav.,* 6:19–26.
64. Raisman, G., and Field, P. M. (1973): Sexual dimorphism in the neuropil of the preoptic area of the rat and its dependence on neonatal androgen. *Brain Res.,* 54:1–29.
65. Ruf, K., and Steiner, F. A. (1967): Steroid-sensitive single neurons in rat hypothalamus and midbrain: Identification by microelectrophoresis. *Science,* 156:667–668.
66. Schutz, G., Killewich, L., Chen, G., and Feigelson, P. (1975): Control of the mRNA for hepatic tryptophan oxygenase during hormonal and substrate induction. *Proc. Natl. Acad. Sci. U.S.A.,* 72:1017–1020.
67. Stern, J. J., and Jankowaik, R. (1972): Effects of actinomycin-D implanted in the anterior hypothalamic-preoptic region of the diencephalon on spontaneous activity in ovariectomized rats. *J. Endocrinol.,* 55:465–466.
68. Sze, P. Y., and Neckers, L. (1974): Requirement for adrenal glucocorticoid in the ethanol-

induced increase in tryptophan hydroxylase activity in mouse brain. *Brain Res.*, 72:375–378.
69. Sze, P. Y., Neckers, L., and Towle, A. C. (1976): Glucocorticoids as a regulatory factor for brain tryptophan hydroxylase. *J. Neurochem.*, 26:169–173.
70. Terkel. A. S., Shryne, J., and Gorski, R. A. (1973): Inhibition of estrogen facilitation of sexual behavior by the intracerebral infusion of actinomycin-D. *Horm. Behav.*, 4:377–386.
71. Toran-Allerand, C. D. (1976): Sex steroids and the development of the newborn mouse hypothalamus and preoptic area *in vitro:* Implications for sexual differentiation. *Brain Res.*, 106:407–412.
72. Wallen, K., Goldfoot, D. A., Joslyn, W. D., and Paris, C. A. (1972): Modification of behavioral estrus in the guinea pig following intracranial cycloheximide. *Physiol. Behav.*, 8:221–223.
73. Whalen, R. E., Gorzalka, B. B., DeBold, J. F., Quadagno, D. M., Kan-Wha Ho, G., and Hough, J. C., Jr. (1974): Studies on the effects of intracerebral actinomycin-D implants on estrogen-induced receptivity in rats. *Horm. Behav.*, 5:337–343.
74. Whalen, R. E., and Rezek, D. L. (1974): Inhibition of lordosis in female rats by subcutaneous implants of testosterone, androstenedione or dihydrotestosterone in infancy. *Horm. Behav.*, 5:125–128.
75. Woo, S. L. C., and O'Malley, B. W. (1976): Hormone inducible messenger RNA. *Life Sci.*, 17:1039–1048.

# Neurophysiological Organization of the Endocrine Hypothalamus

*Leo P. Renaud

Since the previous ARNMD meeting on the hypothalamus, a tremendous body of information has accumulated. A complete review of the neurophysiology related to the endocrine hypothalamus is beyond the scope of this chapter, and the reader is referred to several current reviews (9,19,21, 45). Rather than attempt a comprehensive review, this author has elected to consider some of the recent highlights and trends of development in neurophysiology related to the function of two major hypothalamic-pituitary links, i.e., the *neurohypophysial pathway* for the liberation of oxytocin and vasopressin, and the *tuberoinfundibular pathway* for the release of the various peptide factors considered to regulate adenohypophysial secretion. Both systems are outlined schematically in Fig. 1. Such a schema does not pretend to account for any of the more complicated patterns of connections that now appear to exist (113) among the hypothalamus, the portal vessels, and the posterior pituitary. Figure 1 is instead a working model used to explain currently available neurophysiological data.

## NEUROHYPOPHYSIAL PATHWAYS

Prior to the introduction of microelectrodes, electrical recordings within the hypothalamus relied on data obtained with macroelectrodes; these observations have been adequately reviewed by Cross and Silver (21). In 1959 Cross and Green (20) presented some of the earliest reports on records obtained from single hypothalamic neurons. Subsequent electro-

---

* Division of Neurology, Montreal General Hospital, 1650 Cedar Avenue, Montreal, Quebec, H3G 1A4, Canada

physiological studies by Cross and others have served to characterize the electrophysiology of the neurosecretory neurons related to posterior pituitary, i.e., the magnocellular neurons of the supraoptic and paraventricular nuclei.

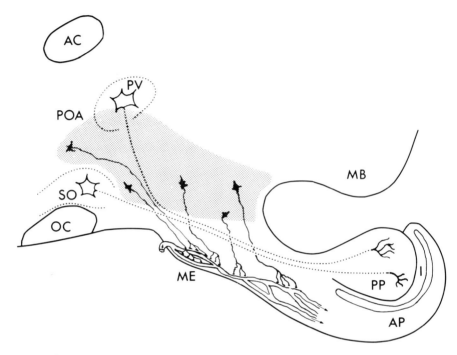

**FIG. 1.** Simplified sketch of the hypothalamus in sagittal section illustrates two classic neurosecretory pathways related to the pituitary: the *neurohypophysial tract* (*interrupted lines*) and the magnocellular paraventricular (PV) and supraoptic (SO) neurons to the posterior pituitary (PP) responsible for liberation of oxytocin and vasopressin; and the *parvicellular tuberoinfundibular* pathway originating from cells in the medial basal hypothalamus (MB) and preoptic area (POA) and terminating on portal capillaries in the median eminence (ME), responsible for elaborating hypothalamic releasing factors that regulate anterior pituitary (AP) secretion. The diagram does not include possible connections between the neurohypophysial system and the portal plexus, nor connections between the tuberoinfundibular system and the posterior pituitary. AC, anterior commissure; I, intermediate lobe; OC, optic chiasm.

## Electrophysiological Identification

Antidromic activation, with stimulating electrodes positioned on the pituitary stalk, has been one of the most useful techniques for the identification of neurosecretory neurons (4,19,25,56,74). Impulses initiated at the axon terminals of neurohypophysial fibers are conducted in retrograde direction to the cell soma where they initiate a fully developed action potential (110). This action potential, recorded with a microelectrode positioned near the cell body, can be differentiated from spontaneous action potentials by the following basic criteria: (a) constant latency all-or-none responses at threshold stimulation strengths for antidromic invasion; (b) ability to follow high-frequency (greater than 150 Hz) stimuli at constant latencies; and (c) cancellation of the antidromic action potential by a spontaneous (or orthodromic) action potential at appropriate interspike intervals, arising from a collision between the antidromic and orthodromic action potentials in the axon. Although the first two criteria can be met in neurons that fail to display any spontaneous activity, and are commonly used to facilitate localization of neurosecretory neurons, the third and probably most definitive of these criteria requires the presence of some form of ongoing activity. In practice, action potentials from supraoptic and paraventricular neurons that are spontaneously active, or are initiated by chemical [microiontophoretic application (53,54) of an excitatory substance such as L-glutamate or DL-homocysteic acid] or natural stimuli (hyperosmolar stimuli, vaginal distension, hemorrhage) are used to trigger the stimulator that in turn excites the axon terminal into producing an action potential at various time intervals after the orthodromic action potential. Provided that the orthodromic action potential truly originates from the same neuron whose terminal is being activated, the antidromic action potential will be observed only if the orthodromic action potential has completely invaded the nerve terminal and repolarization has occurred prior to the initiation of the stimulus. Otherwise, the two impulses will collide somewhere along the axon, and the antidromic action potential will not reach the cell body where the recording microelectrode is positioned (see lower two oscilloscope traces in Fig. 5; also ref. 64).

Recordings obtained from supraoptic and paraventricular neurons identified by antidromic invasion after pituitary stalk stimulation have served to characterize the electrical properties of these neurosecretory cells in a variety of conditions (2,23,28,31,47,61,62,102,103). It is noteworthy that action potentials recorded from these neurons are in all respects similar to those from other neurons comparable in size and shape and located in other areas of the central nervous system of both vertebrates (110) and invertebrates (50). Moreover, the discharge frequency of supraoptic and paraventricular neurons is probably modulated by synaptic excitatory and inhibitory inputs on their cell bodies and dendrites, similar to other neurons within the central nervous system; these electrophysiological changes reflect the neural events that modulate the functions of the neurohypophysis. These neurosecretory cells therefore serve as transducers between brain and endocrine tissue (107). In electrophysiological terms, this means that peptidergic neurosecretory supraoptic and paraventricular neurons behave in a manner identical to any other central neuron. Reassuring as this may be to the neuroendocrinologist, it also signifies that the electrophysiologist cannot rely on action potential configuration to distinguish peptidergic neurons from other central neurons. Antidromic invasion from pituitary stalk stimulation still remains the most definitive criterion for the identification of these specific neurosecretory cells.

### Recurrent Inhibition and Axon Collaterals in the Neurohypophysial Pathway

In 1964 Kandel (50) provided intracellular recordings from analogous neurosecretory neurons in the preoptic nucleus of the goldfish. Not only did he demonstrate action potentials in response to both orthodromic and antidromic stimulation, but he also illustrated the presence of short-latency inhibitory postsynaptic potentials after electrical stimulation of the pituitary stalk. This raised speculation that axons of these neurosecretory cells not only had terminals in the posterior pituitary, but also produced axon collaterals which formed part of a recurrent inhibitory feedback pathway to the parent neuron, either *directly* by a synapse on the neuron of origin, or *indirectly* through transsynaptic excitation of local interneurons whose axons then terminated on the neurosecretory cell. Similar neurophysiological evidence was eventually derived from studies in several vertebrate species (4,25,56,71,74). In extracellular recordings, the presence of an inhibitory postsynaptic potential following pituitary stalk stimulation is suggested by (a) a silent period lasting approximately 80 msec in the spontaneous or glutamate-evoked firing of supraoptic and paraventricular neurons at stimulus intensities below or near threshold for antidromic invasion of the cell under study, and (b) an increase in intensity and duration of the inhibitory period as the stimulus intensity is increased (see Fig. 6C). Intracellular recordings confirmed that inhibitory postsynaptic potentials occur in supraoptic neurons after pituitary stalk stimulation (50,56). From these studies it was concluded that local axon collaterals in the neurohypophysial pathways from both supraoptic and paraventricular neurons participate in this inhibitory event.

Problems of definition of cells considered to represent the local inhibitory interneurons in this recurrent inhibitory pathway have arisen. Thus, Barker and Nicoll (4) were unable to demonstrate local transsynaptic events within these nuclei following pituitary stalk stimulation, and therefore suggested that the axon terminals of neurohypophysial cells synapsed directly on the neurosecretory neurons, although they could not exclude the possibility of local inhibitory interneurons. Koizumi and Yamashita (56) described small Renshaw-type cells in the vicinity of the supraoptic nucleus that responded to stalk stimulation with high-frequency repetitive discharges, and suggested that these could be the local inhibitory interneurons. Other investigators have also described neurons that show transsynaptic activation following stimulation of the neural lobe (71,72).

Morphological evidence for axon collaterals in the neurohypophysial pathway have been provided by both Cajal (16) and Christ (17). On the other hand, Leontovich (58) was unable to observe such collaterals in his Golgi preparation of the dog brain. Recent electron microscopic studies in rat supraoptic nucleus (59,112) indicate that nearly two-thirds of the synapses on these neurons appear to be of intranuclear or otherwise local origin. Presumably some of these terminals arise from interneurons located in and around the supraoptic nucleus and/or intranuclear axon collaterals, or both. However, it is important to realize that there is still no definitive electron microscopic evidence for axon collaterals of neurosecretory neurons; none of the synaptic endings identified within the supraoptic nucleus exhibits the classic neurosecretory granules.

Why this emphasis on recurrent inhibition? Two points emerge from this discussion. Firstly, recurrent inhibition would serve as a functional rate-limiting or negative feedback mechanism that might explain the observed phasic or episodic discharge characteristics of some supraoptic or paraventricular neurons (Figs. 2 and 3). For example, some form of inhibition must account for the long-lasting inhibition in spontaneous discharges that follows osmotically induced augmentation of supraoptic nucleus cellular discharges and for the similar inhibitions observed in paraventricular neurons after the milk ejection reflex (47,61,102). There are of course other explanations for the silence of these neurons and the episodic activity of others, possibly owing to cyclic changes in cell membrane excitability and synaptic activity from other sources. Secondly, if axon collaterals were present in a peptidergic neurosecretory pathway, Dale's hypothesis (22) would suggest that a peptide, i.e., vasopressin or oxytocin, might be liberated and function at central synapses (4,74). However, the transmitter substance involved in recurrent inhibition still remains to be identified. Applied by microiontophoresis, vasopressin does in fact depress the activity of most neurosecretory neurons in the supraoptic nucleus, but only at relatively high microiontophoretic currents (74). The presence of recurrent inhibition in the Brattleboro rat which cannot synthesize vasopressin (32) is a strong argument against vasopressin serving as an inhibitory neurotransmitter in a direct (monosynaptic) recurrent pathway to neurosecretory neurons. Since oxytocin enhances the firing frequency of paraventricular neurosecretory neurons (66), this peptide is also an unlikely neurotransmitter candidate in the recurrent inhibitory pathways. Nevertheless, vasopressin does appear to influence animal behavior and to be important in memory trace consolidation (105), therefore not excluding the possibility that there are central vasopressin- or oxytocin-containing nerve endings that subserve some as yet undefined role in brain tissue separate from their systemic action.

Although physiological evidence attests to their existence, the occurrence of axon collaterals in neurosecretory neurons remains unresolved. Not to be excluded is the possibility that peptidergic neurons can synthesize and release other neurotransmitter agents. The capacity of neurons to secrete more than one agent has been discussed with increasing frequency (13,14,69). This issue will recur in connection with the existence of axon collaterals in the parvicellular tuberoinfundibular system (see below).

## Recurrent Facilitation

Neurophysiological studies in the isolated hypothalamo-hypophysial system of the bullfrog suggest that both recurrent inhibition and facilitation exist in this species (55). Threshold for facilitation was observed to be lower than for inhibition, indicating that cellular activity in the neurohypophysial pathway probably activates the facilitatory mechanism before the inhibitory mechanism. The neural network that would provide for facilitation appears obscure; presumably some form of direct interaction between adjacent neurosecretory neurons is required, through either axon collaterals or contacts between the cell somata, dendrites, or axons. This kind of interaction would provide a powerful means by which numerous neurosecretory cells could be activated synchronously or simultaneously, as, for example, to ensure a sufficiently plentiful output of oxytocin during the milk-ejection reflex (61,102) or vasopressin during hemorrhage (2,103).

## Afferent Pathways

As mentioned above, recent electron microscopic data of afferent connections to the supraoptic nucleus indicate that approximately two-thirds of all synapses on these neurons arise from an intranuclear source (112). Of the extranuclear afferents, the following distribution has been reported: 32.7% from brainstem, 21% from medial basal hypothalamus, 13.5% from amygdala, 13.5% from septum, 8.5% from hippocampus, and the remainder (17%) from the olfactory tubercle and further rostral regions (112). This correlates reasonably well with physiological data that indicate release of vasopressin following stimulation of the medial and lateral mammillary nuclei, midbrain periaqueductal gray, amygdala, hippocampus, olfactory tubercle, and diagonal band of Broca (45,48,71,106). Oxytocin can also be released by stimulation in the cingulate gyrus and posterior hypothalamus (106).

> At the single cell level, septal, amygdala, and reticular formation stimulation can influence the firing frequency of supraoptic and paraventricular neurons transsynaptically (71,72). In the case of the septum, the predominant response from 75% of neurosecretory paraventricular neurons (antidromically identified after pituitary stalk stimulation) was a silent interval at a mean latency of 18 msec and a mean duration of 60 msec. Stimulation in amygdala inhibited 64% of paraventricular neurosecretory cells at a mean latency of 17 msec and over a mean duration of 60 msec. Initial excitatory responses from either stimulus site were unusual. This contrasted sharply with the responses of paraventricular nonneurosecretory neurons that often displayed orthodromic transsynaptic excitatory responses (mean latency 15 msec) to stimulation of the pituitary stalk. These cells, which could function as interneurons in the recurrent inhibitory pathway, responded to septal stimulation with transsynaptic excitation in the majority of instances (94%). The majority of these cells (67%) also displayed orthodromic excitation after amygdala stimulation.

From these observations (71,72) one might suggest that recurrent and afferent inhibition of paraventricular neurosecretory neurons is mediated by similar, and possibly the same, inhibitory interneurons.

## RELATIONSHIP BETWEEN VASOPRESSIN-OXYTOCIN RELEASE AND NEURAL EVENTS

The neurohypophysial tracts from supraoptic and paraventricular nuclei comprise the final neuroendocrine pathway for oxytocin and vasopressin release in response to a variety of stimuli. Generally, in neural tissues that have been studied experimentally, neurotransmitter release at nerve terminals within a particular pathway is correlated with the frequency of action potentials in the system under study. Therefore, it seems appropriate to consider that the neural activity of neurosecretory supraoptic and paraventricular neurons is also correlated with the release of oxytocin and vasopressin in the neural lobe terminals (24,43). Several studies (20,31,33,45) have reported accelerated firing of supraoptic and paraventricular neurons in response to chemical or neural stimuli known to influence the release of oxytocin and/or vasopressin. However, much of the information that was initially derived was indirect and not precisely related to either the stimulus or the release of the hormone, in part because these studies were conducted in anesthetized preparations and in part because techniques for the measurement of these hormones were inadequate (39,40). The following is a summary of the recent attempts by experimental neurophysiologists to circum-

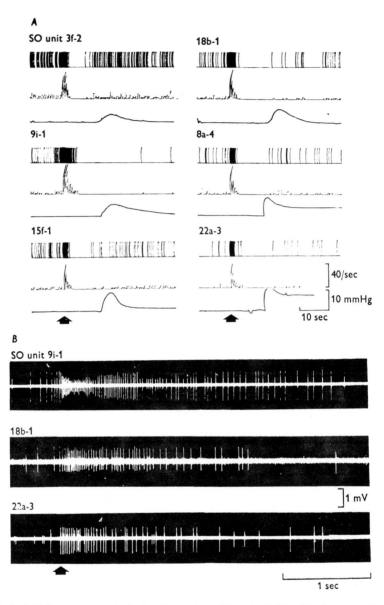

**FIG. 2. A:** Polygraph records derived from six antidromically identified supraoptic neurons illustrate unitary activity correlated with milk ejection. Each vertical deflection in the top traces represents single action potentials, the middle traces represent an integration of discharge frequency, and the lower traces represent intramammary pressure. Note the stereotyped acceleration of activity 10 to 13 sec before a rise in intramammary pressure, and the silent period following the burst of unit activity. **B:** Three examples from three different neurons to illustrate the augmented unit activity in supraoptic neurons preceding a rise in mammary pressure and milk ejection. Adapted from ref. 61 with permission.

vent these problems and to demonstrate directly an association between cellular activity and the release of oxytocin and vasopressin.

### Neural Events Related to Oxytocin Release

The milk-ejection reflex, the response of the myoepithelial cells of the mammary glands that produce a rise in intramammary ductal pressure and expulsion of milk, is considered to be mediated by episodic pulses of circulating oxytocin released from the neurohypophysis. In the lactating rat, there is a close temporal relationship between suckling and milk ejection. Single-unit recordings in the lactating rat have shown a periodic activation of supraoptic and paraventricular neurosecretory neurons related temporally to an increase in intramammary pressure (19.31,61,62,102). A 2- to 4-sec period of activation of both supraoptic and paraventricular neurosecretory cells, characterized by an acceleration in spike discharge frequency up to 30 to 50 spikes/sec in 50% of the neurons, is followed by milk ejection some 12 to 15 sec later (Fig. 2). These pulses of neuronal activity and associated milk release recur uniformly every 4 to 8 min. The afferent stimulus for this periodic activation of "oxytocinergic" neurons during nursing is provided by the suckling activity of the young (62). In the lactating rat, this response persists in the face of deep anesthesia and is dependent in part on the number of suckling young, the degree of mammary distension, and time (62).

This particular pattern of neural response, i.e., a periodic burst of action potentials at high frequency, appears important for neural lobe neurosecretion and the release of oxytocin. Several workers (24,43) have found that electrical stimulation of the pituitary stalk *in vivo* could evoke milk ejection in lactating rabbits or rats only if stimulation frequencies were greater than 20 to 30/sec with 50/sec being optimal. Similarly, *in vitro* studies using incubated rat neurohypophysis have indicated that there was a greater release of oxytocin into the medium at higher frequencies of stimulation and that this neurosecretion was a calcium-dependent mechanism (24). It would seem, therefore, that a rapid succession of action potentials arriving at the nerve terminal achieves the degree of depolarization required for oxytocin release. Outside of the period in which these cells display their characteristic periodic responses, the background activity of these oxytocinergic neurosecretory cells seemingly does not contribute significantly to oxytocin release.

Supraoptic and paraventricular neurosecretory neurons display two general patterns of activity: one characterized by irregular spike discharges at variable frequencies, sometimes randomly distributed (nonphasic), the other characterized by periodic bursts of spikes with periods of 5 to 15 sec separated by intervals with little or no activity (phasic).

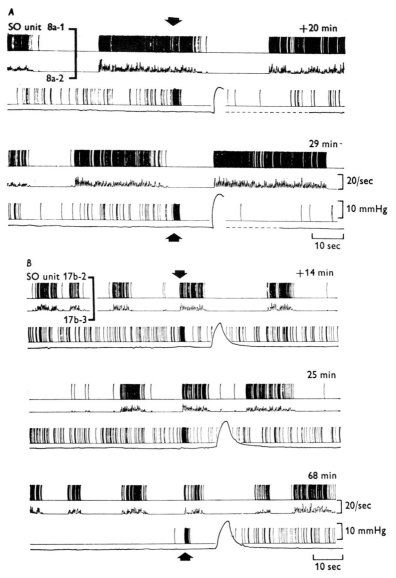

**FIG. 3.** Simultaneous recordings from phasic and nonphasic supraoptic units. In **A**, the activity of a phasic unit illustrated in the upper trace is not correlated with the characteristic activity pattern associated with milk ejection displayed by the lower trace neuron. In **B**, the discharges of a phasic unit shown in the upper trace appear to be synchronized with the characteristic milk-ejection discharge pattern of the lower trace neuron. Adapted from ref. 61 with permission.

Both phasic and nonphasic units in the supraoptic and paraventricular nuclei appear to be involved in the milk-ejection reflex (Fig. 3). Although phasic units may be involved in oxytocin release, the majority of these cells are apparently involved in reflex activity associated with vasopressin release.

In a recent study, Dreifuss et al. (28) demonstrated that vaginal distension was also an effective stimulus to promote an increase in unit activity of supraoptic nucleus neurons in the lactating rat, and that this activity was followed by a rise in intramammary pressure. This suggests oxytocin release, although the magnitude of either the unit activity or the intramammary pressure increase was less than that associated with suckling. These authors pointed out that as many phasically firing neurons as randomly firing neurons were excited by vaginal distension, and they questioned the relative amounts of oxytocin and vasopressin that could be released under these experimental conditions. Perhaps vaginal distension is but one stimulus that can activate both vasopressin- and oxytocin-producing neurosecretory neurons, whereas more specific stimuli are required for selective activation of neurons responsible for the neurosecretion of either oxytocin or vasopressin.

**Neural Events Related to Vasopressin Release**

In their pioneer study on single hypothalamic units, Cross and Green (20) demonstrated that units in the supraoptic nucleus could be activated by intracarotid hyperosmolar stimuli. It has since been demonstrated that osmotic (e.g., dehydration), volumetric (e.g., hemorrhage), and behavioral (e.g., drinking) factors interact to release vasopressin from the neural lobe of the pituitary. However, there has been some difficulty in correlating blood levels of vasopressin with magnocellular neurosecretory neuronal activity, in part owing to conditions of anesthesia which can alter firing patterns of central neural neurons and induce aberrant changes in blood vasopressin levels. Nonetheless, several studies indicate that the activity of supraoptic and paraventricular neurosecretory cells is correlated with circulating levels of vasopressin (2,45,47,103).

Approximately one-fourth of the neurosecretory supraoptic and paraventricular neurons display phasic bursts of activity, a pattern seldom observed from other hypothalamic neurons. Although some phasic neurons respond to suckling and display bursts of activity associated with the milk-ejection reflex (Fig. 3), most of the phasic units do not, and instead are responsive to other forms of stimuli (e.g., hyperosmotic injections, dehydration, carotid occlusion, and hemorrhage) that cause release of vasopressin with little or no oxytocin release. Based on these observations it has been suggested that the phasic neurosecretory supraoptic or paraventricular neuron is a "vasopressinergic" neuron. Additional studies indi-

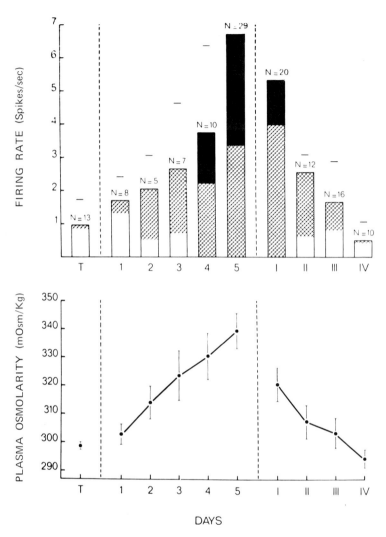

**FIG. 4.** Data obtained from awake, water-deprived, and rehydrated monkey. The upper part of the diagram schematically illustrates patterns and rates of firing of supraoptic neurons compared with normally hydrated monkeys (T). The lower part of the diagram shows plasma osmolarity. With each successive day of water deprivation (1 through 5), the rise of plasma osmolarity was associated with a general increase in firing frequency of supraoptic neurons. The numbers of neurons sampled each day are indicated above the bar graph. Note that with dehydration the irregularly firing neurons (*white area*) become less frequent while phasic discharging cells (*cross-hatched area*) increase in number; at peak dehydration, high-frequency continuously firing neurons appear (*black area*). The sequence appears reversible upon rehydration. Adapted from ref. 2 with permission.

cate that phasic discharges are but one functional state of activity of these neurons. For example, Arnauld et al. (2) demonstrated that the predominant cell firing pattern of supraoptic neurons in the unanesthetized monkey under conditions of dehydration changed from irregular low-frequency discharge patterns (type $i$ cells), to phasic discharges (type $p$ cells), to high-frequency continuously discharging cells (type $c$ cells) in concert with increasing plasma osmolarity; the reverse trend occurred upon rehydration (Fig. 4). As postulated by Wakerley et al. (103), some form of synchronized activity of phasic cells may result in significant discharge of neurohypophysial hormone.

The afferent limb of the reflex activation of vasopressin release is not precisely known. Activity of neurosecretory neurons in the supraoptic nucleus changes in response to vagal or carotid sinus nerve stimulation (4). Osmosensitive neurons in the supraoptic nucleus have been described, but it is still not certain whether the neurosecretory cells themselves, or adjacent cells, behave as the specific osmoreceptor (47). In the latter situation, neurosecretory cells would be responding secondarily to a signal from the true osmoreceptor, which evidently is important in the drinking and behavioral aspects of osmoregulation (45).

## NEUROPHARMACOLOGY OF SUPRAOPTIC AND PARAVENTRICULAR NEURONS

Microiontophoresis, a technique whereby minute quantities of putative neurotransmitters and other agents can be introduced directly into the extracellular environment while extracellular or intracellular excitability characteristics are monitored from neurons (53,54), has been usefully applied to determine the sensitivity (and possible presence of receptor sites) of cells in various CNS regions. Ideally, this method allows for the determination of cellular responsivity to a given compound, or its antagonist, using both systemic and direct administration.

In 1939 Pickford (77) reported on the antidiuretic effect of acetylcholine. Subsequent studies have shown that the supraoptic nucleus contains high levels of choline acetyltransferase and acetylcholinesterase (1,37). Nerve fibers containing acetylcholinesterase have in fact been demonstrated in all regions of the hypothalamus (97). Acetylcholine injected into the carotid artery of unanesthetized dogs has been shown to evoke release of oxytocin and vasopressin. The milk-ejection reflex described earlier is blocked by anticholinergic drugs, and by emotional stress, which involves a central inhibition of oxytocin release (19). Based on an analysis of data from microiontophoresis, it would appear that acetylcholine has a predominantly excitatory action on neurosecretory neurons in the supraoptic (3,26) and paraventricular (68) nuclei, but depresses the firing frequency of adjacent nonneurosecretory neurons that do not send their axons into the pituitary stalk. The excitatory effects of acetylcholine appear to be nicotinic and are antagonized by dihydro-$\beta$-erythroidin; on the other hand, atropine appears to block specifically the acetylcholine-evoked depression (3). Neither intravenous nor microiontophoretically applied atropine has been shown to influence recurrent inhibition in the supraoptic nucleus (74). Thus, acetylcholine does not appear to be the neurotransmitter in the recurrent inhibitory pathway described earlier.

Although acetylcholine can accentuate or depress the release of vasopressin and oxytocin from posterior pituitary, depending on the nature of the receptor activated on the supraoptic or paraventricular neurons, norepinephrine appears to inhibit the release of vasopressin. Norepinephrine applied by microiontophoresis depresses the activity of most neurosecretory neurons, but enhances the discharge frequency of a comparatively greater number of adjacent nonneurosecretory supraoptic and paraventricular neurons (3). Norepinephrine-evoked depression is blocked by $\alpha$-adrenergic blocking agents. Recurrent inhibition in these nuclei persists despite treatment with 6-hydroxydopamine (74), which destroys noradrenergic nerve endings and virtually abolishes all of the norepinephrine fluorescence terminals in the supraoptic nucleus.

According to Fuxe and Hökfelt (38), very few dopaminergic and serotoninergic nerve endings are demonstrable by the fluorescent technique within the paraventricular nuclei. When dopamine and serotonin are applied by microiontophoresis, no consistent responses have been observed among either neurosecretory or nonneurosecretory paraventricular neurons (68); the activity of some neurons is increased, others decreased, and some unchanged. Therefore, the functional significance of the microiontophoretic results is not clear.

The amino acid L-glutamate enhances the activity of virtually all supraoptic and paraventricular neurons tested (68), whereas glycine and gammaaminobutyric acid (GABA) consistently depress their activity. Glycine-evoked depression is antagonized by strychnine, whereas picrotoxin and bicuculline appear to selectively antagonize GABA-evoked depression (74). However, neither of these convulsants administered by microiontophoresis or systemic injection appears to antagonize recurrent inhibition, except for bicuculline which produces partial antagonism when administered systemically in convulsive doses (74).

Angiotensin II, a peptide important in water balance and blood volume regulation, is reported to have a role in brain, and may exert a direct effect on neural tissue (7,76). Neurons in the supraoptic nucleus are excited by iontophoretic application of angiotensin II (73,92), suggesting that these neurons contain specific angiotensin II receptor sites.

## THE "PARVICELLULAR TUBEROINFUNDIBULAR" SYSTEM

We now turn to an electrophysiological analysis of an area that has received comparatively less attention than the neurohypophysial system, i.e., the parvicellular neurosecretory system deemed responsible for regulation of adenohypophysial secretion. Based principally on the pioneering efforts of Harris (42), Green, and others more than two decades ago, a concept has been developed and nourished with regard to the mechanisms for hypothalamic control of adenohypophysis. It is widely believed that a system of tuberoinfundibular neurosecretory neurons located within the medial and basal hypothalamus represents the "final common pathway" for neural

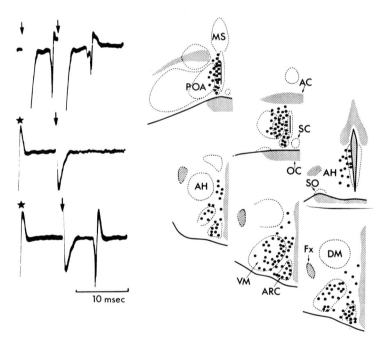

**FIG. 5.** The black dots located throughout the mediobasal hypothalamus and medial preoptic area schematic sections represent the distribution of tuberoinfundibular neurons identified by antidromic invasion following stimulation on the surface of the median eminence. Criteria for antidromic invasion shown on the left for a representative neuron include demonstration of constant-latency responses to paired median eminence stimuli at frequencies beyond 150 Hz (*upper trace*) and collisions at appropriate intervals between spontaneous and antidromic stimuli (*lower traces*). On the right, note the widespread distribution of tuberoinfundibular neurons throughout mediobasal hypothalamus and medial preoptic area. Adapted from ref. 87 with permission.

regulation of adenohypophysial secretion (42,100). The final messengers in this system are hypophysiotropic peptides with release or release-inhibiting potential on the secretion of adenohypophysial cells; these peptides are elaborated into the median eminence portal capillary plexus from nerve endings of these tuberoinfundibular (peptidergic) neurons (8,95). Most workers presume that these peptides are carried in portal blood to the adenohypophysis (78), and their release in response to appropriate neural stimuli requires an increase in neuronal discharges in the tuberoinfundibular system, analogous to events that accompany secretion of posterior pituitary hormones.

"Tuberoinfundibular" neurons are defined *anatomically* as parvicellular neurosecretory neurons located within the hypophysiotrophic area of the mediobasal hypothalamus (41,100). Their fine axons course independently from the larger neurohypophysial fibers and terminate in the surface zone of the median eminence in close relation to the primary capillaries of the portal system (100). These terminals are distinct from those in the neural lobe and contain two vesicular populations: an abundant number of synaptic vesicles of the usual type, and somewhat larger vesicles with dense osmiophilic cores. The *electrophysiological* definition of a "tuberoinfundibular" neuron refers to cells located in the hypothalamus and basal forebrain area that display the criteria for antidromic activation following stimulation applied to the *surface* of the median eminence (63,93).

### Electrical Characteristics of Tuberoinfundibular Neurons

Electrophysiological evidence for hypothalamic tuberoinfundibular neurons was first provided in 1971 (63) with the report that hypothalamic neurons located outside the supraoptic and paraventricular nuclei also displayed antidromic invasion following stimulation applied to the surface of the median eminence. (The location of the stimulating electrode on the surface of the brain is important, since it indicates a preferential termination site in the neighborhood of the portal capillary plexus rather than within the neural lobe.) In this initial study in the rat, the majority of tuberoinfundibular neurons were localized in the arcuate and ventromedial nuclei, while a smaller number were noted in the suprachiasmatic, anterior periventricular, and dorsal premammillary nuclei (63). Subsequent investigators (44,68, 83,84,89) have confirmed and extended these initial observations to include the anterior hypothalamic area and medial preoptic area (Fig. 5). In each instance, electrophysiological recordings have generally been brief, implying the presence of small somata. Tuberoinfundibular neurons appear to comprise less than 10% of the population of recordable neurons within the mediobasal hypothalamus (83). Although it is noteworthy that many of these tuberoinfundibular neurons are recognized within the "hypophysiotrophic area" described by Halász (41), a substantial number of these neurons are located over a more elongated region of the basal forebrain extending into the rostral preoptic area (Fig. 5).

In the studies performed in our laboratory, antidromic invasion latencies for mediobasal tuberoinfundibular neurons have ranged from 0.5 to 14.0 msec, whereas cells located more rostrally in the anterior hypothalamic and medial preoptic areas have displayed antidromic invasion over a latency range of 4.0 to 25.0 msec. Conversion of these latencies to conduction velocities has confirmed earlier reports of impulse conduction and velocities under 2.0 m/sec. Curiously, the tuberoinfundibular neurons with the slowest conduction velocities (under 0.2 m/sec) appear to be located less than 0.5 mm from the ventral surface of the mediobasal hypothalamus, i.e., within the arcuate nucleus and ventral portion of the ventromedial nucleus. One might speculate that these results point to a different population of tuberoinfundibular neurons, possibly the dopaminergic tuberoinfundibular neurons referred to in the literature (60).

> No notable differences have been observed in the spontaneous or antidromic action potentials recorded from tuberoinfundibular neurons, and the action potentials recorded from neighboring hypothalamic or other central neurons. All potentials have displayed an inflexion on their rising phase, which is primarily of positive polarity. All cells have demonstrated the ability to respond to stimuli at frequencies of 200 Hz. Since some neurons can respond to paired stimuli less than 3 msec apart, this would suggest that the absolute refractory period of these neurons is brief. On the other hand, the relative refractory period may be substantially prolonged, since the second antidromic response frequently displays an accentuation of the inflexion on the rising phase of the action potential, coupled with both a reduction in amplitude and prolongation of the spike duration even when tested up to 30 msec after the initial antidromic spike. As with other central neurons, the action potential is presumed to originate at the initial segment, and secondarily invade the cell body and dendrites, thereby forming two components. With paired stimuli, it was often possible to demonstrate dissociation of the antidromic spike into these two separate components and to observe failure of the second (soma-dendritic) component, presumably indicating failure of antidromic invasion into a region with a low safety factor.

While stimulating the median eminence at 1 Hz we have occasionally encountered hypothalamic and preoptic neurons that display antidromic invasion at two or more constant latencies depending on the stimulus intensity (83). Similar observations in the supraoptic and paraventricular nuclei after stimulation of the neural lobe have been reported (4). These results are interpreted to suggest the existence of terminal branching in the tuberoinfundibular pathway. An alternate but less tenable explanation would be the existence of a tortuous pathway through the focus of stimulation for axons of tuberoinfundibular neurons (19).

One of the interesting observations derived from the electrophysiological studies on hypothalamic neurons in general, and on tuberoinfundibular

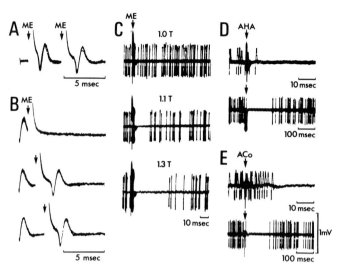

**FIG. 6.** Representative data from a ventromedial nucleus tuberoinfundibular neuron identified by antidromic invasion **(A)** and collision techniques **(B)**, following median eminence stimulation. The sequence in **(C)** illustrates a graded increase in the duration of a silent interval after antidromic invasion [the numbers refer to threshold (T) for antidromic invasion], considered as evidence for recurrent postsynaptic inhibition after median eminence stimuli. **(D,E):** The evoked silent periods following anterior hypothalamic area and amygdala stimulation indicate afferent inhibitory connections on this particular tuberoinfundibular neuron from these areas. Adapted from ref. 87 with permission.

neurons in particular, is the ability of these cells to conduct antidromic impulses (and presumably orthodromic impulses) at frequencies up to 400 to 500 Hz. This appears somewhat surprising for axons whose conduction velocities are under 2.0 m/sec and whose fiber diameters are often less than 1 $\mu$m (10,11). Although spontaneously active tuberoinfundibular cells may not usually discharge for long periods of time at high frequencies, some of these cells do indeed produce bursts of impulses with intraburst frequencies approaching 400 Hz. It remains to be determined whether such activity is a prerequisite for the liberation of releasing factors, similar to the events associated with the liberation of oxytocin and vasopressin described earlier.

## Recurrent Inhibition

Many tuberoinfundibular cells display little or no spontaneous activity in the anesthetized preparation. Most of the active tuberoinfundibular cells display either random discharge patterns or bursts of spikes over widely separated time intervals (83,85). Stimulation of the median eminence is usually followed by a decrease in neuronal excitability lasting 40 to 50 msec, at stimulus intensities subthreshold for antidromic invasion (Fig. 6C). An increase in the stimulus intensity to twice threshold strength usually prolongs the inhibitory period to 100 to 130 msec. Since this inhibition has been observed in both spontaneous and glutamate-evoked activity, one must consider that this represents some form of postsynaptic process analogous to events in supraoptic and paraventricular neurons (83,85,93,94,111).

The presence of recurrent inhibition after median eminence stimulation is further evidence for local hypothalamic axonal arborization in the tuberoinfundibular pathway. This recurrent inhibition could operate either directly, with recurrent axon collaterals terminating on the cells of origin, i.e., tuberoinfundibular neurons (Fig. 9, No. 1), or indirectly through a local inhibitory interneuron (Fig. 9, No. 2). The latter explanation appears more tenable since there are local nontuberoinfundibular neurons that exhibit transsynaptic excitation after median eminence stimulation and could therefore serve as the inhibitory interneuron in the recurrent pathway (83). Also, where spontaneously active tuberoinfundibular neurons demonstrate recurrent inhibition at stimulus intensities subthreshold for antidromic invasion, latencies for the inhibitory period are usually equal to or somewhat greater than the latencies for the antidromic spikes, suggesting a disynaptic pathway.

In accordance with Dale's hypothesis (22), the presence of recurrent inhibition in a putative peptidergic pathway raises the possibility that peptides can be released both in the median eminence portal capillary plexus and at central synapses where they could act as neurotransmitters. At present, this question cannot be adequately answered. However, it is apparent that the action of the transmitter in the recurrent inhibitory pathway is picrotoxin sensitive and may therefore be gamma-aminobutyric acid (111). Should recurrent inhibition utilize a disynaptic pathway (see Fig. 9, No. 2), excitation would be required at the initial synapse. Microiontophoretic studies with the different hypothalamic peptides that have been structurally characterized have in fact demonstrated that these peptides can evoke excitatory or depressant effects on central neurons, although the most frequently observed pattern is a depression of excitability (35,51,52,65,90,91; see also Fig. 10).

> Does recurrent inhibition serve a functional role in the tuberoinfundibular system? It is possible that recurrent inhibition can moderate the background activity in the tuberoinfundibular system or in a particular part of this system. If neurosecretion requires bursts of activity, this inhibitory tone would, however, have to be removed temporarily. We have observed that during bursting activity in spontaneously active tuberoinfundibular neurons, recurrent inhibition does in fact become less effective than during random activity. Thus, some dynamic mechanism may exist, possibly involving extrahypothalamic afferent connections, to alter the bias in the recurrent inhibitory feedback loop conceivably through a direct action on the inhibitory neuron thereby allowing bursting activity to occur.

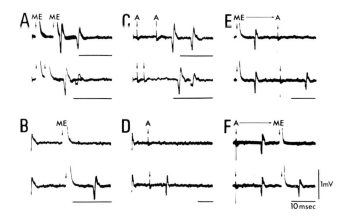

**FIG. 7.** Data that illustrate axon collateral pathways in the tuberoinfundibular system, in this instance to the amygdala. **A,C:** High-frequency responses from both median eminence (ME) and amygdala **(A). B,D:** Cancellations between spontaneous and antidromic spikes from both median eminence and amygdala. **E,F:** Cancellations from antidromic invasion by appropriately timed suprathreshold stimuli to the other branch of this neuron. Reprinted from ref. 85 with permission.

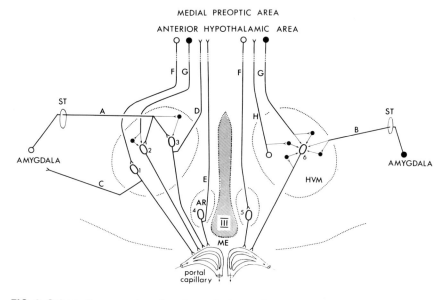

**FIG. 8.** Schematic coronal section through the mediobasal hypothalamus, outlining relative nuclear positions of the ventromedial (HVM) and arcuate (AR) nuclei around the median eminence (ME), and postulated connections of mediobasal tuberoinfundibular neurons with the amygdala, medial preoptic area, and anterior hypothalamic area. Tuberoinfundibular neurons (1–3, 6) within the ventromedial nucleus receive and send fibers to the amygdala and preoptic regions; arcuate nucleus tuberoinfundibular neurons (4, 5) appear related more to rostral than to lateral regions. Pathways presumed to mediate excitatory connections (A, F) originate from open circles, inhibitory connections (B, G) originate from closed circles. The finer continuous lines represent intranuclear recurrent or afferent axon collateral inhibitory pathways. Note the bifurcation in the axons of some tuberoinfundibular neurons (1, 3, 4) with axon collaterals to extrahypothalamic sites. Reprinted from ref. 87 with permission.

## Recurrent Facilitation

In recent studies of the rat tuberoinfundibular system, Sawaki and Yagi (94,111) observed activity considered to represent facilitation after median eminence stimulation. This was considerably enhanced by intravenous picrotoxin, possibly due to the inactivation of a normally dominant recurrent inhibitory system. In their experiments, facilitation was not observed in animals pretreated with $\alpha$-methyl-$p$-tyrosine, suggesting that catecholaminergic neurons may be involved in neural pathways that mediate facilitation of tuberoinfundibular cells after median eminence stimulation.

## Axon Collaterals in the Tuberoinfundibular System

In an attempt to define the electrophysiology of the tuberoinfundibular system, it has been noted that some tuberoinfundibular neurons have exhibited unequivocal antidromic invasion not only from the median eminence, but also from other sites within the brain, i.e., the paraventricular nuclei, the anterior hypothalamic area, the medial preoptic area, thalamic midline structures in and around nucleus medialis dorsalis, and the amygdala (44,82–85,87–89). Evidence provided from cancellation experiments suggests that the axons of these tuberoinfundibular neurons branch very close to their site of origin (87). In each example, the antidromic invasion latencies have been proportionate to the distances separating the recording microelectrode from the stimulus, thereby suggesting that impulse conduction is equivalent in each branch of the axon.

The existence of such widely distributed axon collaterals (Figs. 7 through 9) in a putative peptidergic system heightens speculation as to the role of these axon terminals in intrahypothalamic and extrahypothalamic brain regions. It is plausible that some of these connections are feedback circuits that govern activity in the tuberoinfundibular system, and also inform other brain regions of the status of activity in the tuberoinfundibular system at a particular moment. It is also reasonable to assume that the tuberoinfundibular system is capable of responding to afferent information from extrahypothalamic regions. The axonal arborization patterns described above may be a further indication that these same cells in fact perform very complex integrative roles in neural function through simultaneous reciprocal relations with many brain regions.

Other important problems concern the nature of the neurotransmitter agent(s) contained in the central terminals of these tuberoinfundibular cells, and the possibility that peptides are released at central synaptic sites. For technical reasons these questions may not be easily answered, but some clues have been identified. For example, immunohistochemical and radioimmunoassay studies describe hypothalamic peptides in widespread areas of the brain, and indicate evidence for central peptidergic neural pathways

(5,6,12,29,49,98). Numerous investigators have indicated that hypothalamic peptides exert behavioral effects through a direct action on brain tissues (79,96,105). Subcellular fractionation studies have described hypothalamic peptides localized to synaptosomal fractions of brain (36,81). In certain instances high-affinity receptor binding for these peptides has been observed in brain tissue (15). The hypothalamic peptides thyrotropin releasing hormone (TRH), luteinizing hormone releasing hormone (LHRH), and somatostatin (growth hormone release inhibiting hormone) appear to exert potent depressant effects on central neuronal activity when applied by microiontophoretic techniques (35,51,52,64,65,82,90,91). Thus, there is indirect evidence to suggest that some peptides may in fact influence neuronal behavior. Furthermore, the axon collaterals described by electrophysiologists to exist within a putative peptidergic neural network may be important in the mediation of part of this role for hypothalamic peptides. For the moment,

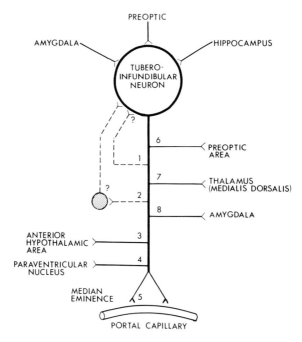

**FIG. 9.** Schematic summary of the connections of mediobasal hypothalamic tuberoinfundibular neurons based on the available experimental data. The principal axon of a tuberoinfundibular neuron is illustrated ending on a median eminence portal capillary, with some terminal branches (No. 5). Other intrahypothalamic axon collaterals mediate recurrent inhibition either directly (1) or indirectly through an inhibitory interneuron (2), and may also reach the anterior hypothalamic area (3) or paraventricular nucleus (4). Extrahypothalamic axon collaterals reach the medial preoptic area (6), the midline thalamic nuclei (7), or amygdala (8). Afferents from the amygdala, preoptic area, and hippocampus are illustrated above.

however, elucidation of the precise role of centrally acting peptides and axonal arborization in putative peptidergic neural networks must await more direct evidence.

### Afferent Connections of Tuberoinfundibular Neurons

Numerous stimulation and lesion experiments have demonstrated that extrahypothalamic regions, notably the amygdala, preoptic area, and hippocampus, can influence patterns of adenohypophysial secretion (60,114). One might anticipate that this influence is exerted through synaptic connections on tuberoinfundibular neurons from these brain areas, and that the functional correlates of these connections could be demonstrable by electrophysiological techniques.

#### *Amygdala*

Electrophysiological data from studies in both the cat (27,70) and rat (85,86,89) indicate that stimulation in the amygdala or on the stria terminalis modifies the excitability of neurons within the ventromedial nucleus. We have recently demonstrated that in the rat, stimulation in either the basolateral or basomedial amygdala, and on the stria terminalis, also exerts short-latency excitatory and/or inhibitory effects on tuberoinfundibular neurons in the mediobasal hypothalamus (84,85; Fig. 6E). This action is selectively directed toward tuberoinfundibular neurons localized within the ventromedial nucleus. Tuberoinfundibular cells located in the arcuate nucleus and periventricular region, as well as in the medial preoptic and anterior hypothalamic area, are generally unresponsive to amygdala stimulation (85). For those tuberoinfundibular cells that display initial excitation, the latencies of these responses evoked by stria terminalis stimulation are generally shorter than those evoked by amygdala stimulation. In spontaneously active tuberoinfundibular cells these initial excitatory responses are followed by silent periods of 70 to 150 msec. A smaller number of tuberoinfundibular neurons mostly located in the dorsal and lateral part of the ventromedial nucleus display initial inhibitory responses suggesting a postsynaptic inhibitory process (Fig. 6E). This inhibition has a longer latency than the excitatory responses, but it is uncertain whether this represents a direct inhibitory connection from the amygdala or whether the longer latencies represent a disynaptic process, arising from activation of local inhibitory interneurons in the amygdalohypothalamic projection to the ventromedial nucleus (Fig. 8).

> In the cat, amygdalofugal fibers to the hypothalamus follow two separate routes, i.e., via the stria terminalis and the ventral amygdalofugal pathway, originating, respectively, from the corticomedial and the basolateral amygdala (27,70). In the rat, lesions of the stria terminalis abolish

most of the amygdala-evoked responses (irrespective of stimulation sites within the amygdala) within the hypothalamic ventromedial nucleus (86). This would suggest, therefore, that there exist species differences in the amygdalofugal projection systems and that in the rat most of the amygdalofugal fibers to the hypothalamus course through the stria terminalis. Such an arrangement would also explain why latencies for the excitatory responses evoked within the ventromedial nucleus are shorter after direct stimulation of the stria terminalis than after amygdala stimulation (85,86).

In accordance with these observations, it has been possible to tentatively outline some of the neural circuitry that interrelates mediobasal hypothalamic neurons, specifically the tuberoinfundibular neurons, with the amygdala in the rat (Fig. 8). It is hoped that this method of analysis will eventually provide a foundation for explaining the role of the amygdala in adenohypophysial secretion. However, at present it is not possible to associate tuberoinfundibular neurons identified by electrophysiology with a specific hypothalamic peptide, and therefore it is not yet possible to define more precisely which aspect of adenohypophysial secretion is likely to be mediated by the neural circuits described above. Based on the information available in this volume and elsewhere (114), it would seem that the amygdala can influence secretory patterns of several adenohypophysial hormones, and thus probably modify the activity of several different types of tuberoinfundibular neurons.

### Anterior Hypothalamic-Medial Preoptic Area

The medial preoptic area is concerned functionally with several different parameters of homeostasis, including temperature regulation (46), water balance (46,101), sexual behavior (75), and endocrine function, in particular the regulation of pituitary gonadotropin secretion (41,64). There is anatomical evidence for both ascending and descending connections between anterior hypothalamic-medial preoptic area (AHA-MPOA) and the mediobasal hypothalamus (18,80,99,100), and some of these have also been defined by electrophysiological studies (34,44,57,64,87).

Dyer and his colleagues (19,34) have reported that 41% of AHA-MPOA neurons project directly to the ventromedial-arcuate nucleus region. In these studies, a stimulating electrode was positioned within the brain in the ventromedial-arcuate complex. Since some of the preoptic tuberoinfundibular neurons, identified by antidromic invasion from stimulation applied to the surface of the median eminence (see Fig. 5), probably send axons through the ventromedial-arcuate nucleus complex, some of these neurons are probably included within the estimate provided by Dyer. Recent studies in our laboratory have demonstrated that tuberoinfundibular neurons in the mediobasal hypothalamus are influenced by AHA-MPOA stimulation (84,87), implying the existence of synaptic pathways whereby these rostral brain regions can influence activity in the caudal parts of the tuberoinfundibular system. A schema of the postulated neural connections between the amygdala and AHA-MPOA and mediobasal tuberoinfundibular neurons is depicted in Fig. 8. This overly simplified model illustrates some of the functional synaptic connections and axonal arborizations that can be identified in the tuberoinfundibular system using currently available electrophysiological techniques.

### Hippocampus

The medial corticohypothalamic tract, originating from the subiculum, emerges from the fornix in the anterior hypothalamic area and ends in the rostral part of the arcuate nucleus (80). Whether this pathway or other fibers that leave the fornix in its passage through the lateral hypothalamus carry impulses to mediobasal tuberoinfundibular neurons remains to be determined. Preliminary observations in our laboratory indicate that stimulation in the dorsal hippocampus can exert orthodromic effects on tuberoinfundibular cells located in the periventricular region, medial part of the ventromedial nucleus, and in the arcuate nucleus (84).

Thus, in summary, there is evidence for synaptic connections on mediobasal hypothalamic tuberoinfundibular neurons from extrahypothalamic regions known to modify different adenohypophysial secretions. In some instances, the excitability of the same tuberoinfundibular neuron can be modified by impulses carried in several different pathways (Fig. 6). With respect to the overall arrangement for the afferents from amygdala, AHA-MPOA, and hippocampus, (a) the amygdala can selectively modify neurons within

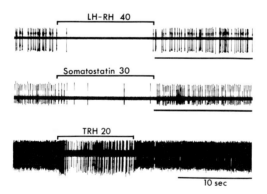

**FIG. 10.** Representative continuous oscillograph traces obtained from central neurons in pentobarbital anesthetized rats to indicate the depressant effects associated with microiontophoretic application of luteinizing hormone releasing hormone (LH-RH, applied to a neuron in the ventromedial hypothalamic nucleus), somatostatin (applied to a parietal cortical neuron), and thyrotropin releasing hormone (TRH, applied to a cerebellar Purkinje neuron). The numbers above each horizontal bar refer to the microiontophoretic currents (in nanoamperes) used to apply each hormone. Time calibration, 10 sec. L. Renaud (*unpublished observations*).

the ventromedial nucleus, (b) the hippocampus can modify cellular activity in the arcuate nucleus and periventricular region, and (c) AHA-MPOA can modify the activity of neurons throughout the mediobasal hypothalamus. We have so far not observed any ascending influences from midbrain periaqueductal gray regions to mediobasal hypothalamic tuberoinfundibular cells.

## NEUROPHARMACOLOGY OF TUBEROINFUNDIBULAR NEURONS

In comparison with the number of studies conducted on magnocellular neurons, there have been relatively few neuropharmacological investigations of tuberoinfundibular neurons. The studies of Yagi and Sawaki (94,111) on recurrent inhibition and facilitation have been described earlier. Monoamines have an important role in the hypothalamic regulation of adenohypophysial secretion (108). Moss and his colleagues (67), reporting on the sensitivity of tuberoinfundibular neurons in the arcuate nucleus, have noted that tuberoinfundibular cells whose activity was enhanced during microiontophoretic application of norepinephrine were either unresponsive or decreased their firing frequency during dopamine microiontophoresis. The reverse pattern applied to tuberoinfundibular cells whose activity was enhanced by dopamine. The amino acid L-glutamate enhanced the activity of most tuberoinfundibular neurons tested. Furthermore, most of the tuberoinfundibular neurons in the arcuate nucleus appear to be unresponsive to hypothalamic peptides TRH or LHRH when applied by microiontophoresis (65,67). This contrasts rather sharply with the sometimes striking effect these peptides have on other hypothalamic or central neurons (35,51,52,65, 90,91; see Fig. 10).

## THE DILEMMA: FUNCTIONAL RELATIONSHIPS

One of the aspirations of numerous investigators interested in neuroendocrinology is to elucidate a causal relationship between action potentials from hypothalamic neurons and the release of pituitary hormones. Most of the successful studies have been related to the neurohypophysial system in connection with oxytocin and vasopressin release. With respect to the tuberoinfundibular system, causal relationships have been established between electrical activity in the basal hypothalamus and gonadotropin release (30,109). However, there are serious limitations in any attempt to define the neural events related to a particular neuroendocrine phenomenon. First, despite a reasonable body of supportive evidence described earlier, it has yet to be firmly established that the electrical activity recorded from so-called tuberoinfundibular elements reflect the "peptidergic" system that regulates adenohypophysial secretion. One approach to this problem would be to demonstrate recordings obtained from neurons identified by specific immunocytochemical localization for a particular peptide releasing (or inhibiting) hormone. It is also important to demonstrate that the same cell projects to the median eminence portal capillary plexus contact zone. Owing to the small size and relative inaccessibility of these structures, this criterion may prove to be a monumental stumbling block. Despite the demonstrable pulsatile patterns in adenohypophysial secretion (104), one is faced with the task of finding a model system where one can relate hormone profiles to neuronal activity.

Although these hormonal fluctuations may occur over a restricted time period (e.g., the change in ACTH is observed within a minute or two after the onset of stress), hormonal measurements are not ideal indices for the neurophysiologist to measure and associate with neuronal activity, since the former depend on a rather laborious and time consuming radioimmunoassay procedure that provides results many hours after the electrophysiological experiment has been completed. Ideally, one would wish to use a neuroendocrine reflex similar to that for oxytocin release during milk ejection. In short, these are some of the major problems that may interfere with a more complete characterization of the role of the tuberoinfundibular system in neuroendocrine regulation.

## CONCLUSIONS AND COMMENTS

Steady progress since the advent of single-unit electrophysiology has led to a considerable appreciation of the role of the hypothalamus in visceral regulation. With respect to neuroendocrinology, the magnocellular neurohypophysial system has provided physiologists with a model with which to explore the nature of the neurosecretory neuron and the process of neurosecretion. This cell has proved to be a true neuroendocrine transducer, transforming electrical events into hormone production (107). Since every neuron whose terminal is engaged in a chemical synapse behaves as a transducer through transformation of electrical impulses into transmitter release, it is not surprising to find that these neurosecretory cells also display electrical properties identical to those of other neurons. This generalization applies also to the parvicellular tuberoinfundibular system. Recent electrophysiological and immunohistochemical evidence suggests that the tuberoinfundibular system may be more than simply a peptidergic neural network strictly involved in adenohypophysial regulation. The widespread localization of certain releasing factors and their apparent influence on neural tissue may provide an early indication that the tuberoinfundibular system is but one of the peptidergic neural networks influencing widespread areas of the central nervous system. One of the major future tasks will be the full physiological characterization of the tuberoinfundibular neuron, a task that will necessarily engage the responsibility of investigators in several different disciplines.

## ACKNOWLEDGMENTS

The author is grateful to the many investigators who granted permission to quote their work. I am indebted to Brian MacKenzie and Robert Nestor for superb technical assistance, Mrs. G. Landrigan and Mrs. M. Walker for secretarial help, and the Canadian Medical Research Council for financial support of the studies conducted in the author's laboratory.

## REFERENCES

1. Abrahams, V. C., Koelle, G. B., and Smart, P. (1957): Histochemical demonstration of cholinesterases in the hypothalamus of the dog. *J. Physiol.*, 139:137-144.
2. Arnauld, E., Dufy, B., and Vincent, J. D. (1975): Hypothalamic supraoptic neurones: Rates and patterns of action potential firing during water deprivation in the unanaesthetized monkey. *Brain Res.*, 100:315-325.
3. Barker, J. L., Crayton, J. W., and Nicoll, R. A. (1971): Noradrenaline and acetylcholine responses of supraoptic neurosecretory cells. *J. Physiol.*, 218:19-32.
4. Barker, J. L., Crayton, J. W., and Nicoll, R. A. (1971): Antidromic and orthodromic responses of paraventricular and supraoptic neurosecretory cells. *Brain Res.*, 33:353-366.
5. Barry, J., and Dubois, M. P. (1975): Immunofluorescence study of LRF-producing neurons in the cat and the dog. *Neuroendocrinology*, 18:290-298.
6. Barry, J., Dubois, M. P., and Carette, B. (1974): Immunofluorescence study of the preoptico-infundibular LRF neurosecretory pathway in the normal, castrated or testosterone-treated male guinea pig. *Endocrinology*, 95:1416-1423.
7. Bennett, J. P., Jr., Arregui, A., and Snyder, S. H. (1976): Angiotensin II as a possible mammalian central neurotransmitter: Synaptic neurochemistry in normal mammalian and Huntington chorea brain tissue. *Neurosci. Abstr.*, 2:775.
8. Blackwell, R. E., and Guillemin, R. (1973): Hypothalamic control of adenohypophyseal secretions. *Annu. Rev. Physiol.*, 35:357-370.
9. Bloom, F. E. (1972): Amino acids and polypeptides in neuronal function. *Neurosci. Res. Program Bull.*, 10:122-251.
10. Brawer, J. (1972): The fine structure of the ependymal tanycytes at the level of the arcuate nucleus. *J. Comp Neurol.*, 145:25-42.
11. Brawer, J., and Sonnenschein, C. (1975): Cytopathological effects of estradiol on the arcuate nucleus of the female rat. A possible mechanism for pituitary tumorigenesis. *Am. J. Anat.*, 144:57-88.
12. Brownstein, M. J., Palkovits, M., Saavedra, J. M., and Kizer, J. S. (1976): Distribution of hypothalamic hormones and neurotransmitters within the diencephalon. In: *Frontiers in Neuroendocrinology*, edited by L. Martini and W. F. Ganong, pp. 1-23. Raven Press, New York.
13. Brownstein, M. J., Saavedra, J. M., Axelrod, J., Zeman, G. H., and Carpenter, D. O. (1974): Coexistence of several putative neurotransmitters in single identified neurons of aplysia. *Proc. Natl. Acad. Sci. U.S.A.*, 71:4662-4665.
14. Burnstock, G. (1976): Do some nerve cells release more than one transmitter? *Neuroscience*, 1:239-248.
15. Burt, D. R., and Snyder, S. H. (1975): Thyrotropin releasing hormone (TRH)—apparent receptor binding in rat-brain membranes. *Brain Res.*, 93:309-328.
16. Cajal, S. R. (1911): *Histologie du Systeme Nerveux de l'Homme et des Vertebres, Vol. 2.* Maloine, Paris.
17. Christ, J. F. (1966): Nerve supply, blood supply and cytology of the neurohypophysis. In: *The Pituitary Gland*, edited by G. W. Harris and B. T. Donovan, pp. 62-130. University of California Press, Berkeley.
18. Conrad, L. C. A., and Pfaff, D. W. (1975): Axonal projections of medial preoptic and anterior hypothalamic neurones. *Science*, 190:1112-1114.
19. Cross, B. A., Dyball, R. E. J., Dyer, R. G., Jones, C. W., Lincoln, D. W., Morris, J. F., and Pickering, B. T. (1975): Endocrine neurons. *Recent Prog. Horm. Res.*, 31:243-294.

20. Cross, B. A., and Green, J. D. (1959): Activity of single neurons in the hypothalamus: Effect of osmotic and other stimuli. *J. Physiol.*, 148:554–569.
21. Cross, B. A., and Silver, I. A. (1966): Electrophysiological studies on hypothalamus. *Br. Med. Bull.*, 22:254–260.
22. Dale, H. A. (1935): Pharmacology and nerve endings. *Proc. R. Soc. Med.*, 28:319–332.
23. Dreifuss, J. J., Harris, M. C., and Tribollett, E. (1976): Excitation of phasically firing hypothalamic supraoptic neurones by carotid occlusion in rats. *J. Physiol.*, 257:337–354.
24. Dreifuss, J. J., Kalnins, I., Kelly, J. S., and Ruf, K. B. (1971): Action potentials and release of neurohypophyseal hormones in vitro. *J. Physiol.*, 215:805–817.
25. Dreifuss, J. J., and Kelly, J. S. (1972): Recurrent inhibition of antidromically identified rat supraoptic neurones. *J. Physiol.*, 220:87–103.
26. Dreifuss, J. J., and Kelly, J. S. (1972): The activity of identified supraoptic neurones and their responses to acetylcholine applied by iontophresis. *J. Physiol.*, 220:105–118.
27. Dreifuss, J. J., Murphy, J. T., and Gloor, P. (1968): Contrasting effects of two identified amygdaloid efferent pathways on single hypothalamic neurons. *J. Neurophysiol.*, 31:237–248.
28. Dreifuss, J. J., Tribollet, E., and Baertschi, A. J. (1976): Excitation of supraoptic neurones by vaginal distention in lactating rats: Correlation with neurohypophyseal hormone release. *Brain Res.*, 113:600–605.
29. Dubois, M. P., and Kolodziejczyk, E. (1975): Centre hypothalamiques du rat sécrétant la somatostatine: répartition des péricaryons en 2 systèmes magno et parvocellulaires (étude immunocytologique). *C.R. Acad. Sci. [D] (Paris)*, 281:1737–1740.
30. Dufy, B., Dufy-Barbe, L., and Poulain, D. (1974): Gonadotropin release in relation to electrical activity in hypothalamic neurons. *J. Neural Trans.*, 35:47–52.
31. Dyball, R. E. J. (1971): Oxytocin and ADH secretion in relation to electrical activity on antidromically identified supraoptic and paraventricular units. *J. Physiol.*, 214:245–256.
32. Dyball, R. E. J. (1974): Single unit activity in the hypothalamo-neurohypophyseal system of Brattleboro rats. *J. Endocrinol.*, 60:135–143.
33. Dyball, R. E. J., and Koizumi, K. (1969): Electrical activity in the supraoptic and paraventricular nuclei associated with neurohypophyseal hormone release. *J. Physiol.*, 201:711–722.
34. Dyer, R. G. (1973): An electrophysiological dissection of the hypothalamic regions which regulate the pre-ovulatory secretion of luteinizing hormone in the rat. *J. Physiol.*, 234:421–442.
35. Dyer, R. J., and Dyball, R. E. J. (1974): Evidence for a direct effect of LRF and TRF on single unit activity in the rostral hypothalamus. *Nature*, 252:486–488.
36. Epelbaum, J., Brazeau, P., Tsang, D., Brawer, J., and Martin, J. B. (1977): Subcellular distribution of radioimmunoassayable somatostatin in rat brain. *Brain Res.*, 126:309–323.
37. Feldberg, W., and Vogt, M. (1948): Acetylcholine synthesis in different regions of the central nervous system. *J. Physiol.*, 107:372–381.
38. Fuxe, K., and Hokfelt, T. (1967): The influence of central catecholamine neurons on the hormone secretion from the anterior and posterior pituitary. In: *Neurosecretion*, edited by F. Stutinsky, pp. 165–175. Springer-Verlag, Berlin.
39. Ginsburg, M. (1968): Production, release, transportation and elimination of the neurohypophyseal hormones. In: *Handbook of Experimental Pharmacology*, edited by B. Berde, pp. 286–371. Springer-Verlag, Berlin.
40. Ginsburg, M., and Brown, L. M. (1956): Effect of anaesthetics and haemorrhage on the release of neurohypophyseal antidiuretic hormone. *Br. J. Pharmacol. Chemother.*, 11:236–244.
41. Halász, B. (1969): The endocrine effects of isolation of the hypothalamus from the rest of the brain. In: *Frontiers in Neuroendocrinology*, edited by W. F. Ganong and L. Martini, pp. 307–342. Oxford University Press, New York.
42. Harris, G. W. (1955): *Neural Control of the Pituitary Gland*. Edward Arnold, London.
43. Harris, G. W., Manabe, Y., and Ruf, K. B. (1969): A study of the parameters of electrical stimulation of unmyelinated fibers in the pituitary stalk. *J. Physiol.*, 203:67–81.
44. Harris, M. C., and Sanghera, M. (1974): Projection of medial basal hypothalamic neurones to the preoptic anterior hypothalamic areas and the paraventricular nucleus in the rat. *Brain Res.*, 81:401–411.
45. Hayward, J. N. (1975): Neural control of the posterior pituitary. *Annu. Rev. Physiol.*, 37:191–210.

46. Hayward, J. N., and Baker, M. A. (1968): Diuretic and thermoregulatory responses to preoptic cooling in the monkey. *Am. J. Physiol.*, 214:843–850.
47. Hayward, J. N., and Jennings, D. P. (1973): Activity of magnocellular neuroendocrine cells in the hypothalamus of unanaesthetized monkeys. II. Osmosensitivity of functional cell types in the supraoptic nucleus and the internuclear zone. *J. Physiol.*, 232:545–572.
48. Hayward, J. N., and Smith, W. K. (1963): Influence of limbic system on neurohypophysis. *Arch. Neurol.*, 9:171–177.
49. Hökfelt, T. (1977): Aminergic and peptidergic brain pathways. In: *The Hypothalamus*, 1976 ARNMD Meeting. Raven Press, New York (*in press*).
50. Kandel, E. R. (1964): Electrical properties of hypothalamic neuroendocrine cells. *J. Gen. Physiol.*, 47:691–717.
51. Kawakami, M., and Sakuma, Y. (1974): Responses of hypothalamic neurons to the microiontophoresis of LH-RH, LH and FSH under various levels of circulating ovarian hormones. *Neuroendocrinology*, 15:290–307.
52. Kawakami, M., and Sakuma, Y. (1976): Electrophysiological evidences for possible participation of periventricular neurons in anterior pituitary regulation. *Brain Res.*, 101:79–94.
53. Kelly, J. S. (1975): Microiontophoretic application of drugs onto single neurons. In: *Handbook of Psychopharmacology, Vol. 2*, edited by L. L. Iversen, S. D. Iversen, and S. H. Snyder, pp. 29–67. Plenum Press, New York.
54. Kelly, J. S., Simmonds, M. A., and Straughan, D. W. (1975): Microelectrode techniques. In: *Methods in Brain Research*, edited by P. B. Bradley, pp. 333–377. John Wiley & Sons, New York.
55. Koizumi, K., Ishikawa, T., and McC. Brooks, C. (1973): The existence of facilitatory axon collaterals in neurosecretory cells of the hypothalamus. *Brain Res.*, 63:408–413.
56. Koizumi, K., and Yamashita, H. (1972): Studies of antidromically identified neurosecretory cells of the hypothalamus by intracellular and extracellular recordings. *J. Physiol.*, 221:683–705.
57. Kreisel, B., Conforti, N., Gutnick, M., and Feldman, S. (1975): Antidromic responses of suprachiasmatic neurons following mediobasal hypothalamic stimulation. *Isr. J. Med. Sci.*, 11:925–927.
58. Leontovich, T. A. (1970): The neurons of the magnocellular neurosecretory nuclei of the dog's hypothalamus. *J. Hirnforsch.*, 11:499–517.
59. Leranth, Cs., Zaborszky, L., Marton, J., and Palkovits, M. (1975): Quantitative studies on the supraoptic nucleus in the rat. I. Synaptic organization. *Exp. Brain Res.*, 22:509–523.
60. Lichtensteiger, W., and Keller, P. J. (1974): Tuberoinfundibular dopamine neurons and the secretion of luteinizing hormone and prolactin: Extrahypothalamic influences, interaction with cholinergic systems and the effect of urethane anesthesia. *Brain Res.*, 74:279–303.
61. Lincoln, D. W., and Wakerley, J. B. (1974): Electrophysiological evidence for the activation of supraoptic neurones during the release of oxytocin. *J. Physiol.*, 242:533–554.
62. Lincoln, D. W., and Wakerley, J. B. (1975): Factors governing the periodic activation of supraoptic and paraventricular neurosecretory cells during suckling in the rat. *J. Physiol.*, 250:443–461.
63. Makara, G. B., Harris, M. C., and Spyer, K. M. (1972): Identification and distribution of tuberoinfundibular neurones. *Brain Res.*, 40:283–290.
64. Moss, R. A. (1976): Unit responses in preoptic and arcuate neurons related to anterior pituitary function. In: *Frontiers in Neuroendocrinology, Vol. 4*, edited by L. Martini and W. F. Ganong, pp. 95–128. Raven Press, New York.
65. Moss, R. L. (1977): Role of hypophysiotropic neurohormones in mediating neural and behavioral events. *Fed. Proc.*, 36:1978–1983.
66. Moss, R. L., Dyball, R. E. J., and Cross, B. A. (1972): Excitation of antidromically identified neurosecretory cells in the paraventricular nucleus by oxytocin applied iontophoretically. *Exp. Neurol.*, 34:95–102.
67. Moss, R. L., Kelly, M., and Riskind, P. (1975): Tuberoinfundibular neurons: Dopaminergic and norepinephrinergic sensitivity. *Brain Res.*, 89:265–277.
68. Moss, R. L., Urban, I., and Cross, B. A. (1972): Microelectrophoresis of cholinergic and aminergic drugs on paraventricular neurons. *Am. J. Physiol.*, 223:310–318.

69. Mroz, E. A., Brownstein, M. J., and Leeman, S. E. (1976): Evidence for substance P in the habenulo-interpeduncular tract. *Brain Res.*, 113:597–599.
70. Murphy, J. T., and Renaud, L. P. (1969): Mechanisms of inhibition in the ventromedial nucleus of the hypothalamus. *J. Neurophysiol.*, 32:85–102.
71. Negoro, H., and Holland, R. C. (1972): Inhibition of unit activity in the hypothalamic paraventricular nucleus following antidromic activation. *Brain Res.*, 42:385–402.
72. Negoro, H., Visessuwan, S., and Holland, R. C. (1973): Inhibition and excitation of units in paraventricular nucleus after stimulation of the septum, amygdala and neurohypophysis. *Brain Res.*, 57:479–483.
73. Nicoll, R. A., and Barker, J. L. (1971): Excitation of supraoptic neurosecretory cells by angiotensin II. *Nature [New Biol.]*, 233:172–174.
74. Nicoll, R. A., and Barker, J. L. (1971): The pharmacology of recurrent inhibition in the supraoptic neurosecretory system. *Brain Res.*, 35:501–511.
75. Numan, M. (1974): Medial preoptic area and maternal behaviour in the female rat. *J. Comp. Physiol. Psychol.*, 87:746–759.
76. Phillips, M. I., and Felix, D. (1976): Specific angiotensin II receptive neurons in the cat subfornical organ. *Brain Res.*, 109:531–540.
77. Pickford, M. (1939): The inhibitory effect of acetylcholine on diuresis in the dog and its pituitary transmission. *J. Physiol.*, 95:226–238.
78. Porter, J. C., Kameri, I. A., and Grazia, Y. R. (1971): Pituitary blood flow and portal vessels. In: *Frontiers in Neuroendocrinology*, edited by L. Martini and W. F. Ganong, pp. 145–175. Oxford University Press, New York.
79. Prange, A. J., Jr., Nemeroff, C. B., Lipton, M. A., Breese, G. R., and Wilson, I. C. (1976): Peptides and the central nervous system. In: *Handbook of Psychopharmacology*, edited by L. L. Iversen, S. D. Iversen, and S. H. Snyder. Plenum Press, New York.
80. Raisman, G., and Field, P. M. (1971): Anatomical considerations relevant to the interpretation of neuroendocrine experiments. In: *Frontiers in Neuroendocrinology*, edited by L. Martini and W. F. Ganong, pp. 3–44. Oxford University Press, London.
81. Ramirez, V. D., Gautron, J. P., Epelbaum, J., Pattou, E., Zamura, A., and Kordon, C. (1975): Distribution of LH-RH in subcellular fractions of the basomedial hypothalamus. *Mol. Cell. Endocrinol.*, 3:339–350.
82. Renaud, L. P. (1975): Electrophysiological evidence to suggest that hypothalamic releasing (inhibiting) peptides may be liberated from nerve terminals in the CNS. *Neurosci. Abstr.*, 1:441.
83. Renaud, L. P. (1976): Tuberoinfundibular neurons in the basomedial hypothalamus of the rat: Electrophysiological evidence for axon collaterals to hypothalamic and extrahypothalamic areas. *Brain Res.*, 105:59–72.
84. Renaud, L. P. (1976): Tuberoinfundibular neurons: Electrophysiological studies on afferent and efferent connections. *Physiologist*, 19:338.
85. Renaud, L. P. (1976): Influence of amygdala stimulation on the activity of identified tuberoinfundibular neurones in the rat hypothalamus. *J. Physiol.*, 260:237–252.
86. Renaud, L. P. (1976): An electrophysiological study of amygdalo-hypothalamic projections to the ventromedial nucleus of the rat. *Brain Res.*, 105:45–58.
87. Renaud, L. P. (1977): Influence of medial preoptic-anterior hypothalamic area stimulation on the excitability of mediobasal hypothalamic neurones in the rat. *J. Physiol.*, 264:541–564.
88. Renaud, L. P., and Hopkins, D. A. (1977): Amygdala afferents from the mediobasal hypothalamus: An electrophysiological and neuroanatomical study in the rat. *Brain Res.*, 121:201–213.
89. Renaud, L. P., and Martin, J. B. (1975): Electrophysiological studies of connections of hypothalamic ventromedial nucleus neurons in rat—evidence for a role in neuroendocrine regulation. *Brain Res.*, 93:145–151.
90. Renaud, L. P., Martin, J. B., and Brazeau, P. (1975): Depressant action of TRH, LH-RH and somatostatin on activity of central neurons. *Nature*, 255:233–235.
91. Renaud, L. P., Martin, J. B., and Brazeau, P. (1976): Hypothalamic releasing factors: Physiological evidence for a regulatory action on central neurons and pathways for their distribution in brain. *Pharmacol. Biochem. Behav.* [Suppl. 1], 5:171–178.
92. Sakai, K. K., Marks, B. H., George, J. M., and Koestner, A. (1974): Specific angiotensin II

receptors in organ-cultured canine supraoptic nucleus cells. *Life Sci.,* 14:1337-1344.
93. Sawaki, Y., and Yagi, K. (1973): Electrophysiological identification of cell bodies of the tuberoinfundibular neurones in the rat. *J. Physiol.,* 230:75-85.
94. Sawaki, Y., and Yagi, K. (1976): Inhibition and facilitation of antidromically identified tuberoinfundibular neurones following stimulation of the median eminence in the rat. *J. Physiol.,* 260:447-460.
95. Schally, A. V., Arimura, A., and Kastin, A. J. (1973): Hypothalamic regulatory hormones. *Science,* 179:341-350.
96. Severs, W. B., and Daniels-Severs, A. E. (1973): Effects of angiotensin on the central nervous system. *Pharmacol. Rev.,* 25:415-449.
97. Shute, C. C. D. (1970): Distribution of cholinesterase and cholinergic pathways. In: *The Hypothalamus,* edited by L. Martini, M. Motta, and F. Fraschini, pp. 167-179. Academic Press, New York.
98. Silverman, A. J. (1976): Distribution of luteinizing hormone-releasing hormone (LH-RH) in the guinea pig brain. *Endocrinology,* 99:30-41.
99. Swanson, L. W. (1976): An autoradiographic study of the efferent connections of the preoptic region in the rat. *J. Comp. Neurol.,* 167:227-256.
100. Szentagothai, J., Flerko, B., Mess, B., and Halasz, B. (1968): *Hypothalamic Control of the Anterior Pituitary.* Akademiai Kiado, Budapest.
101. Van Gemert, M., Miller, M., Carey, R. J., and Moses, A. M. (1975): Polyuria and impaired ADH release following medial preoptic lesioning in the rat. *Am. J. Physiol.,* 228:1293-1297.
102. Wakerley, J. B., and Lincoln, D. W. (1973): The milk ejection reflex of the rat: A 20- to 40-fold acceleration in the firing of paraventricular neurones during oxytocin release. *J. Endocrinol.,* 57:477-493.
103. Wakerley, J. B., Poulain, D. A., Dyball, R. E. J., and Cross, B. A. (1975): Activity of phasic neurosecretory cells during haemorrhage. *Nature,* 258:82-84.
104. Willoughby, J. O., Terry, L. C., Brazeau, P., and Martin, J. B. (1977): Pulsatile growth hormone, prolactin and thyrotropin secretion in rats with hypothalamic deafferentation. *Brain Res. (in press).*
105. Wimersma Greidanus, Tj. B., Bohus, B., and de Weid, D. (1975): The role of vasopressin in memory processes. *Prog. Brain Res.,* 42:135-141.
106. Woods, W. H., Holland, R. C., and Powell, E. W. (1969): Connections of cerebral structures functioning in neurohypophyseal hormone release. *Brain Res.,* 12:26-46.
107. Wurtman, R. J. (1970): Neuroendocrine transducer cells in mammals. In: *The Neurosciences: Second Study Program,* edited by F. O. Schmitt, pp. 530-538. Rockefeller University Press, New York.
108. Wurtman, R. J. (1971): Brain monoamines and endocrine function. *Neurosci. Res. Prog. Bull.,* 9:171-297.
109. Wuttke, W. (1974): Preoptic unit activity and gonadotropin release. *Exp. Brain Res.,* 19:205-216.
110. Yagi, K., Azuma, T., and Matsuda, K. (1966): Neurosecretory cell: Capable of conducting impulse in rats. *Science,* 154:778-779.
111. Yagi, K., and Sawaki, Y. (1974): Recurrent inhibition and facilitation: Demonstration in the tuberoinfundibular system and effects of strychnine and picrotoxin. *Brain Res.,* 84:155-159.
112. Zaborszky, L., Leranth, Cs., Makara, G. B., and Palkovits, M. (1975): Quantitative studies in the supraoptic nucleus in the rat. II. Afferent fiber connections. *Exp. Brain Res.,* 22:525-540.
113. Zimmerman, E. A., and Defendini, R. (1977): Anatomy of neurohypophyseal pathways. In: *The Hypothalamus,* 1976 ARNMD Meeting. Raven Press, New York *(in press).*
114. Zolovick, A. J. (1972): Effects of lesions and electrical stimulation of the amygdala on hypothalamic-hypophyseal regulation. In: *The Neurobiology of the Amygdala,* edited by B. E. Eleftheriou, pp. 643-684. Plenum Press, New York.

# DISCUSSION

*Dr. Sachar:* Have you also found influences of projections of the amygdala on the cells of the arcuate nucleus?

*Dr. Renaud:* Yes. During any penetration through this region, the cells that seem to be specifically responsive are cells in the ventral medial nucleus. Cells in the arcuate nucleus and in the periventricular region are largely unresponsive to stimulation in the amygdala, but there are some cells that do respond, although the latencies are different (generally longer). Few arcuate nucleus neurons seem to receive monosynaptic connections from the amygdala.

*Dr. Bloom:* Can you begin to ascribe a particular neurochemical substance to any of the amygdalohypothalamic pathways that you demonstrated physiologically?

*Dr. Renaud:* Recently we have undertaken a study on the neuropharmacology of GABA and glycine because of the striking inhibition prominent in this region. We have attempted to identify whether or not antagonists to GABA and glycine (i.e., bicuculline, picrotoxin, and strychnine) can antagonize amygdala-evoked synaptic inhibition after their iontophoretic or intravenous administration. I think we have some evidence that picrotoxin and bicuculline diminish the inhibitory periods in that region, but strychnine does not.

*Dr. Snyder:* Did you say that the evidence suggested that the major transmitter of the stria terminalis pathway is GABA?

*Dr. Renaud:* No. I would entertain another possibility. The predominant response on the afferent pathway from stria terminalis seems to be functionally an excitatory connection. Inhibition is secondary, and this may be the putative GABA pathway.

*Dr. Snyder:* Would you give us your ideas as to how the excitatory transmitters of the stria terminalis or of the fornix might be identified?

*Dr. Renaud:* I would speculate that glutamate is probably the neurotransmitter candidate for mediating excitatory, short-latency connections at central synapses. There are possible other candidates, such as aspartic acid and histamine which could also evoke excitation. Histamine causes a brisk onset excitation after microiontophoresis which looks very much like glutamate, but responsive cells are infrequent. The mechanics of transmitter identification in the CNS presents a formidable challenge. On the biochemical side, one could lesion the stria terminalis and look for changes in putative neurotransmitter content, uptake, or binding in the ventromedial hypothalamic nucleus. An alternate method is to attempt to block synaptic excitation by systemic or iontophoretic application of specific antagonists of putative neurotransmitters. This has not yet been attempted in the pathways mentioned. Part of the difficulty here is the lack of specific antagonists to glutamate or aspartate.

*The Hypothalamus,* edited by S. Reichlin,
R. J. Baldessarini, and J. B. Martin,
Raven Press, New York, © 1978.

# The Pineal Gland: A Model of Neuroendocrine Regulation

### *David C. Klein

The pineal gland is considered in this volume because it may be viewed as a prime example of a neuroendocrine control system. In contrast to the state of knowledge of the hypothalamus at the biochemical level, there has been substantial progress in the understanding of cellular and subcellular details regarding neural regulation of synthesis and release of melatonin, the hormonal secretion of the rat pineal. Two fortunate circumstances are responsible for this progress. First, the function of the pineal gland appears to be organized primarily around a single neuroendocrine activity function in that a single neuronal input regulates the production and release of a single hormone from a fairly homogenous cell population. Second, the rat pineal gland can easily be cultured. This has allowed for a productive series of many detailed biochemical studies and some neurophysiological investigations performed within highly defined *in vitro* conditions.

These *in vitro* studies and related *in vivo* studies have made it possible to construct a generally agreed upon description of how the pineal gland functions, a description which spans the full range of biological organization from the level of environmental input to that of molecular interaction within the pineal cell (Fig. 1). Environmental light interacts with a "circadian oscillator" in the suprachiasmatic nucleus, which appears to be responsible for an intrinsic rhythm in melatonin secretion. The neuronal pathway between the suprachiasmatic nucleus and the pineal gland involves central neural structures and terminates with sympathetic neurons. The release of transmitter from these neurons causes a receptor-mediated increase in the production of cyclic AMP in pinealocytes. Cyclic AMP in turn has distinct electrophysiological and biochemical effects on pineal cells, leading to a striking and unique change in pineal metabolism. This change — a 70- to 100-fold increase in the activity of serotonin $N$-acetyltransferase activity within a few hours — is responsible for the daily changes in the production of melatonin. $N$-Acetyltransferase synthesizes the precursor of melatonin

---

* *Present address:* Neuroendocrinology Unit, Laboratory of Developmental Neurobiology, National Institute of Child Health and Human Development, National Institutes of Health, Bethesda, Maryland 20014.

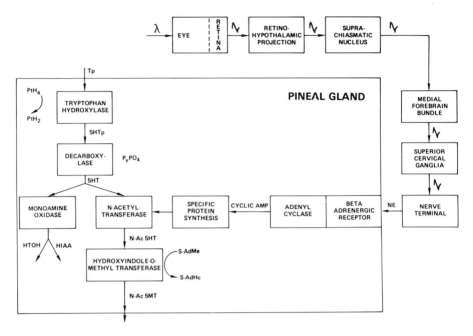

**FIG. 1.** A model of the β-adrenergic regulation of indole metabolism in the pineal gland. Cyclic AMP, adenosine 3′,5′-monophosphate; Tp, tryptophan; PtH₄, tetrahydropteridine; PtH₂, dihydropteridine; 5HTp, 5-hydroxytryptophan; PyPO₄, pyridoxalphosphate; 5HT, 5-hydroxytryptamine, serotonin; HTOH, hydroxytryptophol; HIAA, hydroxyindole acetic acid; N-Ac 5HT, *N*-acetyl 5-hydroxytryptamine, *N*-acetylserotonin; S-AdMe, S-adenosyl methionine; S-AdHe, S-adenosylhomocysteine; N-Ac 5MT, *N*-acetyl 5-methoxytryptamine, melatonin; NE, norepinephrine. From (51).

(Fig. 2); immediate consequences of the increase in enzyme activity are increases in the production and release of melatonin and a subsequent increase in plasma melatonin.

Consideration of the organization and function of this regulatory system suggests that the singular purpose of the entire system (including the neuronal circuitry, the specific transmitter effects, and the intracellular actions of cyclic AMP) is the control of large changes in the activity of a single key *regulating* enzyme. In the following pages, I will describe this system in some detail. It is my belief that analogies to these details exist in hypothalamic function and remain to be discovered and documented. Explorers of the intricacies of hypothalamic function may find it highly profitable to look for these analogies, to search for key *regulating* enzymes, to use what is known about regulation of the pineal gland as a jumping-off point to test hypotheses of hypothalamic function, and to generally regard the pineal gland as a reasonable model of neuroendocrine regulation in the hypothalamus.

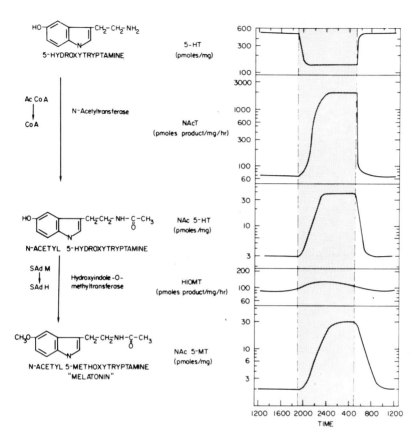

**FIG. 2.** Rhythms in indole metabolism in the rat pineal gland. The metabolic pathway from 5-hydroxytryptamine to melatonin is on the left. The daily variations in the concentrations of metabolites and activities of enzymes are on the right. The shaded portion indicates the dark period of the lighting cycle. The data have been abstracted from reports in the literature. AcCoA, acetyl coenzyme A; CoA, coenzyme A; S AdM, S-adenosyl methionine; S AdH, S-adenosyl homocysteine; 5-HT, 5-hydroxytryptamine, serotonin; NAcT, N-acetyltransferase; HIOMT, hydroxyindole-O-methyltransferase; NAc 5-MT, N-acetyl 5-methyoxytryptamine, melatonin. From 66.

## INDOLE METABOLISM IN THE PINEAL GLAND

### Biochemistry

To understand neural regulation of hormonal output in the pineal gland, one needs first to understand that aspect of indole metabolism which involves the conversion of serotonin to melatonin. Serotonin is synthesized in the pinealocyte from tryptophan taken up from the blood. Synthesis involves 5-hydroxylation by tryptophan hydroxylase followed by decarboxylation by aromatic amino acid decarboxylase (6,22,26,70,71,72).

Serotonin occurs in the pineal gland at concentrations (ca = 0.5 mM) higher than in any other tissue (100,113) (except perhaps the raphe nuclei of the midbrain) because of the high level of tryptophan hydroxylase activity that exists there (70). The conversion of serotonin to melatonin involves two enzymatic steps (Fig. 2). The first is $N$-acetylation by a relatively specific $N$-acetyltransferase enzyme (15). This acetyl group is provided by the cosubstrate acetyl CoA (127). The second enzymatic step is O-methylation by hydroxyindole-O-methyltransferase; the methyl group is provided by S-adenosylmethionine (4).

Melatonin has been found in body fluids of most animals examined including the rat, nonhuman and human primate, sheep, cow, camel, pig, chicken, and iguana (1,36,49,77,93,98,106,116,128). There is a marked daily rhythm in the concentration of melatonin in blood, CSF, and urine with night values in blood being up to 10-fold higher than day values. Light has been shown to have a marked effect on circulating melatonin in certain animals. In the sheep, exposure to light at night results in an extremely rapid drop in circulating melatonin (106). This rapid drop is probably due to a rapid decrease in the rate of release of melatonin from the pineal gland coupled with the rapid inactivation of melatonin, primarily by the liver (38, 68,69).

Based on studies with the rat, it appears that large changes in the release of melatonin coincide with large changes in the rate of production of melatonin, which in turn, coincide with large changes in the rate of production and availability of the precursor of melatonin, $N$-acetylserotonin (14,63,64).

The large increase in the amount of $N$-acetylserotonin synthesized is due to 70- to 100-fold increase in the activity of $N$-acetyltransferase (15, 61). These dynamic changes indicate that this is the key regulating enzyme responsible for the large daily changes in metabolism production in the pineal gland. Evidence that this is a highly specific increase comes from the observation that at night, when melatonin in the rat pineal gland is elevated 10-fold and $N$-acetyltransferase activity is elevated 70- to 100-fold, the other enzyme required for melatonin synthesis, hydroxyindole-O-methyltransferase, does not increase more than 10 to 30%.

Early reports of a many-fold rhythm in the activity of hydroxyindole-O-methyltransferase activity (5,131,132) have been difficult to repeat. Using current technology, it is possible at best to measure only a small daily change in this enzyme (51,55,56,76,101,103,111). In contrast, the large rhythm in rat $N$-acetyltransferase activity has been studied extensively in a number of laboratories and has been found to occur in all mammals examined and in chickens (9,12,21,24,43,90,112).

Several fascinating aspects of the dynamics of $N$-acetyltransferase activity should be pointed out. First, the large changes in the activity of this enzyme appear to cause changes in the concentration of serotonin (14,43, 53,61), the precursor of the enzyme (Fig. 2). Apparently, the rate at which tryptophan is converted to serotonin does not increase sufficiently to replace the serotonin which is consumed in the formation of $N$-acetylated derivatives of melatonin. As a result, the concentrations of pineal serotonin falls. This is consistent with the observation that large changes in melatonin production *in vitro* are not accompanied by large changes in the rate at which tryptophan is hydroxylated (6), and the finding that pineal tryptophan hydroxylase does not increase at night (22).

Second, the daily changes are a reflection of circadian rhythm in that the rhythm persists in constant darkness or in blinded animals kept in constant light (59,61,87,90). This clearly indicates that the rhythm is not dependent upon environmental lighting for its generation and that it is probably driven by an endogenous oscillator. As a result of this persistence in constant darkness of a rhythm in $N$-acetyltransferase activity, there is also a circadian rhythm in the melatonin (102) and a reciprocal rhythm in the serotonin in the pineal gland (119).

A third important characteristic of the dynamics of $N$-acetyltransferase involves the effects of environmental lighting. Exposure to light at night after several hours of darkness results in a rapid decrease in the activity of $N$-acetyltransferase (Fig. 2); within 15 min, values have fallen to about 10% the dark-night value (62). Exposure to darkness during the day does not have the opposite effect, i.e., it does not increase $N$-acetyltransferase activity. During the day, the system appears to be refractory to darkness (13). This clock is only capable of stimulating the pineal gland during the normal dark periods; darkness acts passively to permit increase in $N$-acetyltransferase activity at night. Obviously, light can act to modify the period during which $N$-acetyltransferase is elevated at night in the dark.

Light also entrains the $N$-acetyltransferase rhythm to the rhythm in environmental lighting. In the absence of light or in blinded animals, the rhythm of $N$-acetyltransferase activity within a group of animals becomes both asynchronous with the environmental lighting schedule and asyncronous relative to other animals within a group (21,59,87).

In addition to these daily changes in indoleamine metabolism are photically regulated tonic changes in rat pineal hydroxyindole-O-methyltransferase activity (5,89,131). Dark exposure elevates this enzyme and light exposure depresses it. These changes are evident two weeks after a shift from a normal light:dark lighting schedule to one of either constant light or constant darkness.

It seems highly probable that animals maintained in lighting cycles that provide increasing amounts of darkness will produce proportionately more melatonin owing to the increase in the duration of the period when $N$-acetyltransferase activity is elevated, and to higher tonic levels of hydroxyindole-O-methyltransferase activity. These longer nights would result in longer periods during which circulating melatonin would be elevated and in higher values of melatonin during these periods of accelerated melatonin production. In addition, day values of circulating melatonin might be higher if hydroxyindole-O-methyltransferase activity were higher, even if much less N-acetylserotonin were being produced in the pineal gland. This question has not been thoroughly examined as yet.

## REGULATORY MECHANISMS CONTROLLING PINEAL INDOLE METABOLISM

### Retinohypothalamic Projection

As described above, environmental light acting through the eye can have profound effects on rhythms in pineal indole metabolism. When an animal is exposed to light at night several hours after being in the dark, the low concentrations of serotonin rapidly return to day values (41), the high concentrations of N-acetyltransferase activity drop to day levels (62), and the amount of melatonin in the pineal gland and circulating in blood decreases rapidly (106,128). These profound effects of light appear to be mediated by a recently described neural tract (37,82,88), the retinohypothalamic projection. This important neural pathway is an unmyelinated tract in the optic nerve, probably originating in the ganglion layer of the retina, quite distinct from the primary optic tract and the inferior accessory optic tract (83).

The retinohypothalamic projection terminates in the suprachiasmatic nucleus, and the discovery of both this projection and the importance of the suprachiasmatic nucleus were closely linked (83). The existence of the former became apparent in experiments in which the eyes of rodents were injected with a radioactive amino acid. This type of tracer is known to be taken up by cell bodies and transported down neural processes. Some of the radioactivity injected in the eye was found to be accumulated in terminals in the suprachiasmatic nucleus (37,82,88).

In animals with lesions of the primary optic tracts caudal to the optic chiasm and of the inferior accessory optic tracts, it was found that the rhythm in pineal N-acetyltransferase activity was still entrained to environmental lighting, indicating that this entraining function of light was probably mediated by the only other known visual tract, the retinohypothalamic projection (87). Using similar lesions, we found that the effects of light that cause a continual suppression of the rhythm in pineal N-acetyltransferase activity are also most likely mediated by the retinohypothalamic projection (56). Thus, this unique connection between the eye and the hypothalamus appears to be an important and perhaps the sole route through which light regulates circadian changes in pineal indole metabolism.

Prior to the realization of the importance of the retinohypothalamic projection in the control of pineal function, it was thought that another pathway, the inferior accessory optic tract terminating in the medial terminal nucleus, mediated photoregulation of pineal function (85,86). This conclusion was based on studies in which the chronic effects of light on pineal hydroxyindole-O-methyltransferase activity were used as a measure of pineal function. However, this view may require thorough reevaluation because the results of the initial studies cannot be repeated (83).

The available information suggests that whereas careful and complete lesioning of the primary optic tracts caudal to the optic chiasm of the accessory optic tract and of the medial terminal nucleus do not block the effects of light on pineal hydroxyindole-O-methyltransferase activity, destruction of the optic chiasm does (83). If the lesion of the optic chiasm is effective because of lesioning of the retinohypothalamic tract, which is unclear at present, then it would appear that hydroxyindole-O-methyltransferase might be regulated by the same neural route which regulates pineal $N$-acetyltransferase activity. Alternatively, it is entirely possible that other neural connections mediating the effects of light on pineal hydroxyindole-O-methyltransferase activity might exist, connections which are severed by destruction of the optic chiasm but not the accessory optic tracts on the primary optic tract some distance caudal to the optic chiasm.

The retinohypothalamic projection appears to be important in mediating the effects of light on other neuroendocrine and physiological functions (40,84,117,118).

Studies on the nature of light regulating pineal function have demonstrated that light acts through a sensitive rhodopsin-like photopigment (17,81). In addition, it has been found that the precise structures mediating light are relatively insensitive to its destructive effects as compared to the retina in general (103). Light which produces gross retinal degeneration, as judged by morphological examination, can still function to suppress pineal $N$-acetyltransferase activity. This raises the possibility that either the structures involved in this photodetection may be resistant to light damage, or be located in a portion of the retina which is less likely to be destroyed by light damage, i.e., the most peripheral portions of the retina. Alternatively, the severely reduced retinal function persisting in the light-damaged retina may be sufficient for the control of pineal function.

## The Suprachiasmatic Nucleus

The autoradiographic evidence that led to the discovery of the retinohypothalamic tract suggested that this tract could influence the suprachiasmatic nucleus, a small paired structure located in the basal hypothalamus immediately above the optic chiasm (Fig. 3). The hypothalamus had been previously implicated in regulating circadian rhythms that were destroyed by a large lesion in the basal hypothalamus (34), which, among other structures, could include either the suprachiasmatic nucleus or afferents from the suprachiasmatic nucleus.

To test whether the suprachiasmatic nucleus was involved in regulating

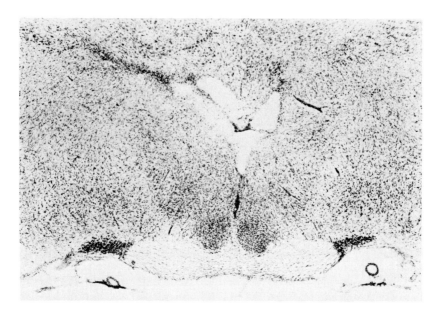

**FIG. 3.** Photomicrograph of a frontal section through the anterior hypothalamus at the level of the suprachiasmatic nucleus. The suprachiasmatic nuclei are seen as the paired dense structures immediately above the optic chiasm. Cresyl violet stain, × 20. From (87).

pineal gland rhythms, we measured $N$-acetyltransferase activity in animals in whom the suprachiasmatic nucleus had been destroyed (56). We found that well placed lesions — those which completely destroyed the suprachiasmatic nucleus — completely abolished the rhythm in $N$-acetyltransferase activity. Incomplete lesions which left part of the suprachiasmatic nucleus and its projections intact also left the rhythm in $N$-acetyltransferase activity intact. This finding was an important step towards establishing the suprachiasmatic nucleus as an endogeneous oscillator — a biological clock regulating the pineal gland (87).

Other evidence supported this postulate. First, the rhythm in $N$-acetyltransferase activity persisted after all known input to the suprachiasmatic nucleus was severed. Second, the rhythm in $N$-acetyltransferase was abolished when neural output from that area was severed (87).

An alternative explanation of the driving force regulating pineal rhythms is that there is a general sympathetic discharge at night when animals are more active and that this increased activity in some way drives pineal rhythms (110,132). This view evolved in part from the finding that lesions of the suprachiasmatic nucleus destroy the rhythm in motor activity (117). This explanation is probably not the case because the increased motor activity associated with swimming either during the day or in the light at night causes an insignificant increase in $N$-acetyltransferase activity relative to that seen in the dark at night when animals are normally active (58,94). The importance of such a general sympathetic discharge appears to be quite minor in regulating pineal N-acetyltransferase activity because the sympathetic discharge occurring during stress, as produced by swimming, immobilization, or cold causes relatively minor increase in N-acetyltransferase activity compared to that which is seen at night in the dark (42,94). The report of a large stress-induced increase in N-acetyltransferase activity in normal animals (74) cannot be confirmed (42,94).

The suprachiasmatic nucleus has some interesting characteristics. As pointed out above, this oscillator apparently cannot stimulate N-acetyltransferase activity during the day, even in darkness. Whereas exposure to darkness at night allows for an increase in N-acetyltransferase activity, a similar exposure during the day does not. Another interesting characteristic (95a) is that if pineal N-acetyltransferase activity is maintained at a low level during the night by light exposure, the clock appears to keep ticking away. As a result if the light period is extended 8 hr and the animal is then exposed to a 10-hr dark period starting at 0300 rather than at 1900, N-acetyltransferase will increase rapidly during the 0300 to 0500 period but will automatically fall at about 0500 to 0700 even in constant darkness. The normal 10 to 12 hr period of elevated N-acetyltransferase activity will not be seen.

Another rhythm in neuroendocrine function, that of hippocampal norepinephrine concentration, may also be generated by the suprachiasmatic nucleus (20,40,84). Whereas this rhythm can be entrained to a cue other than lighting, i.e., water deprivation, the rhythm in pineal N-acetyltransferase activity cannot be entrained to this cue (90).

If it is found that the activity of pineal hydroxyindole-O-methyltransferase is regulated by the same visual pathway which regulates N-acetyltransferase activity (retinohypothalamic projection—suprachiasmatic) which according to Moore, is the most "parsimonious" conclusion (83), then the question must be raised as to why there is no impressive rhythm in hydroxyindole-O-methyltransferase, if it is regulated by a clock. Perhaps the biochemistry of this enzyme is such that large and rapid changes in activity are not possible. That does not exclude the possibility that the activity of this enzyme is increased by light deprivation and decreased by light. However, the slow rate and small magnitude of these changes may preclude large changes being observed during a 24-hr period. As a result, only long-term changes in enzyme activity may be easily detected. The activity of this enzyme at all times of the day would then provide a valuable *averaged* measure of the amount of darkness the animal had been exposed to during the previous few weeks; i.e., with increasingly shorter dark periods, the activity of the enzyme would gradually decrease by a small amount each day.

What emerges as an attractive, working hypothesis is that the rhythm in pineal N-acetyltransferase activity and the tonic changes in hydroxyindole-O-methyltransferase activity are the direct result of a rhythm generated in the suprachiasmatic nucleus. Some insight into the pathway between the suprachiasmatic nucleus and the pineal gland is available, as described in the following sections. The possibility exists that this pathway could in part be involved in regulating both tonic and daily changes in hypothalamic function.

### Central Structures Involved in the Neural Regulation of the Pineal Gland

The neural route by which stimulatory signals apparently reach the pineal gland from the suprachiasmatic nucleus starts with efferent projections directed caudally into the periventricular and ventral tuberal areas of the hypothalamus (122,123). The importance of these projections in generating the rhythm in pineal $N$-acetyltransferase activity has been demonstrated: severing these connections block the rhythm in the activity of this enzyme (87). Neural connections between the ventral tuberal area and the lateral portion of the hypothalamus have also been shown to exist (123), as has a direct projection from the lateral hypothalamus, apparently via the medial forebrain bundle and the midbrain reticular formation, to the upper thoracic intermediolateral cell column (115). Cells in the latter provide preganglionic input to the superior cervical ganglia of the sympathetic nervous system. Functional evidence of the importance of the medial forebrain bundle is provided by experiments which demonstrate that lesions in this area prevent the rhythm in pineal $N$-acetyltransferase activity (87).

This conception of the neural route mediating the effects of the suprachiasmatic nucleus on the pineal gland has been proposed based on studies with the rat. Prior to the general acceptance of this as a generally occuring pathway among mammals, it will be necessary to test this hypothesis further.

### Peripheral Structures Involved in the Neural Regulation of the Pineal Gland

The present concept of the pineal gland being solely innervated by postganglion fibers from the superior cervical ganglia can be traced to the observations of Kappers (46,47). The functional importance of this neural input was first made clear by Fiske (31) who found that removal of these ganglia resulted in blockage of the daily changes in pineal serotonin content. We found that removal of the ganglia also blocked the rhythm in pineal $N$-acetyltransferase activity (65). From these observations, it would appear that sympathetic nerves transmit stimulatory signals which control the pineal gland. Through the discovery that decentralization of the ganglia, accomplished by surgically destroying preganglionic input to the ganglia, also destroys the rhythm in $N$-acetyltransferase (65), it became evident that the endogenous oscillator driving the pineal is not in the ganglia. Evidence supporting the role of the ganglia in regulating $N$-acetyltransferase activity comes from experiments in which electrical stimulation of the ganglia elevated the activity of pineal $N$-acetyltransferase (125).

### Presynaptic Structures and Interactions

The postganglionic neurons that go to the pineal gland form a dense network of nerve fibers, a plexus, characterized not by the formation of the classical synapse, but rather by the frequent appearance of swollen areas of neural processes described as varicosities (2,33,44,79). These varicosities contain dense-cored vesicles within which the transmitter substances or substances which regulate pineal function are thought to reside (2,130).

Norepinephrine, dopamine, serotonin, $\gamma$-aminobutyric acid, histamine, and octopamine are transmitter substances present in the pineal gland (32,45,78,80,99,124,129). Norepinephrine, octopamine, and serotonin are located in the pineal nerves (44,45,99). Denervation of the pineal gland results in a large decrease in both norepinephrine and octopamine and a lesser loss of serotonin which is stored normally in both pineal nerves and cells (119).

Whereas norepinephrine and octopamine can stimulate the activity of $N$-acetyltransferase in cultured glands (64), the other transmitter substances including serotonin, dopamine, $\gamma$-aminobutyryic acid, and histamine have essentially no significant effect (18,64,80). In view of both the location and effects of norepinephrine and octopamine, it is possible that sympathetic stimulation of $N$-acetyltransferase activity may be a reflection of the release of both these compounds.

Treatment of the pineal gland with compounds which accelerate the net release of transmitters from sympathetic neurons produces an increase in pineal $N$-acetyltransferase activity and in melatonin production (3,39,64,94,95). These include compounds which act on neuronal membranes such as cocaine and potassium (39), tyramine, a compound which displaces norepinephrine (3,91), and compounds which prevent the reuptake of norepinephrine, such as desmethylimipramine and imipramine (94,95).

Another set of compounds which have presynaptic effects are the $\alpha$-adrenergic blocking agents (19). This class of compounds can enhance the effects of catecholamines on the stimulation of $N$-acetyltransferase activity, both *in vivo* and *in vitro* (23,64). In addition, these compounds can cause an enhanced response to stress (58,75), apparently by blocking the uptake of circulating catecholamines.

Alpha-blocking agents are known to act presynaptically in other systems (19); the physiological importance of this is not clear. However, norepinephrine has some $\alpha$-adrenergic stimulating potency, and it is possible that the reuptake process, which appears to be critically important in regulating the amount of norepinephrine in extracellular spaces, might be modulated by norepinephrine through an $\alpha$-adrenergic receptor.

### Postsynaptic Events

The initial important event in understanding how indole metabolism is regulated in the pineal gland is the interaction of catecholamines released from nerve terminals with a highly specific receptor on the postsynaptic cell surface. The nature of this receptor has been studied both indirectly, by measuring the effects of adrenergic agonists, and directly, by measuring antagonist binding. The only receptor known to exist on the pinealocyte and to be directly involved in the regulation of indole amine metabolism is the $\beta_1$-adrenergic receptor. Direct measurements of this receptor have found that it has a high affinity for isoproterenol, norepinephrine, and epinephrine (134). Indirect analysis of this receptor has shown that it mediates the adrenergic regulation of adenylate cyclase (126), cyclic AMP (20,120), phosphodiesterase (92), melatonin production (3,64), pineal serotonin (67), N-acetylserotonin production (14), and an increase in the activity of N-acetyltransferase (24,64,96).

One curious aspect of the adrenergic interaction with the pineal cells is the effect adrenergic agonists have on phospholipid metabolism. Norepinephrine appears to stimulate phosphate turnover of certain postsynaptic phospholipids (8,27–29). Remarkably, the effects of norepinephrine are mimicked by an $\alpha$-adrenergic blocking agent (35). The importance of these changes in membrane phospholipid metabolism and the relevance to pineal function is a mystery.

## Plasticity in Responsiveness

The pineal cell does not exhibit a constant level of responsiveness to adrenergic stimulation but exhibits daily changes in sensitivity (107). One of the earliest demonstrations of this plasticity in responsiveness was our finding that constant stimulation with norepinephrine did not produce a constant elevation of N-acetyltransferase activity in pineal organ cultures, but that after about 9 hr of treatment with the agonist, N-acetyltransferase activity fell, even though norepinephrine was continually added (64). In studies with this enzyme by Deguchi and Axelrod (23,25), it was found that a single injection with an adrenergic agonist could produce subsequent subsensitivity, and that supersensitivity would result owing to long periods during which adrenergic stimulation was low. Similar observations by Strada and Weiss (121) on the regulation of cyclic AMP have been reported.

One important contributing factor to this change in sensitivity is that the characteristics of the receptors change with adrenergic stimulation (109). Studies designed to measure the number of adrenergic receptors in the pineal gland and their affinity for adrenergic agonists indicate that these receptors can vary in both number and in affinity, both decreasing following adrenergic stimulation (48,107,109). The factors regulating this phenomenon are not clear. However, plasticity of these receptors is an obvious and important characteristic.

## Intracellular Events

The earliest direct evidence that norepinephrine could alter indole metabolism in the pineal gland was the finding that treatment of cultured pineal glands with norepinephrine resulted in the increased formation and release of melatonin (3). Following that important observation, many investigators sought to explain the events that mediated this event. At first, it was thought that norepinephrine acted by stimulating the activity of hydroxyindole-O-methyltransferase activity (3). However, this has not been proven. In contrast, we repeatedly found that large changes in melatonin production were not accompanied by significant changes in the activity of this enzyme (7,52,60).

A clue to the explanation of melatonin production regulation came from a technical breakthrough, the development of a thin-layer chromatographic analysis procedure which could separate the major derivatives of serotonin (57). Analyses of culture medium from pineal cultures treated with compounds which increased the formation of melatonin from radioactive precursors were performed. These indicated that whenever there was an increase in the formation and release of melatonin, there was also an increase in the formation and release of the precursor of this compound, N-acetylserotonin (7,52,60). This led to our speculation that large changes in melatonin production might result from among other things a change in the activity of N-acetyltransferase (60).

We eventually found that treatment of pineal glands with norepinephrine *in vitro* increased the activity of pineal N-acetyltransferase activity (53), a change closely correlated with an increase in melatonin production (64). It became apparent from this observation that the large changes in melatonin production could be regulated by large changes in the activity of this enzyme. The conventional transition from *in vivo* studies to *in vitro* studies was reversed in this case; the observation that N-acetyltransferase activity was stimulated by adrenergic agonists was made first *in vitro* (53) and was then confirmed *in vivo* (23).

The role of cyclic AMP in the regulation of indoliamine metabolism became apparent when it was shown that a derivative of cyclic AMP, dibutyryl cyclic AMP, could stimulate the activity of N-acetyltransferase (53), and the formation of N-acetylserotonin and melatonin (7), and decrease pineal serotonin (67,68). Prior studies had shown that norepinephrine could stimulate both adenylate cyclase and the formation of cyclic AMP (20,120,126). These results, together with complementary studies using drugs which block the degradation of cyclic AMP, clearly indicated that cyclic AMP was involved as a second messenger of norepinephrine (64).

A clear picture of how cyclic AMP stimulates pineal N-acetyltransferase activity has not emerged. It is known that the effects of this compound depend upon protein synthesis (23,53,64,108). It is not clear whether a new protein is synthesized as a result of cyclic AMP stimulation, or whether cyclic AMP activates a protein, perhaps $N$-acetyltransferase, which is continually being synthesized.

It is highly probable that in addition to requiring protein synthesis, the stimulation of $N$-acetyltransferase activity also requires an activation or protection mechanism. This is made quite clear by the remarkable differences in the rate at which the enzyme appears following adrenergic stimulation, a rate which has a doubling time of 15 to 20 min (24,53), and the far more rapid rate at which the enzyme disappears following cessation of adrenergic stimulation. This has a half-life of about 3 min (62). Recent findings suggest this activation-inactivation process involves cyclic AMP because cessation of adrenergic stimulation causes not only a rapid decrease in $N$-acetyltransferase activity, but also a more rapid decrease in cyclic AMP (54).

It would appear that cyclic AMP might function in several modes. Shortly after adrenergic stimulation starts cyclic AMP spikes to values that are 100-fold higher than baseline. This may trigger the synthesis of mRNA (108) which may be required for the stimulation of N-acetyltransferase if sufficient amounts of the required message is not available. A second effect of cyclic AMP might be to stimulate the synthesis of a new protein. A third would then be to stabilize and maintain $N$-acetyltransferase molecules in an active form (54).

The mode of action of cyclic AMP in this latter regard is not clear. One clue may be found in an important effect of cyclic AMP and norepinephrine. Norepinephrine can hyperpolarize the membrane potential of pinealocytes (114), an action which is mediated by a $\beta$-adrenergic receptor (114) and can be mimicked by cyclic AMP and dibutyryl cyclic AMP (97). Blockage of this hyperpolarization results in blockage of the stimulation of $N$-acetyltransferase (97). Thus, it would appear that the change in membrane potential may be a key event in altering the activity of pineal $N$-acetyltransferase activity.

A clue to the molecular mechanism by which $N$-acetyltransferase activity might be regulated comes from the observation that pineal $N$-acetyltransferase activity is stable in an enzyme cocktail at 37° but highly unstable at 37° in buffer (11). A bit of detective work led to the finding that acetyl CoA, which is present in the enzyme cocktail, was capable of protecting $N$-acetyltransferase against thermal inactivation (11). Analysis of the time course of inactivation indicated that it was remarkably rapid, almost as rapid as that seen in the intact gland when adrenergic stimulation was blocked. Acetyl CoA was found far more active than any other compound, although coenzyme A and the terminal portion of coenzyme A, $\beta$-mercaptoethylamine, were also effective. These observations raise the possibility that cytoplasmic acetyl CoA might be involved in the regulation of the activity of $N$-acetyltransferase.

In view of these observations, we have recently (54) proposed that inactivation of $N$-acetyltransferase activity may involve both acetyl CoA "protection" and cyclic AMP. We theorized that cyclic AMP might act to maintain $N$-acetyltransferase molecules in a stable and active form by encouraging a stabilizing interaction between acetyl CoA and N-acetyltransferase molecules; this might require membrane hyperpolarization. We can only speculate on the nature of this interaction but it could be the cyclic AMP-dependent acetylation of N-acetyltransferase molecules.

### Stress

In view of the previous discussion regarding the regulation of indole metabolism, especially the obvious involvement of catecholamines and of the sympathetic nervous system, it seems reasonable that pineal N-acetyltransferase activity might be elevated by stress due to the associated release of catecholamines by the adrenals and sympathetic nervous system or by direct sympathetic discharge in the pineal gland.

As discussed above, this is not the case. Pineal N-acetyltransferase activity does not increase significantly when rats are stressed by immobilization, cold, or swimming, relative to the increase in activity which occurs during the night (42,58,94). Circulating catecholamines released during stress apparently are not able to stimulate this enzyme significantly and sympathetic discharge in the pineal gland during stress is insignificant as compared to that which occurs at night. However, it has been clearly shown that stress can increase N-acetyltransferase activity if pineal nerves are removed (79,94). The nerves, acting through the well documented reuptake system, protect the pineal gland against circulating catecholamines released by either the adrenal or the sympathetic nervous system during stress (58,94). Further evidence in support of this is that desmethylimipramine, a drug which specifically blocks reuptake, can act both *in vivo* to greatly enhance the stimulation of pineal N-acetyltransferase caused by stress, and can act directly on the pineal nerves *in vitro* to enhance the pinealocyte response to physiological catecholamine (94). Desmethylimipramine is probably acting primarily via this mechanism and is not dependent upon sympathetic discharge in the pineal gland. This was made apparent by the observation that the drug can act in animals with decentralized superior cervical ganglia, i.e., ganglia that send complete innervation from postganglionic cells to the pineal gland but which do not receive central input as a consequence of surgical destruction of preganglionic input (94).

Thus, it would appear that nerve endings are of major importance in regulating the apparent sensitivity of the pineal gland. The clinical importance of presynaptic factors may become more apparent if patients receiving antidepressant treatment with desmethylimipramine are evaluated for their sensitivity to stress. We have shown that this drug enhances *in vitro* melatonin production (95); desmethylimipramine conceivably could act to enhance melatonin production both during the night and during stressful periods.

## Developmental Aspects

The enzymatic capacity to make serotonin in the pineal gland appears to be present at birth in the rat (30,43). The rhythm in serotonin is not apparent, however, until several days later, primarily because the rhythm in pineal N-acetyltransferase activity does not become apparent until the middle of the first week of life (30,43). The developmental appearance of the rhythm in pineal N-acetyltransferase activity at this time is probably due to the ingrowth of sympathetic nerves and the completion of the neuronal circuit from the suprachiasmatic nucleus to the pineal gland (133). The activity of N-acetyltransferase early in the first week of life in the rat is neither continually high nor continually low, but is maintained at a constant, intermediate level, perhaps because catecholamines are continually available to the pineal gland and maintain a constant low level stimulation of the tissue. The ability to respond to norepinephrine is present as early as two days prior to birth (133) indicating that the postsynaptic machinery required to generate a rhythm in pineal N-acetyltransferase activity is developed well before the required presynaptic structures have developed.

The development of the rhythm in pineal melatonin has not been studied. However, it seems reasonable that a rhythm in melatonin in the rat pineal will not be detected prior to the middle or end of the second week of life because this is the stage of development at which hydroxyindole-O-methyltransferase activity is first detectable (55).

The delay in the ability of the pineal gland to produce melatonin raises the question of whether the animal is physiologically naive in regards to this hormone until this time. This is probably not the case. We have found in the rat that [$^3$H]melatonin can cross both the placenta and mammary gland and enter the tissues of the fetal and neonatal animal by these routes (50,104). The first studies on placental transport of melatonin in the rat are currently being confirmed using unanesthetized sheep and an anesthetized non-human-primate (105). It is becoming apparent that even though the fetal or neonatal animal might not be able to synthesize melatonin from serotonin, it is possible that a rhythm in circulating melatonin could be established, one that is generated maternally and transmitted via normal nutritional pathways. This provides a theoretical mechanism by which the fetus receives information about environmental light, and might be an important means of introducing the developing fetus to 24-hr periodicity.

## Cell Separation Studies

One characteristic of the pineal gland which has enabled substantial advances in the understanding of its function is that all the cells appear to be of one type—the pinealocyte. This has been the tacit assumption of biochemists. The morphologist, however, can subdivide this one cell type into two: the light and the dark pinealocyte. No clear ultrastructural characteristic can be used to distinguish these; the only major difference in the cell type is the density of the nuclear and cytoplasmic material as evidenced by staining techniques. These two subtypes of pinealocytes might represent cells in different stages of activity or cells of distinct function (2,3,120).

A second type of cell in the pineal, comprising about 10% of the cells, is a supporting glia-like cell described as an interstitial cell, or an astrocyte (2,130).

One of our current interests is to determine if we can separate these three cell types, the light pinealocyte, the dark pinealocyte and the interstitial cell. Our first attempts (16) have successfully separated distinct cell types as characterized by their position on a density gradient (Fig. 4). The ability of norepinephrine to stimulate $N$-acetyltransferase activity is restricted to one band of cells. Electronmicroscopic examination of these cells has not been completed; it is hoped that this will indicate that distinct cell types have been isolated. It may then be possible to use these purified cells for more specific studies of the cellular distribution of receptors, enzymes, and second messengers and to provide a more precise picture of pineal function on a subcellular basis. It might be possible, for example, to identify

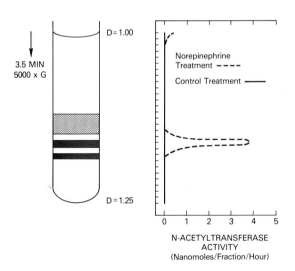

**FIG. 4.** Separation of pineal cells on a density gradient. Cells were prepared from 2-day-old pineal glands by trypsinization and applied to the top of a density gradient. The gradient was spun for 5 min at 5000 × g. Three bands of material were detected visually as indicated. The gradient was fractionated using a fraction collector; each fraction was diluted, and the cells were collected by centrifugation. Each fraction was divided into several samples which were incubated for 24 hr, and then treated with norepinephrine or with no drug for 6 hr. Norepinephrine stimulated $N$-acetyltransferase activity in fractions corresponding to the middle band of material. From (16).

only one cell type that specifically responds to adrenergic stimulation by an increase in cyclic AMP. Conceivably one cell type might be found to contain primarily γ-aminobutyric acid and another to contain none; one cell type might contain N-acetyltransferase and another hydroxyindole-O-methyltransferase; one cell might have β-adrenergic receptors and another α-adrenergic receptors.

This kind of approach to the study of tissue function would permit substantial progress in understanding the function of specific hypothalamic nuclei on a cellular and subcellular level. It will be helpful to isolate specific homogenous groups of cells enriched with one specific hypothalamic releasing factor and then use these to study how the production and release of these hormones is regulated. A growing number of studies have demonstrated that glia cells have the ability to respond to agonists with an increase in the concentration of a second messenger. Thus, it will be necessary to separate glia cells from cells of specific function before it will be possible to determine if metabolic changes in metabolic function measured are associated with the cell responsible for the release and production of a hormone.

The cell fractionation method we have used provides one approach to prepare pure cells. Other methods will no doubt have to be applied; methods which make use of receptors on cells to remove cells which contain only these specific receptors, methods which make use of electrical differences in cells so that cells can be electrophoresed, and methods which make use of unique membrane proteins so that cells can be immunologically separated, perhaps by immobilized antibodies. The application of these techniques will certainly lead to a more precise understanding of how the hypothalamus functions and how specific hypothalamic nuclei are regulated. With these available, it should be possible to construct an integrated concept of function for each type of hypothalamic cell, a concept which involves central neural circuitry, pre- and post-synaptic events, and intracellular mechanisms of regulation including specific biochemical and electrophysiological changes in function organized to control a single key function, perhaps through a specific *regulating* enzyme.

## REFERENCES

1. Arendt, J., Paunier, L., and Sizonenko, P. D. (1975): Melatonin radioimmunoassay. *J. Clin. Endocrinol. Metab.* 40:347–350.
2. Arstila, A. U. (1967): Electron microscopic studies on the structure and histochemistry of the pineal gland of the rat. *Neuroendocrinology, (Suppl. 6,)*, 2:1–11.
3. Axelrod, J., Shein, H. M., and Wurtman, R. J. (1969): Stimulation of $C^{14}$-melatonin synthesis from $C^{14}$-tryptophan by noradrenaline in rat pineal in organ culture. *Proc. Natl. Acad. Sci., USA*, 62:554–559.
4. Axelrod, J. and Weissbach, H. (1960): Enzymatic O-methylation of N-acetylserotonin to melatonin. *Science*, 131:1312–1313.
5. Axelrod, J., Wurtman, R. J., and Snyder, S. H. (1965): Control of hydroxyindole-O-

methyltransferase activity in the rat pineal gland by environmental lighting. *J. Biol. Chem.,* 240:949–954.
6. Bensinger, R. E., Klein, D. C., Weller, J. L., and Lovenberg, W. M. (1974): Radiometric assay of total tryptophan hydroxylation by intact cultured pineal glands. *J. Neurochem.,* 23:111–117.
7. Berg, G. R. and Klein, D. C. (1971): Pineal gland in organ culture. II. Role of adenosine 3′,5′-monophosphate in the regulation of radiolabelled melatonin production. *Endocrinology,* 89:453–461.
8. Berg, G. R. and Klein, D. C. (1972): Norepinephrine stimulates $^{32}$P incorporation into a specific phospholipid fraction of postsynaptic pineal membranes. *J. Neurochem.,* 19:2519–2532.
9. Binkley, S. and Geller, E. B. (1975): Pineal enzymes in chickens: Development of daily rhythmicity. *Gen. Comp. Endocrinol.,* 27:424–429.
10. Binkley, S., Klein, D. C., and Weller, J. (1974): Dark induced increase in pineal serotonin N-acetyltransferase activity: A refractory period. *Experientia,* 29:1339–1340.
11. Binkley, S., Klein, D. C., and Weller, J. L. (1976): Pineal serotonin N-acetyltransferase activity: Protection of stimulated activity by acetyl CoA and related compounds. *J. Neurochem.,* 26:51–55.
12. Binkley, S. A., MacBride, S., Klein, D. C., and Ralph, C. (1973): Pineal enzymes: Regulation of avian melatonin synthesis. *Science,* 181:273–275.
13. Binkley, S. A., MacBride, S., Klein, D. C., and Ralph, C. L. (1975): Regulation of pineal rhythms in chickens: Refractory period and nonvisual light perception. *Endocrinology,* 96:848–853.
14. Brownstein, M., Saavedra, J. M., and Axelrod, J. (1973): Control of pineal N-acetylserotonin by a beta-adrenergic receptor. *Molec. Pharmacol.,* 9:605–611.
15. Buda, M. and Klein, D. C. (1977): Function and regulation of pineal N-acetyltransferase activity. In: *Biochemistry and Function of Amine Enzymes,* edited by E. Usdin and N. Weiner, Pergamon Press, Elmsford, New York (*in press*).
16. Buda, M. and Klein, D. C. (*unpublished data*).
17. Cardinali, D. P., Larin, F., Wurtman, R. J. (1971): Action spectra for effects of light on hydroxyindole-O-methyltransferases in rat pineal, retina, and harderian gland. *Endocrinology,* 91:877–886.
18. Chiou, C. Y., Trzeciakowski, J., and Klein, D. C. (1976): Histamine mediation of nicotine effects on postganglionic sympathetic junctions. *Neuropharmacology,* 15:689–893.
19. Cubeddu, L. X., Langer, S. Z., and Weiner, N. (1974): The relationships between alpha receptor block, inhibitions of norepinephrine uptake and the release and metabolism of $^3$H-norepinephrine. *J. Pharmacol. Exp. Ther.,* 188:368–385.
20. Deguchi, T. (1973): Role of the beta adrenergic receptor in the elevation of adenosine cyclic 3′5′-monophosphate and induction of serotonin N-acetyltransferase in rat pineal glands. *Mol. Pharm.,* 9:184–190.
21. Deguchi, T. (1977): Circadian rhythms of enzyme and running activity under ultradian lighting schedule. *Amer. J. Physiol.,* 232:375–381.
22. Deguchi, T. (1977): Tryptophan hydroxylase in pineal gland of rat: postsynaptic localization and absence of circadian change. *J. Neurochem.,* 28:667–668.
23. Deguchi, T. and Axelrod, J. (1972): Induction and superinduction of serotonin N-acetyltransferase by adrenergic drugs and denervation in the rat pineal. *Proc. Natl. Acad. Sci., USA,* 69:2208–2211.
24. Deguchi, T. and Axelrod, J. (1972): Control of circadian change in serotonin N-acetyltransferase activity in the pineal organ by the β-adrenergic receptor. *Proc. Natl. Acad. Sci., USA,* 69:2547–2550.
25. Deguchi, T. and Axelrod, J. (1973): Superinduction of serotonin N-acetyltransferase and supersensitivity of adenyl cyclase to catecholamines in denervated pineal glands. *Mol. Pharm.,* 9:612–619.
26. Deguchi, T. and Barchas, J. (1973): Comparative studies on the effect of parachloropheylalanine on hydroxylation of tryptophan in pineal and brain of rat. In: *Serotonin and Behavior,* edited by J. Barchas, and E. Usdin, pp. 33–47. Academic Press, New York.
27. Eichberg, J. and Hauser, G. (1974): Stimulation by local anesthetics of the metabolism of acidic phospholipids in the rat pineal gland. *Biochem. Biophys. Res. Commun.,* 60:1460–1467.

28. Eichberg, J., Shein, H. M., and Hauser, G. (1973): Effect of neurotransmitters and other pharmacological agents on the metabolism of phospholipids in pineal gland cultures and cloned neuronal and glial cells. *Biochem. Soc. Trans.*, 1:352–359.
29. Eichberg, J., Shein, H. M., Schartz, M., and Hauser, G. (1973): Stimulation of $^{31}$Pi incorporation into phosphatidylinositol and phosphatidylglycerol by catecholamines and $\beta$-adrenergic receptor-blocking agents in rat pineal organ cultures. *J. Biol. Chem.*, 235:3615–3622.
30. Ellison, N., Weller, J. L., and Klein, D. C. (1972): Development of a circadian rhythm in pineal N-acetyltransferase. *J. Neurochem.*, 19:1335–1341.
31. Fiske, V. M. (1964): Serotonin rhythm in the pineal organ: control by the sympathetic nervous system. *Science*, 146:253–254.
32. Giarman, N. J. and Day, M. (1958): Presence of biogenic amines in the bovine pineal body. *Biochem. Pharmacol.*, 1:235–237.
33. Hakanson, R., Lombard Des Gouttes, M. N., and Owman, C. (1967): Activities of tryptophan hydroxylase, DOPA decarboxylase and monoamine oxidase as correlated with the appearance of monoamines in the developing rat pineal. *Life Sci.*, 6:2577–2583.
34. Halász, B. (1969): Effects of isolation of the hypothalamus from the rest of the brain. In: *Frontiers in Neuroendocrinology*, edited by W. F. Ganona and L. Martini, pp. 307–342. Oxford Press, New York.
35. Hauser, G., Shien, H. M., and Eichberg, J. (1974): Relationship of $\alpha$-adrenergic receptors in rat pineal gland to drug induced stimulation of phospholipid metabolism. *Nature*, 252:482–483.
36. Hedlund, L., Lischko, M., Rollag, M. D., Niswender, G. D. (1977): Melatonin: daily cycle in plasma and cerebrospinal fluid of calves. *Science*, 195:686–687.
37. Hendrickson, A. E., Wagoner, N., and Cowan, W. M. (1972): An autoradiographic and electron microscopic study of retino-hypothalamic connections. *Z. Zellforsch.*, 135:1–26.
38. Hirata, F., Hayaishi, O., Tokuyama, T., and Serioh, S. (1974): *In vitro* and *in vivo* formation of two new metabolites of melatonin. *J. Biol. Chem.*, 249:1611–1616.
39. Holz, R. W., Deguchi, T., and Axelrod, J. (1974): Stimulation of serotonin N-acetyltransferase in pineal organ culture by drugs. *J. Neurochem.*, 22:205–209.
40. Ibuka, N. and Kawamura, H. (1975): Loss of circadian rhythm in sleep-wakefulness cycle in the rat by suprachiasmatic nucleus lesions. *Brain Res.*, 96:76–81.
41. Illnerová, H. (1971): Effect of light on the serotonin content of the pineal gland. *Life Sci.*, 10:955–961.
42. Illnerová, H. (1975): The effects of immobilization of the activity of serotonin N-act in the rat epiphysis. In: *Catecholamines and Stress*, edited by Usdin, Kvetnansky, and Kopin, p. 129. Pergamon Press, New York.
43. Illnerov, A. H. and Skopou, A. J. (1976): Regulation of the diurnal rhythm in rat pineal serotonin-N-acetyltransferase activity and serotonin content during ontogenesis. *J. Neurochem.*, 26:1051–1052.
44. Jaim-Etcheverry, G. and Zieher, L. M. (1974): Localizing serotonin in central and peripheral nerves. In: *Neurosciences*, 3rd edition, edited by F. O. Schmidt, pp. 917–923. MIT Press, Cambridge.
45. Jaim-Etcheverry, G. and Zieher, L. M. (1975): Octopamine probably coexists with noradenaline and serotonin in vesicles pineal adrenergic nerves. *J. Neurochem.*, 25:915–918.
46. Kappers, J. A. (1960): The development, topographical relations and innervation of the epiphysis cerebri in the albino rat. *Z. Zellforsch.*, 52:163–215.
47. Kappers, J. A. (1965): Survey of the innervation of the epiphysis cerebri and the accessory pineal organs of vertebrates. In: *The Structure and Function of the Epiphysis Cerebri*, Progress in Brain Research, Volume 10, edited by J. Ariëns Kappers and J. P. Schadé, Elsevier, Amsterdam.
48. Kebabian, J. W., Zatz, M., Romero, J. A., and Axelrod, J. (1975): Rapid changes in rat pineal beta-adrenergic receptor: Alterations in L-($^3$H)alprenolol binding and adenylate cyclase. *Proc. Natl. Acad. Sci., USA*, 72:3735–3739.
49. Kennaway, D. J., Frith, R. G., Phillipou, G., Matthews, C. D., and Seamark, R. F. (1977): A specific radioimmunoassay for melatonin in biological tissue and fluids and its validation by gas chromatography-mass spectrometry. *Endocrinology*, 101:119–127.

50. Klein, D. C. (1972): Evidence for the placental transfer of $^3$H acetyl-melatonin. *Nature,* 237:118–119.
51. Klein, D. C. (1974): Circadian rhythms in indole metabolism in the rat pineal gland. In: *The Neurosciences. Third Study Programme,* edited by F. O. Schmidt. pp. 509–515. MIT Press, Cambridge, Massachusetts.
52. Klein, D. C. and Berg, G. R. (1970): Pineal gland: Stimulation of melatonin production by norepinephrine involves cyclic AMP-mediated stimulation of N-acetyltransferase. *Advances in Biochem. Psychopharmacol.,* 3:241–263.
53. Klein, D. C., Berg, G. R., and Weller, J. (1970): Melatonin synthesis: Adenosine 3′, 5′-monophosphate and norepinephrine stimulate N-acetyltransferase. *Science,* 168: 979–980.
54. Klein, D. C., Buda, M., Kappor, C. L., and Krishma, G. (1977): Pineal serotonin N-acetyltransferase activity: An abrupt disease in cyclic AMP may be the signal for "Turn Off". *Science (in press).*
55. Klein, D. C. and Lines, S. V. (1969): Pineal hydroxyindole-O-methyltransferase activity in the growing rat. *Endocrinology,* 89:1523–1525.
56. Klein, D. C. and Moore, R. Y. *(unpublished observations).*
57. Klein, D. C. and Notides, A. (1969): Thin-layer chromatographic separation of pineal gland derivatives of serotonin-$^{14}$C. *Analytical Biochem.,* 31:480–483.
58. Klein, D. C. and Parfitt, A. (1976): A protective role of nerve endings in stress-stimulated increase in pineal N-acetyltransferase activity. In: *Catecholamines and Stress,* edited by R. Kvetnansky and E. Usdin, pp. 119–128. Pergamon Press, New York.
59. Klein, D. C., Reiter, R. J., and Weller, J. L. (1971): Pineal N-acetyltransferase activity in blinded and anosmic rats. *Endocrinology,* 89:1020–1023.
60. Klein, D. C. and Rowe, J. (1970): Pineal gland in organ culture. I. Inhibition by harmine of serotonin-$^{14}$C oxidation, accompanied by stimulation of melatonin-$^{14}$C production. *Mol. Pharmacol.,* 6:164–171.
61. Klein, D. C. and Weller, J. L. (1970): Indole metabolism in the pineal gland: a circadian rhythm in N-acetyltransferase. *Science,* 169:1093–1095.
62. Klein, D. C. and Weller, J. L. (1972): A rapid light-induced decrease in pineal serotonin N-acetyltransferase activity. *Science,* 177:532–533.
63. Klein, D. C. and Weller, J. L. (1972): The role of N-acetylserotonin in the regulation of melatonin production. In: *International Congress Series* No. 256, p. 52. Excerpta Medica, Amsterdam.
64. Klein, D. C. and Weller, J. L. (1973): Adrenergic-adenosine 3′,5′-monophosphate regulation of serotonin N-acetyltransferase activity and the temporal relationship of serotonin N-acetyltransferase activity to synthesis of $^3$H-N-acetylserotonin and $^3$H-melatonin in the cultured rat pineal gland. *J. Pharmacol. Exp. Ther.,* 186:516–527.
65. Klein, D. C., Weller, J. L., and Moore, R. Y. (1971): Melatonin metabolism: neural regulation of pineal serotonin N-acetyltransferase activity. *Proc. Natl. Acad. Sci., USA,* 68:3107–3110.
66. Klein, D. C. and Yuwiler, A. (1973): Beta-adrenergic regulation of indole metabolism in the pineal gland. In: *Frontiers in Catecholamine* Research, edited by E. Usdin and S. Snyder, p. 321. Pergamon Press, London.
67. Klein, D. C., Yuwiler, A., Weller, J. L., and Plotkin, S. (1973): Postsynaptic adrenergic-cyclic AMP control of the serotonin content of cultured rat pineal glands. *J. Neurochem.,* 21:1261–1272.
68. Kopin, J. I., Pare, C. M. B., Axelrod, J., and Weissbach, H. (1961): The fate of melatonin in animals. *J. Biol. Chem.,* 236:11.
69. Kveder, S. and McIsaac, W. M. (1961): The metabolism of melatonin (N-acetyl-5-methoxytryptamine) and 5-methoxytryptamine. *J. Biol. Chem.,* 236:12.
70. Lovenberg, W., Jequier, E., and Sjoerdsma, A. (1967): Tryptophan hydroxylation: measurements in pineal gland, brain stem, and carcinoid tumor. *Science* 155:217–218.
71. Lovenberg, W., Jequier, E., and Sjoerdsma, A. (1968): Tryptophan hydroxylation in mammalian systems. *Adv. in Pharmacol.,* 6A:21–29.
72. Lovenberg, W., Weissbach, H., and Udenfriend, S. (1962): Aromatic L-amino acid decarboxylase. *J. Biol. Chem.,* 237:89–92.
73. Lynch, W. J. (1971): Diurnal oscillations in pineal melatonin content. *Life Sciences,* 10:791–795.

74. Lynch, H. J., Eng, J. P., and Wurtman, R. J. (1973): Control of pineal indole biosynthesis by changes in sympathetic tone caused by factors other than environmental lighting. *Proc. Natl. Acad. Sci., USA*, 70:1704–1708.
75. Lynch, H. J., Ho, M., and Wurtman, R. J. (1977): The adrenal medulla may mediate the increase in pineal melatonin synthesis caused by stress, but not that caused by exposure to darkness. *J. Neural Transm.*, 40:87–97.
76. Lynch, H. J. and Ralph, C. L. (1970): Diurnal variation in pineal melatonin and its non-relationship to HIOMT activity. *Am. Zool.*, 10:300.
77. Lynch, H. J., Wurtman, R. J., Moskowitz, M. A., Archer, M. C., and Ho, M. H. (1975): Daily rhythm in human urinary melatonin. *Science*, 187:169–170.
78. Machado, A. B. M., Faleiro, L. C., and Dasilva, D. A. (1965): Study of mast cell and histamine contents of the pineal body. *Z. Zellforsch. Mikrosk. Anat.*, 65:521–525.
79. Machado, C. R. S., Wragg, L. E., and Machado, A. B. M. (1968): A histochemical study of sympathetic innervation and 5-hydroxytryptamine in the developing pineal body of the rat. *Brain Res.*, 8:310–318.
80. Mata, M. M., Schrier, B. K., Klein, D. C., Weller, J. L., and Chiou, C. Y. (1976): On GABA function and physiology in the pineal gland. *Brain Res.*, 118:383–394.
81. Minneman, R. P., Lynch, H., and Wurtman, R. J. (1974): Relationship between environmental light intensity and retina-mediated suppression of rat pineal serotonin N-acetyltransferase. *Life Sci.*, 15:1791–1796.
82. Moore, R. Y. (1973): Retinohypothalamic projection in mammals: a comparative study. *Brain Res.*, 49:403–409.
83. Moore, R. Y. (1977): The innervation of the mammalian pineal gland. In: *The Pineal and Reproduction,* edited by R. J. Reiter. S. Karger, Basel (*in press*).
84. Moore, R. Y., and Eichler, V. B. (1972): Loss of a circadian adrenal corticosterone rhythm following suprachiasmatic lesions in the rat. *Brain Res.*, 42:201–206.
85. Moore, R. Y., Heller, A., Bhatnagar, R. K., Wurtman, R. J., and Axelrod, J. (1968): Central control of the pineal gland: visual pathways. *Arch. Neurol.*, 18:208–218.
86. Moore, R. Y., Heller, A., Wurtman, R. J., and Axelrod, J. (1967): Visual pathways mediating pineal response to environmental light. *Science*, 155:220–223.
87. Moore, R. Y. and Klein, D. C. (1974): Visual pathways and the central neural control of a circadian rhythm in pineal serotonin N-acetyltransferase activity. *Brain Res.*, 71:17–33.
88. Moore, R. Y., and Lenn, N. J. (1972): A retinohypothalamic projection in the rat. *J. Comp. Neurol.*, 146:1–14.
89. Moore, R. Y. and Rapport, R. L. (1970): Pineal and gonadal function in the rat following cervical sympathectomy. *Neuroendocrinology*, 7:361–374.
90. Moore, R. Y. and Traynor, M. E. (1976): Diurnal rhythms in pineal N-acetyltransferase and hippocampal norepinephrine: effects of water deprivation, blinding, and hypothalamic lesions. *Neuroendocrinology*, 20:250–259.
91. Muscholl, E. (1966): Indirectly acting sympathomimetic amines. *Pharmacol. Rev.*, 18:551–559.
92. Neff, N. H. and Oleshansky, M. A. (1975): Rat pineal adenosine cyclic 3′,5′-monophosphate phosphodiesterase activity: modulation in vivo by a beta adrenergic receptor. *Mol. Pharmacol.*, 11:552–557.
93. Ozaki, Y., Lynch, H. J., and Wurtman, R. J. (1976): Melatonin in rat pineal, plasma, and urine: 24-hour rhythmicity and effect of chlorpromazine. *Endocrinology*, 98:1418–1424.
94. Parfitt, A. G. and Klein, D. C. (1976): Sympathetic nerve endings protect against acute stress-induced increase in N-acetyltransferase (E.C. 2.3.1.5.) activity. *Endocrinology*, 99:840–854.
95. Parfitt, A. and Klein, D. C. (1977): Desmethylimipramine causes an increase in the production of [$^3$H]melatonin by isolated pineal glands. *Biochem. Pharmacol.*, 26:906–907.
95a. Parfitt, A. and Klein, D. C. (*Unpublished observations.*)
96. Parfitt, A., Weller, Joan L., and Klein, D. C. (1976): Beta adrenergic-blockers decrease adrenergically stimulated N-acetyltransferase activity in pineal glands in organ culture. *Neuropharmacology*, 15:353–358.
97. Parfitt, A., Weller, J. L., Sakai, K. K., Marks, B. H., and Klein, D. C. (1975): Blockade

by ouabain or elevated potassium ion concentration of the adrenergic and adenosine cyclic 3′,5′-monophosphate-induced stimulation of pineal serotonin N-acetyltransferase activity. *Mol. Pharmacol.*, 11:241–255.
98. Pelham, R. W., Vaughan, G. M., Sandock, K. L., and Vaughan, M. K. (1973): Twenty-four hour cycle of a melatonin-like substance in the plasma of human males. *J. Clin. Endocrinol. Metab.*, 37:341–344.
99. Pelligrino de Iraldi, A. and Zieher, L. M. (1966): Noradrenaline and dopamine content of normal decentralized, and denervated pineal glands of the rat. *Life Sci.*, 5:149–154.
100. Quay, W. B. (1963): Circadian rhythm in rat pineal serotonin and its modification by estrous cycle and photoperiod. *Gen. Comp. Endocrinol.*, 3:473–479.
101. Quay, W. B. (1967): Lack of rhythm and effect of darkness in rat pineal content of N-acetylserotonin-O-methyltransferase. *Physiologist*, 10:286.
102. Ralph, C. L., Hull, D., Lynch, H. J., and Hedlund, L. (1971): A melatonin rhythm persists in rat pineals in darkness. *Endocrinology*, 89:1361–1366.
103. Reiter, R. J. and Klein, D. C. (1971): Observations on the pineal glands, the harderian glands, the retinas, and the reproductive organs of adult female rats exposed to continuous light. *J. Endocrinol.*, 51:117–125.
104. Reppert, S. and Klein, D. C. (1977): Milk transport and biological fate of melatonin in the neonatal rat. *Endocrinology (in press)*.
105. Reppert, S., Klein, D. C., and Chez, R. (*unpublished observation*).
106. Rollag, M. D. and Niswender, G. D. (1976): Radioimmunoassay of serum concentrations of melatonin in sheep exposed to different lighting regimens. *Endocrinology*, 98:482–488.
107. Romero, J. A. and Axelrod, J. (1974): Pineal β-adrenergic receptor: diurnal variation in sensitivity. *Science*, 184:1091–1092.
108. Romero, J. A., Zatz, M., and Axelrod, J. (1975): Beta-adrenergic stimulation of pineal N-acetyltransferase: adenosine 3′:5′-cyclic monophosphate stimulates both RNA and protein synthesis. *Proc. Natl. Acad. Sci., USA*, 72:2107–2111.
109. Romero, J. A., Zata, M., Kebabian, J. W., and Axelrod, J. (1975): Circadian cycles in binding of ³H-alprenolol to beta-adrenergic receptor sites in rat pineal. *Nature*, 258:435–436.
110. Roth, W. D. (1965): Metabolic and morphologic studies on the rat pineal organ during puberty. In: *Structure and Function of the Epiphysis Cerebri*, Progress in Brain Research, Volume 10, edited by J. Ariëns Kappers and J. P. Schadé, pp. 552–563. Elsevier, Amsterdam.
111. Rowe, J. W., Richert, J. R., Klein, D. C., and Reichlin, S. (1970): Relation of the pineal gland and environmental lighting to thyroid function in the rat. *Neuroendocrinology*, 6:247–254.
112. Rudeen, P. K., Reeler, R. S., and Vaughan, M. K. (1975): Pineal serotonin N-acetyltransferase activity in four mammalian species. *Neuroscience Letters*, 1:225–229.
113. Saavedra, J. M., Brownstein, M., and Axelrod, J. (1973): A specific and sensitive enzymatic-isotopic microassay for serotonin in tissues. *J. Pharmacol. Exp. Ther.*, 186:508–515.
114. Sakai, K. K. and Marks, B. H. (1972): Adrenergic effects on pineal cell membrane potential. *Life Sci.*, 11:285–291.
115. Saper, C. B., Loewy, A. D., Swanson, L. W., and Cowan, W. M. (1976): Direct hypothalamoautonomic connections. *Brain Res.*, 117:305–312.
116. Smith, J. A., Padmick, D., Mee, T. S., Minneman, K. P., and Bud, E. D. (1977): Synchronous nyctohemeral rhythms in human blood melatonin and in human post-mortem pineal enzyme. *Clin. Endocrinol.* 6:219–225.
117. Stephan, F. K. and Zucker, I. (1972): Circadian rhythms in drinking behavior and locomotor activity of rats are eliminated by hypothalamic lesions. *Proc. Natl. Acad. Sci., USA*, 69:1583–1586.
118. Stephan, F. K. and Zucker, I. (1972): Rat drinking rhythms. Central visual pathways and endocrine factors mediating responsiveness to environmental illumination. *Physiol. Behav.*, 8:315–326.
119. Snyder, S. H., Zweig, M., Axelrod, J., and Fischer, J. E. (1965): Control of the circadian rhythm in serotonin content of the rat pineal gland. *Proc. Natl. Acad. Sci., USA*, 53:301–303.
120. Strada, S., Klein, D. C., Weller, J., and Weiss, B. (1972): Norepinephrine stimulation of

cyclic adenosine monophosphate in cultured pineal glands. *Endocrinology,* 90:1470–1476.
121. Strada, S. J. and Weiss, B. (1974): Increased response to catecholamines of the cyclic AMP system of rat pineal gland induced by decreased sympathetic activity. *Arch. Biochem. Biophys.,* 160:197–204.
122. Swanson, L. W., and Cowan, W. M. (1975): The efferent connections of the suprachiasmatic nucleus of the hypothalamus. *J. Comp. Neurol.,* 160:1–12.
123. Szentagothai, J., Flerko, B., Mess, B., and Halasz, B. (1960): Hypothalamic control of the anterior pituitary. Akademia Kiado, Budapest.
124. Vellan, E. L., Glessing, L. R., and Stalsberg, H. (1970): Free amino acids in the pineal and pituitary glands of human brain. *J. Neurochem.,* 17:699–701.
125. Volkman, P. H., and Heller, A. (1971): Pineal N-acetyltransferase: effect of sympathetic stimulation. *Science,* 173:839–840.
126. Weiss, B. and Costa, E. (1968): Selective stimulation of adenyl cyclase of rat pineal gland by pharmacologically active catecholamines. *J. Pharmacol. Exp. Ther.,* 161:310–319.
127. Weissbach, H., Redfield, B. G., and Axelrod, J. (1961): The enzymatic acetylation of serotonin and other naturally occurring amines. *Biochim. Biophys. Acta,* 54:190–192.
128. Wilkinson, M., Arendt, J., and de Ziegler, D. (1977): Determination of a dark-induced increase in pineal N-acetyltransferase activity and simultaneous radioimmunoassay of melatonin in pineal, serum, and pituitary tissue of the male rat. *J. Endocrinol.,* 72:243–244.
129. Wolfe, D., Potter, D., Richardson, K., and Axelrod, J. (1962): Localizing norepinephrine in sympathetic axon by electron microscopic autoradiography. *Science,* 138:440–442.
130. Wolfe, D. E. (1965): The epiphyseal cell: An electron microscopic study of its intercellular relationships and intracellular morphology in the pineal body of the albino rat. In: *Structure and Function of the Epiphysis Cerebri,* Progress in Brain Research, Volume 10, edited by J. Ariëns Kappers and J. P. Schadé, pp. 332–386. Elsevier, Amsterdam.
131. Wurtman, R. J., Axelrod, J., and Phillips, L. (1963): Melatonin synthesis in the pineal gland: Control by light. *Science,* 142:1071–1072.
132. Wurtman, R. J. and Mokowitz, M. A. (1977): The pineal organ (First of Two Parts). *New Engl. J. Med.,* 296:1329–1333.
133. Yuwiler, A., Klein, D. C., Buda, M., and Weller, J. L. (1977): Pineal N-acetyltransferase activity: Developmental aspects. *Am. J. Physiol. (in press).*
134. Zatz, M., Kebabian, J. W., Romero, J. A., Lefkowitz, R. J., and Axelrod, J. (1976): Pineal beta adrenergic receptor: correlation of binding of 3H-1-alprenolol with stimulation of adenylate cyclase. *J. Pharmacol. Exp. Ther.,* 196:714–722.

## DISCUSSION

*Dr. Baldessarini:* Dr. Klein, could you comment just briefly on what you think is going on during the rapid inactivation phase? I assume you think there is a protein phosphorylation change mediated by the beta receptors?

*Dr. David Klein:* This is currently a popular idea, i.e., that protein kinase is involved. Unfortunately, there are no facts supporting this. However, we do have the following facts: When we homogenize pineal glands containing high levels of N-acetyltransferase, put the homogenate in a test tube, and incubate it at 37°, the activity spontaneously decreases with an impressively short halving time (3 to 4 min). This can be blocked completely by low concentrations of acetyl-coenzyme A (Co-A). We do not know if this really is a reflection of the physiological regulatory mechanism. We do know, however, that even the terminal portion of acetyl-CoA, mercaptoethylamine, also can act at impressively low concentrations to block the inactivation.

One of my collaborators, Dr. Aryan Namboodiri, is presently investigating the role of protein kinase and phosphoprotein phosphatase. He is using inhibitors and activators of both these enzymes to try to elucidate the role of protein phosphorylation in the rapid inactivation of N-acetyltransferase activity. He is also trying to tie this in with the acetyl-CoA story.

*Dr. Reichlin:* I would like to comment about the resemblance of the pineal to the hypothalamus. I think the pineal is a model for the hypothalamologists, not only because the suprachiasmatic nucleus influences the function of the superior cervical ganglia cells which then impinge back on the pineal. I think you can consider the pineal the model system of a biogenic amine neurotransmitter regulating protein synthesis in another cell, in this case the melatonin-producing cell. Then you have worked out the molecular biology of this system to show how receptor activation can influence protein synthesis.

My own view is that this is very similar to what is happening inside the hypothalamus where central biogenic amine pathways are mediated by serotonin, dopamine, and norepinephrine, in turn, impinging on the function of peptidergic pathways, which have protein synthetic machinery. In the pineal such machinery is in the open and can be examined, whereas in the hypothalamus, it is concealed. The value of the pineal system is that it can be stripped bare so that the essentials of control can be studied.

*Dr. Martin:* Dr. Bosches has a question for Dr. Klein: "How do you reconcile the pineal gland studies with the work of Albert Wolfson on gonadal growth of the premigratory bird when it is exposed to light?"

*Dr. Klein:* I think that our hypothesis fits in with this just perfectly. We think pineal function is increased by longer periods of darkness. With more light there is less function and the inhibitory effect of the pineal gland is decreased. Thus, light exposure would allow gonadal growth in the premigratory bird.

# Neuroendocrine Organization of Growth Hormone Regulation

*Joseph B. Martin, Paul Brazeau, Gloria S. Tannenbaum, John O. Willoughby, Jacques Epelbaum, L. Cass Terry, and Dominique Durand

The secretion of growth hormone (GH) is precisely regulated by neural influences, both stimulatory and inhibitory. This control is achieved by at least two hypothalamic hormones: GH releasing factor (GRF), the structure of which is still unknown, and GH release-inhibiting factor (SRIF, GIF, or somatostatin[1]), which has now been isolated and structurally identified. The secretion of these hormones is in turn regulated by monoamines, notably dopamine (DA), norepinephrine (NE), and serotonin, which act at a neural level, perhaps as neurotransmitters.

The purpose of this chapter is to review recent evidence for neural control of GH secretion. Several other reviews on various aspects of GH control have been published during the last few years (8,47,50,65).

## HYPOTHALAMIC REGULATION OF GH SECRETION

The hypothalamus exerts a predominantly stimulatory effect on GH secretion in the mammal. Hypothalamic destruction in man causes growth failure and defective GH responses to hypoglycemia, L-DOPA, and arginine (47,50,65). Sleep-associated GH release may also be blocked by hypothalamic lesions. Small lesions of the median eminence and midline basal hypothalamus block insulin-induced (1) and stress-mediated (9) GH release in the squirrel monkey. Lesions of the hypothalamic ventromedial nucleus (VMN) in young female rats result in growth retardation and a fall in plasma and pituitary GH levels (23,24). Lesions outside the VMN and arcuate (infundibular) nucleus are not associated with GH deficiency (52) unless caused by thyrotropin failure. Growth retardation *without* decreased plasma GH levels results from lesions of the dorsal medial hypothalamic nuclei (4). The defect in these animals appears to be caloric deficiency secondary to hypophagia.

---

* *Present address:* Montreal Neurological Institute, 3801 University Street, Montreal, Quebec H3A 2B4, Canada
[1] The terms SRIF, GIF, and somatostatin are used interchangeably in this chapter.

Electrical stimulation of the medial basal hypothalamus in the pentobarbital-anesthetized rat elicits prompt GH secretion (25,46). Effective stimulation sites in the hypothalamus are strictly confined to the region of the VMN and arcuate; stimulation of the lateral or anterior hypothalamus, the supraoptic or paraventricular nuclei, and the mammillary bodies has no effect on GH. Stimulation of the preoptic area causes inhibition of GH (55), a response which is probably mediated by the preoptic-anterior hypothalamic-somatostatinergic system (see below). Clinical and experimental evidence also suggests that this region exerts inhibitory control on GH secretion (8,47,65).

The rise in plasma GH levels induced by hypothalamic stimulation occurs after termination of the stimulus, as a postinhibitory rebound surge of secretion, the peak of the response occurring 10 to 15 min after termination of the stimulus; longer periods of stimulation cause a further delay in the GH rise (50). Presumably, hypothalamic stimulation elicits somatostatin release resulting in inhibition of GH secretion, followed by a postinhibitory rebound in GH, the latter possibly due to release of GRF. Similar postinhibitory GH surges have been described after hypothalamic stimulation in the sheep (45) and dog (W. Ganong, *personal communication*).

Hypothalamic electrical stimulation experiments provide evidence that a considerable degree of anatomic specificity exists for regulation of individual hormones. Hypothalamic sites selective for TSH, LH, FSH, and GH release have now been identified in the rat.

### Episodic GH Secretion

Plasma levels of GH measured at frequent intervals throughout the day and night show striking variations in man and experimental animals. Individual surges of GH secretion in man may reach plasma levels of 20 to 60 ng/ml, the largest bursts occurring during the first 2 hr of night sleep (22,50, 79). Finkelstein et al. (22) showed that the number and magnitude of spontaneous GH bursts in man are age dependent; transition from early puberty to adolescence was associated with increasing frequency of GH surges, up to eight times in a 24-hr period. This study showed a close correlation between age and total 24-hr secretory rates of GH with highest levels during adolescence, followed by a decline in young adults.

A number of studies have been undertaken to determine the basis of these physiologic variations in GH secretion. Factors such as sleep, exercise, and stress account for certain of these fluctuations, but available data indicate that many of the surges are spontaneous. As reviewed elsewhere in detail (50,65), it does not appear that the abrupt changes in plasma GH levels are caused by or related to variations in glucose, amino acids, or free fatty acids. Physiologic changes in these metabolites have minimal effects on GH secretion. Hyperglycemia only temporarily suppresses daytime surges of GH secretion, and nocturnal GH rises are not affected by either fasting or hyperglycemia, although more frequent GH secretory pulses do occur during fasting (27,56).

The profile of the secretory bursts of GH and their nonsuppressibility by potential metabolic regulators of GH secretion suggest that the surges are the result of primary activation of GH secretion induced by neural mechanisms.

Recent experimental observations support this hypothesis. In the rat, GH secretion is characterized by high-amplitude secretory bursts of GH which may reach levels of 400 to 800 ng/ml (Fig. 1). The bursts show a striking acute profile with a rapid rise in plasma GH, abrupt termination of secretion, and decline in plasma consistent with the known half-life of the hormone. The surges of GH secretion show regular 3- to 4-hr interpeak intervals in

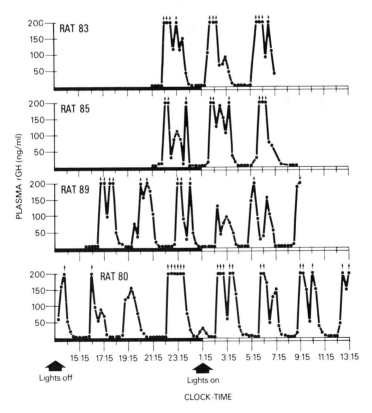

**FIG. 1.** Episodic rGH secretion in 4 male rats sampled for 6 to 24 hr. Surges of rGH rise from undetectable levels (< 1ng/ml) to exceed 200 ng/ml. The timing of the surges indicates entrainment to the light-dark cycle. (From ref. 81 with permission.)

the male rat which are entrained to the light-dark cycle (81). The bursts of GH release occur independently of fluctuations in corticosterone (86), prolactin (88), TSH (88), and glucose and insulin (82), and they are not related to stress.

Pulsatile GH secretion in the rat is significantly reduced by bilateral lesions of the hypothalamic VMN (54) and persists, albeit in somewhat altered form, in rats with hypothalamic islands (88). Since it is not consistently affected by feeding or glucose infusions (82), episodic secretion of GH in the rat, as in man, is not primarily determined by requirements for glucose homeostasis. The timing of the bursts shows no correlation with sleep stages or with activity; we have concluded, therefore, that the secretory GH pattern is regulated by an intrinsic CNS rhythm which is cued by the light-dark cycle but unaffected by sleep-wake rhythms (87).

## HYPOTHALAMIC PEPTIDES AND GH SECRETION

### Peptides that Stimulate GH Secretion

#### GRF

Despite considerable physiologic and biochemical evidence for the existence of a GRF, attempts to identify this substance have thus far been unsuccessful. Crude and semipurified extracts of hypothalamus are effective in stimulating GH release both *in vivo* and *in vitro* (26,44,78,84). All of these studies have used bioassays that do not totally exclude the possibility that the GH releasing activity may have been due in part (or in total) to contamination with vasopressin, TRH, or other peptides which may cause GH release, but which are generally not believed to be the specific GRF. A list of these peptides is given in Table 1.

TABLE 1. *Peptides that Release Growth Hormone*

| | |
|---|---|
| 1. Vasopressin | 7. Neurotensin |
| 2. TRH | 8. Metenkephalin |
| 3. LHRH | 9. $\beta$-Endorphin |
| 4. $\alpha$-MSH | 10. Cholera Enterotoxin |
| 5. Glucagon | 11. Basic Myelin Protein |
| 6. Substance P | |

#### TRH and LHRH

TRH is ineffective in causing GH release in normal human subjects but is remarkably potent in acromegaly (33) and in patients with renal failure (28). TRH stimulates GH release in the cow and in the urethane-anesthetized rat, and is effective *in vitro* in causing GH secretion from incubated bovine, ovine, and rat pituitaries (see 12 for review). TRH-induced GH release in acromegaly is not inhibited by somatostatin. LHRH also causes GH release in certain acromegalic subjects (21), and this response is also not affected by somatostatin.

### Vasopressin

Vasopressin releases GH in man, monkey, and rat. This effect was considered to be of little physiologic significance prior to the observations of Zimmerman et al. (90) that vasopressin (and neurophysin) is present in extremely high concentrations in the pituitary portal blood of surgically stressed rhesus monkeys. This has raised the possibility that vasopressin might act as a "GRF" to stimulate GH release during stress. That vasopressin cannot be the only GRF is demonstrated by the finding that rats with hereditary vasopressin deficiency (Brattleboro strain) have normal plasma GH levels. Furthermore, Brattleboro rats show normal GH release after hypothalamic electrical stimulation (J. B. Martin, *unpublished observations*), and stimulation of the supraoptic nuclei in normal rats has no effect on GH release (50).

### $\alpha$-MSH

Several papers have appeared which suggest that ACTH and $\alpha$-MSH may be effective in releasing GH in man (77). The specificity and significance of this response will require further study.

### Glucagon

Glucagon is well known for its capacity to stimulate release of GH in man (19). The site of action of glucagon is unknown, but its effects are not believed to be secondary to changes in plasma glucose, a point still debated by some investigators. The effect of glucagon is potentiated by $\beta$-receptor blockade, suggesting that it acts on hypothalamic noradrenergic pathways (47).

### Substance P

Substance P, an undecapeptide which was isolated and structurally identified by Chang and Leeman (15), is widely distributed in the central and peripheral nervous system. Immunocytochemical and immunoassay studies have localized the peptide to dorsal root ganglia and the dorsal horn of the spinal cord and to hypothalamus, brainstem, cortex, and other brain regions (30).

Intravenous injection of substance P in doses of 5 or 50 $\mu$g/100 g body weight causes both GH and prolactin release in urethane-anesthetized rats (37). The response is blocked by L-DOPA and by nicotine. Substance P is effective in rats with hypothalamic destruction and after injection into the pituitary portal veins, implying a direct effect at the pituitary level (37). Rivier and co-workers (68) have also shown that substance P releases GH in the urethane-anesthetized rat but could detect no direct effects on monolayer cultures of pituitary. They postulate that the effect is indirect perhaps via the hypothalamus. They further reported that the effect was blocked by diphenhydramine, an antihistaminic (and anticholinergic) agent. These results suggest that histamine or acetylcholine may be implicated in the pathway of action of the peptide.

## Neurotensin

Neurotensin was also identified in hypothalamic extracts by Leeman and Carraway (13). Brown and collaborators (10) have recently reported that the peptide is present in high concentrations in the median eminence and that, like substance P, it acts to induce release of both GH and prolactin when administered to urethane-anesthetized rats. The minimum effective dose for GH release was 10 $\mu$g. Neurotensin was also ineffective in causing GH release *in vitro* and its action was blocked by diphenhydramine. These workers proposed that its stimulatory effects act via extrapituitary mechanisms.

## Opioid Peptides

The identification of the pentapeptide methionine enkephalin by Hughes and co-workers (32) and the subsequent elucidation of several larger peptides, the endorphins (see chapter by Guillemin, *this volume*), the amino acid sequences of which are contained within $\beta$-lipotropin (Fig. 2), have led to extensive investigation of their role as endogenous opiates and of their potential neuroendocrine effects.

Morphine sulfate is itself a potent stimulant of GH release in the anesthetized and unanesthetized rat (39,51), doses as small as 5 $\mu$g/kg being effective in the unanesthetized animal (Fig. 3). The effects are blocked by somatostatin and by naloxone, a specific morphine antagonist. Morphine has no direct effect on the pituitary *in vitro* (51), but large hypothalamic lesions that spare the median eminence do not abolish the response. These findings suggest that morphine acts directly on the median eminence.

Methionine enkephalin causes GH release when given intravenously in large doses (68); it is more effective when administered intraventricularly or intracisternally, and the response can be enhanced by concomitant administration of antiserum to somatostatin (17). The larger opioid peptide $\beta$-endorphin is also effective in releasing GH (68). The GH response to

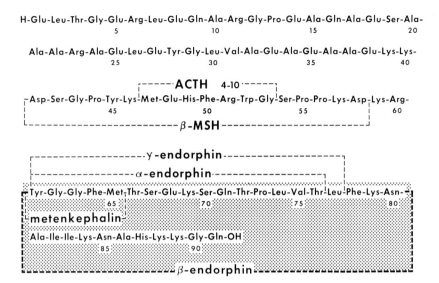

**FIG. 2.** Structures of opioid and other peptides contained within the molecule of β-lipotropin.

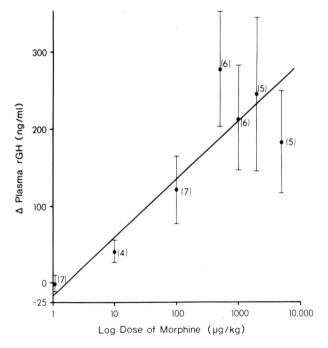

**FIG. 3.** Effects of morphine on plasma rGH levels in male rats. Each point represents the mean ± SE of the increment in rGH 15 min after administration of morphine. A log-dose relationship is evident between 1 and 1,000 µg/kg. In this and subsequent figures, the number of animals in each group is shown in parentheses. (From ref. 51 with permission.)

both methionine enkephalin and β-endorphin is blocked by naloxone. This is in contrast to the stimulatory effects on GH of TRH, substance P, and neurotensin, which in the urethane-anesthetized rat are not affected by the specific morphine antagonist.

## Cholera Enterotoxin GRH

Rappaport and Grant (64) reported a potent GH stimulating substance of microbial origin. This material, a protein of molecular weight 84,000, was isolated from cholera enterotoxin and shown to be active *in vitro* in eliciting GH release by stimulation of cyclic AMP. Its effects were blocked by somatostatin.

GRF activity has also been reported in human lung tumor tissue (3), indi-

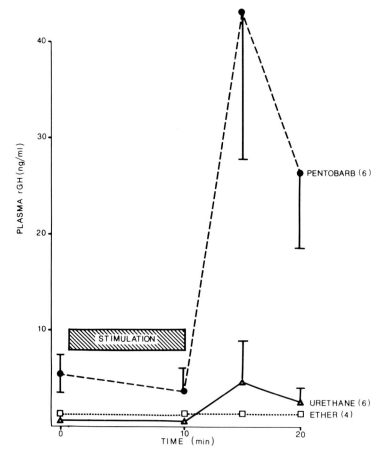

**FIG. 4.** Failure of hypothalamic stimulation to elicit rGH release in urethane- or ether-anesthetized rats compared to effects in pentobarbital-anesthetized rats. Note that rGH rise in latter group occurs after termination of the stimulation.

cating the possibility that the peptide may be synthesized in neoplastic tissue as has been documented for other peptide hormones.

These results are an embarrassment of riches. It is apparent that a host of peptides of various sizes and structures ranging from TRH to cholera enterotoxin are effective in causing GH secretion. In many cases the stimulatory effect is observed only in urethane-anesthetized rats. This anesthetic, which is lethal, results in low basal GH secretion, presumably by suppression of episodic GH release and perhaps by sustained somatostatin release. Although urethane anesthesia provides convenient stable low baseline GH levels, the interpretation of results with the use of this agent are fraught with difficulty. For example, electrical stimulation of the hypothalamus is ineffective in eliciting GH release in urethane-anesthetized rats, in contrast to pentobarbital anesthesia, which produces less suppression of GH secretion (Fig. 4). Moreover, the use of urethane anesthesia has resulted in incorrect conclusions with respect to the function of monoamines in GH regulation (see below). Most workers still believe that the *specific* GRF remains to be identified.

## Peptides that Inhibit GH Secretion

### Somatostatin

The elucidation of the structure of somatostatin (7) and the development of immunological techniques for radioimmunoassay and immunocytochemistry have led to a rapid extension of knowledge with respect to the function, distribution, and significance of this tetradecapeptide. In addition, its discovery has complicated the understanding of GH regulation, necessitating a consideration of both stimulatory and inhibitory influences. Somatostatin has widespread inhibitory effects on GH secretion. It inhibits GH secretion in the rat induced by electrical stimulation (48), pentobarbital (6), morphine (51), and chlorpromazine (36). It prevents GH release to L-DOPA in man (71), baboon (69), and dog (43), and GH secretion stimulated by insulin-hypoglycemia (29,89) and arginine in man. Sleep-associated GH secretion is also prevented by somatostatin (59). Somatostatin has no effect on basal prolactin, FSH, LH, or ACTH secretion but prevents TRH-induced TSH release (72) without affecting TRH-mediated prolactin release. It does not block LHRH-induced LH or FSH release. It is effective in suppressing elevated plasma GH levels in acromegaly and in diabetes.

The onset of effect of somatostatin is rapid and the duration of action brief. After cessation of infusion, GH levels tend to rebound. This has been demonstrated in several species, including man. Evidence of such rapid postinhibitory rebound secretion after termination of somatostatin administration raises the important question of whether such a mechanism might be important in physiologic regulation of episodic GH secretion. To investigate this possibility, we have studied the effects of intermittent somatostatin infusions in unanesthetized rats. In intact rats, repeated administration of somatostatin at hourly intervals caused transient suppression of GH secretion during the infusion. Rebound secretion, however, occurred only when spontaneous surges of secretion were evident. Administration of somatostatin during trough periods of low GH secretion resulted in no evidence of postinhibitory rebound (Fig. 5), the effects of somatostatin being indistinguishable from those of saline administration. Further studies in rats with hypothalamic VMN lesions (which abolish episodic GH release) support these conclusions. Administration of somatostatin to such rats is not followed by GH rebound, excluding a direct pituitary effect (Fig. 6). On the other hand, superfused *isolated* rat pituitaries do show postinhibitory release of GH, indicating that temporary cessation of GH secretion by somatostatin may be followed by enhanced pituitary GH secretion in the absence of hypothalamic control (12,76). In the studies of Stachura (76), increased duration of somatostatin inhibition was followed by progressively larger postinhibitory rebound GH responses.

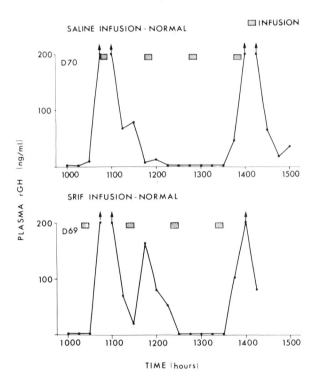

**FIG. 5.** Comparison of effects of saline infusion (*top*) and SRIF infusion (*bottom*) on pulsatile rGH secretion in 2 intact rats. Postinhibitory rebound of rGH after SRIF occurs only at expected time of rGH surge. Infusion of SRIF during trough period fails to produce rebound.

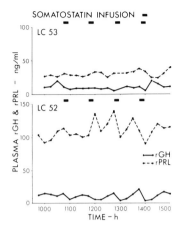

**FIG. 6.** Plasma rGH and rPRL in 2 rats with hypothalamic VMN lesions. Infusions of somatostatin fail to produce postinhibitory rGH rebound secretion. rPRL levels are elevated as occurs with hypothalamic lesions which interfere with normal inhibitory control and are unaffected by somatostatin.

Administration of specific antiserum to somatostatin to unanesthetized rats elevates basal GH secretion and partially reverses stress-induced GH inhibition (83). These findings indicate that somatostatin is likely released during stress in the rat, and support a role of somatostatin in physiologic episodic GH secretion.

After consideration of the available physiologic evidence, it can be concluded that compelling arguments can be made for the necessity of a specific, as yet unidentified GRF. These may be summarized as follows: the predominant hypothalamic control of GH secretion is stimulatory. The effect of destruction and of interruption of hypothalamic-pituitary continuity is a decrease in GH secretion, and this is evident experimentally as a failure of the episodic release mechanism. If hypothalamic control of GH secretion were due to tonic somatostatin secretion with intermittent rebound release, then hypothalamic lesions which abolish episodic GH secretion should result in a rise in GH, as occurs with prolactin. This is not the case. A second line of evidence for a GRF is that stimulation of extrahypothalamic regions such as the hippocampus and amygdala elicits rapid GH secretion that is stimulus bound and cannot be accounted for by postsomatostatin rebound effects (see below). The rise in GH after electrical stimulation of the basolateral amygdala results in a three- to five-fold rise in plasma GH within 5 min of the onset of stimulation, a response that strongly suggests a GRF releasing mechanism. In contrast, stimulation of other regions such as the preoptic area and corticomedial amygdala causes GH inhibition without rebound, implying activation of a somatostatinergic control system. Finally, evidence of the mechanism of pulsatile GH secretion in the rat also requires a GRF drive. Hypothalamic VMN lesions obliterate pulsatile secretion (54) and result in base-line *low* GH levels, whereas hypothalamic islands, which cause somatostatin depletion, result in enhanced GH episodic release (85).

## Nonpituitary Effects of Somatostatin

In addition to its effects on GH and TSH, somatostatin inhibits insulin and glucagon secretion by direct effects on pancreatic islets, inhibits gastrin secretion in normal human subjects and in patients with excessive gastrin secretion, inhibits ACTH in patients with ACTH-secreting pituitary adenomas, and inhibits secretin, renin, calcitonin, and parathormone secretion under certain conditions (see 50 for review). The physiologic significance of these effects will require further study.

## Brain Distribution of Somatostatin

Bioassay and radioimmunoassay studies show that somatostatin is localized to a number of brain regions outside the hypothalamus. Highest concentrations of extrahypothalamic somatostatin have been reported in the preoptic area, amygdala, brainstem, and spinal cord (11,20,61), but it is also present in cerebral cortex, thalamus, cerebellum, pineal gland (62), and peripheral nerve (30). Subcellular distribution studies indicate that somatostatin in the hypothalamus, preoptic area, and amygdala is localized predominantly in synaptosomes or nerve terminals (20) (Figs. 7 through 9). This localization is consistent with a function of the peptide as a synaptic neurotransmitter or modulator.

Determination of the site of origin of somatostatin in various brain regions is important to the understanding of its function. Electrophysiologic studies have provided evidence that tuberoinfundibular (hypophysiotrophic) cells have branching axon collaterals, one of which terminates on the portal vessels while the others end in various other regions of the hypothalamus, the preoptic area, or the limbic system (see chapter by Renaud, *this volume*). These observations permit speculation that peptidergic neurons, like

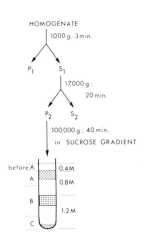

**FIG. 7.** Procedure for purification of synaptosomes. Band B on sucrose gradient was synaptosome-enriched when examined by electron microscopy.

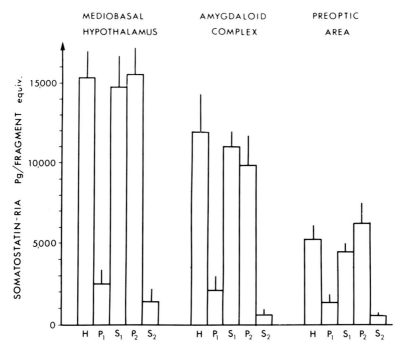

**FIG. 8.** Recovery of radioimmunoassayable SRIF during preparation of synaptosomes. The $P_2$ component contains the bulk of SRIF in tissue from the hypothalamus, the amygdala, and the preoptic area. Vertical lines at top of each bar indicate SE of 6 or more experiments.

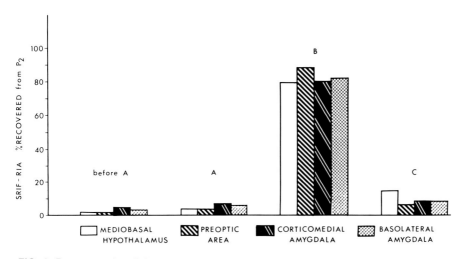

**FIG. 9.** Recovery of radioimmunoassayable SRIF on sucrose gradient. The bulk of the activity is in Band B, corresponding to synaptosomes.

dopaminergic, noradrenergic, and serotonergic neurons, may give rise to complex neural pathways, which terminate in widespread areas of the neural axis. Moreover, recent anatomic studies of the efferent connections of the hypothalamic VMN in the rat have shown terminations in a variety of extrahypothalamic brain regions including the preoptic area, thalamus, amygdala, septum, and brainstem (70).

To investigate the possibility that somatostatinergic pathways might arise from the medial basal hypothalamus or preoptic area, we made selective lesions in these regions and measured the content of somatostatin by radioimmunoassay in the amygdala and cortex. Bilateral anterior hypothalamic-periventricular lesions reduced somatostatin content in the median eminence to less than 10% of control levels (Fig. 10). However, such lesions had no effect on somatostatin concentration in the preoptic area, amygdala, or cortex (Fig. 10). Combined preoptic and VMN lesions also had no effect on amygdaloid or cortex somatostatin content. Hypothalamic islands that preserved the ventromedial-arcuate region attached to the pituitary stalk result in a reduction in somatostatin content in the ME and the preoptic area but have no effect on levels in the amygdala and cortex (Fig. 11). These findings argue against a somatostatinergic pathway arising in the mediobasal hypothalamus which is distributed to extrahypothalamic regions.

Immunohistochemical studies indicate that the primary localization of cell perikarya of the hypothalamic-median eminence somatostatinergic pathway is the anterior periventricular nucleus (2) (see chapter by Hökfelt, *this volume*). In addition, such studies have shown neuronal cell bodies that are

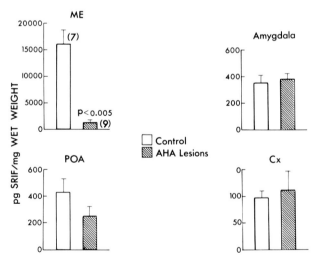

**FIG. 10.** Effect of anterior hypothalamic-periventricular lesions on SRIF content in median eminence (ME), amygdala, preoptic area (POA), and cortex (Cx). Lesions extended into the anterior hypothalamic area and partially destroyed the anterior periventricular region.

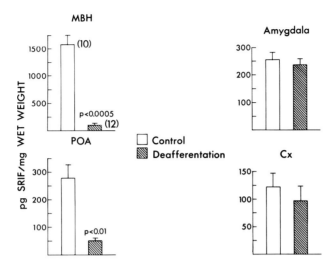

**FIG. 11.** Effect of complete hypothalamic deafferentation on SRIF concentration in the medial basal hypothalamus (MBH), amygdala, preoptic area (POA), and cortex (Cx).

immunoreactive for somatostatin in the preoptic area, amygdala, and dorsal root ganglia. These observations indicate that somatostatin, like substance P, neurotensin, and the opioid peptides, is distributed throughout various neuronal pathways in the central nervous system.

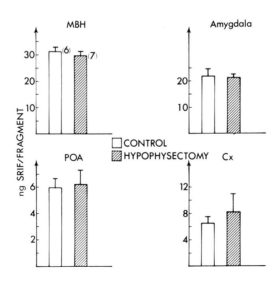

**FIG. 12.** Lack of effect of hypophysectomy on content of SRIF in various brain regions.

There is little information presently available to indicate the precise role of somatostatin in neuronal function. Brain levels of somatostatin are not affected by hypophysectomy (Fig. 12). Somatostatin depresses electrical activity in neurons (66) and inhibits calcium release from synaptosomes (80). These findings, together with a predominant synaptosome localization of the peptide, implicate a function as a synaptic transmitter or modulator.

Somatostatin is detectable in the cerebrospinal fluid (CSF). In the report of Patel et al. (60), elevated levels were described in patients with certain intracranial tumors (pinealoma, medulloblastoma) and in meningitis, indicating that in certain diseases somatostatin leaks into the CSF.

## SUPRAHYPOTHALAMIC REGULATION OF GH SECRETION

Physiologic studies demonstrating the effects of stress and sleep on GH secretion point to a probable role of extrahypothalamic structures in GH control. The isolated hypothalamus appears capable of maintaining near normal basal secretion of GH, including pulsatile release (88), but selective hypothalamic cuts indicate that disconnection of specific inputs to the medial basal hypothalamus can have differential excitatory or inhibitory effects on GH secretion, at least in the rat.

The anatomic pathways connecting the limbic system to the medial basal hypothalamus, in general, and the VMN-arcuate complex in particular, are well defined. Direct monosynaptic inputs from the hippocampus reach the arcuate nucleus via the medial corticohypothalamic tract (63). The amygdala has a large monosynaptic connection from its corticomedial subdivision which reaches the VMN via the stria terminalis (18). The connection(s) of the basolateral amygdala to the medial basal hypothalamus has not been as clearly defined. Electrophysiologic studies in the cat (58) indicated an inhibitory pathway from the corticomedial amygdala via the stria terminalis to the VMN and a complementary excitatory pathway from the basolateral amygdaloid complex to the hypothalamic VMN. However, studies in the rat have not shown a similar clear distinction (see chapter by Renaud, *this volume*).

Electrical stimulation of the hippocampus causes GH release, whereas stimulation of the amygdala can elicit either a rise or a fall in plasma GH depending on the precise site stimulated (46,53). Stimulation of the basolateral amygdala causes prompt GH release which is entrained to the stimulus, plasma levels increasing within 5 min of the onset of stimulation and declining after its termination (49). This response is blocked by bilateral hypothalamic VMN lesions indicating that the GH release effects are mediated through the medial basal hypothalamus. Stimulation of the corticomedial amygdala causes a fall in plasma GH levels comparable to that observed with preoptic stimulation. Since an important component of the efferent system of the corticomedial amygdala travels in the stria terminalis

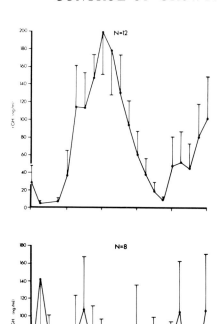

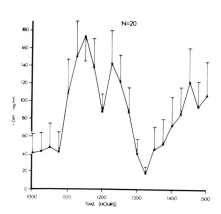

**FIG. 13.** Effects of anterior hypothalamic deafferentation on light-entrainment of pulsatile rGH secretion (*middle*). Posterior deafferentation (*top*) has no effect, compared to normal rats (*bottom*). (From ref. 88 with permission.)

to terminate in the septum and preoptic area (18), it remains to be shown whether the corticomedial inhibitory response is mediated via connections to these areas. It is significant that coronal cuts through the anterior hypothalamus cause an increase in growth in the rat (57) and an elevation in plasma GH levels (16). Such cuts interrupt light-entrainment of episodic

GH secretion (Fig. 13). Complete medial basal hypothalamic deafferentation in which the VMN and arcuate are isolated from the overlying brain results in an elevation in plasma GH levels, an effect that is due to enhanced episodic secretion (88). These observations suggest that extrahypothalamic inhibitory inputs to medial basal hypothalamus are important to maintain tonic inhibition of GH secretion in the rat. Further support for this is derived from experiments in which destruction of the preoptic area was shown to block stress-induced GH inhibition in this species (67).

## MONOAMINES AND GH REGULATION

Growth hormone is regulated by a dual system of hypothalamic hypophysiotrophic hormones, one inhibitory and the other excitatory. Release of these hypothalamic hormones is, in turn, regulated by monoaminergic neurons. The finding that many pharmacologic stimuli affect GH secretion in man has led to extensive investigations of the role of monoamines, in particular norepinephrine, dopamine, and serotonin, in normal and abnormal GH secretory states. Additional studies have also examined the effects of melatonin and of specific receptor agonists and antagonists on GH secretion. Such studies have led to the development of important insights into basic control mechanisms and to the discovery of new therapeutic approaches to disorders of GH secretion.

Norepinephrine, dopamine, and serotonin each have a stimulatory effect on GH secretion in the primate. Oral administration of L-DOPA, the precursor of both NE and DA, causes release of GH in man, monkey, and dog (50). Evidence that the L-DOPA effect is mediated through its conversion to NE is derived from experiments showing that L-DOPA-induced GH release is blocked by phentolamine, an $\alpha$-receptor blocking agent.

Administration of subemetic doses of apomorphine, a centrally active dopamine receptor stimulating agent, also releases GH in man, indicating an additional dopaminergic control mechanism independent of NE conversion (40). The effects of apomorphine are blocked by pimozide, a dopamine antagonist, but not by methysergide, a serotonin antagonist (S. Lal, *personal communication*). GH release induced by L-DOPA and apomorphine is attenuated by prior glucose administration, indicating that glucoreceptor stimulation can partially override catecholaminergic stimuli for GH release.

Additional evidence for a role of $\alpha$-adrenergic receptors in GH control is provided by the report of Lal and co-workers (41) that clonidine, a centrally active $\alpha$ agonist, is effective in stimulating GH release in man.

Indeed, most of the stimuli that cause GH release appear to act via central $\alpha$-adrenergic receptors (47). In man, GH release induced by insulin hypoglycemia, arginine, exercise, vasopressin, L-DOPA, and certain stresses is prevented by phentolamine (50). On the other hand, propranolol, a $\beta$-adrenergic blocker, enhances GH release induced by glucagon, vasopressin, and L-DOPA. The $\beta$-agonist isoproterenol is reported to inhibit GH release. These studies indicate that GH secretion is facilitated by both dopaminergic and noradrenergic $\alpha$-receptor stimulation and inhibited by noradrenergic $\beta$-receptor stimulation. Pharmacologic blockade of GH release by drugs such as chlorpromazine, pimozide, and haloperidol is thought to occur at a hypothalamic level, probably by competitive blockade of dopaminergic receptor sites.

Serotonin also appears to be involved in GH release. Oral administration of L-tryptophan or 5-hydroxytryptophan (5HTP) in man and monkey causes GH release, although the response is not great (14,34). The release of GH to hypoglycemia is blocked by the serotonin antagonists methysergide and cyproheptadine (5). Melatonin is reported both to facilitate GH release and to block release to insulin hypoglycemia (74,75). These apparent conflicting effects of melatonin were explained by differences in the time course of action of the substance.

Physiologic GH release, as in sleep, may, however, be regulated by another mechanism. Neither $\alpha$- nor $\beta$-adrenergic nor dopaminergic receptor blockade with chlorpromazine has any effect on nocturnal sleep-associated GH release. Block of serotonin receptors with methysergide has been reported to result in an increase in GH release during sleep, whereas other reports have shown either opposite effects or no effect. Further studies are required to resolve this issue.

Certain species differences seem to exist with respect to GH responses to pharmacologic stimuli. In the rhesus monkey, intravenous clonidine and 5-OH-tryptophan are potent stimulants for GH release, whereas apomorphine is ineffective except at doses that induce vomiting (and, therefore, probably acting as a nonspecific stress effect) (14). In the studies by Chambers and Brown (14), administration of dihydroxyphenylserine, a precursor of norepinephrine, also was effective in releasing GH. Similarly, in the dog, systemic administration of L-DOPA and clonidine causes GH release whereas apomorphine is ineffective. These findings suggest that species differences exist and that there may not be an available animal model that precisely duplicates GH control in man.

Our studies in the rat have attempted to define the role of biogenic amines in episodic or pulsatile GH secretion. The spontaneous surges of GH secretion are abolished by pretreatment with $\alpha$-methyl-$p$-tyrosine 250 mg/kg i.p. (Fig. 14), an agent that inhibits synthesis of dopamine and norepinephrine. Growth hormone secretion can be restored in $\alpha$-MT-treated rats by administration of clonidine (Fig. 15) but not by apomorphine. Furthermore, pimozide and butaclamol (dopamine antagonists) have minimal inhibitory effects on pulsatile GH release, whereas phenoxybenzamine causes complete suppression of secretion. These findings indicate that the $\alpha$-adrenergic mechanisms have a facilitatory effect on episodic GH secretion in this species, as is probably also true in man.

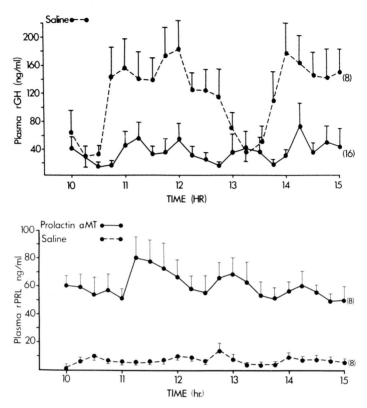

**FIG. 14.** Suppression of rGH (*upper part*) and elevation of rPRL (*lower part*) in rats treated with α-methyl-p-tyrosine (α-MT). α-MT-treated group is shown by solid line, saline-treated group by dashed line.

There is also evidence of a serotonergic facilitatory mechanism for episodic GH secretion in the rat. Growth hormone secretion is suppressed by *p*-chlorophenylalanine, an inhibitor of serotonin synthesis, and by methysergide, a serotonin antagonist. On the other hand, GH secretion can be stimulated in the rat by administration of 5-OH tryptophan (73).

It should be emphasized that systemic or intraventricular administration of these pharmacologic agents provides little direct evidence of their physiologic role or site of action in GH control. It is probable that biogenic amine agonists and antagonists act at a hypothalamic level to modulate GRF or somatostatin release. Intrahypothalamic infusions of NE are effective in releasing GH in the baboon (47). Moreover, cerebral intraventricular or hypothalamic injection of systemically ineffective doses of phentolamine prevent hypoglycemic-induced GH release in this species. There is no convincing experimental evidence to indicate that catecholamines or serotonin acts at the pituitary somatotrope level, either directly or in synergism with hypothalamic hormones. In electrical stimulation experiments in the rat,

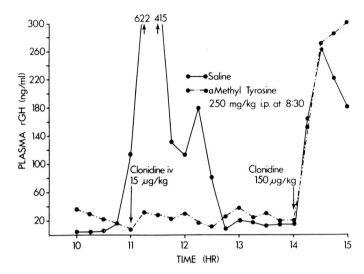

**FIG. 15.** Stimulatory effect of clonidine on suppressed rGH levels in α-MT-treated rat. (From ref. 18a, with permission.)

we have shown that GH release after VMN stimulation is not blocked by agents which alter DA, NE, and serotonin content in rat brain. On the other hand, GH release induced by amygdaloid and hippocampal stimulation is prevented by α-methyl-p-tyrosine and by p-chlorophenylalanine, results consistent with the interpretation that monoamines may function as neurotransmitters in the relay of these responses from higher neural centers to hypothalamic GRF or somatostatinergic neurons. Amygdaloid-induced release of GH was prevented by α- but not by β-adrenergic blockade (Fig. 16).

These observations provide a degree of uniformity of GH regulation that has otherwise been lacking in studies using anesthetized rats. For example, intravenous administration of pharmacological agents to urethane-anesthetized rats suggested that β-receptors are important in facilitation of GH release (38). Intravenous chlorpromazine, a dopamine antagonist, elicited GH release, and this effect was blocked by propranolol. Furthermore, L-DOPA caused GH suppression under urethane and blocked GH release to TRH and substance P. These observations are valid probably only under the effects of urethane, which may be accompanied by excess somatostatin release.

Alterations in monoaminergic control may play a role in certain disease states. In acromegaly there is usually a "paradoxical" suppression of elevated GH levels by L-DOPA, apomorphine (42), and bromoergocryptine, a long-acting dopaminergic receptor stimulating agent that is currently widely used for its inhibitory effects on prolactin secretion (42). At least in

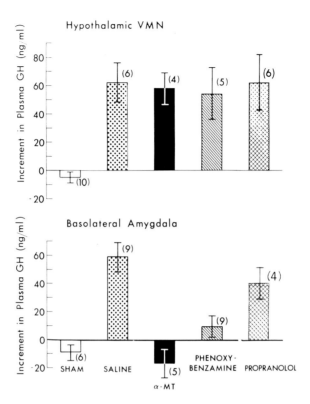

**FIG. 16.** Comparison of pharmacologic pretreatment on rGH release induced by hypothalamic versus basolateral amygdala electrical stimulation. $\alpha$-MT, phenoxybenzamine, and propranolol have no effect on hypothalamic-induced GH increment (*top*), whereas $\alpha$-MT and phenoxybenzamine (but not propranolol) partially block the GH rise after basolateral amygdala stimulation.

some cases, this abnormality in GH regulation seems to be a secondary and not a primary disorder. In one case successful total removal of a GH-secreting adenoma resulted in a return to normal of L-DOPA-, apomorphine-, and TRH-stimulated GH release (31). In one case of the maternal deprivation syndrome, $\beta$-blockade with propranolol resulted in restoration of the plasma GH responses to hypoglycemia (35). Excessive $\beta$-receptor stimulation might have induced chronic suppression of GH in this subject.

## CONCLUSIONS

An analysis of the role of the brain in GH secretion is useful because it provides a tool by which to examine brain mechanisms. Abnormalities of GH secretion commonly result from disturbances in hypothalamic control. Aberrant GH responses to pharmacologic stimuli in certain diseases may provide clues to more widespread disorders of brain function. The tools for complete analysis of such a regulatory system must await the discovery of GRF and the development of methods to assay levels of the hypothalamic peptides in blood and CSF. New approaches to the study of neurologic disease should include the possibility that low molecular weight peptides may serve as markers of neurologic dysfunction.

## ACKNOWLEDGMENTS

The authors thank Dr. Albert Parlow and the NIAMDD for generous gifts of radioimmunoassay materials for rat GH and prolactin. The gift of somatostatin from Ayerst Laboratories, Montreal, is gratefully acknowledged. Ms. J. Audet, A. Saunders, and W. Gurd provided technical assistance, and Mrs. A. D'Amato typed the manuscript. G.S.T., J.O.W., and L.C.T. were supported by fellowships, P.B. by a scholarship, and J.B.M. by an associateship from the Medical Research Council of Canada. J.E. was supported by a Killam fellowship and D.D. by a fellowship from the Republic of France. MRC grants to P.B. and J.B.M. provided funds to carry out this research.

## REFERENCES

1. Abrams, R. L., Parker, M. L., Blanco, S., Reichlin, S., and Daughaday, W. H. (1966): Hypothalamic regulation of growth hormone secretion. *Endocrinology*, 78:605–613.
2. Alpert, L. C., Brawer, J. R., Patel, Y. C., and Reichlin, S. (1976): Somatostatinergic neurons in anterior hypothalamus: Immunohistochemical localization. *Endocrinology*, 98:255–258.
3. Beck, C., Larkins, R. G., Martin, T. J., and Burger, H. G. (1973): Stimulation of growth hormone release from superfused rat pituitary by extracts of hypothalamus and of human lung tumours. *J. Endocrinol.* 59:325–333.
4. Bernardis, L. L., and Goldman, J. K. (1972): Growth and metabolic changes in weanling rats with lesions in the dorsomedial hypothalamic nuclei. *Exp. Brain Res.*, 15:424–429.
5. Bivens, C. H., Lebovitz, H. E., and Feldman, J. M. (1973): Inhibition of hypoglycemia-induced growth hormone secretion by the serotonin antagonists cryproheptadine and methysergide. *N. Engl. J. Med.*, 289:236–239.
6. Brazeau, P., Rivier, J., Vale, W., and Guillemin, R. (1974): Inhibition of growth hormone secretion in the rat by synthetic somatostatin. *Endocrinology*, 94:184–187.
7. Brazeau, P., Vale, W., Burgus, R., Ling, N., Butcher, M., Rivier, J., and Guillemin, R. (1973): Hypothalamic polypeptide that inhibits the secretion of immunoreactive pituitary growth hormone. *Science*, 179:77–79.
8. Brown, G. M., and Reichlin, S. (1972): Psychological and neural regulation of growth hormone secretion. *Psychosom. Med.*, 34:45–61.
9. Brown, G. M., Schlach, D. S., and Reichlin, S. (1971): Hypothalamic mediation of growth hormone and adrenal stress response in the squirrel monkey. *Endocrinology*, 89:694–703.
10. Brown, M., Lazarus, L., Ling, N., Rivier, J., Kobayashi, R., and Vale, W. (1977): Central nervous system distribution of neurotensin (NT), NT binding sites and NT degradative activity *Clin. Res. (in press)*.
11. Brownstein, M., Arimura, A., Sato, H., Schally, A. V., and Kizer, J. S. (1975): The regional distribution of somatostatin in the rat brain. *Endocrinology*, 96:1456–1461.
12. Carlson, H. E., Mariz, I. K., and Daughaday, W. H. (1974): Thyrotropin-releasing hormone stimulation and somatostatin inhibition of growth hormone secretion. *Endocrinology*, 94:1709–1913.
13. Carraway, R., and Leeman, S. E. (1975): The amino acid sequence of a hypothalamic peptide neurotensin. *J. Biol. Chem.*, 250:1907–1911.

14. Chambers, J. W., and Brown, G. M. (1976): Neurotransmitter regulation of growth hormone and ACTH in the rhesus monkey: Effects of biogenic amines. *Endocrinology,* 98:420–428.
15. Chang, M. M., and Leeman, S. E. (1970): Isolation of a sialogogic peptide from bovine hypothalamic tissue and its characterization as substance P. *J. Biol. Chem.,* 245:4784–4790.
16. Collu, R., Jéquier, J.-C., Letarte, J., Leboeuf, G., and Ducharme, J. R. (1973): Effect of stress and hypothalamic deafferentation on the secretion of growth hormone in the rat. *Neuroendocrinology,* 11:183–190.
17. Cusan, L., Dupont, A., Garon, M., and Coy, D. H. (1976): Stimulatory effect of met-enkephalin (D-Ala$^2$) met-enkephalin and $\beta$-endorphine on growth hormone secretion. *Clin. Res.,* 24:656A.
18. de Olmos, J. S., and Ingram, W. R. (1972): The projection field of the stria terminalis in the rat brain: An experimental study. *J. Comp. Neurol.,* 146:303–334.
18a. Durand, D., Martin, J. B., and Brazeau, P. (1977): Evidence for a role of $\alpha$-adrenergic mechanisms in regulation of episodic growth hormone secretion in the rat. *Endocrinology.*
19. Eddy, R. L., Gilliland, P. F., Ibarra, J. D., McMurray, J. F., Jr., and Thompson, J. Q. (1974): Human growth hormone release–comparison of provocative test procedures. *Am. J. Med.,* 56:179–185.
20. Epelbaum, J., Brazeau, P., Tsang, D., Brawer, J., and Martin, J. B. (1977): Subcellular distribution of radioimmunoassayable somatostatin in rat brain. *Brain Res.,* 126:309–323.
21. Faglia, G., Beck-Peccoz, P., Travaglini, P., Paracchi, A., Spada, A., and Lewin, A. (1973): Elevations in plasma growth hormone concentration after luteinizing hormone-releasing hormone (LRH) in patients with active acromegaly. *J. Clin. Endocrinol. Metab.,* 37:336–340.
22. Finklestein, J. W., Roffwarg, H. P., Boyar, R. M., Kream, J., and Hellman, L. (1972): Aged-related changes in the twenty-four hour spontaneous secretion of growth hormone. *J. Clin. Endocrinol. Metab.,* 35:665–670.
23. Frohman, L. A., and Bernardis, L. L. (1968): Growth hormone and insulin levels in weanling rats with ventromedial hypothalamic lesions. *Endocrinology,* 82:1125–1132.
24. Frohman, L. A., Bernardis, L. L., Burck, L., Maran, J. W., and Dhariwal, A. P. S. (1972): Hypothalamic control of growth hormone secretion in the rat. In: *Growth and Growth Hormone,* edited by A. Pecile and E. E. Muller, pp. 271–282. Excerpta Medica, Amsterdam.
25. Frohman, L. A., Bernardis, L. L., and Kant, K. (1968): Hypothalamic stimulation of growth hormone secretion. *Science,* 162:580–582.
26. Frohman, L. A., Maran, J. W., and Dhariwal, A. P. S. (1971): Plasma growth hormone responses to intrapituitary injections of GH RF in the rat. *Endocrinology,* 88:1483–1488.
27. Glick, S. M., and Goldsmith, S. (1968): The physiology of growth hormone secretion. In: *Growth Hormone,* edited by A. Pecile and E. E. Müller, pp. 84–88. Excerpta Medica, Amsterdam.
28. Gonzalez-Barcena, D., Kastin, A. J., Schalch, D. S., Torres-Zamora, M., Perez-Pasten, E., Kato, A., and Schally, A. V. (1973): Responses to thyrotropin-releasing hormone in patients with renal failure and after infusion in normal men. *J. Clin. Endocrinol. Metab.,* 36:117–120.
29. Hall, R., Besser, G. M., Schally, A. V., Coy, D. H., Evered, D., Goldie, D. J., Kastin, A. J., McNeilly, A. S., Mortimer, C. H., Phenekos, C., Tunbridge, W. M. G., and Weightman, D. (1973): Action of growth-hormone-release inhibitory hormone in healthy men and in acromegaly. *Lancet,* 2:581–586.
30. Hökfelt, T., Elde, R., Johansson, O., Luft, R., Nilsson, G., and Arimura, A. (1976): Immunohistochemical evidence for separate populations of somatostatin-containing and substance P-containing primary afferent neurons in the rat. *Neuroscience,* 1:131–136.
31. Hoyte, K., and Martin, J. B. (1975): Recovery from paradoxical GH responses in acromegaly after transsphenoidal selective adenonectomy. *J. Clin. Endocrinol. Metab.,* 41:656–659.
32. Hughes, J., Smith, T., Kosterlitz, H. W., Fothergill, L. A., Morgan, B. A., and Morris, H. R. (1975): Identification of two related pentapeptides from the brain with potent opiate agonist activity. *Nature,* 258:577–579.

33. Irie, M., and Tsushima, T. (1972): Increase of serum growth hormone concentration following thyrotropin-releasing hormone injection in patients with acromegaly or gigantism. *J. Clin. Endocrinol. Metab.*, 35:97-100.
34. Imura, H., Nakai, Y., and Yoshimi, T. (1973): Effect of 5-hydroxytryptophan (5-HTP) on growth hormone and ACTH release in man. *J. Clin. Endocrinol. Metab.*, 36:204-206.
35. Imura, H., Yoshimi, T., and Ikekubo, K. (1971): Growth hormone secretion in a patient with deprivation dwarfism. *Endocrinol. Jpn.*, 18:301-304.
36. Kato, Y., Chihara, K., Ohgo, S., and Imura, H. (1974): Effects of hypothalamic surgery and somatostatin on chlorpromazine-induced growth hormone release in rats. *Endocrinology*, 95:1608-1613.
37. Kato, Y., Chihara, K., Ohgo, S., Iwasaki, Y., Abe, Y., and Imura, H. (1976): Growth hormone and prolactin release by substance P in rats. *Life Sci.*, 19:441-446.
38. Kato, Y., Dupre, J., and Beck, J. C. (1973): Plasma growth hormone in the anesthetized rat: Effects of dibutyryl cyclic AMP, prostaglandin E, adrenergic agents, vasopressin, chlorpromazine, amphetamine and L-dopa. *Endocrinology*, 93:135-145.
39. Kokka, N., and George, R. (1974): Effects of narcotic analgesics, anesthetic and hypothalamic lesions on growth hormone and adrenocorticotropic hormone secretion in rats. In: *Narcotics and the Hypothalamus*, edited by E. Zimmerman and R. George, pp. 137-157. Raven Press, New York.
40. Lal, S., de la Vega, C., Sourkes, T. L., and Friesen, H. G. (1973): Effect of apomorphine on growth hormone, prolactin, luteinizing hormone and follicle-stimulating hormone levels in human serum. *J. Clin. Endocrinol. Metab.*, 37:719-724.
41. Lal, S., Tolis, G., Martin, J. B., Brown, G. M., and Guyda, H. (1975): Effects of clonidine on growth hormone, prolactin, luteinizing hormone, follicle-stimulating hormone and thyroid-stimulating hormone in the serum of normal men. *J. Clin. Endocrinol. Metab.*, 41:703-708.
42. Liuzzi, A., Chiodini, P. G., Botalla, L., Silvestrini, F., and Müller, E. E. (1974): Growth hormone (GH)-releasing activity of TRH and GH-lowering effect of dopaminergic drugs in acromegaly: Homogeneity in the two responses. *J. Clin. Endocrinol. Metab.*, 39:871-876.
43. Lovinger, R., Boryczka, A. T., Schackelford, R., Kaplan, S. L., Ganong, W. F., and Grumbach, M. M. (1974): Effect of synthetic somatotropin release inhibiting factor on the increase in plasma growth hormone elicited by L-Dopa in the dog. *Endocrinology*, 95:943-946.
44. Malacara, J. M., Valverde, R., and Reichlin, S. (1972): Elevation of plasma radioimmunoassayable growth hormone in the rat induced by porcine hypothalamic extract. *Endocrinology*, 91:1189-1198.
45. Malven, P. V. (1974): Altered release of GH, LH and prolactin (PRL) induced by electrical stimulation of the median eminence (ME) in unanesthetized sheep. Progr. 56th Annual Meeting Endocrine Society, p. A127.
46. Martin, J. B. (1972): Plasma growth hormone (GH) response to hypothalamic or extrahypothalamic electrical stimulation. *Endocrinology*, 91:107-115.
47. Martin, J. B. (1973): Neural regulation of growth hormone secretion. Medical Progress Report. *N. Engl. J. Med.*, 288:1384-1393.
48. Martin, J. B. (1974): Inhibitory effect of somatostatin (SRIF) on the release of growth hormone (GH) induced in the rat by electrical stimulation. *Endocrinology*, 94:497-503.
49. Martin, J. B. (1974): The role of hypothalamic and extrahypothalamic structures in the control of GH secretion. In: *Advances in Human Growth Hormone Research*, edited by S. Raiti, pp. 223-249. NIH Publication No. 74-612, Washington, D.C.
50. Martin, J. B. (1976): Brain regulation of growth hormone secretion. In: *Frontiers in Neuroendocrinology, Vol. 4*, edited by L. Martini and W. F. Ganong, pp. 129-168. Raven Press, New York.
51. Martin, J. B., Audet, J., and Saunders, A. (1975): Effect of somatostatin and hypothalamic ventromedial lesions on GH release induced by morphine. *Endocrinology*, 96:839-847.
52. Martin, J. B., and Jackson, I. (1974): Neural regulation of TSH and GH secretion. In: *Anatomical Neuroendocrinology*, edited by W. E. Stumpt and L. D. Grant, pp. 343-353. S. Karger, Basel.
53. Martin, J. B., Kontor, J., and Mead, P. (1973): Plasma GH responses to hypothalamic,

hippocampal and amygdaloid electrical stimulation: Effects of variation in stimulus parameters and treatment with α-methyl-ρ-tyrosine (α-MT). *Endocrinology,* 92:1354–1361.
54. Martin, J. B., Renaud, L. P., and Brazeau, P. (1974): Pulsatile growth hormone secretion: Suppression by hypothalamic ventromedial lesions and by long-acting somatostatin. *Science,* 186:538–540.
55. Martin, J. B., Tannenbaum, G., Willoughby, J. O., Renaud, L. P., and Brazeau, P. (1975): Functions of the central nervous system in regulation of pituitary GH secretion. In: *Hypothalamic Hormones; chemistry, physiology, pharmacology, and clinical uses,* edited by M. Motta, P. G. Crosignani, and L. Martini, pp. 217–236. Academic Press, New York.
56. Merimee, T. J., and Pulkkinen, A. J. (1976): Body composition and the metabolic responses of fasting. Progr. 58th Annual Meeting Endocrine Society, p. 296.
57. Mitchell, J. A., Hutchins, M., Schindler, W. J., and Critchlow, V. (1973): Increases in plasma growth hormone concentration and naso-anal length in rats following isolation of the medial basal hypothalamus. *Neuroendocrinology,* 12:161–173.
58. Murphy, J. T., and Renaud, L. P. (1969): Mechanism of inhibition in the ventromedial nucleus of the hypothalamus. *J. Neurophysiol.,* 32:85–102.
59. Parker, D. C., Rossman, L. G., Siler, T. M., Rivier, J., Yen, S. S. C., and Guillemin, R. (1974): Inhibition of the sleep-related peak in physiologic human growth hormone release by somatostatin. *J. Clin. Endocrinol. Metab.,* 38:496–499.
60. Patel, Y. C., Rao, K., and Reichlin, S. (1977): Somatostatin in human cerebrospinal fluid. *N. Engl. J. Med. (in press).*
61. Patel, Y. C., Weir, G. C., and Reichlin, S. (1975): Anatomic distribution of somatostatin (SRIF) in brain and pancreatic islets as studied by radioimmunoassay (RIA). Progr. 57th Annual Meeting Endocrine Society, p. A127.
62. Pelletier, G., Labrie, F., Arimura, A., and Schally, A. V. (1974): Electron microscopic immunohistochemical localization of growth hormone-release inhibiting hormone (somatostatin) in the rat median eminence. *Am. J. Anat.,* 140:445–450.
63. Raisman, G., and Field, D. M. (1971): Anatomical considerations relevant to the interpretation of neuroendocrine experiments. In: *Frontiers in Neuroendocrinology,* edited by L. Martini and W. F. Ganong, pp. 1–44. Oxford University Press, London.
64. Rappaport, R. S., and Grant, N. H. (1974): Growth hormone releasing factor of microbial origin. *Nature,* 248:73–75.
65. Reichlin, S. (1974): Regulation of somatotrophic hormone secretion. In: *Handbook of Physiology-Endocrinology, Vol. IV, Part 2,* edited by E. Knobil and W. H. Sawyer, pp. 405–447. Williams & Wilkins Co., Baltimore.
66. Renaud, L. P., Brazeau, P., and Martin, J. B. (1975): Depressant action of TRH, LH-RH and somatostatin on the activity of central neurons. *Nature,* 255:233–235.
67. Rice, R. W., and Critchlow, V. (1976): Extrahypothalamic control of stress-induced inhibition of growth hormone secretion in the rat. *Endocrinology,* 99:970–976.
68. Rivier, C., Brown, M., and Vale, W. (1977): Effect of neurotensin, substance P and morphine sulfate on the secretion of prolactin and growth hormone in the rat. *Endocrinology,* 100:529–533.
69. Ruch, W., Koerker, D. J., Carino, M., Johnson, S. D., Webster, B. R., Ensinck, J. W., Goodner, C. J., and Gale, C. C. (1974): Studies on somatostatin (somatotropin release inhibiting factor) in conscious baboons. In: *Advances in Human Growth Hormone Research,* edited by S. Raiti, pp. 271–293. DHEW Publication No. (NIH) 74-612, Washington, D.C.
70. Saper, C. B., Swanson, L. W., and Cowan, W. M. (1976): The efferent connections of the ventromedial nucleus of the hypothalamus of the rat. *J. Comp. Neurol.,* 169:409–442.
71. Siler, T. M., VanderBerg, G., and Yen, S. S. C. (1973): Inhibition of growth hormone release in humans by somatostatin. *J. Clin. Endocrinol. Metab.,* 37:632–634.
72. Siler, T. M., Yen, S. S. C., Vale, W., and Guillemin, R. (1974): Inhibition by somatostatin on the release of TSH induced in man by thyrotropin-releasing factor. *J. Clin. Endocrinol. Metab.,* 38:742–745.
73. Smythe, G. A., Brandstater, J. F., and Lazarus, L. (1975): Serotoninergic control of rat growth hormone secretion. *Neuroendocrinology,* 17:245–257.
74. Smythe, G. A., and Lazarus, L. (1974): Growth hormone responses to melatonin in man. *Science,* 184:1373–1374.

75. Smythe, G. A., and Lazarus, L. (1974): Suppression of human growth hormone secretion by melatonin and cryproheptadine. *J. Clin. Invest.,* 54:116–121.
76. Stachura, M. E. (1976): Influence of synthetic somatostatin upon growth hormone release from perifused rat pituitaries. *Endocrinology,* 99:678–683.
77. Strauch, G., Girault, D., Rifai, M., and Bricaire, H. (1973): Alpha-MSH stimulation of growth hormone release. *J. Clin. Endocrinol. Metab.,* 37:990–993.
78. Szabo, M., and Frohman, L. A. (1975): Effects of porcine stalk median eminence and prostaglandin $E_2$ on rat growth hormone secretion *in vivo* and their inhibition by somatostatin. *Endocrinology,* 96:955–961.
79. Takahashi, Y., Kipnis, D. M., and Daughaday, W. H. (1968): Growth hormone secretion during sleep. *J. Clin. Invest.,* 47:2079–2090.
80. Tan, A. T., Tsang, D., Renaud, L. P., and Martin, J. B. (1977): Effect of somatostatin on calcium transport in guinea pig cortex synaptosomes. *Brain Res.,* 123:193–196.
81. Tannenbaum, G. S., and Martin, J. B. (1976): Evidence for an endogenous ultradian rhythm governing growth hormone secretion in the rat. *Endocrinology,* 98:540–548.
82. Tannenbaum, G. S., Martin, J. B., and Colle, E. (1976): Ultradian growth hormone rhythm in the rat; effects of feeding, hyperglycemia and insulin-induced hypoglycemia. *Endocrinology,* 99:720–727.
83. Terry, L. C., Willoughby, J. O., Brazeau, P., Martin, J. B., and Patel, Y. (1976): Antiserum to somatostatin prevents stress-induced inhibition of growth hormone secretion in the rat. *Science,* 192:565–567.
84. Wilber, J. F., Nagel, T., and White, W. F. (1971): Hypothalamic growth hormone releasing activity (GRA): Characterization by the in vitro rat pituitary and radioimmunoassay. *Endocrinology,* 89:1419–1424.
85. Willoughby, J. O., Epelbaum, J., Brazeau, P., and Martin, J. B. (1977): Effect of brain lesions and hypothalamic deafferentation on somatostatin distribution in the rat brain. Submitted to *Endocrinology*.
86. Willoughby, J. O., and Martin, J. B. (1976): Episodic GH secretion: Evidence for a hypothalamic dopaminergic mechanism. In: *Hypothalamus and Endocrine Functions,* edited by F. Labrie, J. Meites, and G. Pelletier, pp. 303–320. Plenum Press, New York.
87. Willoughby, J. O., Martin, J. B., Brazeau, P. B., and Renaud, L. P. (1976): Pulsatile growth hormone: Failure to demonstrate a correlation to sleep phases in the rat. *Endocrinology,* 98:593–598.
88. Willoughby, J. O., Terry, L. C., Brazeau, P., and Martin, J. B. (1977): Pulsatile growth hormone, prolactin and thyrotropin secretion in rats with hypothalamic deafferentation. *Brain Res.,* 127:137–152.
89. Yen, S. S. C., Siler, T. M., and De Vane, G. W. (1974): Effect of somatostatin in patients with acromegaly. Suppression of growth hormone, prolactin, insulin and glucose levels. *N. Engl. J. Med.,* 290:935–938.
90. Zimmerman, E. A., Carmel, P. W., Husain, M. K., Ferin, M., Tannenbaum, M., Frantz, A. G., and Robinson, A. G. (1973): Vasopressin and neurophysin: High concentrations in monkey hypophyseal portal blood. *Science,* 182:925–927.

## DISCUSSION

*Dr. Martin:* We have begun to examine the neuropharmacology of the spontaneous GH secretory mechanism. I think it is important to emphasize that in the adult human, episodic GH secretion gradually attenuates with age so that by age 50, there may be little or no remaining evidence of episodic GH secretion, perhaps a manifestation of aging.

The Finkelstein work emphasizes that episodic GH secretion during the growth period is the predominant mechanism of GH release. In this sense, the rat model becomes a useful way of asking the kinds of questions that Dr. Sachar has emphasized. We have, therefore, spent a considerable amount of time trying to assess the pharmacologic basis of the surges.

Briefly stated, we find the following: Dopamine and norepinephrine depletion, with α-methyl-p-tyrosine, completely abolishes the GH surges. They can be restored by the administration of clonidine, an α-agonist, but not by apomorphine, a dopamine agonist. Secondly, serotonin antagonists, either by blocking synthesis of serotonin with p-chlorophenylalanine or by receptor blockade with methysergide, also abolish the surges. So it appears that there is a combined facilitatory α-adrenergic and serotonergic mechanism for GH secretion in the rat, as there is in man.

*Dr. Plum:* Dr. Martin, would you comment briefly on why growth hormone has all these extrahypothalamic inputs. What is the biological advantage to the organism to have representation in the limbic system, from the amygdala and from a variety of other telencephalic levels?

*Dr. Martin:* The one brain regulator for GH that has been identified, namely, somatostatin, has probably been coopted, in Dr. Reichlin's terms, by the hypothalamus for use in neuroendocrine control. It is probable that its presence in other brain regions is for other more "primitive" reasons, many of which are probably not directly related to growth hormone regulation at all. Extrahypothalamic brain regions are doubtlessly important in mediation of stress-related, sleep-associated, or circadian-determined hormone changes.

I might add one comment concerning why the brain secretes peptides episodically. I think this suggests that brain rhythms represent a fundamental process in neuroendocrine regulation. Homeostasis is achieved in peripheral tissues by the long-acting cascade effects of target hormone secretion, which act to smooth out episodic events. In the case of growth hormone, this target substance is the group of peptides called "somatomedins," of which at least three have now been partially identified. What we can learn from study of growth hormone secretion is the manner in which brain rhythmic function occurs. The physiologic effects at the peripheral tissue level are accomplished by the long-acting effects of the induced secretion of the somatomedins.

*Dr. Sachar:* Dr. Martin has done another masterly review of a complex field. I would like to make one short comment and then ask a question. You have underscored the role of monoaminergic influences in the regulation of growth hormone secretion. For psychiatrists, this approach appears to have some value in studying the presumed monoamine functional depletion in depressive states in that some of these monoaminergic stimuli such as insulin-induced hypoglycemia and dextroamphetamine produce deficient growth hormone responses in patients with endogenous depressive illness. Have you had a chance, Dr. Martin, to look at the responses of lesioned animals, particularly the animals with the isolation of the hypothalamus, to some of these pharmacological stimuli? You mentioned the effects of stress. What happens to growth hormone secretion in the animals when you give them insulin to produce hypoglycemia, or when you give them monoaminergic agonists, for example?

*Dr. Martin:* Hypoglycemia in the rat, as Dr. Tannenbaum has reported, tends to inhibit the GH surges as do other stresses. We believe this is due to somatostatin release. We have not examined pharmacologic responses in rats with lesions or hypothalamic deafferentation.

*Dr. Krieger:* Would you like to speculate as to where you think the locus is of the paradoxical responses you see, both to the stimulatory effect of TRF and to the inhibitory effect of bromocryptine. Do you think these are hypothalamic receptors or pituitary receptors?

*Dr. Martin:* My guess is that the paradoxic response to glucose changes is hypothalamic. Several reports indicate that anterior hypothalamic lesions are associated with paradoxical responses to glucose. However, I think that the paradoxic inhibitory response to L-DOPA and the TRH stimulation of growth hormone which occurs in acromegaly are most likely due to a direct pituitary effect. One can demonstrate, at least in a significant percentage of such patients, that the responses revert to normal after complete adenomectomy. I think tumor cells become responsive to these agents because of dedifferentiation and by alteration in receptor sites on the surfaces of the cells in the appearance of receptors that are active locally.

*Dr. Guillemin:* Most of the evidence at hand can be interpreted by proposing that acromegaly is a disease of pituitary somatotroph receptors. For instance, it appears that any stimulus will stimulate secretion of growth hormone. Wylie Vale did several experiments over the last couple of years in which he obtained fragments of the pituitary of several acromegalic patients, placed them in tissue cultures, and then after a few days of culture added TRF or LRF, which normally do not induce secretion of growth hormone. Most, although not all of these tissues will now respond to TRF or LRF in releasing growth hormone. In several cases, the patient came back with evidence of incomplete hypophysectomy; administration to the patient of the particular peptide (TRF or LRF) that we had observed to release growth hormone *in vitro* did also release growth hormone *in vivo,* which normally does not happen. It seems to me therefore that this is indicative that the fundamental problem may well be at the level of the pituitary somatotroph receptors. I am aware that it could always be argued that what is abnormal is not the receptor but one step or another of what follows after the receptor has recognized its endogenous ligand. In this latter proposal, however, it would mean that normally the receptor is promiscuous, accepting TRF, LRF, or GRF, which is not in keeping with much of the evidence for selectivity of receptors as has been shown in many tissues.

*The Hypothalamus,* edited by S. Reichlin,
R. J. Baldessarini, and J. B. Martin,
Raven Press, New York, © 1978.

# The Hypothalamic Regulation of LH and FSH Secretion in the Rhesus Monkey*

## E. Knobil and T. M. Plant

The following account reviews our current understanding of the control systems which govern the secretion of the gonadotropic hormones during the menstrual cycle of the rhesus monkey, a representative member of our own suborder amenable to extensive experimental manipulations.

The conceptual basis for the investigation of the neuroendocrine regulation of the menstrual cycle, a relatively recent endeavor in the experimental sense, is deeply embedded in one of the principal cornerstones of neuroendocrinology: the elucidation of the control systems which govern the estrous cycle and ovulation in the rat (3,36,37). Not surprisingly, however, it has become apparent that rather salient differences exist between the neuroendocrine mechanisms which direct the brief ovarian cycle of this rodent and those which control the 28- to 36-day menstrual cycle of the higher primates.

The time courses of the circulating gonadotropic hormones (LH and FSH) during the menstrual cycle of the rhesus monkey (Fig. 1) may be viewed as being the resultant of tonic secretion interrupted, once every 28 days on the average, by an abrupt discharge of these hypophysial hormones followed by ovulation some 36 hr later (26,39,40). These secretory patterns, which closely resemble those in the human female (34), are controlled by negative and positive feedback loops involving two major components: the ovary and the mediobasal hypothalamico-hypophysial apparatus.

The role of the ovarian steroids in these feedback loops has been reviewed in detail previously (18,19) and may be summarized as follows. Tonic gonadotropin secretion, which is characterized by pulsatile, rhythmic circhoral discharges of LH and FSH, is principally controlled by a classical, negative feedback action of estradiol. The preovulatory gonadotropin surge is the consequence of a positive feedback action of this same steroid when its concentration in the serum surpasses a well-defined threshold and remains elevated for approximately 36 hr. The time course of gonadotropin secre-

---

Department of Physiology, University of Pittsburgh School of Medicine, Pittsburgh, Pennsylvania 15261

* The substance of this paper will also appear as part of a chapter in *Frontiers of Neuroendocrinology, Vol. 5,* edited by L. Martini and W. F. Ganong, Raven Press, New York, 1978 (*in press*).

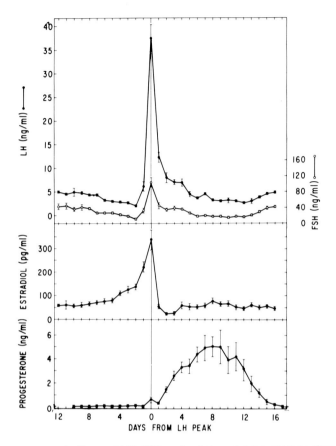

**FIG. 1.** Serum concentrations of LH, FSH, estradiol, and progesterone throughout the menstrual cycle of the rhesus monkey normalized to the day of the preovulatory gonadotropin surge (day 0). Each point represents the mean ± SE of at least 7 observations. From (18) with permission.

tion observed during the entire menstrual cycle can be replicated experimentally in ovariectomized animals by superimposing increments in serum estrogen concentration, produced by the injection of estradiol benzoate, on constant levels of the steroid released from Silastic capsules (Fig. 2).

The foregoing observations permit the conclusion that the time courses of LH and FSH secretion throughout the menstrual cycle are controlled by the patterns of ovarian estrogen secretion during the cycle and that the ovary may be considered as the "Zeitgeber" for the timing of ovulation in the rhesus monkey. By contrast, the preovulatory LH surge in the rat is timed by a neural signal transduced by a discharge of a hypothalamic gonadotropin releasing hormone (LHRH) into the pituitary portal circulation (35). Nevertheless, in the rhesus monkey, as in the rat, LHRH plays an essential role in the control of gonadotropin secretion.

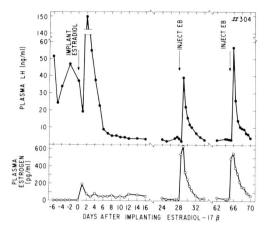

**FIG. 2.** Replication of the secretory pattern of LH, as observed during the menstrual cycle, in an ovariectomized rhesus monkey by superimposing increments in circulating estrogen levels [subcutaneous injections of estradiol benzoate (EB) at arrow] on constant estrogen concentrations achieved by a subcutaneous estradiol-containing Silastic capsule implanted on day 0. From (17) with permission.

Placement of large radiofrequency lesions in the mediobasal hypothalamus (MBH) of ovariectomized monkeys, which do not infarct the adenohypophysis, results in a profound and permanent inhibition of LH and FSH secretion (32). Passive immunization of ovariectomized animals with specific antisera to LHRH (25) produces a similar, albeit transient and less pronounced, inhibition of gonadotropin secretion (Fig. 3). Immunoreactive LHRH has been demonstrated in the hypothalamus (4,43) and in hypophysial portal blood (9,30) of the rhesus monkey, and, although the macaques seem to be less responsive than other species to single injections of the synthetic decapeptide (2,11,22,24,27), this peculiarity of the genus appears to be of quantitative rather than qualitative significance.

That the secretion of LHRH in the monkey is, at least in part, under the control of catecholaminergic neurons is suggested by the finding that in ovariectomized animals single intravenous injections of $\alpha$-adrenergic blocking agents, which have been reported not to inhibit pituitary responsiveness to exogenous LHRH (38), are followed, within minutes, by a cessation of pulsatile gonadotropin discharges and a resultant fall in the plasma concentration of these hormones (5). This finding further suggests that the signals which initiate the circhoral pulsatile discharges of gonadotropin originate within the CNS. In this regard, Carmel et al. (9) have reported that the concentration of immunoreactive LHRH in pituitary stalk blood from ovariectomized monkeys appears to fluctuate with a frequency not incompatible with that of circhoral LH oscillations in the peripheral circulation (10). On the other hand, the same group of workers has also described a single experiment in which pulsatile LH release was seen following pituitary

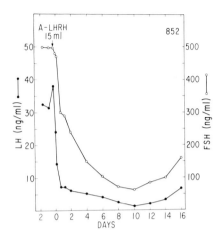

**FIG. 3.** Abrupt suppression of serum LH and FSH concentrations in an ovariectomized rhesus monkey following a single intravenous injection, on day 0, of a specific antiserum to LHRH. From (25) with permission.

stalk section, and intercalation of a Silastic barrier, in an ovariectomized animal receiving a constant infusion of LHRH (12). The latter observation suggests that the signal for circhoral release of gonadotropin originates within the pituitary itself, but it is difficult to conceive of an integrating mechanism, resident within the hypophysis, which could cause the synchronous discharge of gonadotropins by a vast number of adenohypophysial cells.

In the rhesus monkey, as in the rat (6,14,15), the neural component of the control system which governs tonic gonadotropin secretion is resident in the MBH. Complete surgical disconnection of this structure from the remainder of the brain does not significantly interfere either with the circhoral discharges of LH and FSH or with the negative feedback inhibition of their secretion by estradiol (21).

In striking contrast to its effect in the rat (7,14,15,29), however, surgical disconnection of the MBH in the monkey does not appear to interfere with the initiation of estrogen-induced gonadotropin surges nor with spontaneous ovulatory discharges of LH and FSH (21). The MBH "islands" produced in the monkey (Fig. 4) were judged to be complete by examination of serial coronal sections of the hypothalamus and were anatomically similar to those described by Halasz and others in the rat (7,14,15). They included the median eminence and arcuate nucleus as well as portions of the ventromedial nucleus, the premamillary region, and the mamillary bodies, but not the suprachiasmatic nucleus and preoptic area. It was concluded from these observations that the neural components of the control system which govern the preovulatory gonadotropin surge in the rhesus monkey are also located within the MBH (21).

This conclusion, however, was difficult to reconcile with the interpretation by Norman et al. (31) of their own finding that some lesions in the rostral hypothalamus of the rhesus monkey blocked the positive feedback

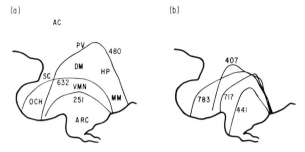

**FIG. 4.** Reconstructions, from serial coronal sections, of complete MBH "islands" produced by surgical disconnection in 7 intact female rhesus monkeys superimposed on diagrammatic parasagittal sections of the hypothalamus, in whom **(a)** spontaneous and **(b)** estradiol benzoate-induced gonadotropin surges were observed. Hypothalamic nuclei: *ARC*, arcuate; *DM*, dorsomedial; *VMN*, ventromedial; *HP*, posterior hypothalamic; *PV*, paraventricular; *SC*, suprachiasmatic; *MM*, mammillary body. The optic chiasm and anterior commissure are labeled *OCH* and *AC*, respectively. From (21) with permission.

action of estradiol on gonadotropin release. These lesions caused extensive bilateral damage of the ventromedial preoptic area–anterior hypothalamic area (POA-AHA)[1] and thus differed from the complete surgical disconnections described above which left this region of the brain essentially intact.[2] The possibility remained, therefore, that in response to an elevation in estradiol, the ventromedial POA-AHA could produce a neurotransmitter, or LHRH itself, which is then transported to the MBH by a non-neural pathway. The recent description of LHRH in the organum vasculosum of the lamina terminalis (OVLT) in the rhesus monkey (30, 43) provided substance to such speculation. Alternatively, the ventromedial POA-AHA could generate a neural signal in response to estradiol that is relayed to the MBH by nerve fibers capable of rapid functional regeneration. If nerve regeneration is the explanation for the continued functioning of the pituitary-ovarian axis in monkeys with MBH "islands," one may wonder why such regeneration should be limited to nerve fibers involved in the control of estrogen-induced gonadotropin discharges since abnormalities in cortisol, thyroxine, and growth hormone secretion as well as unremitting diabetes insipidus, which are also observed in these animals, persist (8,21, 23).

These alternative explanations were directly tested by aspirating all neural tissue dorsal and anterior to the optic chiasm, including the OVLT

---

[1] Corpora lutea or elevated serum progesterone levels, however, were observed in two animals several months after placement of bilateral lesions in the POA-AHA. The authors suggested that these sequelae of ovulation may be a consequence of "neural reorganization" (31).

[2] It should be noted, however, that following anterior surgical disconnection of the MBH in female rhesus monkeys extensive areas of necrosis were produced in the POA and AHA without interfering with the positive feedback action of estradiol (21).

and the suprachiasmatic nucleus, in a series of ovariectomized female monkeys and examining the capacity of these preparations to discharge LH and FSH in response to the positive feedback action of estrogen (16). The MBH "peninsulae" that remained following decerebration were continuous with the brainstem and contained the median eminence, arcuate nucleus, mamillary bodies, and varying portions of the ventromedial, dorsomedial, supraoptic, and paraventricular nuclei (Fig. 5). The administration of estradiol benzoate immediately upon completion of the ablation procedure resulted in massive gonadotropin discharges 18 to 24 hr later (Fig. 6). Since there is no reason to believe that the mechanisms underlying estradiol-induced gonadotropin discharges in monkeys with MBH "peninsulae" differ from those in animals with intact nervous systems, it must be concluded that the positive feedback action of this steroid is demonstrable in the absence of any neural or neurohumoral inputs from regions anterior to the MBH.

We have also attempted to repeat the experiments of Norman et al. (31) by placing radiofrequency lesions in the ventromedial POA-AHA of four intact female monkeys but without the aid of intraventricular radiopaque dyes (33). These lesions extended from the lamina terminalis to the caudal aspect of the optic chiasm and encompassed the suprachiasmatic nucleus and, with one exception, all of the OVLT. Laterally, they extended from midline toward the medial aspects of the supraoptic nuclei. In the dorsoventral plane, the area of destruction reached from the superior surface of the optic chiasm to within approximately 2 mm of the anterior commissure. In one of these animals, a spontaneous

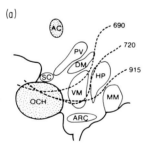

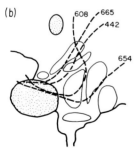

**FIG. 5.** Reconstructions, from serial coronal sections, of the MBH "peninsulae" produced by aspiration in 3 estrogen-treated **(a)** and 4 control **(b)** ovariectomized rhesus monkeys superimposed on diagrammatic parasagittal sections of the hypothalamus. The dashed lines demarcate the anterior and dorsal limits of the residual neural tissue. Hypothalamic nuclei indicated as in Fig. 4. From (16).

ovulatory gonadotropin surge was observed 27 days after placement of the lesion (Fig. 7). In two of the other three lesioned animals, characteristic surges of LH and FSH were elicited in response to the injection of estradiol benzoate 1 to 2 months after surgery. Although the remaining animal failed to respond to the positive feedback action of estradiol, so did one of five control animals given repeated injections of estradiol benzoate. We are currently unable to provide a compelling explanation for the discrepancy between our own studies and the observations reported by Norman et al.

That the arcuate nucleus is the primary structure within the MBH responsible for the hypothalamic control of both tonic and surge gonadotropin secretion in the rhesus monkey is suggested by the findings (32) that placement of relatively discrete radiofrequency lesions in the arcuate region of ovariectomized animals caused a rapid fall in serum LH and FSH to undetectable levels and abolished the positive feedback action of estradiol (Fig. 8). Placement of larger lesions, which only partially destroyed the arcuate region (10 to 60%), had lesser effects (Fig. 9). When the arcuate region was spared entirely by the lesions, LH and FSH secretion did not differ from control (32). The striking deficits in gonadotropin secretion that follow the placement of lesions in the arcuate region cannot be attributed to a generalized reduction in hypophysiotropic stimulation because basal adrenocortical and thyroid function did not appear to be grossly influenced, growth hormone discharge in response to insulin hypoglycemia was not abolished, and elevations in prolactin secretion[3] were not observed (32).

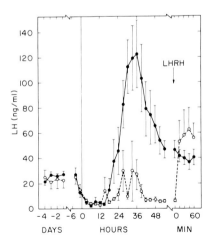

**FIG. 6.** Time courses of serum LH concentrations before and after ablation of all neural tissue rostral to the MBH in 4 estrogen-treated (*closed circles*) and 4 control (*open circles*) monkeys. The completion of the ablation procedure and the simultaneous administration of estradiol benzoate at time 0 are indicated by the vertical line. The severe surgical procedure resulted in a transient inhibition of LH secretion. The intravenous administration of LHRH indicated by the arrow, immediately prior to the termination of the experiments, induced a release of gonadotropin in the control animals but not in the estrogen-treated group. Each point represents the mean ± SE of 4 observations. FSH responded in like manner. From (16).

---

[3] The finding that relatively discrete lesions of the arcuate region, which abolish gonadotropin secretion in the rhesus monkey, do not influence basal serum prolactin levels suggests that the areas of the MBH involved in the regulation of gonadotropin release and those which control prolactin secretion are anatomically distinct in this species (32).

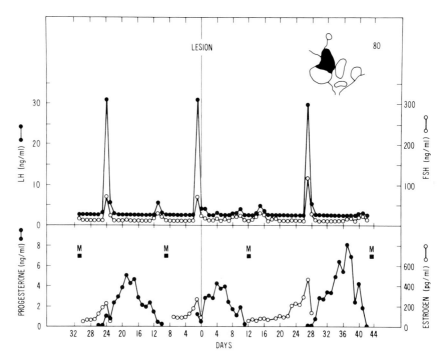

**FIG. 7.** Time courses of serum gonadotropins and ovarian steroid hormones in an intact rhesus monkey before and after placement of a large radiofrequency lesion in the ventromedial POA-AHA on day 0. The reconstruction of the lesion from serial coronal sections of the brain is superimposed on a diagrammatic parasagittal section of the hypothalamus. Note that a spontaneous gonadotropin surge, followed by ovulation, occurred at the appropriate time after the placement of this large lesion. Menstrual periods are indicated by M. For identity of hypothalamic nuclei, see Fig. 4.

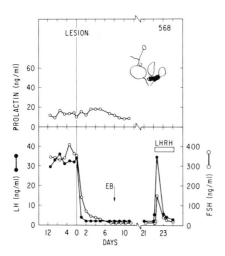

**FIG. 8.** Reduction in serum LH and FSH concentrations and abolition of the positive feedback action of estradiol in an ovariectomized rhesus monkey following placement of a radiofrequency lesion (on day 0) in the arcuate region and the dorsal aspect of the posterior median eminence. The injection of estradiol benzoate (42 μg/kg BW) on day 8 is indicated by the arrow. The continuous intravenous infusion of LHRH (*horizontal bar*) at a dose of 6.8 μg/hr resulted in a transient discharge of gonadotropic hormones. Note that this lesion did not result in a rise in serum prolactin levels. From (32).

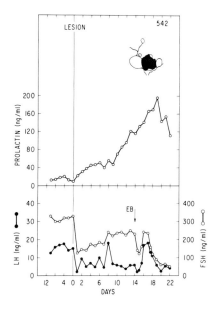

**FIG. 9.** Time courses of serum LH, FSH, and prolactin concentrations is an ovariectomized rhesus monkey before and after placement of a large radiofrequency lesion in the MBH that spared the ventral aspects of the arcuate region. An injection of estradiol benzoate (42 µg/kg BW) on day 14 is indicated by the arrow. Note that gonadotropin secretion was only partially suppressed while circulating prolactin levels increased. From (32).

It seems reasonable to conclude, therefore, that the seemingly specific abolition of gonadotropin secretion following arcuate lesions is the consequence of damage to neurons that either secrete LHRH or control the secretion of this releasing factor. Whether the majority of the perikarya of these neurons reside primarily within the region of the arcuate nucleus or whether their axons course through this region to the median eminence cannot be definitively answered at present. Although perikarya containing immunoreactive LHRH have been identified in the arcuate nucleus of the rhesus monkey, such cell bodies are also diffusely distributed throughout the hypothalamus of this species (4,43).

The feedback actions of estradiol on gonadotropin secretion may occur at the level of the MBH, there controlling LHRH release, or at the level of the adenohypophysis, there modulating the response of the gonadotrophs to hypophysiotropic stimulation, or at both of these levels.

Ferin et al. (13) observed that the pulsatile LH secretion in ovariectomized monkeys was inhibited by the microinjection of estradiol at various sites in the hypothalamus. These same workers, however, later reported that acute intravenous injections of estradiol did not suppress immunoreactive LHRH levels in pituitary portal blood (9). Neill and his colleagues (30) have proposed that the negative feedback action of estradiol is exerted primarily at the neural level, on the basis of their findings that the concentration of LHRH in portal blood collected from a limited number of ovariectomized animals was greater than that in early follicular phase females. Similarly elevated levels of LHRH were observed in animals undergoing estrogen-induced gonadotropin surges, which led Neill et al. (30) to suggest that the positive feedback action of estradiol, like its inhibitory effect, is exerted primarily at the neural level. On the other hand, less direct evidence based on the augmentation of the gonadotropin response to LHRH administration in women (41,42) has suggested that the principal site of the positive action of estrogen may be at the level of the pituitary (42).

This problem has recently been reexamined in ovariectomized rhesus monkeys bearing hypothalamic lesions, which abolish gonadotropin secretion, but in whom essentially normal production of LH and FSH was reestablished by the chronic administration of synthetic LHRH (28). Initial attempts to restore gonadotropin secretion in such animals by the continuous intravenous infusion of LHRH were markedly unsuccessful. This replacement regimen, although causing an immediate discharge of LH and FSH lasting for 24 to 36 hr, did not result in a sustained elevation of these hormones despite the continued administration of the decapeptide. Unaccountably, the pituitary seemed to become refractory to the sustained hypophysiotropic stimulus with a resultant fall in serum gonadotropins to undetectable levels (see Fig. 8). When the LHRH was administered in a pulsatile mode, however, using constant infusion pumps programmed to deliver a 6-min pulse once per hour, serum LH and FSH concentrations gradually rose from undetectable levels to approximately those observed during the prelesion control period. The administration of estradiol to such animals, while continuing the LHRH replacement, resulted in a profound decline in circulating gonadotropin levels followed by an unambiguous discharge of LH and FSH (Fig. 10). The time course of this biphasic pattern of gonadotropin secretion is remarkably similar to that observed in response to the negative and positive feedback actions of estradiol in otherwise intact ovariectomized animals (18). Since endogenous LHRH production is abolished in this experimental preparation, these results demonstrate that, in the rhesus monkey, estradiol can exert both its negative and positive feedback actions on

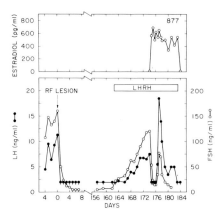

**FIG. 10.** Cessation of tonic gonadotropin secretion in an ovariectomized rhesus monkey by placement of a radiofrequency lesion in the arcuate region on day 0, and its subsequent restoration by a chronic intermittent intravenous infusion of synthetic LHRH (*horizontal bar*) initiated on day 162 (1 µg/min for 6 min every hour). The subcutaneous implantation of estradiol-containing Silastic capsules 12 days later (day 174) resulted in an abrupt decline of LH and FSH levels followed by a discharge of these hormones. The time course of this biphasic gonadotropin response to estrogen administration was similar to that observed in an ovariectomized animal with an intact central nervous system [see (18)].

LH and FSH secretion at the level of the pituitary gland. These experiments, however, do not exclude feedback actions of the steroid at the neural level in the context of a physiological setting.

These findings are in harmony with our earlier observations that passive immunization with antisera to LHRH (25) or administration of a variety of neuroactive drugs (18), both of which inhibit tonic gonadotropin secretion (see above) and block ovulation in the rat (1,20,36), do not interfere with estradiol-induced gonadotropin surges in the rhesus monkey. They are also consonant with the view that in this species, LHRH may be a permissive hormone whose principal role is to maintain the functional integrity of the gonadotrophs while the "control" of gonadotropin secretion is exerted principally at the adenohypophysial level by estradiol.

## ACKNOWLEDGMENTS

The work conducted in this laboratory has been generously supported by grants from the National Institutes of Health and the Ford Foundation. Dr. Plant is a Postdoctoral Research Fellow of the National Institute of Child Health and Human Development.

## REFERENCES

1. Arimura, A., Debeljuk, L., and Schally, A. V. (1974): Blockade of the preovulatory surge of LH and FSH and of ovulation by anti-LH-RH serum in rats. *Endocrinology,* 95:323–325.
2. Arimura, A., Spies, H. G., and Schally, A. V. (1973): Relative insensitivity of rhesus monkeys to the LH-releasing hormone (LH-RH). *J. Clin. Endocrinol. Metab.,* 36:372–374.
3. Barraclough, C. A. (1973): Sex steroid regulation of reproductive neuroendocrine processes. In: *Handbook of Physiology, Section 7, Endocrinology, Vol. 11. Female Reproductive System, Part I,* edited by R. O. Greep, pp. 29–56. American Physiological Society, Washington, D.C.
4. Barry, J., Girod, C., and Dubois, M. P. (1975): Topographie des neurones elaborateurs de LRF chez les primates. *Bull Assoc. Anat.,* 59:103–110.
5. Bhattacharya, A. N., Dierschke, D. J., Yamaji, T., and Knobil, E. (1972): The pharmacologic blockade of the circhoral mode of LH secretion in the ovariectomized rhesus monkey. *Endocrinology,* 90:778–786.
6. Blake, C. A., and Sawyer, C. H. (1974): Effects of hypothalamic deafferentation on the pulsatile rhythm in plasma concentrations of luteinizing hormone in ovariectomized rats. *Endocrinology,* 94:730–736.
7. Blake, C. A., Weiner, R. I., Gorski, R. A., and Sawyer, C. H. (1972): Secretion of pituitary luteinizing hormone and follicle stimulating hormone in female rats made persistently estrous or diestrous by hypothalamic deafferentation. *Endocrinology,* 90:855–861.
8. Butler, W. R., Krey, L. C., Espinosa-Campos, J., and Knobil, E. (1975): Surgical disconnection of the medial basal hypothalamus and pituitary function in the rhesus monkey. III. Thyroxine secretion. *Endocrinology,* 96:1094–1098.
9. Carmel, P. W., Araki, S., and Ferin, M. (1976): Pituitary stalk portal blood collection in rhesus monkeys: Evidence for pulsatile release of gonadotropin-releasing hormone (GnRH). *Endocrinology,* 99:243–248.
10. Dierschke, D. J., Bhattacharya, A. N., Atkinson, L. E., and Knobil, E. (1970): Circhoral oscillations of plasma LH levels in the ovariectomized rhesus monkey. *Endocrinology,* 87:850–853.

11. Ehara, Y., Ryan, K. J., and Yen, S. S. C. (1972): Insensitivity of synthetic LRF in LH-release of rhesus monkeys. *Contraception,* 6:465–478.
12. Ferin, M., Carmel, P. W., and Vande Wiele, R. L. (1974): The neuroendocrine regulation of LH secretion by estrogens in rhesus monkeys. *Adv. Biosci.,* 15:223–234.
13. Ferin, M., Carmel, P. W., Zimmerman, E. A., Warren, M., Perez, R., and Vande Wiele, R. L. (1974): Location of intrahypothalamic estrogen-responsive sites influencing LH secretion in the female rhesus monkey. *Endocrinology,* 95:1059–1068.
14. Halasz, B., and Gorski, R. A. (1967): Gonadotrophic hormone secretion in female rats after partial or total interruption of neural afferents to the medial basal hypothalamus. *Endocrinology,* 80:608–622.
15. Halasz, B., and Pupp, L. (1965): Hormone secretion of the anterior pituitary gland after physical interruption of all nervous pathways to the hypophysiotrophic area. *Endocrinology,* 77:553–562.
16. Hess, D. L., Wilkins, R. H., Moossy, J., Chang, J. L., Plant, T. M., McCormack, J. T., Nakai, Y., and Knobil, E. (1977): Estrogen-induced gonadotropin surges in decerebrated female rhesus monkeys with medial basal hypothalamic peninsulae. *Endocrinology (in press).*
17. Karsch, F. J., Dierschke, D. J., Weick, R. F., Yamaji, T., Hotchkiss, J., and Knobil, E. (1973): Positive and negative feedback control by estrogen of luteinizing hormone secretion in the rhesus monkey. *Endocrinology,* 92:799–804.
18. Knobil, E. (1974): On the control of gonadotropin secretion in the rhesus monkey. *Recent Prog. Horm. Res.,* 30:1–36.
19. Knobil, E., Dierschke, D. J., Yamaji, T., Karsch, F. J., Hotchkiss, J., and Weick, R. F. (1972): Role of estrogen in the positive and negative control of LH secretion during the menstrual cycle of the rhesus monkey. In: *Gonadotropins,* edited by B. B. Saxena, G. G. Belling, and H. M. Gandy, pp. 72–86. John Wiley & Sons, New York.
20. Koch, Y., Chobsieng, P., Zor, U., Fridkin, M., and Lindner, H. R. (1973): Suppression of gonadotropin secretion and prevention of ovulation in the rat by antiserum to synthetic gonadotropin releasing hormone. *Biochem. Biophys. Res. Commun.,* 55:623–629.
21. Krey, L. C., Butler, W. R., and Knobil, E. (1975): Surgical disconnection of the medial basal hypothalamus and pituitary function in the rhesus monkey. I. Gonadotropin secretion. *Endocrinology,* 96:1073–1087.
22. Krey, L. C., Butler, W. R., Weiss, G., Weick, R. F., Dierschke, D. J., and Knobil, E. (1973): Influences of endogenous and exogenous gonadal steroids on the actions of synthetic LRF in the rhesus monkey. In: *Hypothalamic Hypophysiotropic Hormones,* edited by C. Gaul and E. Rosemberg, pp. 39–47. Excerpta Medica, Amsterdam.
23. Krey, L. C., Lu, K.-H., Butler, W. R., Hotchkiss, J., Piva, F., and Knobil, E. (1975): Surgical disconnection of the medial basal hypothalamus and pituitary function in the rhesus monkey. II. GH and cortisol secretion. *Endocrinology,* 96:1088–1093.
24. Levitan, D., Beitins, I. Z., Milton, G., Barnes, A., and McArthur, J. W. (1977): Insensitivity of bonnet monkeys to (D-Ala$^6$, Des-Gly$^{10}$) LHRH ethylamide, a potent new luteinizing hormone releasing hormone analogue in rats and mice. *Endocrinology,* 100: 918–922.
25. McCormack, J. T., Plant, T. M., Hess, D. L., and Knobil, E. (1977): The effect of luteinizing hormone releasing hormone (LHRH) antiserum administration on gonadotropin secretion in the rhesus monkey. *Endocrinology,* 100:663–667.
26. Monroe, S. E., Atkinson, L. E., and Knobil, E. (1970): Patterns of circulating luteinizing hormone and their relation to plasma progesterone levels during the menstrual cycle of the rhesus monkey. *Endocrinology,* 87:453–455.
27. Mori, J., and Hafez, E. S. E. (1973): Release of LH by synthetic LH-HR in the monkey, *Macaca fascicularis. J. Reprod. Fertil.,* 34:155–157.
28. Nakai, Y., Plant, T. M., Hess, D. L., Keogh, E. J., and Knobil, E. (1977): On the sites of the negative and positive feedback actions of estradiol in the control of gonadotropin secretion in the rhesus monkey. Presented at 59th Annual Meeting of the Endocrine Society, Abstract #298.
29. Neill, J. D. (1972): Sexual differences in the hypothalamic regulation of prolactin secretion. *Endocrinology,* 90:1154–1159.
30. Neill, J. D., Dailey, R. A., Tsou, R. C., Patton, J., and Tindall, G. (1976): Control of the

ovarian cycle in the monkey. In: *Ovulation in the Human,* edited by P. G. Crosignani and D. R. Mishell, pp. 115–125. Academic Press, London.
31. Norman, R. L., Resko, J. A., and Spies, H. G. (1976): The anterior hypothalamus: How it affects gonadotropin secretion in the rhesus monkey. *Endocrinology,* 99:59–71.
32. Plant, T. M., Krey, L. C., Moossy, J., McCormack, J. T., Hess, D. L., and Knobil, E. (1977): The arcuate nucleus and the control of gonadotropin and prolactin secretion in the female rhesus monkey (*Macaca mulatta*). *Endocrinology (in press).*
33. Plant, T. M., Moossy, J., Hess, D. L., Nakai, Y., McCormack, J. T., and Knobil, E. (1977): The rostral hypothalamus and gonadotropin secretion in the rhesus monkey. Proceedings of the Annual Meeting of the Society for Study of Reproduction, Austin, Texas.
34. Ross, G. T., Cargille, C. M., Lipsett, M. B., Rayford, P. L., Marshall, J. R., Strott, C. A., and Rodbad, D. (1970): Pituitary and gonadal hormones in women during spontaneous and induced ovulatory cycles. *Recent Prog. Horm. Res.,* 26:1–48.
35. Sarkar, D. K., Chiappa, S. A., and Fink, G. (1976): Gonadotropin-releasing hormone surge in pro-oestrous rats. *Nature,* 264:461–463.
36. Sawyer, C. H. (1969): Regulatory mechanisms of secretion of gonadotrophic hormones. In: *The Hypothalamus,* edited by W. Haymaker, E. Anderson, and W. J. H. Nauta, pp. 389–430. Charles C Thomas, Springfield, Ill.
37. Sawyer, C. H. (1975): Some recent developments in brain-pituitary-ovarian physiology. *Neuroendocrinology,* 17:97–124.
38. Spies, H. G., and Norman, R. L. (1975): Interaction of estradiol and LHRH on LH release in rhesus females: Evidence for a neural site of action. *Endocrinology,* 97:685–692.
39. Weick, R. F., Dierschke, D. J., Karsch, F. J., Butler, W. R., Hotchkiss, J., and Knobil, E. (1973): Periovulatory time courses of circulating gonadotropic and ovarian hormones in the rhesus monkey. *Endocrinology,* 93:1140–1147.
40. Yamaji, T., Peckham, W. D., Atkinson, L. E., Dierschke, D. J., and Knobil, E. (1973): Radioimmunoassay of rhesus monkey follicle-stimulating hormone (RhFSH). *Endocrinology,* 92:1652–1659.
41. Yen, S. S. C., Lasley, B. L., Wang, C. F., Leblanc, H., and Siler, T. M. (1975): The operating characteristics of the hypothalamic-pituitary system during the menstrual cycle and observations of biological action of somatostatin. *Recent Prog. Horm. Res.,* 31:321–357.
42. Young, J. R., and Jaffe, R. B. (1976): Strength-duration characteristics of estrogen effects on gonadotropin response to gonadotropin releasing hormone in women. II. Effects of varying concentrations of estradiol. *J. Clin. Endocrinol. Metab.,* 42:432–442.
43. Zimmerman, E. A., and Antunes, J. L. (1976): Organization of the hypothalamic-pituitary system: Current concepts from immunohistochemical studies. *J. Histochem. Cytochem.,* 24:807–815.

## DISCUSSION

*Dr. Reichlin:* I want to congratulate Dr. Knobil on bringing the gonadotrophic regulation system back to the pituitary and ovary. Your data strongly suggest that all that is needed from the hypothalamus is a steady source of LHRH, but the cyclicity of the system is regulated by the ovary and pituitary.

Dr. Michael Besser from London, in unpublished work, has done the same thing in man that you have in monkeys. He maintained women with hypogonadotrophic hypogonadism on sustained injections of LHRH over a period of many months. He found in these women, as you find in your monkeys, that without any change in cyclicity of LHRH injection, there is induced cyclicity of ovulation.

*Dr. Knobil:* In defense of the hypothalamus, I would like to point out that: a) the region of the arcuate nucleus is absolutely essential for normal gonadotropin secretion, and b) these experiments do not exclude a hypothalamic site for the positive feedback action of estrogen. They just demonstrate that estrogen *can* act at the level of the pituitary to initiate gonadotropin secretion. The next step in these investigations is to assess the physiological significance of this experimental model in the initiation of the preovulatory gonadotrophin surge.

*Dr. Guillemin:* For those interested in the intimate physiological mechanism of this complicated system for controlling the cyclicity of the secretion of the pituitary gonadotrophins and ovarian hormones, the systematic experiments by Dr. Knobil and his collaborators are a remarkable achievement of classic physiology.

Can you briefly explain your understanding of the mechanisms involved in the various types of estrogen effects in terms of cellular biology?

*Dr. Knobil:* Unfortunately, I can provide no explanations at the cellular level. Clearly, this is where the explanations for the negative and positive feedback actions of steroids are going to be. Unfortunately, these descriptive terms tell us nothing about mechanism. We are limited for the moment to descriptions of the sequelae of increasing serum estradiol concentrations: There is first a decline in pituitary gonadotrophin secretion, followed by a discharge. I hope that the mechanisms whereby estrogens elicit these responses will become available within the next decade.

*Dr. Michael:* Dr. Knobil, in what way do you think the primate hypothalamus might be involved in the induction of cycles in the immature animal, that is, the control of puberty, which presumably will be rather different in primates and rodents? How do you conceptualize the mechanism by which exteroceptive factors influence gonadotrophin regulation?

*Dr. Knobil:* Let me deal with the second question first. Even though we are obliged to conclude that the basic, primary mechanisms which govern tonic and surge secretion of gonadotrophins are within the medial basal hypothalamus, perhaps even within the arcuate nucleus, there can be absolutely no doubt that this control system is modulated by a great variety of extrahypothalamic inputs. That is completely clear.

The first question is more difficult and is one that we asked several years ago. We performed the MBH deafferentation procedure in a number of sexually immature animals to determine whether the advent of sexual maturation would be delayed or accelerated. In fact, complete surgical disconnection of the MBH did nothing. Although we have every reason to believe that the cuts were complete on physiologic grounds, it is difficult to interpret these negative results with confidence. There is a possibility of some nerve regeneration during the several years required to complete this experiment, and there is also the possibility of diffusion of neuroactive substances into the hypothalamic island.

*The Hypothalamus*, edited by S. Reichlin, R. J. Baldessarini, and J. B. Martin, Raven Press, New York, © 1978.

# Sleep-Related Endocrine Rhythms

## *Robert M. Boyar

The study of sleep has intensified during the past 25 years because of the development of precise methods for assessing the qualitative and quantitative aspects of this physiologic state. Sleep has been shown to be a dynamic process characterized by rapid changes in electrophysiologic events. The major insights into neuroendocrine mechanisms associated with sleep have paralleled methodologic advances in the measurement of protein and steroid hormones. Sleep-associated hormone secretory events have been studied in an attempt to gain insight into basic neuroendocrine mechanisms that govern pituitary hormone secretory activity.

The observation that sleep consists of alternating, recurrent cyclic phenomena was largely responsible for the great upsurge of interest in sleep and related biologic phenomena. Aserinsky and Kleitman (1) noted that recurrent periods of eye movement activity occurring approximately every 90 min could be seen beneath the closed eyelids during sleep. Dement and Kleitman (11) defined five electroencephalographic stages of sleep. Stages 2 through 4 (non-rapid eye movement) represent a progressive deepening of sleep, whereas stage 1-REM is associated with rapid eye movements and desynchronized EEG activity. As a subject proceeds from wakefulness to sleep, the muscles relax and the EEG gives way to a stage 1 pattern of low voltage, mixed frequency. Stage 2 is characterized by sleep spindles or k complexes. The latter are characterized by high-amplitude activity with a sharply delineated negative spike followed by a positive wave. Stages 3 and 4 are characterized by high-voltage, synchronized slow waves of 1 to 2 cycles per second. Stage 4 occurs when more than half of a 30-second epoch consists of this slow-wave activity, whereas lesser amounts, 20 to 50%, are classified as stage 3. During REM sleep, the EEG pattern is similar to that in stage 1 but is accompanied by episodic rapid eye movements, a low-amplitude fast EEG pattern, penile erections, pontine-geniculate-occipital spikes, and electromyographic muscle relaxation.

Normal adults generally have four to six sleep cycles per night depending on the total length of sleep. The mean cycle length is approximately 90 to 100 min and shortens toward the latter part of the sleep period. In normal

---

* Department of Internal Medicine, University of Texas Health Science Center at Dallas, Southwestern Medical School, 5323 Harry Hines Boulevard, Dallas, Texas 75235

subjects, stages 3 and 4 predominate at the beginning of the night, decreasing in amount toward the end of the night, whereas REM periods increase in frequency and duration as the night progresses.

It is not unreasonable to have predicted a relationship between sleep stage sequencing throughout the night and pituitary hormone secretory activity. This is based on the fact that there is strong evidence that peptidergic neurons that synthesize the hypothalamic releasing and inhibiting hormones are themselves influenced by monoaminergic neurons in the central nervous system. These findings, coupled with the evidence that specific sleep stages are related to changes in the activity of monoaminergic neurons in particular brain regions, suggest a common mechanism that could link pituitary hormone secretory activity to the sleep-wake cycle. The classic studies of Jouvet (13) showed that the periods of EEG synchronization or slow-wave sleep result from a reduction in ascending reticular activity which requires serotonergic mechanisms for its initiation and maintenance. On the other hand, REM sleep or desynchronization results from neural activity in the nucleus locus ceruleus and requires a serotonergic mechanism for its initiation and noradrenergic and cholinergic mechanisms for its maintenance. These studies provide a rationale for suggesting that neurotransmitters involved in the synthesis and release of hypothalamic releasing and inhibiting hormones may also be important in the initiation and maintenance of the REM-non-REM sleep cycle.

## GROWTH HORMONE

In 1966 Quabbe et al. (21) noted nocturnal peaks of growth hormone in human subjects who were studied during a 24-hr fast. Takahashi et al. (25) measured growth hormone levels during sleep and found a significant release persisting for 2 to 4 hr in association with slow-wave sleep. This study showed for the first time that a pituitary hormone was secreted in association with polygraphically defined sleep. The important role of sleep rather than the time at which sleep occurred was shown convincingly by Sassin et al. (24). These authors showed that the sleep-related secretion of growth hormone could be reversed acutely by a sleep-wake reversal study (Fig. 1). This study showed that nocturnal growth hormone secretory activity was sleep related and not a circadian rhythm independent of this state. These authors also showed that on 18 of 21 nights growth hormone was secreted within 90 min of sleep onset, and that on 32 of 37 occasions this rise was associated with slow-wave sleep. However, the relationship between growth hormone secretion during sleep and slow-wave sleep is not completely fixed; slow-wave sleep can occur without growth hormone secretion, and growth hormone release during sleep can occur in the absence of polygraphically defined slow-wave sleep. In a subsequent study, Pawel et al. (20) showed that the initiation of sleep-related growth hormone secretory activity was

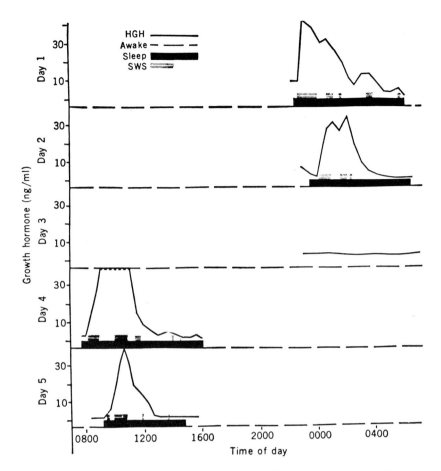

**FIG. 1.** Plots of five experimental sessions on consecutive days of a study of the reversal of the sleep-waking cycle from a representative subject. Peak concentration of HGH on the morning of 4th day reached 40 ng/ml. Time according to 24-hr clock. Graph illustrates absence of HGH secretion on 3rd night, when subject was kept awake, and resumption of secretion on 4th and 5th mornings, when he was asleep. Base-line and postreversal release show relation to slow-wave sleep. SWS, slow-wave sleep. (Reprinted with permission from ref. 24.)

related to the onset of polygraphically defined stage 3 sleep. Further studies to determine if the growth hormone secretory activity during sleep was related to alterations in metabolic substrates such as glucose, amino acids, or FFA failed to show any convincing evidence that would indicate that changes in these substrates serve as a trigger for sleep-related growth hormone release. Recently, Parker et al. (18) were able to inhibit growth hormone secretion during sleep by somatostatin without affecting the initiation of slow-wave sleep.

Finally, it is important to state that the secretion of growth hormone is not constant throughout life. Finkelstein et al. (12) showed that prepubertal children secrete growth hormone almost exclusively during the sleep period. However, with the onset of sexual maturation, there is a marked increase in the frequency and amplitude of growth hormone secretory activity (Fig. 2). Additional studies by Carlson et al. (9) in elderly individuals showed decreased growth hormone secretion during the nocturnal sleep period, suggesting that aging may be associated with a loss of sleep-associated growth hormone release.

In an attempt to gain insight into the neural mechanism controlling sleep-

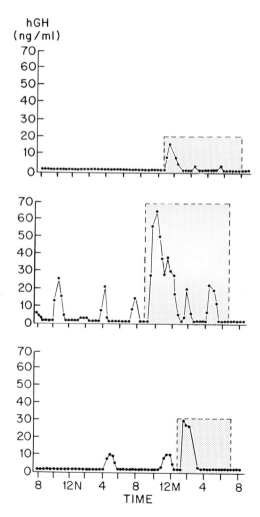

**FIG. 2.** Growth hormone secretory patterns in prepubertal (*top panel*), pubertal (*middle panel*), and young adult male subjects (*bottom panel*). Shaded area indicates the period of nocturnal sleep. (Reprinted with permission from ref. 12.)

related growth hormone release, investigators have carried out neuropharmacologic studies using relatively specific pharmacologic agents. In one study, Mendelson et al. (16) showed that methysergide, a potent serotonin antagonist, inhibited growth hormone secretory activity in response to insulin hypoglycemia; however, sleep-related growth hormone release was significantly increased. These data suggested that sleep-related growth hormone secretion is under the influence of an inhibitory serotonergic mechanism. Of additional interest was the finding that different effects on growth hormone secretion were obtained in response to two stimuli with this pharmacologic agent. Chihara et al. (10) showed suppression or delay of sleep-related growth hormone release after administration of another serotonin blocker, cyproheptadine. The latency required for this effect suggested that cyproheptadine inhibited the neural mechanism responsible for sleep-related growth hormone release. These investigators also showed that there was no change in the duration from sleep onset to the first period of slow-wave sleep, suggesting that the inhibition of growth hormone release was independent of any effect on slow-wave sleep. Recently, Malarkey et al. (15), using parachlorophenylalanine (PCPA), a tryptophan hydroxylase inhibitor, failed to show any significant effect on the sleep-related release of growth hormone. It is apparent that with three different antiserotonin agents three completely different results were obtained. These differences point up the difficulty in studying physiologic events with agents that have multiple effects on neurotransmitter function. For example, in addition to its antiserotonin effect, methysergide also has dopaminergic activity and cyproheptadine has antihistaminic and anticholinergic activity.

## PROLACTIN SECRETION DURING SLEEP

In 1972 Sassin and colleagues (23) showed that prolactin secretion was augmented during nocturnal sleep. Prolactin levels increased 60 to 90 min after sleep onset, achieving maximal levels in the early morning between 5 and 7 a.m. (Fig. 3). Subsequent studies by Sassin et al. (22) using acute sleep-wake reversal and by Parker et al. (19) using daytime naps showed that this enhanced nocturnal prolactin secretory activity was associated with sleep itself and was not the result of a circadian rhythm. Studies in pregnancy during the first, second, and third trimesters showed that the sleep-related augmentation of prolactin persists, albeit at much higher levels than normal (Fig. 4). This effect is probably related to the rising levels of estrogen during pregnancy and their effect on the pituitary lactotrope (4). Studies in patients with prolactin-secreting pituitary tumors showed a loss of the sleep-related increase in prolactin secretion (Fig. 5), such that the waking levels were equivalent to those during sleep, although pulsatile prolactin secretory activity was still present (5).

In an attempt to gain insight into the mechanisms involved in the aug-

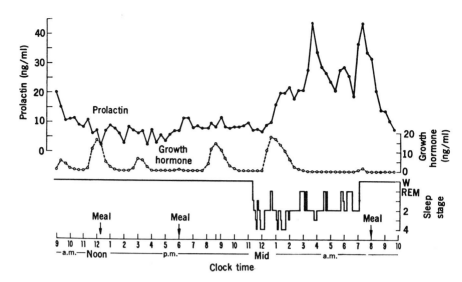

**FIG. 3.** Prolactin, growth hormone, and sleep stage in a male subject, age 23. The results of this 24-hr catheter study are representative of 24-hr pattern seen in normal subjects of both sexes. (Reprinted with permission from ref. 23.)

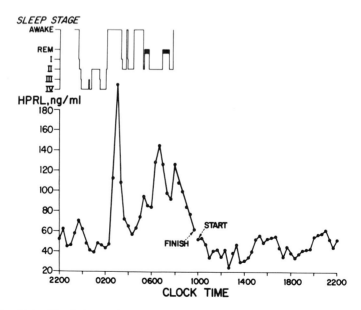

**FIG. 4.** The 24-hr, 20-min interval plasma PRL levels in a young woman during pregnancy. The sleep histogram is depicted above the nocturnal sleep period. (Reprinted with permission from Boyar, R. M., et al. (1976): *J. Clin. Endocrinol. Metab.*, 43:1418.)

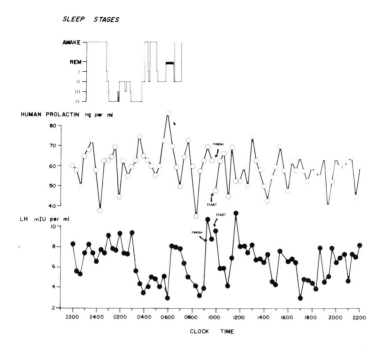

**FIG. 5.** Plasma LH and prolactin in a young woman with hyperprolactinemia and galactorrhea. Note the absence of the normal sleep-related augmentation of plasma prolactin. (Reprinted with permission from ref. 5.)

mentation of plasma prolactin levels during sleep, Mendelson et al. (16) showed that prolactin release during sleep was significantly decreased, after methysergide compared with studies during a control period. These studies suggested that serotonin has a stimulatory effect on prolactin secretion during sleep. However, recent studies by Malarkey et al. (15) showed that PCPA had no effect on prolactin secretion when given in doses that resulted in a reduction in urinary 5-hydroxyindoleacetic acid and plasma serotonin. The physiologic significance of plasma prolactin in the nonpregnant or nonlactating individual is not known. Preliminary studies have shown that sleep-related release is present from late prepuberty to old age, and therefore this secretory pattern does not undergo ontogenetic change. Whether this sleep-related rhythm has any physiologic significance will have to await future investigations.

## TSH SECRETION AND SLEEP

The hypothalamic-pituitary-thyroid axis has always been considered a system that tends to maintain the hormonal milieu within narrow limits. The relative insensitivity of thyroxine and triiodothyronine to stress and various other interventions suggests resistance to external modulation.

The difficulty in studying TSH secretory patterns during the 24-hr sleep-wake cycle has related to the limited sensitivity of most available radioimmunoassays. However, recent improvements in the TSH radioimmunoassay that have resulted in enhanced sensitivity have permitted recent studies of TSH dynamics during the 24-hr sleep-wake cycle. Measurements of plasma T4 and T3 have also been made to determine if correlations exist between these hormones. Recently, Azukizawa et al. (2) showed that plasma TSH levels show a characteristic nocturnal rise between 2300 and 0100 hr in male subjects who were studied during the 24-hr sleep-wake cycle. During the sleep period, it was shown that TSH secretory activity was generally inhibited. The measurement of plasma T4 showed some variability, but no diurnal variation or response to the rise in plasma TSH was evident. Plasma T3 levels showed less fluctuation; however, these were generally synchronized with those of T4. The factors responsible for the pre-sleep augmentation of TSH secretion and suppression during sleep have not been identified. Earlier investigations suggested a reciprocal relationship between cortisol secretory activity and TSH (17), but recent studies using multiple sampling have failed to confirm the finding that plasma cortisol modulates the pulsatile, circadian variation of TSH (26).

## GONADOTROPIN SECRETION DURING PUBERTY

Plasma luteinizing hormone (LH) and to a lesser extent follicle-stimulating hormone (FSH) are secreted in pulses or episodes characterized by abrupt rises of hormone concentration and slower declines that approximate the half-lives of these hormones. Studies of LH secretion in normal adult males showed episodic secretory activity occurring throughout the 24-hr period with no difference between the mean LH concentration asleep compared with waking (6). When studies of LH secretory activity were carried out in children during varying stages of sexual maturation, an interesting set of findings emerged. In "late" prepubertal and pubertal children, LH secretion and to a lesser extent FSH secretion become augmented synchronously with sleep (3). The increase in pulsatile LH secretion occurs coincident with sleep onset with a seeming relationship to the number of sleep cycles (Fig. 6). In "late" prepubertal and early pubertal subjects, there is inhibition or cessation of LH secretory activity during waking. Progression of the normal pubertal process is associated with the onset of LH secretory activity during waking, but still of lesser magnitude than during sleep. After the completion of sexual maturation, there is no significant effect of sleep on LH secretory activity.

The biologic importance of this augmented LH secretory activity in pubertal boys was shown by finding simultaneous augmentation of testosterone secretion (7). With sleep delay or reversal, there is a corresponding delay and reversal of the augmented LH and testosterone secretory activity. This demonstrates that LH secretory activity in pubertal subjects

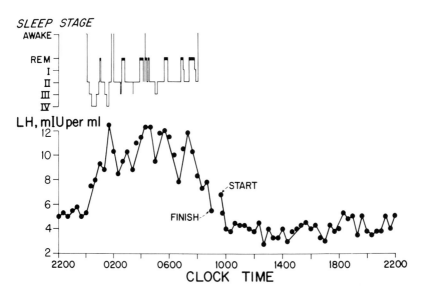

**FIG. 6.** Plasma LH concentration sampled every 20 min around the clock in an early pubertal boy. The sleep histogram is shown above the 8-hr period of nocturnal sleep. (Reprinted with permission from Boyar, R. M., et al. (1972): *N. Engl. J. Med.*, 287:582.)

is related not to the time sleep occurs but to sleep itself. These observations show that puberty can be identified before the onset of clinical evidence of this event by the occurrence of pulsatile LH secretory activity during sleep. The factors responsible for this augmentation during sleep have not been identified.

Recent studies in patients with hypogonadotropic hypogonadism, in which luteinizing hormone-releasing factor (LH-RH) was infused during sleep and waking periods, showed no significant difference in the amount of LH released during the sleep or waking periods (8). These findings suggest that the pituitary is not sensitized to LH-RH during sleep and that the onset of pulsatile LH secretory activity is related to increased secretion of LH-RH. There have been no measurements of plasma LH-RH activity in pubertal subjects, a time when the pituitary is relatively insensitive to LH-RH, and therefore an optimal time to find measurable amounts of plasma LH-RH. It is of interest that even though LH secretory activity in adult males is randomly distributed throughout the 24-hr sleep-wake cycle, there is frequently a nocturnal increase in plasma testosterone that may or may not be preceded by a significant rise in plasma LH. This augmentation of testosterone secretion during sleep in normal adult males results in a circadian variation, with the highest levels occurring during the early morning hours and the lowest levels in the late afternoon.

Studies of the 24-hr LH and FSH secretory patterns in pubertal girls have shown findings similar to those in pubertal boys except that augmented

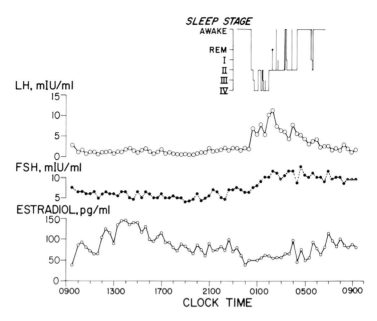

**FIG. 7.** Plasma LH, FSH, and estradiol measured every 20 min in a normal early pubertal girl. (Reprinted with permission from Boyar, R. M., et al. (1976): *J. Clin. Endocrinol. Metab.*, 43:1418.)

FSH secretion appears more commonly in girls. Studies of LH, FSH, and estradiol levels in 9 pubertal girls showed that although LH and FSH levels are augmented during sleep, plasma estradiol levels do not rise until 10 to 12 hr later (Fig. 7). This lag in the rise of plasma estradiol may result from the requirement for two coordinate cell types within the ovary to synthesize estrogen. Another explanation for this finding may be that plasma prolactin rises during sleep and may compete with LH and FSH for binding sites within the ovary. In any event, the absence of a rise in plasma estradiol during sleep allows the pituitary to respond to gonadotropin releasing hormone with a synchronous release of LH and FSH during sleep.

Finally, studies of LH and FSH secretory activity in normal cycling women suggest that there is random LH secretory activity during the 24-hr sleep-wake cycle similar to that in adult males; however, there appears to be inhibition of LH secretion 2 to 3 hr after sleep onset (14). The significance of this inhibition is not known, nor is the reason for its occurrence. There have been no reported studies of the CNS mechanisms controlling pulsatile LH secretory activity during sleep in pubertal subjects.

## THE ACTH-CORTISOL CIRCADIAN RHYTHM

Although ACTH-cortisol has a circadian rhythm, a clear relationship to the sleep-wake cycle has not been delineated. The total daily production

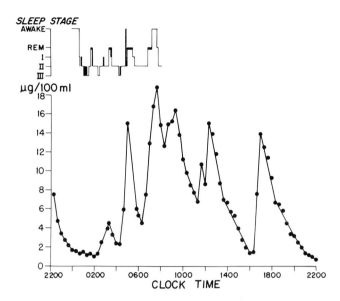

**FIG. 8.** Plasma cortisol measured at 20-min intervals in a normal adult male. Note the near undetectable cortisol levels at the time of sleep onset (lights out).

of cortisol occurs as a result of discrete secretory episodes that cluster around the 0300 to 0900 time period (Fig. 8). During the period of 2300 to 0300 hours, the plasma cortisol concentration approaches near undetectable levels (28). Since this suppression of ACTH-cortisol secretory activity occurs in close proximity to sleep onset (lights out), an inhibitory effect of sleep has been suggested. Recently, Weitzman et al. (27) studied subjects in isolation and demonstrated that lights out and sleep time progressively shifted. In all instances where this occurred, there was a corresponding shift in the time of inhibition of cortisol secretion which persisted for several hours. These findings suggest that the 24-hr ACTH-cortisol rhythm may have two components: an endogenous circadian rhythm and a sleep or lights-out inhibition. Further studies will be required to define precisely this important hormonal rhythm and its relationship to the sleep-wake cycle.

## CONCLUSIONS

This review has attempted to summarize the progress that has been made over the past 10 years concerning the relationship of pituitary hormone secretion to the sleep-wake cycle. It has been clearly established that growth hormone, prolactin, and pubertal LH secretory activities are closely linked to the sleep-wake cycle. The changes with regard to TSH are more complex and seem to involve a pre-sleep maxima which is followed by a decrease or suppressive effect of sleep. The biologic importance of this augmentation of TSH secretion prior to sleep is unclear, since no corresponding change in plasma T4 or T3 has been demonstrated. Studies of ACTH-cortisol suggest a relationship of sleep (lights out); however, the dominant factor appears to be an endogenous circadian rhythm which can be shifted only after prolonged shifts of the sleep-wake cycle. Studies of the neurohumoral basis of pulsatile and episodic secretion of pituitary hormones are in their infancy. This area of research in humans is complicated by the inability to measure hypothalamic hormones and neurotransmitters. Studies have attempted to delineate the neurochemical basis of the sleep-related release of growth hormone and prolactin using neuropharmacologic agents; however, the results obtained have been conflicting. These differences may be related to differences in the route of administration, duration of drug administration, or the multiple effects of these drugs. These difficulties will be important to consider in future studies using neuropharmacologic agents designed to elucidate the factors responsible for the stimulation of pituitary hormone secretion during sleep. In this area future studies will be needed to shed light on the neurochemical correlates of sleep-related endocrine rhythms.

## REFERENCES

1. Aserinsky, E., and Kleitman, N. (1953): Regularly occurring periods of eye motility and concomitant phenomena during sleep. *Science*, 118:273–274.
2. Azukizawa, M., Pekary, A. E., Hershman, J. M., and Parker, D. C. (1976): Plasma thyrotropin, thyroxine, and triiodothyronine relationships in man. *J. Clin. Endocrinol. Metab.*, 43:533–542.
3. Boyar, R. M., Finkelstein, J. W., David, R., Roffwarg, H., Weitzman, E., and Hellman, L. (1973): Twenty-four hour patterns of plasma luteinizing hormone and follicle-stimulating hormone in sexual precocity. *N. Engl. J. Med.*, 289:282–286.
4. Boyar, R. M., Finkelstein, J. W., Kapen, S., and Hellman, L. (1975): Twenty-four hour prolactin (PRL) secretory patterns during pregnancy. *J. Clin. Endocrinol. Metab.*, 40:1117–1120.
5. Boyar, R. M., Kapen, S., Finkelstein, J. W., Perlow, M., Sassin, J. F., Fukushima, D. K., Weitzman, E. D., and Hellman, L. (1974): Hypothalamic-pituitary function in diverse hyperprolactinemic states. *J. Clin. Invest.*, 53:1588–1598.
6. Boyar, R., Perlow, M., Hellman, L., Kapen, S., and Weitzman, E. (1972): Twenty-four hour pattern of luteinizing hormone secretion in normal men with sleep stage recording. *J. Clin. Endocrinol. Metab.*, 35:73–81.
7. Boyar, R. M., Rosenfeld, R. S., Kapen, S., Finkelstein, J. W., Weitzman, E. D., and Hellman, L. (1974): Human puberty: Simultaneous augmented secretion of luteinizing hormone and testosterone during sleep. *J. Clin. Invest.*, 54:609–618.
8. Boyar, R. M., Wu, R. H. K., Kapen, S., Hellman, L., Weitzman, E. D., and Finkelstein, J. W. (1976): Clinical and laboratory heterogeneity in idiopathic hypogonadotropic hypogonadism. *J. Clin. Endocrinol. Metab.*, 43:66–73.
9. Carlson, H. E., Gillin, J. C., Gorden, P., and Snyder, F. (1972): Absence of sleep-related growth hormone peaks in aged normal subjects and in acromegaly. *J. Clin. Endocrinol. Metab.*, 34:1102–1112.
10. Chihara, K., Kato, Y., Maeda, K., Matsukura, S., and Imura, H. (1976): Suppression by cyproheptadine of human growth and cortisol secretion during sleep. *J. Clin. Invest.*, 57:1393–1402.
11. Dement, W. C., and Kleitman, N. (1957): Cyclic variations in EEG during sleep and their

relation to eye movements, body motility, and dreaming. *Electroencephalogr. Clin. Neurophysiol.,* 9:673–690.
12. Finkelstein, J. W., Roffwarg, H. P., Boyar, R. M., Kream J., and Hellman, L. (1972): Age-related change in the twenty-four hour spontaneous secretion of growth hormone.*J. Clin. Endocrinol. Metab.,* 35:665–670.
13. Jouvet, M. (1969): Biogenic amines and the states of sleep. *Science,* 163:32–41.
14. Kapen, S., Boyar, R., Hellman, L., and Weitzman, E. D. (1973): Episodic release of luteinizing hormone at mid-menstrual cycle in normal adult women. *J. Clin. Endocrinol. Metab.,* 36:724–729.
15. Malarkey, W. B., and Mendell, J. R. (1976): Failure of a serotonin inhibitor to effect nocturnal GH and prolactin secretion in patients with Duchenne muscular dystrophy. *J. Clin. Endocrinol. Metab.,* 43:889–892.
16. Mendelson, W. B., Jacobs, L. S., Reichman, J. D., Othmer, E., Cryer, P. E., Trivedi, B., and Daughaday, W. H. (1975): Methysergide: Suppression of sleep-related prolactin secretion and enhancement of sleep-related growth hormone secretion. *J. Clin. Invest.,* 56:690–697.
17. Nicoloff, J. T., Fisher, D. A., and Appleman, M. A. (1970): The role of glucocorticoids in the regulation of thyroid function in man.*J. Clin. Invest.,* 49:1922–1929.
18. Parker, D. C., Rossman, L. G., Siler, T. M., Rivier, J., Yen, S. S. C., and Guillemin, R. (1974): Inhibition of the sleep-related peak in physiologic human growth hormone release by somatostatin. *J. Clin. Endocrinol. Metab.,* 38:496–499.
19. Parker, D. C., Rossman, L. G., and VanderLaan, E. F. (1973): Sleep-related nuchthemeral and briefly episodic variation in human plasma prolactin concentrations. *J. Clin. Endocrinol. Metab.,* 36:1119–1124.
20. Pawel, M. A., Sassin, J. F., and Weitzman, E. D. (1972): The temporal relation between HGH release and sleep stage changes at nocturnal sleep onset in man. *Life Sci.,* 11:587–593.
21. Quabbe, H. J., Schilling, E., and Helge, H. (1966): Pattern of growth hormone secretion during a 24-hour fast in normal adults.*J. Clin. Endocrinol. Metab.,* 26:1173–1177.
22. Sassin, J. F., Frantz, A. G., Kapen, S., and Weitzman, E. D. (1973): The nocturnal rise of human prolactin is dependent on sleep. *J. Clin. Endocrinol. Metab.,* 37:436–440.
23. Sassin, J. F., Frantz, A. G., Weitzman, E. D., and Kapen, S. (1972): Human prolactin: 24-hour pattern with increased release during sleep. *Science,* 177:1205–1207.
24. Sassin, J. F., Parker, D. C., Mace, J. W., Gotlin, R. W., Johnson, L. C., and Rossman, L. G. (1969): Human growth hormone release: Relation to slow-wave sleep and sleep-waking cycles. *Science,* 165:513–515.
25. Takahashi, Y., Kipnis, D. M., and Daughaday, W. H. (1968): Growth hormone secretion during sleep. *J. Clin. Invest.,* 47:2079–2090.
26. VanCauter, E., LeClercq, R., Vanhaelst, L., and Golstein, J. (1974): Simultaneous study of cortisol and TSH daily variations in normal subjects and patients with hyperadrenalcorticism. *J. Clin. Endocrinol. Metab.,* 39:645–652.
27. Weitzman, E. D., Czeisler, C. A., and Moore Ede, M. C. (1976): Relationship of cortisol, GH and body temperature to sleep in man living in an environment free of time cues. Abstracts of the 58th Meeting of the Endocrine Society, San Francisco, p. 252.
28. Weitzman, E. D., Fukushima, D., Nogeire, C., Roffwarg, H., Gallagher, T. F., and Hellman, L. (1971): Twenty-four hour pattern of the episodic secretion of cortisol in normal subjects. *J. Clin. Endocrinol. Metab.,* 33:14–22.

## DISCUSSION

*Dr. Sachar:* Regarding the complexity of the neuropharmacological control of prolactin secretion during sleep, I would like to point out that children who are treated with amphetamine for minimal brain dysfunction show suppression of the nocturnal rise in prolactin secretion.

*Dr. Guillemin:* Dr. Boyar, have you ever given somatostatin in any of your studies in these patients or normal individuals? Don Parker showed that in normal individuals, infusion of somatostatin during several sleep cycles consistently inhibits the cyclic secretion of growth hormone while not modifying at all the EEG recording of the particular sleep stage of this patient. After the end of the somatostatin infusion, an immediate release of growth hormone secretion usually took place at the inappropriate EEG stage of sleep.

*Dr. Krieger:* I would like to go back to some of the studies to which you referred on the neuropharmacology of growth hormone release during sleep and the diverse effects of the three agents supposedly affecting serotonin secretion. I believe that there is a real lack of specificity of the number of the agents that are presumably used as specific antogonists or depleters. Many of the conclusions drawn from the use of a number of neuropharmacological agents should be tempered, not only because such agents are not specific, but also because a substance that one thinks has an effect on a specific neurotransmitter may cause compensatory changes in levels of another transmitter at the same time.

*Dr. Getzow:* Have you studied fasting women as controls for your anorexia nervosa patients?

*Dr. Boyar:* We have not studied the effect of fasting on LH secretory activity in patients with anorexia nervosa. The patients were in the hospital and had varying degrees of reduced caloric intake. Some were fasting, some were eating minimally, some were eating well, although body weight was low in all patients at the time of the study. I would conclude from the spectrum of patients that we studied that LH secretion is not affected by fasting, although this has not been critically studied. Fasting affects growth hormone secretion, and that may explain the controversy concerning growth hormone secretory activity in patients with anorexia nervosa. I think one has to be careful about caloric intake when assessing growth hormone levels in patients with this disorder.

Patients with primary anorexia nervosa have LH secretory patterns that are similar to those of normal prepubertal or pubertal girls.[1] After psychologic improvement and achievement of normal body weight, the LH secretory pattern "matures" to one that is characteristic of normal adult women. These findings suggest that body composition plays an important role in the maturational state of the LH secretory "program."

---

[1] Boyar, R. M., Katz, J., Finkelstein, J. W., Kapen, S., Weiner, H., Weitzman, E. D., and Hellman, L. (1974): Anorexia nervosa: Immaturity of the 24-hour luetinizing hormone secretory pattern. *N. Engl. J. Med.,* 291:861–865.

*The Hypothalamus,* edited by S. Reichlin,
R. J. Baldessarini, and J. B. Martin,
Raven Press, New York, © 1978.

# Newer Understanding of Human Hypothalamic-Pituitary Disease Obtained Through the Use of Synthetic Hypothalamic Hormones

## Lawrence A. Frohman

The structure of thyrotropin releasing hormone (TRH), the first of the hypothalamic hormones to be identified, was reported in 1969 (8). Two years later, the structure of luteinizing hormone releasing hormone (LRH) was determined (31), and shortly thereafter somatostatin (SRIF) was identified and characterized (6). Within a short period of time synthetic preparations of these hypothalamic hormones became widely available to investigators throughout the world and clinical studies were initiated. The literature pertaining to the results of these studies has grown voluminous and includes many significant contributions to our understanding of the effects of these peptides and their mechanism of action, of the physiological processes which they regulate, and of their usefulness as diagnostic agents in the evaluation of clinical neuroendocrine disease. Their use can no longer be considered experimental, but rather constitutes an integral part of the endocrinologists' armamentarium of diagnostic procedures. This chapter will review the results of some of the studies that have helped to define human neuroendocrine physiology and pathophysiology.

According to classic endocrine principles, the role of a proposed hormone in normal and altered physiological conditions must be supported by four types of evidence:

1. production of a deficit in function by extirpation of the source of the hormone;
2. repair of the deficit by administration of exogenous hormone;
3. measurement of hormone levels in tissue and, more importantly, in plasma; and
4. alteration in secretion of the hormone accompanying change in hormone-dependent function.

With respect to the hypothalamic hormones, numerous problems have been encountered in attempting to provide the classic proofs of hormone

---

*Present address:* Division of Endocrinology and Metabolism, Michael Reese Medical Center, Department of Medicine, University of Chicago, Pritzker School of Medicine, Chicago, Illinois 60616

secretion. The sites of production are not localized to one specific anatomic area for any of the hypothalamic hormones characterized to date. Administration of exogenous hormone has been carried out and will of course constitute the major emphasis in this review; but the well-recognized fact that the hypothalamic hormones have actions at extrapituitary sites and the impracticality of administering these peptides selectively to the pituitary have raised the question of whether some of the effects observed may actually not be directly on the pituitary but rather on other tissues, notably the CNS. Specific evidence for this phenomenon will be presented. The measurement of releasing hormone levels in circulating plasma has also met with considerable difficulty. Although there are reports of TRH immunoreactivity in plasma from several laboratories (11,12,41), the levels from different laboratories show considerable variability and have not been demonstrated to vary under physiologic and pathologic circumstances where changes might have been expected. LRH has been reported in peripheral plasma by several laboratories (1,29,54) using radioimmunoassay techniques and by one group (57) using bioassay. Somatostatin has not been identified in peripheral plasma to date. However, in contrast to the usual circumstances in which the systemic circulation is used to transport hormones from one organ to another, there is no requirement that hypothalamic hormones be present in systemic circulation since their intended effects are at the pituitary and the small quantities detected in the systemic circulation may represent overflow hormone. LRH has been identified in portal blood of several laboratory animals, but the conditions of collection were not compatible with observing normal physiologic fluctuations. Obviously, stimulation and suppression of hormone secretion cannot be accomplished until the above problems can be overcome. Thus, a major contribution to the understanding of hypothalamic pituitary disease in humans relates to the information gained from the administration of the synthetic releasing hormones. As each of the hormones is reviewed, this presentation will discuss, in order, effects of the hypothalamic hormones on pituitary hormone secretion in normal subjects, the factors that modify pituitary responses, effects of the hypothalamic hormones in diseases of known pathophysiology, and, finally, the effects of the hypothalamic hormones in diseases of uncertain etiology in which the results have provided new insight.

## TRH

In normal subjects, pituitary TSH secretion is stimulated by TRH administered either as a bolus injection or as a constant infusion (Fig. 1). The response to a single injection is rapid, peaking at 15 to 30 min, then returning to base line subsequent to that time. Repeated injections result initially in reproducible responses, but after successive stimulations TSH responses diminish. This has been shown to be due to an elevation in en-

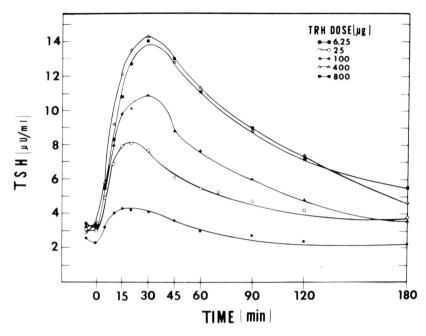

**FIG. 1.** Time course of serum TSH response in 8 normal subjects to the i.v. injection of 5 doses of TRH. (Courtesy of Snyder, P. J., and Utiger, R. D. [1972]: Response to thyrotropin releasing hormone [TRH] in normal man. *J. Clin. Endocrinol. Metab.,* 34:380–385.)

dogenous triiodothyronine ($T_3$) and thyroxine ($T_4$) levels, which exert an inhibitory feedback effect on the pituitary TSH response to TRH (62). Administration of either $T_3$ or $T_4$ completely abolishes the TSH response to TRH (Fig. 2) As might be expected from this observation, TSH responses in hyperthyroidism are depressed and TSH responses in hypothyroidism are enhanced when compared with normal subjects. Of interest, and yet unexplained, is a depressed or absent TSH response to TRH in subjects with euthyroid Graves' disease, i.e., subjects with normal $T_3$ and $T_4$ levels but in whom thyroid function is autonomous (19). Other hormones also influence the TSH response to TRH. The TSH response is enhanced by estrogens, being increased both in women and by pretreatment of men with estrogens; and high-dose glucocorticoid therapy suppresses the TSH response to TRH (44).

In patients with known pituitary disease, the TSH response to TRH is dependent on the extent of preservation of the thyrotrope population. Thus, two groups of subjects with pituitary insufficiency can be defined: those with a normal TSH response and those in whom the response is diminished or absent (Fig. 3) (16). The latter group includes subjects clinically euthyroid but with a reduced thyrotropin reserve and also subjects clinically hypothyroid on the basis of TSH deficiency. One may also occasionally encounter subjects with no evidence of organic pituitary disease in whom an

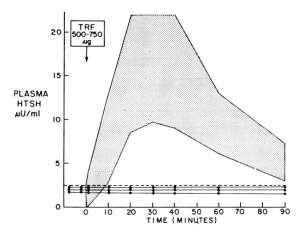

**FIG. 2.** Suppression of plasma TSH response to TRH by 0.3 mg l-thyroxine. The shaded area indicates the range of responses in normal subjects not receiving l-thyroxine. (Courtesy of Fleischer, N., Burgus, R., Vale, W., Dunn, T., and Guillemin, R. [1970]: Preliminary observations on the effect of synthetic thyrotropin releasing factor on plasma thyrotropin levels in man. *J. Clin. Endocrinol. Metab.,* 31:109–112.)

isolated TSH deficiency is present and in whom TRH does not elicit a TSH response (Fig. 4) (40).

In children with idiopathic hypopituitarism (either uni- or multihormonal in type) in whom the disease is manifested primarily as growth hormone deficiency, several groups (10,18) have noted that the TSH response to TRH

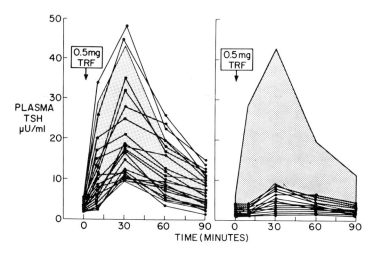

**FIG. 3.** Effect of TRH on plasma TSH levels in euthyroid subjects with pituitary tumors. The stippled area represents the range of TSH levels in control subjects receiving TRH. Two groups of subjects can be distinguished, those with normal responses (*left*) and those in whom the response is impaired (*right*). (From ref. 16 with permission.)

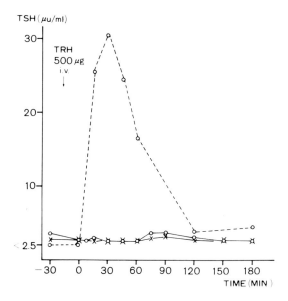

**FIG. 4.** Absence of serum TSH response to TRH in 2 siblings (*solid lines*) with familial isolated thyrotropin deficiency. A normal response (*broken line*) was observed in the patients' mother. The results indicate a pituitary etiology for the TSH deficiency in these subjects. (From ref. 40 with permission.)

was intact and in some instances prolonged (Fig. 5). It has been postulated therefore that the defect in such children, at least with respect to TSH secretion, is due to TRH deficiency and that the deficiencies of the other pituitary hormones are presumably due to impaired secretion of their re-

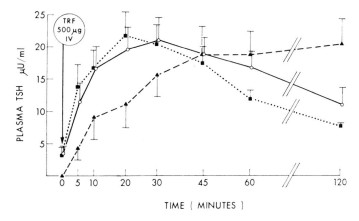

**FIG. 5.** TSH responses to TRH in normal children (*closed squares*), patients with isolated GH deficiency (*closed triangles*), and patients with idiopathic hypopituitary dwarfism and TSH deficiency (*open circles*). The last group exhibited the delayed and prolonged response which has been observed in hypothalamic disorders. (From ref. 10 with permission.)

spective releasing hormones. Although this view must remain inferential until more specific evidence becomes available, the demonstration of normal or enhanced TSH responses to TRH in subjects with known hypothalamic disease even in the absence of hypothyroidism (14,56) has provided support for this concept. It is of interest to note that the postulated deficiency of TRH does not impair the pituitary response to TRH stimulation. This is in contrast to the impaired gonadotropin response to LRH and the diminished adrenocortical response to ACTH after long periods of LRH or ACTH deficiency. The term "hypothalamic hypothyroidism" has been used to describe those conditions in which diminished TSH secretion is presumably due to a deficiency of TRH (46,58).

Inasmuch as the negative feedback of thyroid hormones appears to be on the pituitary rather than on the hypothalamus, one can reasonably ask why TRH is needed at all. The answer may well be that acute alterations in TRH secretion do not occur as part of the feedback mechanism but that the response threshold for TSH secretion due to alterations in circulating thyroid hormone levels is modified by the presence or absence of TRH. One line of evidence from studies in rats using antiserum to TRH provides additional insight into the role of TRH. We have recently reported that the injection of anti-TRH serum into rats inhibits the TSH response to cold environment, a response which is neurally mediated and which presumably requires TRH secretion (66). Martin et al. (48) have shown that the chronic elevation of TSH after thyroidectomy is partially suppressed by destruction of the "thyrotropic" area of the hypothalamus and anti-TRH serum has also been shown to decrease plasma TSH levels in hypothyroid rats (25a). Thus, the role of TRH appears to be twofold: that of a mediator of acute neurally stimulated rises in TSH secretion, and chronically as a modulator of the set point and extent of the TSH response to the negative feedback effect of thyroid hormones.

In addition to its stimulatory effects on TSH secretion, TRH is a potent releaser of prolactin. The initial observation was unexpected (68) since the physiologic control of prolactin was presumed to be mediated by an inhibiting rather than a stimulating factor. Within a very short period, however, there followed widespread clinical application. When prolactin and TSH levels were measured in response to increasing infusion rates of TSH, the response threshold in normal subjects appeared to be the same with respect to the two pituitary hormones (Fig. 6) (42). In other words, TRH is as effective a releaser of prolactin as of TSH. Kinetic studies following bolus injection (27) rather than infusion reveal similar secretory patterns, or possibly even a more rapid prolactin than TSH response (Fig. 7). Many of the factors that modulate the TSH response to TRH have similar effects on the prolactin response. The prolactin response to TRH is greater in women than in men, and the response in men can be increased by pretreatment with estrogen. TRH produces an enhanced prolactin response in hypothyroidism and a diminished response in hyperthyroidism. On a quantitative basis,

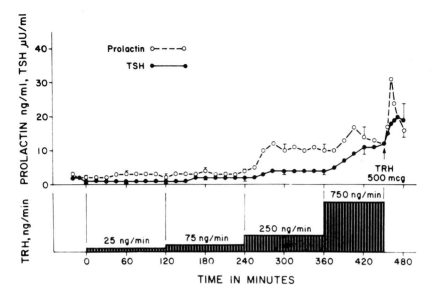

**FIG. 6.** Prolactin and TSH responses to a continuous infusion of TRH at increasing rates in 5 normal men. The smallest increases in TRH concentration capable of producing an increase in TSH concentrations also increased prolactin concentrations. (From ref. 42 with permission.)

however, the modulating effects of thyroid hormones appear to be greater on the TSH response, whereas those of estrogen are greater with respect to prolactin secretion. In contrast to TSH, the prolactin response to TRH is not suppressed by glucocorticoids (47).

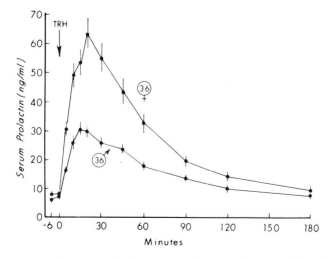

**FIG. 7.** Time course of serum prolactin responses in normal male and female subjects to an i.v. injection of 400 μg TRH. (From ref. 27 with permission.)

The prolactin response to TRH is frequently altered in patients with pituitary disorders (Table 1). With destruction of the anterior pituitary, the prolactin response to TRH is diminished or absent, whereas it remains intact in most patients with the empty sella syndrome presumably due to persistent pituitary tissue (50). Hyperprolactinemia is present in 40 to 70% of patients with documented chromophobe adenomas of the pituitary. The prolactin response to TRH in such patients is occasionally absent and does not necessarily correlate with basal prolactin levels. In patients with craniopharyngiomas in which hypothalamic involvement is common, plasma prolactin levels are normal in half ot the subjects and a response to TRH is usually observed (28). Consequently, the use of TRH to distinguish between pituitary tumors and other sources of hyperprolactinemia has not proved to be entirely reliable.

TABLE 1. *Prolactin responses to thyrotropin releasing hormone in pituitary disease*

| | |
|---|---|
| Anterior pituitary destruction: | Decreased to absent |
| Empty sella syndrome: | Normal |
| Chromophobe adenoma: | Normal to absent |
| Craniopharyngioma: | Normal |

In children with idiopathic growth hormone deficiency, the prolactin response to TRH appears to be subnormal in the presence of intact thyroid function and greater than normal if the subject is hypothyroid (17). Treatment of such subjects with thyroid hormone is effective in suppressing the enhanced prolactin response. Thus, despite the fact that prolactin is under an inhibitory CNS control and that interruption of hypothalamic-pituitary integrity by stalk section results in hyperprolactinemia, it appears that diseases of the hypothalamus can occur in which other pituitary hormone deficiencies are present but in which there is no hyperprolactinemia. In fact, the physiologic regulation of prolactin secretion in subjects with idiopathic hypopituitarism is probably intact since subjects with isolated growth hormone deficiency are capable of normal postpartum lactation (51).

Although dopamine is not yet accepted as a hypothalamic hormone, a few comments should be made concerning its role in prolactin secretion. There is considerable controversy as to whether the prolactin inhibiting factor (PIF) is a peptide or dopamine itself. The identification of dopamine in portal hypophysial blood (3), its effectiveness in suppressing prolactin secretion when injected into portal vessels (67), and the existence of dopamine receptors in the pituitary (30) have all favored this concept. Dopamine and/or the PIF which it controls is capable of inhibiting the prolactin response to TRH (Fig. 8) (43), and thus may constitute a tonic inhibitory tone on prolactin secretion which may then be modified by the stimulatory effects of TRH or of a separate prolactin releasing factor (5,65).

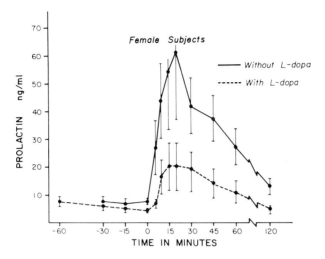

**FIG. 8.** Suppressive effect of L-DOPA pretreatment on plasma prolactin responses to TRH in normal women. L-DOPA was administered by mouth 60 and 30 min prior to TRH injection. (From ref. 43 with permission.)

In addition to its effects on TSH and prolactin secretion, TRH has been demonstrated to stimulate the release of growth hormone under certain specific conditions (Table 2). These include anorexia nervosa (34,38), depression (37), protein calorie malnutrition (2), hypothyroidism (25), uremia (22), and acromegaly (26,56). The response characteristics, with one exception, are similar to those observed for TSH and prolactin under normal conditions, suggesting a pituitary site of action. In acromegalic patients (Fig. 9), one could explain these observations by postulating an abnormal receptor mechanism in tumor tissue. However, in some patients the abnormal response to TRH persists after selective adenomectomy and a return of normal basal growth hormone levels (55). Further studies will be required to clarify the mechanism of this effect. From a practical standpoint, the growth hormone responses can be of assistance in establishing the diagnosis of acromegaly and anorexia nervosa in patients in whom other studies are inconclusive.

TABLE 2. *Conditions in which thyrotropin releasing hormone stimulates the release of growth hormone*

|  |
| --- |
| Anorexia nervosa |
| Depression |
| Hypothyroidism |
| Protein calorie deprivation |
| Uremia |
| Acromegaly |

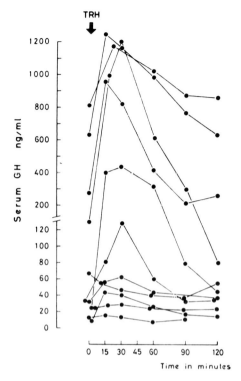

**FIG. 9.** Serum GH responses to 500 μg TRH in patients with acromegaly or gigantism. Responses were noted in 8 of 11 patients. (From ref. 26 with permission.)

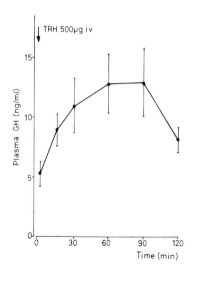

**FIG. 10.** Plasma GH responses to 500 μg TRH in 11 patients with anorexia nervosa. The GH response is delayed in comparison with that observed in acromegaly. (From ref. 38 with permission.)

The growth hormone responses to TRH in anorexia nervosa exhibit a response pattern different from that observed in the other disorders or from that of TSH or prolactin (Fig. 10). A delayed response, peaking at 60 to 90 min after TRH injection, is suggestive of a centrally mediated effect rather than one directly on the pituitary (37).

Evidence for inhibitory effects of TRH on growth hormone secretion has also been provided (36) by studies in which a TRH infusion was shown to inhibit the growth hormone release response to L-DOPA (Fig. 11). Similar inhibitory effects on the growth hormone response to other stimuli have also been shown and raise the intriguing possibility that TRH might act within the CNS to stimulate somatostatin secretion. One might also raise the question as to whether those disorders in which growth hormone secretion is characteristically depressed or absent, such as obesity, might be due to an excessive turnover of TRH.

Lastly, TRH has been reported to increase ACTH and cortisol levels in patients with Cushing's disease and with pituitary adenomas and Nelson's syndrome (35). It remains to be determined whether the abnormal response to TRH originates with tumor tissue or whether it is a response of the entire corticotrope population.

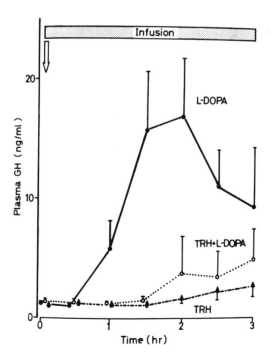

**FIG. 11.** Suppressive effect of TRH infusion (1 mg) on the GH response to L-DOPA (600 mg). (From ref. 36 with permission.)

## LRH

The hormonal responses to LRH are somewhat more complicated than are those to TRH. LRH stimulates the release of both LH and FSH, and in men and cycling women, the LH response is considerably greater than that of FSH. This fact along with other experimental observations has suggested the hypothesis that LRH is primarily a physiologic stimulator of LH secretion and that there exists a separate FSH releasing factor. However, the more widely held view is that there is a single releasing factor and that qualitative and quantitative differences in the FSH and LH response to LRH are determined largely by alterations in the levels of circulating gonadal hormones.

The acute LH response to a single injection of LRH in normal subjects is similar to that observed for TRH, i.e., with a peak occurring 30 to 45 min after injection (73) (Fig. 12). The onset of the FSH response, though, is delayed as compared with that of LH. The relatively slower return of FSH values to preinjection levels is due in part to the fact that its metabolic clearance rate is only one-fourth that of LH. Estrogen dosage level and duration of treatment modify the LH and FSH responses to LRH. Low-dose estrogen administration enhances the LH response and generally the FSH response in women, whereas at a high dose the response is characteristically inhibited. Progestinic drugs either exhibit no effect or actually enhance the estrogen stimulation (23,32). In contrast, estrogen treatment in males tends to suppress rather than augment the response to LRH. More recently, it has been appreciated that LH secretion occurs from at least two pools in the pituitary (7) and that the effects of LRH on each of these pools are probably affected differently by the various gonadal steroids (31).

In studies performed at different phases of the menstrual cycle, there appears to be an increased responsiveness of LH late in the follicular phase, when there are high circulating estradiol levels (Fig. 13). This reflects an increase in the secretory capacity rather than in the sensitivity to LRH (32). Under conditions of elevated estrogen levels but not in its absence, a priming pulse of LRH results in a bigger response to a second dose with respect to both FSH and LH (Fig. 14). The augmentation is enhanced further by progesterone. This phenomenon can be demonstrated with either individual pulses or a priming infusion and has formed the basis on which to interpret the responses to LRH in hypogonadal subjects. In this regard, the LRH effect on LH secretion resembles more the previously mentioned ACTH-cortisol relationship than the TRH effect on TSH where priming does not appear to have similar effects.

Prior to puberty, the LH response is minimal and less pronounced than is the FSH response (52). This increase in LH sensitivity to LRH at puberty is presumed due to the initiation of endogenous LRH secretion and its sensitizing effect. In postmenopausal women in whom LH and FSH levels are

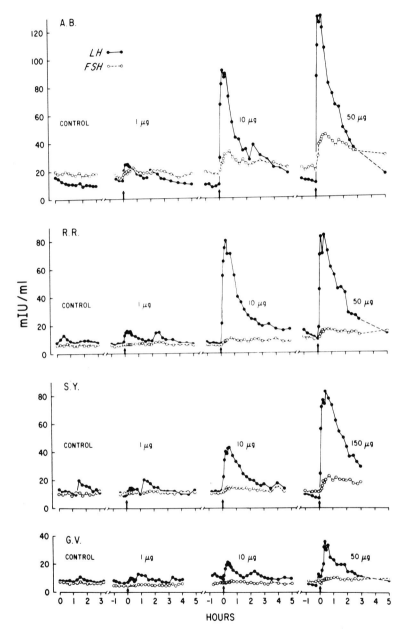

**FIG. 12.** Serum LH and FSH responses to graded doses of LRH in 4 normal males. The LH response is greater in magnitude and peaks at an earlier time than the FSH response. (From ref. 73 with permission.)

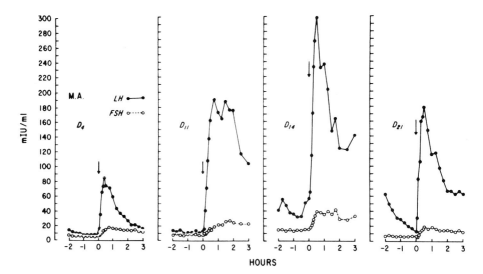

**FIG. 13.** Variations in the LH and FSH responses to the same dose of LRH during different phases of the menstrual cycle in a normal female. The peak response is observed at mid-cycle (day 14). (From ref. 73 with permission.)

both elevated, the FSH response to LRH is also frequently greater than is the LH response. Disproportionately high FSH responses relative to those of LH occur in settings of germ cell deficiency such as prepuberty, castration, and menopause, and have been attributed to loss of the inhibitory effects on the pituitary of inhibin, a newly described peptide secretion of germinal elements in testis and ovary.

In some hypogonadal subjects with gonadotrope cell destruction due to pituitary tumors, responses to LRH are impaired or absent, but a surprisingly large number have responses in the normal range. In subjects with delayed sexual development with or without association of other pituitary hormone deficiencies and where the clinical findings suggest a hypothalamic rather than a pituitary etiology, LH and FSH levels may or may not respond to LRH injection (Fig. 15) (52). Failure to respond is usually due to LRH deficiency; responsivity can be restored by frequent doses of LRH.

In children with premature puberty the LH responses to LRH are greater than those seen in normal prepubertal children. This heightened pituitary sensitivity is probably due to premature activation of the neural mechanisms which normally initiate LRH secretion during puberty. The responses to LRH are not increased in girls with precocious thelarche or precocious adrenarche, suggesting that the mechanisms of these premature maturational processes are not the same as that of premature puberty (49).

There has recently been considerable interest in the gonadotropin responses in women with anorexia nervosa. This disorder, in which amenor-

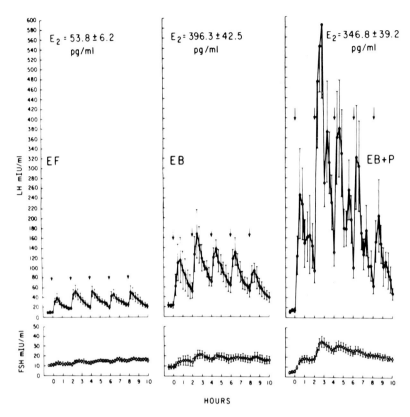

**FIG. 14.** Augmentation of LH and FSH responses to repeated injections of LRH (10 μg) following estradiol benzoate alone (EB) and in combination with progesterone (EB + P) administration to normal subjects during the early follicular phase of the menstrual cycle compared with responses to the same stimulus without treatment (EF). Estradiol ($E_2$) concentrations prior to LRH injections are also shown. (From ref. 32 with permission.)

rhea frequently but not always parallels the weight loss, is accompanied by changes in basal LH and FSH secretion which suggest a reversion to a prepubertal pattern (4). Although the results reported by different groups vary slightly (59,71), a characteristic alteration in the LRH response pattern is evident (Fig. 16). With increasing weight loss there is a progressive decrease in the LH response and an increase in the FSH response to LRH, a pattern which resembles that of the prepubertal state. In some patients with severe cachexia, the FSH response also decreases but invariably remains greater than the LH response. Changes in the relative LH and FSH responses to weight loss may be secondary to cachexia-induced changes in hypothalamic function, although an enhanced FSH response has been seen in some women with secondary amenorrhea of presumed CNS origin in whom no other endocrine-metabolic disturbances have been found (31).

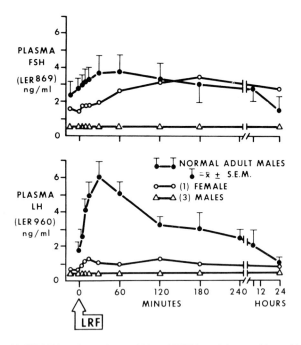

**FIG. 15.** Effect of LRH (100 μg) on plasma LH and FSH in subjects with multiple hypothalamic hormone deficiency and sexual infantilism. (Courtesy of Roth, J. C., Kelch, R. P., Kaplan, S. C., and Grumbach, M. M. [1973]: Patterns of LH, FSH and testosterone release stimulated by synthetic LRF in prepubertal, pubertal and adult subjects, and in patients with gonadotropin deficiency and XO gonadal dysgenesis. In: *Hypothalamic Hypophysiotropic Hormones,* edited by C. Gual and E. Rosemberg, pp. 236–246 Excerpta Medica, Amsterdam.)

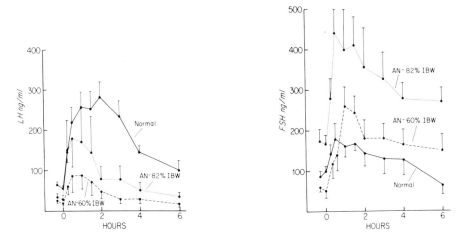

**FIG. 16.** Serum LH (*left*) and FSH (*right*) responses to LRH (100 μg) in patients with anorexia nervosa at 60% and 82% of ideal body weight and in 8 normally cycling females during the midfollicular phase. (From ref. 59 with permission.)

Women with the amenorrhea-galactorrhea syndrome associated with high prolactin values have gonadotropin levels compatible with a follicular phase and show an intact LH response to LRH (Fig. 17). It is assumed that the elevated prolactin levels in some way inhibit the ovulatory surge of LRH (63). One patient in whom ovarian failure coexisted, however, was fully capable of elevating her basal LH levels to values which in the normal female are seen at ovulation. One must presume, therefore, that the CNS locus required for the postmenopausal gonadotropin elevation is distinct from that responsible for the ovulatory surge, and that the latter is inhibited in hyperprolactinemic states.

Gonadotropin responses to LRH, and in particular that of LH, are greatest under those conditions in which endogenous LRH is being secreted. The use of exogenous LRH to distinguish between pituitary and central nervous system disorders has proven useful in some patients, although the

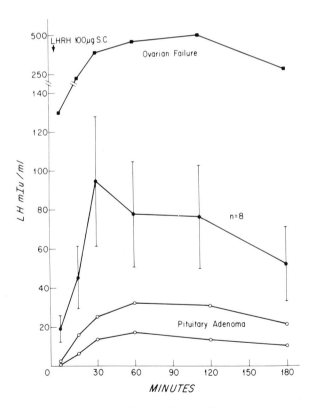

**FIG. 17.** Plasma LH responses to LRH in patients with galactorrhea-amenorrhea and hyperprolactinemia in whom clinical evidence of a pituitary tumor was present in two and ovarian failure in one. Intact responses to LRH were present in all except the patients with demonstrable pituitary tumors. (From ref. 63 with permission.)

marked variability in the gonadotropin responses associated with changes in circulating gonadal steroids as well as endogenous LRH secretion has made the elucidation of specific pathophysiological processes somewhat difficult.

Considerably less information is available in males with disorders of gonadotropin secretion. In subjects with oligospermia an increased response of both LH and FSH to LRH administration has been shown, and similar changes have been reported in patients with Klinefelter's disease in which testosterone therapy has been shown to reduce the response to normal levels (70).

As with TRH, LRH injection causes atypical secretory responses in patients with pituitary tumors. Growth hormone secretion in subjects with acromegaly has been induced by LRH (13,53), and recently, prolactin responses to LRH have also been observed in this disease (9). The responses are rapid and believed to be due to direct pituitary effects. In order to determine whether normal or tumor tissue is responsible for this effect, studies will have to be performed in patients cured of their disease by selective removal of tumor in whom normal growth hormone and prolactin secretory control mechanisms have been reestablished.

## SOMATOSTATIN

The discovery of somatostatin has complicated attempts to clarify the neuroendocrine control of growth hormone secretion in both normal and disease states. Inasmuch as there is little convincing evidence for feedback control of growth hormone secretion at the level of the pituitary, the previously held concept that changes in growth hormone secretion were a direct result of parallel changes in the secretion of growth hormone releasing factor has had to be modified to include the possibility of reciprocal changes in somatostatin secretion.

Administration of somatostatin to normal subjects either as a single injection or as an infusion results in a lowering of basal growth hormone levels and a suppression of the growth hormone response in normal subjects to whatever stimulus is employed (Fig. 18) (60). The spontaneous rises in growth hormone levels associated with the onset of deep sleep are also inhibited by somatostatin (45).

The inhibitory effects of somatostatin on growth hormone secretion are seen in patients with acromegaly as well as in normal subjects (Fig. 19) (24,74). The suppression of growth hormone levels occurs immediately after initiation of a somatostatin infusion and rebounds immediately upon its cessation. The rebound secretion is actually quantitatively similar to or even greater than the amount of suppressed secretion if the somatostatin

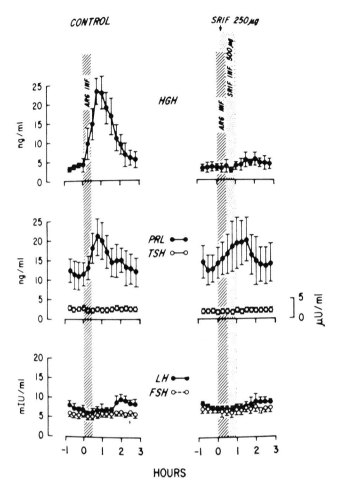

**FIG. 18.** Suppression of arginine-stimulated GH release by somatostatin in 4 normal adult subjects. Note absence of suppression of prolactin response. (From ref. 60 with permission.)

infusion is kept relatively short (64). This observation initially suggested that the episodic secretory spikes of growth hormone, which are even more prominent in the rat than in humans, might be due to an intermittent cessation of somatostatin secretion rather than to stimulation by a releasing factor. This theory appears to be incorrect since the administration of somatostatin antiserum to rats (15), while elevating growth hormone levels, does not interfere with the rhythmic secretion of the hormone. It also tends to confirm the previous hypothesis that the rises in growth hormone levels in response to various CNS-mediated stimuli occur as a result of activation by a growth hormone releasing factor. The role of somatostatin in growth hormone con-

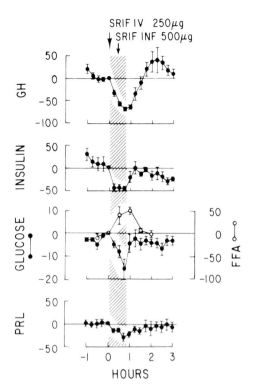

**FIG. 19.** Suppression of GH, prolactin, and insulin levels by somatostatin administration in 5 acromegalic patients. (From ref. 74 with permission.)

trol therefore is probably one of a tonic inhibitory tone. Whereas a deficiency in somatostatin secretion could contribute to the elevated growth hormone levels in acromegaly, it does not appear sufficient to explain the qualitatively abnormal responses to physiologic and pharmacologic stimuli exhibited in this disease (20).

Somatostatin is also an inhibitor of TSH secretion in normals, in hypothyroid subjects, and following TRH administration (Fig. 20) (33,61,72). Although there is presently insufficient evidence on which to establish a physiologic role of somatostatin in TSH secretion, it may exert a tonic inhibitory effect since treatment of rats with antisomatostatin serum increases the TSH response to cold exposure, and in thyroidectomized rats partially replaced with $T_4$, increases TSH levels.

Although somatostatin exhibits inhibitory effects on the secretion of numerous other hormones unrelated to the pituitary, including insulin, glucagon, gastrin, and renin, the other pituitary hormones (LH, FSH, ACTH, and prolactin) do not appear to be affected by somatostatin under normal conditions. In patients with pituitary tumors, however, there have been several

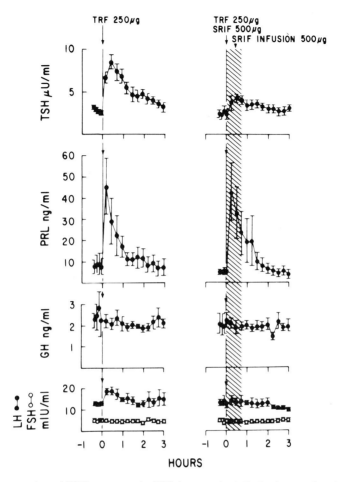

**FIG. 20.** Suppression of TSH response to TRH by somatostatin in 4 normal males. (From ref. 61 with permission.)

reported actions of somatostatin which are not observed in normals. In at least some acromegalics, somatostatin is capable of suppressing prolactin secretion (74), and in contrast to its effects on other stimuli, the growth hormone responses to TRH and LRH are not suppressed (21). Furthermore, it has recently been reported (69) that the elevated ACTH levels in subjects with basophilic adenomas of the pituitary and Nelson's syndrome are suppressed during somatostatin infusion (Fig. 21). Once again, it is not clear whether this represents a response limited to tumor tissue or whether these effects would also be demonstrable subsequent to tumor removal. In addition, the relationship of this observation, if any, to the pathogenesis of Cushing's syndrome is not known.

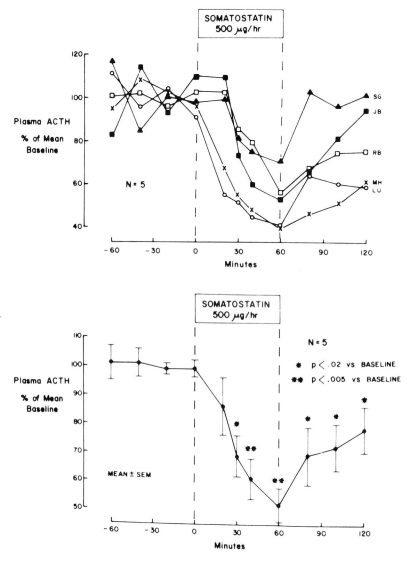

**FIG. 21.** Suppression of plasma ACTH levels in patients with Nelson's syndrome by somatostatin. (From ref. 69 with permission.)

## SUMMARY

In summary, clinical studies using the hypothalamic hormones over the past 5 years have provided important evidence for their tonic effects on the pituitary hormonal responses to both negative and positive feedback mechanisms. The studies have provided evidence of qualitative alterations in the pituitary hormonal responses as a result of the endogenous secretion of hypothalamic hormones. Further clarification of the hypothalamic hormone secretory patterns will require additional development and application of sensitive and specific assays for their measurement in plasma.

The pathophysiology of certain hypothalamic-pituitary disorders has been proposed on the basis of the use of the hypothalamic hormones in selected deficiency states, and suggestive evidence has been presented for the etiologic role of TRH and LRH deficiencies in specific disorders. The evidence for altered somatostatin secretion, however, remains unconvincing. Additional insight into the hypothalamic disorders mentioned, into others such as idiopathic growth hormone deficiency, Cushing's syndrome, the amenorrhea-galactorrhea complex, and anorexia nervosa, and into a large number of neurologic and psychiatric diseases including Huntington's chorea, Parkinson's disease, schizophrenia, and tardive dyskinesia, in which abnormalities of neuroendocrine responses are now being recognized, will also require the use of more indirect techniques involving neuropharmacologic agents in conjunction with pituitary hormone measurements. This approach, together with the eventual measurement of endogenous hypothalamic hormone levels, as well as continued studies with exogenous hormone administration, will hopefully permit a distinction among metabolic disorders of the hormone secreting cells, defects in hormonal biosynthesis, abnormalities of feedback mechanisms, and alterations in neurotransmitter physiology, any or all of which could be responsible for producing disordered hypothalamic hormone secretion.

## REFERENCES

1. Arimura, A., Sato, H., Kumasaka, T., Worobec, R. B., Debeljuk, L., Dunn, J., and Schally, A. V. (1973): Production of antiserum to LH-releasing hormone (LH-RH) associated with gonadal atrophy in rabbits: Development of radioimmunoassays for LH-RH. *Endocrinology*, 93:1092–1103.
2. Becker, D., Kronheim, S., and Pimstone, B. (1975): Serum growth hormone responses to thyrotropin-releasing hormone in children with protein-calorie malnutrition. *Horm. Metab. Res.*, 7:358–359.
3. Ben-Jonathan, N., Oliver, C., Mical, R. S., and Porter, J. C. (1976): Hypothalamic secretion of dopamine into hypophyseal portal blood. *Fed. Proc.*, 35:306.
4. Boyar, R. M., Katz, J., Finkelstein, J. W., Kaplan, S., Weiner, H., Weitzman, E. D., and Hellman, L. (1974): Anorexia nervosa: Immaturity of 24-hour luteinizing hormone secretory pattern. *N. Engl. J. Med.*, 291:861–865.
5. Boyd, A. E., III, Spencer, E., Jackson, I. M. D., and Reichlin, S. (1976): Prolactin-releasing factor in porcine hypothalamic extract distinct from TRH. *Endocrinology*, 99: 861–871.
6. Brazeau, P., Vale, W., Burgus, R., Ling, N., Butcher, M., Rivier, J., and Guillemin, R. (1973): Hypothalamic polypeptide that inhibits the secretion of immunoreactive pituitary growth hormone. *Science*, 179:77–79.
7. Bremmer, W. J., and Paulsen, C. A. (1974): Two pools of luteinizing hormone in the human pituitary: Evidence from constant administration of luteinizing hormone-releasing hormone. *J. Clin. Endocrinol. Metab.*, 39:811–815.
8. Burgus, R., Dunn, T. F., Desiderio, D., and Guillemin, R. (1969): Structure moléculaire

du facteur hypothalamique hypophysiotrope TRF d'origine ovine: Mise en evidence par spectrometre de masse de la sequence PCA-His Pro-NH$_2$. *C. R. Acad. Sci. [D] (Paris)*, 269:1870–1873.
9. Catania, A., Cantalamessa, L., and Reschini, E. (1976): Plasma prolactin response to luteinizing hormone-releasing hormone in acromegalic patients. *J. Clin. Endocrinol. Metab.*, 43:689–691.
10. Costom, B. H., Grumbach, M. M., and Kaplan, S. L. (1971): Effect of thyrotropin-releasing factor on serum thyroid-stimulating hormone. *J. Clin. Invest.*, 50:2219–2225.
11. Emerson, C. H., and Utiger, R. D. (1975): Plasma thyrotropin-releasing hormone concentrations in the rat. Effect of thyroid excess and deficiency and cold exposure. *J. Clin. Invest.*, 56:1564–1570.
12. Eskay, R. L., Oliver, C., Warberg, J., and Porter, J. C. (1976): Inhibition of degradation and measurement of immunoreactive thyrotropin releasing hormone in rat blood and plasma. *Endocrinology*, 98:269–277.
13. Faglia, G., Beck-Peccoz, P., Travaglini, P., Paracchi, A., Spada, A., and Lewin, A. (1973): Elevations in plasma growth hormone concentration after luteinizing hormone-releasing hormone in patients with active acromegaly. *J. Clin. Endocrinol. Metab.*, 37:338–340.
14. Faglia, G., Peccoz, P. B., Ambrosi, B., Ferrari, C., and Neri, V. (1971): Prolonged and exaggerated elevations of plasma thyrotropin after thyrotropin-releasing factor in patients with pituitary tumors. *J. Clin. Endocrinol. Metab.*, 33:999–1002.
15. Ferland, L., Labrie, F., Jobin, M., Arimura, A., and Schally, A. V. (1976): Physiological role of somatostatin in the control of growth hormone and thyrotropin secretion. *Biochem. Biophys. Res. Commun.*, 68:149–154.
16. Fleischer, N., Coscia, A. M., and Lorente, M. (1973): The effect of thyrotropin releasing factor on plasma TSH in subjects with pituitary tumors. In: *Hypothalamic Hypophysiotropic Hormones*, edited by C. Gual and E. Rosemberg, pp. 326–337. Excerpta Medica, Amsterdam.
17. Foley, T. P., Jr., Jacobs, L. S., Hoffman, W., Daughaday, W. H., and Blizzard, R. M. (1972): Human prolactin and thyrotropin concentrations in the serum of normal and hypopituitary children before and after the administration of synthetic thyrotropin-releasing hormone. *J. Clin. Invest.*, 51:2143–2150.
18. Foley, T. P., Jr., Owings, J., Hayford, J. T., and Blizzard, R. M. (1972): Serum thyrotropin responses to synthetic thyrotropin-releasing hormone in normal and hypopituitary patients. *J. Clin. Invest.*, 51:431–437.
19. Franco, P. S., Hershman, J. M., Haigler, E. D., Jr., and Pittman, J. A., Jr. (1973): Response to thyrotropin-releasing hormone compared with thyroid suppression tests in euthyroid Graves' disease. *Metabolism*, 22:1357–1365.
20. Frohman, L. A., and Stachura, M. E. (1975): Neuropharmacologic control of neuroendocrine function in man. *Metabolism*, 24:211–234.
21. Giustina, G., Reschini, E., Peracchi, M., Cantalamessa, L., Cavagnini, F., Pinto, M., and Bulgheroni, P. (1974): Failure of somatostatin to suppress thyrotropin releasing factor and luteinizing hormone releasing factor-induced growth hormone release on acromegaly. *J. Clin. Endocrinol. Metab.*, 38:906–909.
22. Gonzalez-Barcena, D., Kastin, A. J., Schalch, D. S., Torres-Zamora, M., Perez-Pasten, E., Kato, A., and Schally, A. V. (1973): Response to thyrotropin-releasing hormone in patients with renal failure and after infusion in normal men. *J. Clin. Endocrinol. Metab.*, 36:117–120.
23. Gual, C., Lichtenberg, R., Schally, A. V., Ortiz, A., Perez-Palacios, G., and Midgley, A. R., Jr. (1973): Effects of sex steroids on the pituitary responsiveness to synthetic LH- and FSH-releasing hormone in women. In: *Hypothalamic Hypophysiotropic Hormones*, edited by C. Gual and E. Rosemberg, pp. 230–235. Excerpta Medica, Amsterdam.
24. Hall, R., Besser, G. M., Schally, A. V., Coy, D. H., Evered, D., Goldie, D. J., Kastin, A. J., McNielly, B. S., Mortimer, C. H., Phenekos, C., Tunbridge, W. M. G., and Weightman, D. (1973): Action of growth hormone-release inhibitory hormone in healthy men and in acromegaly. *Lancet*, 2:581–584.
25. Hamada, N., Uoi, K., Nishizawa, Y., Okamoto, T., Hasegawa, K., Morii, H., and Wada, M. (1976): Increase of serum GH concentration following TRH injection in patients with primary hypothyroidism. *Endocrinol. Jpn.*, 23:5–10.

25a. Harris, A., Christianson, D., Smith, M. S., Braverman, L., and Vagenakis, A. (1977): The physiological role of TRH in the regulation of TSH and prolactin secretion in the rat. *Clin. Res.,* 25:463A.
26. Irie, M., and Tsushima, T. (1972): Increase of serum growth hormone concentration following thyrotropin-releasing hormone injection in patients with acromegaly or gigantism. *J. Clin. Endocrinol. Metab.,* 35:97–100.
27. Jacobs, L. S., Snyder, P. J., Utiger, R. D., and Daughaday, W. H. (1973): Prolactin response to thyrotropin-releasing hormone in normal subjects. *J. Clin. Endocrinol. Metab.,* 36:1069–1073.
28. Jenkins, J. S., Gilbert, C. J., and Ang, V. (1976): Hypothalamic-pituitary function in patients with craniopharyngiomas. *J. Clin. Endocrinol. Metab.,* 43:394–399.
29. Keye, W. R., Jr., Kelch, R. P., Gordon, D., Niswender, G. D., and Jaffe, R. B. (1973): Quantitation of endogenous and exogenous gonadotropin releasing hormone by radioimmunoassay. *J. Clin. Endocrinol. Metab.,* 36:1263–1267.
30. Kimura, H., and MacLeod, R. M. (1975): Dopamine receptors and the regulation of prolactin secretion. Progr. of the 57th Meeting, Endocrine Society, Abstr. No. 87.
31. Koninckx, P., DeHertogh, R., Heyns, W., Meulepas, E., Brosens, I., and DeMoor, P. (1976): Secretion rates of LH and FSH during infusions of LH-FSH/RH in normal women and in patients with secondary amenorrhea: Suggestive evidence for two pools of LH and FSH. *J. Clin. Endocrinol. Metab.,* 43:159–167.
32. Lasley, B. L., Wang, C. F., and Yen, S. S. C. (1975): The effects of estrogen and progesterone on the functional capacity of the gonadotrophs. *J. Clin. Endocrinol. Metab.,* 41:820–826.
33. Lucke, C., Höffken, B., and von zur Mühlen, A. (1975): The effect of somatostatin on TSH levels in patients with primary hypothyroidism. *J. Clin. Endocrinol. Metab.,* 41:1082–1084.
34. Lundberg, P. O., Walinder, J., Werner, I., and Wide, L. (1972): Effects of thyrotrophin-releasing hormone on plasma levels of TSH, FSH, LH, and GH in anorexia nervosa. *Eur. J. Clin. Invest.,* 2:150–153.
35. Luria, M., and Krieger, D. T. (1976): Response of plasma ACTH to TRF, vasopressin or hypoglycemia in Cushing's disease and Nelson's syndrome. *Clin. Res.,* 24:274A.
36. Maeda, K., Kato, Y., Chihara, K., Ohgo, S., Iwasaki, I., and Imura, H. (1975): Suppression by thyrotropin releasing hormone of human growth hormone release induced by l-dopa. *J. Clin. Endocrinol. Metab.,* 41:408–411.
37. Maeda, K., Kato, Y., Ohgo, S., Chihara, K., Yoshimoto, Y., Yamaguchi, N., Kuromaru, S., and Imura, H. (1975): Growth hormone and prolactin release after injection of thyrotropin-releasing hormone in patients with depression. *J. Clin. Endocrinol. Metab.,* 40:501–505.
38. Maeda, K., Kato, Y., Yamaguchi, N., Chihara, K., Ohgo, S., Iwasaki, Y., Yoshimoto, Y., Moridera, K., Kuromaru, S., and Imura, H. (1976): Growth hormone release following thyrotrophin-releasing hormone injection into patients with anorexia nervosa. *Acta Endocrinol. (Kbh.),* 81:1–8.
39. Matsuo, H., Baba, Y., Nair, R. M. G., Arimura, A., and Schally, A. V. (1971): Structure of the proposed LH- and FSH-releasing hormone. I. The proposed amino acid sequence. *Biochem. Biophys. Res. Commun.,* 43:1334–1339.
40. Miyai, K., Azukizawa, M., and Kumuhara, Y. (1971): Familial isolated thyrotropin deficiency with cretinism. *N. Engl. J. Med.,* 285:1043–1048.
41. Montoya, E. M., Seibel, M. J., and Wilber, J. F. (1975): Thyrotropin-releasing hormone secretory physiology: Studies by radioimmunoassay and affinity chromatography. *Endocrinology,* 96:1413–1418.
42. Noel, G. L. Dimond, R. C., Wartofsky, L., Earll, J. M., and Frantz, A. G. (1974): Studies of prolactin and TSH secretion by continuous infusion of small amounts of thyrotropin-releasing hormone. *J. Clin. Endocrinol. Metab.,* 39:6–17.
43. Noel, G. L., Suh, H. K., and Frantz, A. G. (1973): L-dopa suppression of TRH-stimulated prolactin release in man. *J. Clin. Endocrinol. Metab.,* 36:1255–1258.
44. Otsuki, M., and Sakoda, M. (1973): TRF-induced TSH response in several endocrine disorders. In: *Hypothalamic Hypophysiotropic Hormones,* edited by C. Gual and E. Rosemberg, pp. 332–337. Excerpta Medica, Amsterdam.
45. Parker, D. C., Rossman, L. G., Siler, T. N., Rivier, J., Yen, S. S. C., and Guillemin, R.

(1974): Inhibition of sleep-related peak in physiologic human growth hormone release by somatostatin. *J. Clin. Endocrinol. Metab.*, 38:496-499.
46. Pittman, J. A., Jr., Haigler, E. D., Jr., Hershman, J. M., and Pittman, C. S. (1971): Hypothalamic hypothyroidism. *N. Engl. J. Med.*, 285:844-845.
47. Re, R. M.. Kourides, I. A., Ridgway, E. C., Weintraub, B. D., and Maloof, F. (1976): The effect of glucocorticoid administration on human pituitary secretion of thyrotropin and prolactin. *J. Clin. Endocrinol. Metab.*, 43:338-346.
48. Reichlin, S. (1971): Neuroendocrine-pituitary control. In: *The Thyroid*, edited by S. C. Werner and S. H. Ingbar, pp. 95-111. Harper & Row, New York.
49. Reiter, E. O., Kaplan, S. L., Conte, F. A., and Grumbach, M. M. (1975): Responsivity of pituitary gonadotropes to luteinizing hormone-releasing factor in idiopathic precocious puberty, precocious thelarche, precocious adrenarche, and in patients treated with medroxyprogesterone acetate. *Pediatr. Res.*, 9:111-116.
50. Ridgway, E. C., Kourides, I. A., Kliman, B., Bigos, T., and Maloof, F. (1975): Thyrotropin and prolactin pituitary reserve in the "empty sella syndrome." *J. Clin. Endocrinol. Metab.*, 41:968-973.
51. Rimoin, D. L.. Holzman, G. B., Merimee, T. J., Rabinowitz, D., Barnes, A. C., Tyson, J. E. A., and McKusick, V. A. (1968): Lactation in the absence of human growth hormone. *J. Clin. Endocrinol. Metab.*, 28:1183-1188.
52. Roth, J. C., Kelch, R. P., Kaplan, S. L., and Grumbach, M. M. (1972): FSH and LH response to luteinizing hormone-releasing factor in prepubertal and pubertal children, adult males and patients with hypogonadotropic and hypergonadotropic hypogonadism. *J. Clin. Endocrinol. Metab.*, 35:926-930.
53. Rubin, A. L., Levin, S. R., Bernstein, R. I., Tyrrell, J. B., Noacco, C.. and Forsham, P. H. (1973): Stimulation of growth hormone by luteinizing hormone-releasing hormone in active acromegaly. *J. Clin. Endocrinol. Metab.*, 37:160-162.
54. Saito, S., Musa, K., Oshima, I., Yamamoto, S., and Funato, T. (1975): Radioimmunoassay for luteinizing hormone releasing hormone in plasma. *Endocrinol. Jpn.*, 22:247-253.
55. Samaan, N. A., Leavens, M. E., and Jesse, R. H., Jr. (1974): Serum growth hormone and prolactin response to thyrotropin-releasing hormone in patients with acromegaly before and after surgery. *J. Clin. Endocrinol. Metab.*, 38:957-963.
56. Schalch, D. S., Gonzalez-Barcena, D., Kastin, A. J., Schally, A. V., and Lee, L. A. (1972): Abnormalities in the release of TSH in response to thyrotropin-releasing hormone in patients with disorders of the pituitary, hypothalamus and basal ganglia. *J. Clin. Endocrinol. Metab.*, 35:609-615.
57. Seyler, L. E., Jr., and Reichlin, S. (1973): Luteinizing hormone releasing factor in plasma of postmenopausal women. *J. Clin. Endocrinol. Metab.*, 37:197-203.
58. Shenkman, L., Mitsuma, T., Suphavai, A., and Hollander, C. S. (1972): Hypothalamic hypothyroidism. *J.A.M.A.*, 222:480-481.
59. Sherman, B. M., Halmi, K., and Zamudio, R. (1975): LH and FSH response to gonadotropin-releasing hormone in anorexia nervosa: Effect of nutritional rehabilitation. *J. Clin. Endocrinol. Metab.*, 41:135-142.
60. Siler, T. N., Vandenberg, G., Yen, S. S. C., Brazeau, P., Vale, W., and Guillemin, R. (1973): Inhibition of growth hormone release in humans by somatostatin. *J. Clin. Endocrinol. Metab.*, 37:632-634.
61. Siler, T. N., Yen, S. S. C., Vale, W., and Guillemin, R. (1974): Inhibition by somatostatin of the release of TSH induced in man by thyrotropin-releasing factor. *J. Clin. Endocrinol. Metab.*, 38:742-745.
62. Snyder, P. J., and Utiger, R. D. (1973): Repetitive administration of thyrotropin-releasing hormone results in small elevations of serum thyroid hormones and in marked inhibition of thyrotropin response. *J. Clin. Invest.*, 52:2305-2312.
63. Spark, R. F., Pallotta, J., Naftolin, F., and Clemens, R. (1976): Galactorrhea-amenorrhea syndromes: Etiology and treatment. *Ann. Intern. Med.*, 84:532-537.
64. Stachura, M. E. (1976): Influence of synthetic somatostatin upon growth hormone release from perfused rat pituitaries. *Endocrinology*, 99:678-682.
65. Szabo, M., and Frohman, L. A. (1976): Dissociation of prolactin-releasing activity from thyrotropin-releasing hormone in porcine stalk median eminence. *Endocrinology*, 98:1451-1459.
66. Szabo, M., and Frohman, L. A. (1976): Suppression of cold-stimulated TSH secretion in

the rat by anti-TRH serum. Prog. of the 58th Meeting of the Endocrine Society, Abstr. No. 265.
67. Takahara, J., Arimura, A., and Schally, A. V. (1974): Suppression of prolactin release by a purified porcine PIF preparation and catecholamines infused into a rat hypophysial portal vessel. *Endocrinology,* 95:462–465.
68. Tashjian, A. H., Jr., Borowsky, N. J., and Jensen, D. K. (1971): Thyrotropin-releasing hormone: Direct evidence of prolactin production by pituitary cells in culture. *Biochem. Biophys. Res. Commun.,* 43:516–523.
69. Tyrrell, J. B., Lorenzi, M., Gerich, J. E., and Forsham, P. H. (1975): Inhibition by somatostatin of ACTH secretion in Nelson's syndrome. *J. Clin. Endocrinol. Metab.,* 40:1125–1127.
70. Wagner, H., Bockel, K., Hrubesch, M., and Grote, G. (1973): Examination of the pituitary-gonadal relationship in man with synthetic LH/FSH releasing hormone. In: *Hypothalamic Hypophysiotropic Hormones,* edited by C. Gual and E. Rosemberg, pp. 257–270. Excerpta Medica, Amsterdam.
71. Warren, M. P., Jewelewicz, R., Dyrenfurth, I., Ans, R., Khalaf, S., and VandeWiele, R. L. (1975): The significance of weight loss in the evaluation of pituitary response to LH-RH in women with secondary amenorrhea. *J. Clin. Metab.,* 40:601–611.
72. Weeke, J., Hansen, A. P., and Lundbaek, K. (1975): Inhibition by somatostatin of basal levels of serum thyrotropin in normal men. *J. Clin. Endocrinol. Metab.,* 41:168–171.
73. Yen, S. S. C., Rebar, R., VandenBerg, G., Naftolin, F., Judd, H., Ehara, Y., Ryan, K. J., Rivier, J., Amoss, M., and Guillemin, R. (1973): Clinical studies with synthetic LRF. In: *Hypothalamic Hypophysiotropic Hormones,* edited by C. Gual and E. Rosemberg, pp. 217–229. Excerpta Medica, Amsterdam.
74. Yen, S. S. C., Siler, T. N., and DeVane, G. W. (1974): Effect of somatostatin in patients with acromegaly. *N. Engl. J. Med.,* 290:935–938.

*The Hypothalamus*, edited by S. Reichlin,
R. J. Baldessarini, and J. B. Martin.
Raven Press, New York, © 1978.

# Nonendocrine Diseases and Disorders of the Hypothalamus

## *Fred Plum and Robert Van Uitert

Here in this well-concealed spot, almost to be covered with a thumb-nail lies the very main-spring of primitive existence — vegetative, emotional, reproductive — on which with more or less success, man has come to superimpose a cortex of inhibitions.

H. Cushing, 1929 (57)

The 1939 session of this Association gave almost its entire attention to the anatomy and physiology of the normal hypothalamus (HT) as well as to the coarser manifestations of the diseases or dysfunctions of that tiny structure. The limits of knowledge of the times meant that no one directed attention to neurochemistry and neurotransmitters, and even neuroendocrinology barely received mention except in reference to the neurohypophysis. The present volume, by contrast, has dealt mainly with the hypothalamus as a special endocrine gland, and has devoted little time to describing how workers over the past 37 years have illuminated the anatomy, physiology, and function of this tiny bit of brain whose mere 4 g integrates almost all higher physiological functions. This anchor chapter in the volume reviews some nonendocrine aspects of hypothalamic function with emphasis on human disease.

## GENERAL FACTORS INFLUENCING EXPRESSION OF HYPOTHALAMIC DISEASE IN MAN

Before turning to a detailed description of specific disorders, it is useful to emphasize certain general aspects of hypothalamic disease.

### Locus of Function and Site of Lesion

The hypothalamus is characterized by a reticular organization similar to that of the reticular formation of the lower brainstem; like those of the

---

* *Present address:* Cerebrovascular Disease Research Center, Department of Neurology, New York Hospital-Cornell Medical Center, 525 East 68th Street, New York, New York 10021

TABLE 1. *Functions and related disorders of hypothalamic regions*

| POAHT | Tuberal HT | Post. HT |
|---|---|---|
| Receptor: Thermal, osmotic | (?) Caloric | Thermal setpoint |
| Integrates: Endocrine, thermal, autonomic | Cognition, endocrine, autonomic, caloric balance, fluid balance | Consciousness, cognition, complex endocrine, autonomic, thermal |
| Contains: Sleep-inducing mechanism, telencephalic parasympathetic paths | Thermoregulatory paths, final common endocrine paths | Reticular activating system, thermoregulatory effectors |
| Acute lesions: Insomnia? hyperthermia, DI, or SIADH | Hyperthermia, DI, HT endocrine disorders | Hypersomnia, emotional disorders, poikilothermia, autonomic storm |
| Chronic lesions: Insomnia, complex endocrine changes (e.g., precocious puberty), final common path endocrine changes, hypothermia, absent thirst | Medial: Memory loss, emotional disorders, hyperphagia and obesity, final common path endocrine disorders<br>Lateral: Emotional disorders, emaciation, loss of thirst | Memory loss, emotional disorders, poikilothermia, incoordination of autonomic reflexes, complex endocrine disorders (e.g., precocious puberty) |

Abbreviations: POAHT, preoptic anterior hypothalamic area; HT, hypothalamus; DI, diabetes insipidus; SIADH, syndrome of inappropriate ADH secretion.

latter structure, some hypothalamic functions have a relatively precise anatomic localization whereas others appear to depend on the activities of more diffusely distributed neurons. For example, only the magnocellular neurons of the supraoptic and paraventricular nuclei manufacture antidiuretic hormones; by contrast, neurons scattered over a large region of the preoptic anterior hypothalamus (POAHT) serve as thermodetectors, and a similarly widespread distribution appears to characterize other vegetative receptor systems. At least in regard to their integrating activity, some areas and even neurons (e.g., in the ventromedian area) undoubtedly participate in more than one hypothalamic function.

Accepting the difficulties of assigning a precise neuroanatomy for each hypothalamic function, one still can identify certain main activities represented in the structure's major subdivisions. Table 1 presents a brief review of the normal functions and related disorders attributed to each large HT region in man. The subsequent text will provide details on many of the points presented in the table.

## Rate of Progression of the Lesion

In the hypothalamus, as in most other regions of the brain, slow-growing lesions can reach a considerable size before causing symptoms, although suddenly or rapidly developing abnormalities of even a relatively small size can produce striking clinical changes, especially in strategic locations

(Table 1). Most disturbances of consciousness of hypothalamic origin, for example, are caused by acute or subacute damage. Similarly, sustained hyperthermia, as well as almost all severe disturbances of cardiovascular, pulmonary, or gastrointestinal function that accompany hypothalamic disease, reflect acute diencephalic damage rather than more insidious or chronic processes. Only two exceptions break the rule that profound autonomic abnormalities reflect acute damage or huge lesions. One is diencephalic epilepsy, a disorder characterized by recurrent paroxysms of autonomic hyperactivity. The other occurs when an HT lesion interrupts the final common pathway of a function. Examples of the latter include diabetes insipidus following damage to the supraoptic-paraventricular median eminence system and poikilothermia resulting from destruction of the final thermoregulatory outflow in the posterior HT.

In contrast to the acute lesions of the HT which tend to produce autonomic abnormalities, chronic damage to the structure tends to interfere with cognition and complex homeostatic functions such as caloric balance, fluid balance, and endocrine regulation. These complex functions require the integration of neural activities originating in multiple and widespread regions of the brain and their transduction by hypothalamus into either endocrine or neural effector mechanisms. Other parts of the brain cannot readily replace these synthetic, integrative, and transductive activities even though the component parts of each function may be produced without an intact HT. Chronic HT disease characteristically advances slowly and often spares at least part of the anatomical region it involves. The resulting signs and symptoms are those of chronic disruptions of function: for example, one observes empty-minded wakefulness (i.e., dementia) rather than sleep-like coma and obesity-emaciation rather than the severe autonomic disturbances that characteristically accompany equal-sized acute hypothalamic destruction.

### Bilaterality of Lesions

Unilateral HT damage seldom causes symptoms in man. With the exception of efferent sympathetic pathways, hypothalamic projections are not lateralized as are somatic sensorimotor tracts. Furthermore, either the right or left half of the HT can modulate and coordinate the complex general behavior that the region regulates, and only slight or transient functional loss occurs with unilateral damage. Thus, transient memory loss may occasionally follow unilateral limbic system lesions in man, but all available evidence indicates that sustained memory failure requires bilateral damage. Similarly, unilateral destruction of the ventromedial (VM) hypothalamus may produce a slight degree of obesity in animals, but a selective effect of VM destruction in man has not been demonstrated (220).

Because damage to the hypothalamus must be bilateral in order to cause symptoms does not mean that the lesion must be symmetrical. Corresponding functional areas or pathways can be interrupted at different levels in the hypothalamus and still produce symptoms. Cognitive functions are especially dependent on multiple telencephalic inputs to the HT and can be deranged by injury to different pathways converging on either half of the structure.

The fact that representative areas on both sides of the HT must be damaged or destroyed before a function disappears explains why disorders referable to disease or damage of the closely adjacent medial HT structures is more commonly seen than disorders of the lateral hypothalamus. The high incidence of neuroendocrine disorders in hypothalamic disease is at least partly due to the architectural accident of the juxtaposition of the paired neurosecretory tracts in the median eminence.

## Influence of Age on Hypothalamic Function

The nonendocrine functions of the HT change with the age of the organism. Newborns possess immature hypothalamic physiologic mechanisms (54,152,160) and require external assistance to maintain almost every homeostatic function that the adult hypothalamus regulates. Accordingly, disease or damage to the HT in the newborn may not be readily apparent. The morphology of the HT also changes with age. In the neonate, hypothalamic nuclei appear more compact than in the adult, mostly due to incomplete myelination of axon bundles traversing the structures (53).

Nisbett et al. (193) have postulated that deterioration of the nonendocrine functions of the ventromedial nuclear region of the HT may explain some of the characteristics of aging. They note that both aging men and VM-lesioned animals gain weight and accumulate centripetal fat, show decreased motor activity, have an increased incidence of sleep disturbance and psychological disorders, and demonstrate finicky eating habits. Similarly, the well-established susceptibility of the aged to environmentally induced hypothermia and the loss of thermal discomfort perception in these individuals may reflect a reduction in efficiency of hypothalamic heat regulating mechanisms.

## Causes of Hypothalamic Disease

Many diseases can damage the HT; the most frequently reported causes are neoplasms, with craniopharyngiomas the most common offender. Astrocytomas and suprasellar dysgerminomas, although much less frequent, are the next most common neoplasms. Because of the developmental origin of craniopharyngioma as well as some of the other tumors, the majority of reported HT disorders have occurred in patients under the age of 25 years. Table 2 lists the commonly reported causes of HT disease by age group and

TABLE 2. *Causes of hypothalamic disease*

I. Premature infants and neonates:
   Intraventricular hemorrhage
   Meningitis: Bacterial
   Tumors: Glioma, hemangioma
   Trauma
   Hydrocephalus, hydranencephaly, kernicterus

II. 1 month–2 years:
   Tumors: Glioma, especially optic glioma
      Histiocytosis X
      Hemangiomas
   Hydrocephalus, meningitis
   "Familial" disorders: Laurence-Moon-Biedl, Prader-Labhart-Willi

III. 2–10 years:
   Tumors: Craniopharyngioma
      Glioma, dysgerminoma, hamartoma, histiocytosis X, leukemia
      Ganglioneuroma, ependymoma, medulloblastoma
   Meningitis: Bacterial
      Tuberculous
   Encephalitis: Viral and demyelinating
      Various viral encephalitides and exanthematous demyelinating encephalitides
      Disseminated encephalomyelitis
   "Familial" Disorders: Diabetes insipidus, etc.

IV. 10–25 years:
   Tumors: Craniopharyngioma
      Glioma, hamartoma, dysgerminoma
      Histiocytosis X, leukemia
      Dermoid, lipoma, neuroblastoma
   Trauma
   Subarachnoid hemorrhage, vascular aneurysm, arteriovenous malformation
   Inflammatory diseases, meningitis, encephalitis, sarcoid, tuberculosis
   Associated with midline brain defects: agenesis of corpus callosum
   Chronic hydrocephalus or increased intracranial pressure

V. 25–50 years:
   Nutritional: Wernicke's disease
   Tumors: Glioma, lymphoma, meningioma
      Craniopharyngioma, pituitary tumors
      Angioma, plasmacytoma, colloid cysts
      Ependymoma, sarcoma, histiocytosis X
   Inflammatory: Sarcoid
      Tuberculosis, viral encephalitis
   Subarachnoid hemorrhage, vascular aneurysms, arteriovenous malformation
   Damage from pituitary radiation therapy

VI. >50 years:
   Nutritional: Wernicke's disease
   Tumors: Sarcoma, glioblastoma, lymphoma
      Meningioma, colloid cysts, ependymoma, pituitary tumors
   Vascular: Infarct, subarachnoid hemorrhage
      Pituitary apoplexy
   Infectious: Encephalitis, sarcoid, meningitis

gives the approximate order of frequency with which they occur. It can be seen that some illnesses, such as encephalitis, attack the various age groups equally often whereas others are restricted to specific ages of development.

### Functions of the Hypothalamus

The hypothalamus influences at least six major and fairly distinct nonendocrine functions: (a) consciousness and sleep; (b) cognition; (c) emotional behavior and affect; (d) autonomic balance; (e) caloric balance; and (f) water-osmolar balance. Each of these has several subfunctions, and certain behaviors, for example, feeding and aggression, sometimes arise in such close concert that one suspects that they may share the same neurological substrates. Nevertheless, these relatively large functional subdivisions provide a convenient framework on which to review our present knowledge.

## THE HYPOTHALAMUS AND HIGHER NERVOUS FUNCTIONS

### Consciousness and Sleep-Wake Behavior

#### Physiology and Studies in Animals

By 1939 investigators already knew that stimulation of the anterior hypothalamus produced behavioral unresponsiveness in animals and man (266), and that excitation of the posterior HT in animals evoked arousal. Conversely, anterior hypothalamic destruction in animals induced wakefulness, whereas posterior HT destruction in both animals and man (depending on size and acuteness of the lesion) precipitated drowsiness, stupor, or unresponsive coma. From these observations, Ranson (210) proposed a posterior HT "wake center," and Nauta (191) later postulated an anterior HT "sleep center."

The intervening years have seen the development of a theory of consciousness based on the excitatory influence of the brainstem reticular activating system (RAS) on forebrain structures (187). Sleep may be regarded as a form of normal behavior requiring the active participation of fairly specific pontine structures as well as the anterior HT. The decreased level of consciousness in normal sleep behavior may be rapidly reversed by stimulation of the animal. In contrast, coma results from a loss of RAS activity without the activation of other neuronal circuits and is clinically distinguishable from sleep by the inability of strong stimuli to produce arousal. General arousal is generated from the posterior HT and mesencephalic tegmentum and appears to require a lesser level of RAS activity than is necessary to maintain fully conscious mental activity. Normal sleep-wake cycles are determined by the anterior HT "sleep center" and the posterior HT "wake center" superimposed on influences from pontine and other brainstem nuclei.

According to Moruzzi (187), the posterior HT "wake center" depends on a functionally intact brainstem RAS for maintenance of normal awake behavior. Stimulation of this HT region produces EEG desynchronization and arousal. Large lesions that acutely damage the mesencephalic RAS immediately caudal to the HT (Fig. 1) result in unresponsive coma that may persist for several weeks before spontaneous sleep-wake cycles are restored (166). Similarly, in experimental animals sleep-wake cycles require several weeks to develop following isolation of the HT from the mesencephalon by midcollicular decerebration (104). Destruction of the posterior HT with sparing of the adjacent midbrain reticular formation leads to hypersomnolence. The hypersomnolence simulates physiologic sleep behavior in that the animal may be aroused with strong stimuli (213), but it is unclear whether brain metabolism in this state lies closer to the normal levels seen in sleep or to the depressed level of metabolism that exists in coma. In either case, the occurrence of EEG synchronization with posterior HT lesions indicates only that the RAS has been disturbed and does not differentiate between the decreased level of RAS activity associated with physiologic sleep behavior and the more complete loss of RAS function that initiates coma. Sleep-wake cycles, accompanied by a concurrent decline in hypersomnia, recover more

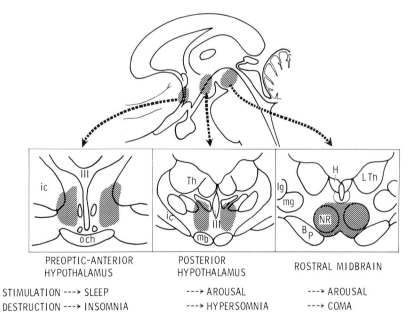

**FIG. 1.** Hypothalamic and midbrain regions critical to consciousness and sleep-wake behavior. Stimulation or destruction of the cross-hatched regions produces the noted results. BP, basis pedunculi; H, habenula; ic, internal capsule; lg, lateral geniculate; mb, mammillary body; mg, medial geniculate; NR, red nucleus; och, optic chiasm; th, thalamus; III, third ventricle.

rapidly after posterior HT insults than after mesencephalic lesions (179).

The anterior HT "sleep center" appears to function by inhibiting the RAS, thereby inducing EEG synchronization, and producing behavioral phenomena characteristic of true sleep. Destruction of this region in animals produces not only prolonged wakefulness (180,191) but also increased RAS activity, possibly accounting for the frequently noted hyperkinesis with such lesions.

Jouvet (144) and his school have gathered evidence that the induction of different stages of true sleep is mediated by different neurotransmitters. Slow-wave sleep is associated with the excitation of serotonergic pathways originating in caudal pontine nuclei, and rapid eye movement (REM) sleep with norepinephrine pathways arising in the locus ceruleus. Despite these known specific transmitter pathways, however, experimental studies fail to clarify whether hypersomnolence resulting from hypothalamic damage represents a selective neurotransmitter imbalance or merely the massive failure of many rostrally aimed arousal pathways irrespective of their neurotransmitter.

*Studies in Man*

Although clinical studies of sleep and consciousness in man have not independently produced any new understanding of consciousness and coma since 1939, the findings have provided strong support for the discoveries in experimental animals. The observation that precise posterior HT stimulation produces EEG desynchronization and arousal in animals has been repeated in man by Sano et al. (230) with similar results. Similarly, small stereotaxically placed, posterior HT lesions in man produce temporary hypersomnolence (230,241).

Examples of naturally occurring HT lesions in man and the disorders of consciousness they have produced are listed in Table 3. Von Economo (80) made the first clinical association of hypersomnolence with posterior HT-rostral mesencephalic lesions during his studies of encephalitis lethargica. Subsequent cases have demonstrated the separation of hypersomnic phenomena into unarousable coma produced by mesencephalic lesions (157, 223,258) and a more arousable hypersomnic state simulating natural sleep produced by posterior HT damage (67,68,168,264). A few examples of intermittent hypersomnia somewhat resembling idiopathic narcolepsy have also been associated with HT neoplasms, but the location of the offending tumors was no different from that of lesions causing a more sustained hypersomnia (42,96,240).

Von Economo (79) first noted insomnia in man as a consequence of anterior HT encephalitis. Although rare, other examples of this phenomenon have been reported with anterior HT lesions, but because of their size, neither the anatomy nor physiology of the lesions has lent itself to precise

TABLE 3. *Examples of hypothalamic lesions in man associated with changes in consciousness or sleep-wake cycle*

A. HT lesions producing arousable hypersomnolence

| Case | Age/sex | Diagnosis | ICP | HT lesion | Mesencephalic lesion | Symptoms |
|---|---|---|---|---|---|---|
| Davison and Demuth (67) | 52 M | Meningioma | NI | Posterior HT (R) compressed (L) tumor invasion | (L) None (R) slight compression | Hypersomnolence, arousable |
| Davison and Selby (68) | 31 M | Angioma | NI | Anterior HT to caudal mammillaries | None | Hypersomnolence, arousable |
| Lipsett et al. (168) | 6 M | Post-varicella encephalitis | NI | Infiltrated with lymphocytes throughout HT | None | Hypersomnolence, arousable |
| Weitzman and Triedman (264) | 27 F | Nonlipid histiocytosis | No autopsy: hydrocephalus | Anterior HT to mammillaries | None | Hypersomnolence, arousable |

B. HT lesions producing intermittent hypersomnolence

| Case | Age/sex | Diagnosis | ICP | HT lesion | Mesencephalic lesion | Symptoms |
|---|---|---|---|---|---|---|
| Castaigne and Escourole (42) | 38 F | Craniopharyngioma | NI | Anterior HT to posterior mammillaries | None | Intermittent hypersomnia |
| Fulton and Bailey (96) | 28 F | Lymphoma | NI | Anterior HT to posterior mammillaries | None | Intermittent progressing to sustained hypersomnia |
| Souques et al. (240) | 37 F | Metastatic epithelioma | NI | Anterior HT to posterior mammillaries | Between peduncles only | Intermittent progressing to sustained arousable hypersomnia |

TABLE 3 (Continued)

| | | | C. Lesions producing unarousable hypersomnolence (coma) | | | | |
|---|---|---|---|---|---|---|---|
| Case | Age/sex | Diagnosis | ICP | HT lesion | Mesencephalic lesion | Symptoms | |
| Koeppen et al. (157) | 70 M | Wernicke's | NI | Mammillary and perimammillary necrosis | Periaqueductal hemorrhage and gliosis | Hypersomnolence, unarousable | |
| Rosenblum and Feigin (223) | 57 M | Wernicke's | NI | Mammillary region necrotic | Hemorrhagic necrosis extending into tegmentum | Comatose; responsive after several days | |
| Walter et al. (258) | 62 F | Perithelial sarcoma | NI | Posterior HT | Rostral midbrain level to red nucleus | Hypersomnolence, difficult to arouse | |

| | | | D. Lesions producing hyposomnolence or altered sleep-wake cycles | | | |
|---|---|---|---|---|---|---|
| Case | Age/sex | Diagnosis | ICP | HT lesion | Mesencephalic lesion | Symptoms |
| Davison and Demuth (67) | 45 F | Craniopharyngioma | ? | Anterior HT, later invading posterior HT | None | Insomnia followed by arousable hypersomnia preterminal |
| Davison and Demuth (67) | 20 F | Postencephalitis lethargica (chronic) | NI | Neuronal depletion of caudal HT, slight loss in SO nuclei | Substantia nigra calcification | Altered sleep-wake cycles; parkinsonism |
| Killeffer and Stern (153) | 13 F | Craniopharyngioma | NI | Anterior HT to caudal mammillaries | None | Altered sleep-wake cycles; periods of hypersomnolence |
| Miyasaki et al. (184) | 45 F | Trauma | NI | Anterior and tuberal HT | None | Insomnia |

E. Lesions destroying hypothalamus without altering consciousness

| Case | Age/sex | Diagnosis | ICP | HT lesion | Mesencephalic lesion | Symptoms |
|---|---|---|---|---|---|---|
| Avioli et al. (13) | 12 F | Histiocytosis X | ? | Optic chiasm to mammillaries | ? | Alert |
| Bastrup-Madsen and Greisen (22) | 2 M | Acute leukemia | ? | Infiltrates throughout HT | Infiltrates throughout brainstem | Alert |
| Cussen (60) | 13 months M | Astrocytoma | ? | Optic chiasm to mammillaries | None | Alert, hyperactive |
| Daly and Nabarro (64) | 22 F | Dysgerminoma | Nl | Optic chiasm to mammillaries | None | Alert, until terminal hyperosmolar coma |
| Dods (74) Cases A, B, C | 2–19 months M and F | Astrocytoma | ? | Optic chiasm to mammillaries | None | Alert, hyperactive |

ICP, intracranial pressure; Nl, normal; ?, unknown.

study (67,184). Hypothalamic lesions associated with altered or reversed diurnal sleep-wake patterns also have tended to be large and poorly localized (67,153).

Large, slowly progressing lesions occasionally invade all regions of the HT without disturbing consciousness or sleep-wake patterns (13,22,60, 64,74). It must be assumed in such patients either that the mesencephalic portion of the RAS and at least part of the HT "sleep" and "wake centers" remain functionally intact or that more dorsal diencephalic or rostral telencephalic regions possess the capacity to establish a diurnal rhythm of consciousness. Presently at least, no direct evidence suggests the latter possibility.

### Memory

The close anatomic association between the hypothalamus and structures of known importance to memory, such as the reticular formation of the brainstem, the hippocampus, and the limbic system, would lead one to predict that the functions of memory recall also would include hypothalamic pathways, and experimental and clinical evidence supports this supposition. Gold and Proulx (109) demonstrated in animals that the acquisition of new short-term memory traces requires an intact ventromedian hypothalamus. Similarly, escape behavior to aversive stimuli could not be conditioned in rats with lateral hypothalamic (LH) lesions (33). Although the phenomenon initially was regarded as due to the loss of affective responses to sensation, Schwartz and Teitelbaum (233) demonstrated that it must be a more specific inability to learn new associations, since the animals retained previously learned aversion responses. The apparent short-term memory loss caused by LH lesions has been ascribed by these investigators to the disruption of laterally lying amygdalofugal pathways projecting to the region of the VM nucleus. As noted below, similar anatomical distributions of diencephalic lesions are found in patients with memory failure.

Analysis of cases with large destructive lesions helps little in defining the anatomy of the pathways subserving memory. In a few instances, however, spontaneously occurring diseases have damaged the brain fairly precisely and these examples are instructive.

Claude and L'Hermitte (48) first correlated memory loss in man with neoplastic damage to the mammillary bodies. Subsequent reports confirmed the association of isolated HT lesions causing short-term memory loss, but in each case the subjects examined had relatively large invasive tumors destroying the VM regions of the HT in addition to the mammillary bodies (18,99,147,269). Many workers have described damage to the region of the mammillary bodies in the Wernicke-Korsakoff syndrome and attributed the memory loss to this abnormality (35,216). Victor, Adams, and Collins (256) disputed this conclusion because they found that such memory failure correlated better with degeneration of the dorsomedial nucleus of the thalamus than with degeneration of the mammillary bodies. A more precise localization of the HT area critical to memory loss is suggested by the study of the patient described by Reeves and Plum (214). This young woman with striking memory loss and confusion had a well-localized hamartoma that damaged the ventromedian and premammillary region of the HT but entirely spared the mammillary bodies themselves.

Most human examples of memory failure with hypothalamic disease manifest the classic Korsakoff syndrome, displaying a loss of short-term memory but a relative preservation of immediate and long-term recall. Some unusual variants have been described, however. Roeder and Müller (220) reported that one of their patients subjected to a stereotaxic lesion of the VM developed selective visual memory loss as a result. Kahn and Crosby (147) described a young man with a dysgerminoma destroying the posterior hypothalamus, mammillary bodies, and optic tracts. The subject not only suffered the blindness expected with such a lesion but also experienced a poor auditory short-term memory with retained tactile memory.

From the above it appears that memory loss frequently follows ventromedian hypothalamic destruction in man, but less consistently occurs when lesions fall immediately outside this precise region. The hippocampus is known to be essential to short-term memory (201), but the surgeon can section the fornix (a major hippocampal projection to the hypothalamus) without producing memory loss (101). It may be postulated that the brain compensates for the loss of the fornix of hippocampal projections that travel to critical hypothalamic sites via the amygdalofugal paths traversing the lateral hypothalamus to terminate in the VM nuclear region. According to this view, memory loss from hypothalamic lesions would occur only when disease destroyed either this critical, terminal VM zone for memory or both the amygdalofugal and the fornix projections to it.

### Other Mental Functions

Standard tests of "intelligence" depend heavily on the ability of the subject to remember instructions, to be motivated to carry out the test, and to behave in a relatively restrained manner. Consequently, memory failure, apathy, or unrestrained emotional outbursts that may result from HT disease can limit any assessment of higher cognitive functioning. Descriptions of mental function in a few patients with restricted ventromedian hypothalamic disease (4,174,214) have been reported despite these difficulties. The patients display severe mental abnormalities, some appearing demented and others retaining almost no contact with the environment and perhaps hallucinating as well. The latter mental states more closely resemble acute delirium than quieter behavior characteristic of the degenerative dementias.

## Emotion and Behavior

Bard's 1928 experiments demonstrating "sham rage" (19) showed that the hypothalamus has a major influence on emotions. Experimental work and human observation in this area has produced a huge literature, too extensive for this review. As a summary of current evidence, the hypothalamus appears to exert its effects on behavior in three spheres: it coordinates the motor, autonomic, and endocrine components of behavior; it produces the behavior appropriate to the affective state; and it influences the intensity of each behavior (145). Hypothalamic activity also may influence the intrinsic affective state of the organism. At least this is suggested by accounts from patients and by the behavioral reinforcement conferred by self-stimulation of the medial forebrain bundle of the HT in experimental animals (196). Lesions of the HT produce pathologic emotional states by altering one or more of these basic functions.

## Rage and Fear

The emotions of rage and fear represent the most commonly reported behavioral disturbance due to HT disease in human adults and in children older than 2 years; they less frequently characterize HT disease in infants. Rage and fear represent the affective correlates to the fight-or-flight behaviors associated with sympathetic autonomic discharge.

Characteristically, hypothalamic-engendered rage and fear in man are accompanied by fully coordinated behavioral responses that tend to occur as episodic outbursts with normal behavior in between. Such outbursts, as might be expected, generally include an intense autonomic component. Many such patients are aware that their behavior is abnormal and may even apologize for it. Although inappropriately intense, the rage and fear usually burst forth in response to a threatening stimulus, such as restraint or a delay in feeding (22,214).

Attacks of rage and aggression also can result from lesions of the medial temporal lobes and orbitofrontal cortex as well as of the HT (206). The destructive lesions of the human HT that produce these behaviors all involve the basal parts of the structure (4,17,22,153,174,214), and the location of the smaller of these lesions suggests that the critical sites involve either pathways descending from the cerebral cortex to the hypothalamus (273) or the VM nuclear region where these inputs converge (214). In either case the inhibitory influence of the telencephalic inputs on HT behavioral control is disrupted. Rage is the usual accompanying emotional expression produced by destructive VM lesions in man. In contrast, stimulation of the posterior HT regions that results in sympathetic responses in man also elicits fear reactions and the emotion of horror rather than rage (117,195, 230).

## Apathy

Apathy is the emotional and behavioral antithesis of rage and fear. Lesions of the lateral hypothalamus or posterior sympathetic tracts in animals produce an apathetic hypoactive state. It has been more difficult to document this locus in man, but what evidence exists suggests a similar anatomical distribution (26,149,254,264). Stereotactic lesions placed either in the medial posterior hypothalamus (230,232) where sympathetic responses were elicited on stimulation or in the caudolateral region of the structure (241) have produced apathy and hypoactivity in patients who previously displayed attacks of aggressive behavior.

### Euphoria

Euphoria and the accompanying appearance of well-being are unusual and only rarely considered as manifestations of HT dysfunction in the adult (4,174). Anterior HT tumors in infants under the age of 2 years, however, are regularly accompanied by inappropriately cheerful behavior (32,60,73, 74). The apparent lack of concern is associated with hyperactivity and usually disappears or evolves into aggressive behavior after that age (100).

Speculation on the pathogenesis of this behavior disorder centers on the disruption of frontocortical-thalamo-HT pathways by anterior HT lesions. Such lesions would produce the equivalent of a frontal lobotomy, but such a suggestion hardly explains the age-linked predominance of euphoric behavior in infants.

### Sexual Behavior

Excessive or uncontrollable sexual behavior represents a rare and questionably established manifestation of limbic system disease in man. Although medial temporal lobe lesions produce hypersexual behavior associated with aggression, Poeck and Pilleri (206) point out that caudal HT lesions can produce hypersexual behavior divorced from rage or excessive desire.

Loss of libido represents a frequent symptom of HT disease, especially in the male. Almost all appropriately evaluated cases have been associated with gonadotropin deficiency, presumably due to loss of LHRH secretion in the diseased HT (184,214,254). Gonadotropin measurements have been omitted in the few patients described with normal levels of sex steroid hormones and loss of libido (7,194,275). Similarly, reports of loss of libido following unilateral stereotactic lesioning of the VM nucleus in homosexuals have not included gonadotropin levels in the affected subjects (40,220), although hypogenitalism has not been noted in these patients.

## AUTONOMIC FUNCTIONS

### Heat Regulation

*Experimental Studies*

Aronsohn and Sachs in 1885 discovered the hypothalamic influence on thermoregulation. By 1913 Meyer (183) had developed his classic concept of separate anterior HT heat loss and posterior HT heat production centers. Ranson's 1939 presentation before this Association (211) provided a wealth of new experimental detail but differed relatively little in its conclusions from Meyer's view, except to recognize that the posterior hypothalamus contained mechanisms for heat production and heat loss even though the latter seldom manifested themselves since gross damage to the area usually resulted in hypothermia.

Several excellent reviews summarize the main recent advances in understanding the physiology and pharmacology of heat regulation (25,97,121, 129,143). Although this section examines hypothalamic disorders of heat regulation in man, these are better understood if one briefly recapitulates the pertinent experimental evidence (Fig. 2).

Many investigators have demonstrated that the preoptic anterior hypothalamus contains both warm receptors that increase their frequency of firing with elevated temperatures and activate heat dissipation processes and cold receptors that respond to cold with an increased firing rate and thereby excite heat production mechanisms. Warm receptors appear to predominate in the POAHT, but both the warm and cold POAHT receptors are important thermodetectors, and their efferent signals, along with those from lower CNS thermoreceptors, provide a major component of what is sensed elsewhere in the HT as deep body temperature. No clear evidence

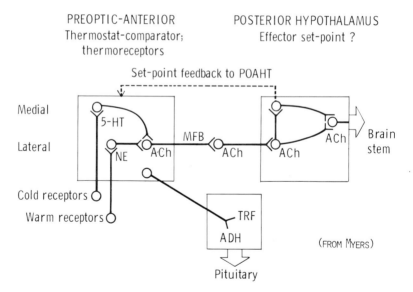

**FIG. 2.** Hypothalamic thermoregulation depends on preoptic-anterior thermoreception and integration and posterior effectors and setpoint, as described in the text. Injection of 5-HT into the POAHT of the cat elevates body temperature; NE injection inhibits this response. Synapses between thermoregulatory interneurons and effectors in both heat gain and heat dissipation pathways appear to be cholinergic. ACh, acetylcholine; ADH, antidiuretic hormone; 5-HT, 5-hydroxytryptamine (serotonin); MFB, medial forebrain bundle; TRF, thyrotropin releasing factor. [Redrawn from Myers (190) with permission.]

indicates that the posterior HT normally contributes a thermoreceptive function that influences physiologic temperature regulation, although local posterior HT receptors may alter behavioral responses to thermal stress (1,129).

Homeothermic animals maintain a constant body temperature by activating their heat production and heat loss mechanisms in a manner that resembles the functioning of a thermostat regulating around a predetermined thermal setpoint. The HT contains this "biologic thermostat" and thus determines the body temperature. Whether the setpoint function is mediated by a functionally distinct and anatomically specific group of neurons or merely represents the hypothalamic integration of the various independent but thermosensitive heat loss and heat gain mechanisms is an area of controversy. In either case, stimulation of pyrogen receptors present in both the POAHT and the more caudal brainstem results in fever by elevating this setpoint function to a higher base line (189,253).

The HT integrates the thermal information generated by HT thermoreceptors, pyrogen receptors, and setpoint function with data acquired from other CNS and peripheral thermoreceptors and also with nonthermal stimuli that influence temperature control. These include among others the level of arousal, the diurnal rhythm, and the stage of the menstrual cycle. Hensel (129) concludes that the setpoint for thermal regulatory control reflects a combination of these various inputs and not just the local temperature at any given specific central or peripheral site.

The hypothalamus coordinates and activates appropriate heat dissipation and heat production mechanisms by the following systems: (a) affective sensations of thermal discomfort (1,124); (b) behavioral responses whereby the organism attempts to alter its immediate or remote environment (41); and (c) physiological responses. These latter may be either autonomic, such as changes in vascular tone, sweating or panting, or endocrine, such as TSH changes in lower animals and perhaps human infants (215). In addition, changes mediated by the sympathetic nervous system occur in the organism's metabolic rate which require POAHT receptor integrity and provide the predominant source of thermogenesis during prolonged cold exposure (94).

The clarification of the HT mechanisms governing thermoregulation allows one better to understand not only the effects of more gross animal experiments but also hypothalamic-engendered disorders of temperature control in man. The evidence is interpreted to indicate that the POAHT generates thermal signals, integrates these with other thermally dependent signals, and sends an effector signal to the posterior HT via the medial forebrain bundle. Gross stimulation of the POAHT activates mainly the numerically more abundant warm receptors, leading thereby to heat dissipation and a fall in body temperature. Conversely, gross POAHT or LH destruction causes transient hyperthermia that subsides gradually over several days to weeks (212) as more caudally placed and peripheral thermoreceptors take charge (9). According to Squires and Jacobson (242), more selective medial POAHT destruction in experimental animals results in a chronic hypothermia at room temperature because this area contains most of the cold receptors which initiate heat production.

The posterior HT is the final integration site for thermoregulation and generates the effector signals for physiologic, affective, and behavioral responses. Myers (190) believes that the perimammillary area also contains a site essential to the setpoint function. Selective stimulation of dorsomedial-caudal HT and rostral mesencephalic sites activates heat production mechanisms. In contrast, ventrolateral-caudal HT stimulation inhibits heat production and activates heat dissipation. As Hardy (125) points out, these sites are the caudalmost loci at which reticular neurons can influence thermoregulation, for thermal information in the lower brainstem exists only in axonal tracts. Accordingly, large posterior HT lesions destroy both heat loss and heat production mechanisms and result in poikilothermia.

## Human Disorders of Heat Regulation

### Sustained Hypothermia

Human HT diseases result in sustained hypothermia either by destroying heat production mechanisms or by establishing an abnormally low setpoint. Under most environmental conditions, basal heat production is sufficient to maintain body temperature above that of the environment. When the ambient temperature equals or exceeds thermal neutrality, heat production mechanisms become superfluous or even burdensome to thermoregulation. It follows that the detection of hypothermia due to impaired heat production mechanisms requires an ambient temperature lower than the normal body temperature. In contrast, an abnormally low setpoint would establish a low body temperature that maintains a new constancy independent of environmental fluctuations.

Sustained hypothermia is rare but may occur with anterior hypothalamic lesions in man as in animals. Fox et al. (90) reported a young man who developed varying degrees of hypothermia when exposed to low ambient temperatures but activated a fully coordinated heat dissipation response when his body temperature was elevated to 37°C. Although the patient could not activate nonshivering thermogenesis, components of other heat production mechanisms less dependent on the POAHT region were retained. These included shivering with profound hypothermia and thermal discomfort. Autopsy disclosed distortion and extensive gliosis of the medial POAHT of unknown cause (Fig. 3). As with Squires and Jacobson's (242) experimental study, the case demonstrates the independence of hypothalamic heat loss and heat gain mechanisms in the anterior HT and pre-

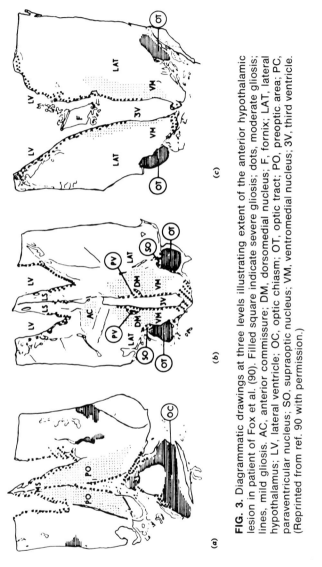

FIG. 3. Diagrammatic drawings at three levels illustrating extent of the anterior hypothalamic lesion in patient of Fox et al. (90). Filled square indicate severe gliosis; dots, moderate gliosis; lines, mild gliosis. AC, anterior commissure; DM, dorsomedial nucleus; F, fornix; LAT, lateral hypothalamus; LV, lateral ventricle; OC, optic chiasm; OT, optic tract; PO, preoptic area; PC, paraventricular nucleus; SO, supraoptic nucleus; VM, ventromedial nucleus; 3V, third ventricle. (Reprinted from ref. 90 with permission.)

sumably reflects a disproportionate destruction of cold receptors in that region. Hypothermia due to a chronically lowered setpoint function has been well documented in man by Hockaday et al. (138). Both of Hockaday's patients had central nervous system disease. An autopsy of one of the cases showed infarcts that involved hypothalamus, but other neurologic abnormalities were too widespread to allow the firm conclusions to be drawn that the disturbance was due to hypothalamic damage.

*Paroxysmal Hypothermia*

Paroxysmal hypothermia consists of attacks of lowered body temperature that vary widely in frequency from daily to more than a decade apart. The attacks usually begin abruptly and with no consistent precipitating cause or change in the environment. They last from minutes to days and are characterized by the sudden onset of sweating, flushing of the skin, and a fall in body temperature, usually to 32°C or lower. Fatigue, decreased mental responsiveness, hypoventilation, hypotension, cardiac arrhythmias, ataxia, lacrimation, and asterixis may accompany the temperature decline. The attacks subside either slowly (hours to days) or rapidly with shivering and peripheral vasoconstriction. During hypothermia, both heat production and heat dissipation mechanisms respond normally in fully coordinated fashion to thermal stress, although around a lower temperature setpoint. Some of the patients have had abnormally low body temperatures between attacks.

Table 4 lists the 13 patients with paroxysmal hypothermia reported in the literature; Gowers (1907) (114) and Cunliffe et al. (1971) (56) described apparently similar subjects but omitted mention of body temperature. Of the tabulated patients, only two were studied post-mortem. Penfield's (200) patient had a cholesteatoma of the third ventricle compressing but not invading the HT. The patient of Noël et al. (194) had gliosis and cell loss in the premammillary and arcuate nuclear regions of the HT. Other cases, however, have manifested circumstantial evidence for HT disease, such as diabetes insipidus (119), and at least five patients have had agenesis of the corpus callosum (119,194,229,235).

Although many aspects of paroxysmal hypothermia suggest the occurrence of an epileptic discharge that resets a central thermostat, most of the patients have not responded to anticonvulsant drugs. Animal and human observations on the pathogenesis of the disorder coincide fairly closely, however. The localization of gliosis in the patient of Noël and co-workers reportedly involved the same premammillary area that Myers (190) has implicated as mainly responsible for setpoint functions in animals; regrettably, no illustrations of the lesions were provided.

*Sustained Hyperthermia*

Abnormalities that theoretically could result in sustained hyperthermia of HT origin include: (a) loss of heat dissipation mechanisms; (b) stimulation

TABLE 4. Paroxysmal hypothermia—characteristics of the documented cases reported in the literature

| Case | Age/sex | Base-line setpoint (°C) | Episodic setpoint (°C) | Duration of episodes | Frequency of episodes | Associated paroxysmal symptoms | Dx or other signs and symptoms | Response to anticonvulsants |
|---|---|---|---|---|---|---|---|---|
| Penfield (200) | 41 F | 37–37.8 | 36.1 | 5–12 min | q.12 hr | Salivation, low spontaneity, tearing, hiccuping, pupil dilation | Cholesteatoma | Improvement |
| Hines and Bannick (135) | 22 M | 37.2 | 32.8–33.3 | 1 hr | q.2 hr for 4–6 weeks per year | Nausea, vomiting, fatigue | None | Excellent responses |
| Hoffman and Pobirs (140) | 43 M | 36.7 | 33.9–34.7 | 1–2 hr | Daily for 3–4 months/years | Weakness | None | Controlled episodes |
| Lennox (163) | 10 months F | 36.1 | 33.3 | 8–10 hr | Monthly | Sinking feelings, pallor | Changed to hyperthermic episodes at 18 months | Hyperthermic episodes controlled |
| Duff et al. (76) Case #1 | 30 M | 36.1–36.7 | 28.3–29.4 | 4–5 hr | Daily | Unresponsiveness, bradycardia, atrial fibrillation, hypotension, hyporeflexia, sluggish pupils, EEG slowing | None | ? |
| Case #2 | 21 F | 37 | 30.6–32.8 | 4 days–4 weeks | 1 per several months | Hypotension, bradycardia, slow "cerebration," EEG slowing | Generalized seizures | Hyperthermia started while on diphenylhydantoin (Dilantin®); no response |
| Shapiro et al. (235) Case #1 | 46 F | 35–37 | 33.6 | ½–2 hr | 1–4 days | Fatigue, lethargy, low spontaneity, dyspnea, bradycardia, hypotension, EEG slowing | Agenesis of corpus callosum | No response |
| Case #2 | 34 F | 34–36 | 30 | Minutes–hours | 1 week–several months | Stupor, mute, bradycardia, hypotension, EEG slowing | Agenesis of corpus callosum | No response |

TABLE 4 (*Continued*)

| Case | Age/sex | Base-line setpoint (°C) | Episodic setpoint (°C) | Duration of episodes | Frequency of episodes | Associated paroxysmal symptoms | Dx or other signs and symptoms | Response to anticonvulsants |
|---|---|---|---|---|---|---|---|---|
| Guihard et al. (119) | 7 M | 35 | 31 | "Brief" | 3 months | Bradycardia, hypotension, lethargy | Agenesis of corpus callosum, precocious puberty, polydipsia-polyuria | No response |
| Noël et al. (194) | 41 M | 34.4 | 29 | Several hours | 1–2 days | Low spontaneity, bradycardia, hypotension, EEG slowing | Agenesis of corpus callosum, gliosis of arcuate and premammillary HT, hypogonadism | ? |
| Fox et al. (91) | 41 F | 37 | 34 | 10–20 min | Daily – q.2 hr | Slight confusion, hypotension, tachycardia | None | No response |
| Thomas and Green (250) | 39 F | 35.5–36 | 33 with daily drops to 31 | 1–4 weeks | ? q.several months | Lethargy, ataxia, asterixis, bradycardia | Generalized seizures; amenorrhea | No response |
| Sadowsky and Reeves (229) | 33 M | 34.7 | 33.7 | ? hours | ? | Lethargy, fatigue | Agenesis of corpus callosum | No response |

of heat production; and (c) elevation of the setpoint. The latter two mechanisms would stimulate heat production in the presence of normal ambient temperatures and may be impossible to differentiate clinically. The first of these mechanisms, loss of heat dissipation, would not be clinically apparent at room temperature but would require ambient temperatures of 37° or more. This may be one reason why no case of isolated loss of heat dissipation due to central nervous system disease has been reported.

All evidence indicates that sustained hyperthermia due to neuronal damage occurs only as a consequence of an acute pathological process, and even then is a self-limited phenomenon. The most frequent causes include craniotomy, trauma, or bleeding into the region of the anterior hypothalamus or third ventricle (58,86,113,148,184). A few cases of sustained hyperthermia have been reported in association with damage of the tuberal region of the HT, but with sparing of the posterior HT (276). With sustained hyperthermia, the body temperature rises to potentially fatal levels of 106° or higher as a result of active heat production. As noted by Cushing (58), cardiovascular changes normally accompanying fever are disproportionately small with hyperthermia due to HT lesions.

If one restricts the evidence to cases in which post-mortem or equivalent detailed examination has ruled out more likely causes of fever, hypothalamic hyperthermia persists for no more than 2 weeks after the acute event. Cases of hyperthermia reported as extending for longer periods almost certainly have reflected the presence of undetected pneumonia, malignancy, or some other systemic problem. The literature of the preantimicrobial era reports patients with brain tumors in whom multiple brief hyperthermic episodes were attributed to repeated nonfatal decompensations of the tumor itself including hemorrhage or necrosis (66,276), but infection or pulmonary infarctions were not ruled out. Lipsett et al. (168) report prolonged fever with HT encephalitis, but in such instances persistent infection was as likely a cause as direct neuronal damage.

*Paroxysmal Hyperthermia*

Neurogenic paroxysmal hyperthermia is a rare condition characterized by brief episodes of shaking chills, high spiking fever, and occasionally other autonomic phenomena that resolves either quickly with sweats and flushing or more slowly over several hours. The disorder differs from repeated episodes of sustained neurogenic hyperthermia by being of much briefer duration. More important is the differentiation of the disorder from the many other and more frequent causes of intermittent fever, and the diagnosis of neurogenic hyperthermia must be made by exclusion.

Although the disorder is usually nonrhythmically paroxysmal in nature, one patient reported by Wolff et al. (272) had episodes of hyperpyrexia occurring in regular 3-week cycles. The studies of metabolism and thermoregulatory function in this patient and in others with the syndrome have failed to demonstrate any consistent abnormality.

The HT has been incriminated in patients with paroxysmal hyperthermia mostly because other clinical signs suggested HT disease; few cases have been examined pathologically. The autopsied patients of Reeves and Plum (214) and Anderson et al. (7), however, both had intermittent hyperthermia accompanying lesions involving the posterior tuberal region and extending anteriorly along the floor of the third ventricle. The presence of perimammillary lesions in these two patients and the paroxysmally intense nature of the symptoms suggest that the syndrome is caused by a seizure-type of disorder akin to that which readjusts the setpoint in paroxysmal hypothermia. Lennox (163) describes a girl whose symptoms support this concept: her paroxysmal illness started with well-documented attacks of hypothermia which subsequently shifted to hyperthermia, finally controlled by phenytoin.

*Relative Poikilothermia*

Poikilothermia is defined as the fluctuation in body temperature of greater than 2°C with changes in ambient temperature (129). The disorder results from impairment or loss of both heat production and heat dissipation mechanisms and is the most common central neurogenic abnormality of heat regulation in man. The severity of impaired thermal regulation varies widely from patient to patient, depending on the nature and degree of the hypothalamic defect.

Poikilothermia in man results from destruction or dysfunction of the final integrator and effector sites for thermoregulation in the posterior HT and rostral mesencephalon. Thermal discomfort, coordinated autonomic function, and behavioral regulation all disappear with large posterior HT lesions. As a result, many patients with poikilothermia are unaware of their condition and show no signs of discomfort or behavioral regulations with thermal stress. Furthermore, although nonthermal stimuli may evoke appropriate autonomic reflexes in these patients, thermal stress does not. Brainstem lesions destroying the tracts descending from the posterior HT likewise produce poikilothermia, as does metabolic depression (e.g., by anesthesia or myxedema) of the hypothalamic interneurons integrating and effecting thermoregulation.

Large destructive lesions of the posterior HT cause severe poikilothermia, marked by a large decline in body temperature when the environmental temperature is in the usual range of 20 to 25°C. Consequently, most cases of poikilothermia present themselves with lowered body temperature. The few patients reported with hyperthermia have been observed in environments with high ambient temperatures and most have been children, who normally have higher metabolic rates than adults (12,153). Almost the entire HT is sometimes destroyed in patients with this syndrome (66,153, 247). The hypothermia that accompanies the posterior HT-rostral mesencephalic destruction of Wernicke's disease is most likely poikilothermic in nature (157,204,223).

Clinical evidence indicates that the lesion producing poikilothermia must be bilateral. As proof of this point, Branch et al. (31) reported a patient with sarcoid granuloma unilaterally involving the posterior HT who remained normothermic until a small infarct damaged the contralateral side. Only after the infarct did body temperature fall toward ambient levels.

Inefficient thermoregulation affects both the newborn and the senescent even in the absence of acquired HT disease. Neonates, especially when premature, are normally moderately poikilothermic and become even more so when insults such as intraventricular hemorrhage damages the HT (54). The body temperature of the elderly also tends to fall abnormally in the face of a cool environment. MacMillan et al. (171) report that elderly survivors of hypothermic episodes have inadequate heat-generating mechanisms and often lack thermal discomfort.

**Sympathetic and Parasympathetic Functions of the Hypothalamus**

The centripetal influences of the HT on sympathetic and parasympathetic function in man have stimulated many thoughtful studies which involve not only the neurological disciplines, but studies in cardiology, respiration, gastroenterology, and psychosomatic medicine as well. Even to cite the informative reviews on the subject could strain the length of our bibliography. Furthermore, studies of this aspect of hypothalamic function in man and animals are so extensive and well established that one suspects that even the details have become common knowledge to neurological clinicians and biologists. Accordingly, this section is relatively brief.

The classic studies of Ranson, Hess, and others interpreted the HT as the "head ganglion of the autonomic nervous system" (131,146,259). The POAHT was considered to control overall parasympathetic tone, and the posterior HT, sympathetic tone. In 1939, however, no one believed that the HT directly influenced specific autonomic reflexes.

Experimental studies of the last 20 years have revised some of these concepts and indicate that the notion of a "head ganglion" is accurate only in that the HT functions as a critical point where more rostrally originating, telencephalic autonomic influences converge and are integrated (158). Gross electrical stimulation of the anterior HT results in parasympathetic responses due to the large number of descending telencephalic pathways in the region that mediate parasympathetic functions. Gross stimulation of the posterior HT yields sympathetic responses partly emanating from the diffuse sympathetic reticular core that lies within the region but also arising from telencephalic originating tracts, such as the stria terminalis, that terminate in the tuberal and posterior HT and that produce sympathetic responses when stimulated.

Stimulation at any anterior-posterior level of the HT can produce either sympathetic or parasympathetic responses depending mainly on the transverse locus of stimulation but partly on the nature of the stimulus as well. Sympathetic responses tend to follow ventral and medial stimulation (16, 118) but occur with stimulation of structures as far lateral as the medial forebrain bundle (MFB) (224). Parasympathetic responses result from either periventricular or lateral HT stimulation. Both animal and man tend to respond similarly to stimulation in anatomically homologous regions. Sano et al. (230), for example, evoked in humans symptoms of tachycardia, tachypnea, hypertension, and pupillary dilation when they stimulated the posteromedial HT. More lateral stimulation evoked hypotension, bradycardia, and pupillary constriction.

The more refined experimental approaches of recent years have disclosed that contrary to earlier beliefs, the HT autonomic functions are not diffuse but highly specific, even to the degree that stimulation of an isolated fiber can engender a specific response in specific peripheral organs (158). A similar precision applies to afferent pathways: it now is known, for example, that the HT receives direct specific sensory inputs from peripheral autonomic receptors (134,251).

Despite the capacity of the HT for individual autonomic responses, however, its predominant and singular role is to coordinate these autonomic functions with each other and with the rest of the organism's behavior (274). Without the HT, one can still obtain individually normal base-line autonomic activities, but the complex coordination of autonomic and behavioral responses that assures homeostasis is lost.

## *Diencephalic Epilepsy*

Paroxysmal, spontaneous, abnormal neuronal hyperactivity arising either within the hypothalamus itself or in structures that directly project to it can involve hypothalamic efferent autonomic functions and produce seizures that include coordinated patterns of intense autonomic hyperactivity. When such autonomic hyperactivity produces the initiating symptoms of the seizure, one appropriately terms the condition diencephalic epilepsy. For the clinician, such attacks often provide a difficult problem in diagnosis because psychophysiologic responses to unexpressed or subconscious emotional drives sometimes can induce similar episodes of nonepileptic origin. The unfortunate result has been to make the concept of diencephalic epilepsy something of a wastebasket diagnosis. It deserves better.

Autonomic hyperactivity can accompany many nonhypothalamic seizures in addition to diencephalic epilepsy. Striking cardiovascular and respiratory changes routinely accompany generalized convulsions. Gastrointestinal, respiratory, and cardiovascular alterations contribute prominently to the patterns of temporal lobe epilepsy and may reflect seizure discharges from that structure projecting to the hypothalamus. Intrinsic lesions of the brain that either lie in regions closely adjacent to the HT or destroy part of that structure can produce epileptic attacks characterized by functional alterations in cardiovascular, respiratory, heat-regulating, gastrointestinal, and glandular systems (163,235). These autonomic alterations form a pattern that is excessive or inappropriate to the present environmental stimuli but could be considered to be a well-coordinated response to a different environmental stress. For example, the activation of heat dissipation mechanisms in paroxysmal hypothermia, as mentioned above, would be a well-coordinated response to a hot environment, although it is inappropriate at normal ambient temperatures. In most well-studied patients with organic lesions associated with such spontaneous paroxysmal attacks, consciousness has been preserved and the subjects have reported the behavioral and affective responses appropriate to the altered autonomic response (170).

Penfield (200) coined the term diencephalic epilepsy to describe attacks of autonomic hyperactivity occurring in a patient with a cholesteatoma lying within the third ventricle. Others have described convincing examples of the disorder in association with the third ventricular dilatation of hydrocephalus or accompanying small tumors of the thalamus (83). Still other patients have had paroxysms associated with tumors directly invading and destroying the hypothalamus (170,239), although all anatomically studied instances have spared at least part of the posterior, effector regions of the structure.

The diagnosis of diencephalic epilepsy is best reserved for patients in whom one finds well-documented evidence of hypothalamic dysfunction. Most such patients have other evidences of diencephalic abnormality by clinical or laboratory tests. The EEG in such patients often has an abnormally slow frequency, but one seldom finds the paroxysmal dysrhythmias that characterize generalized or focal cerebral epilepsy. Currently available anticonvulsants control the paroxysms of only some of these patients, and their effect is not reliable as a diagnostic aid. One must be especially careful not to overdiagnose diencephalic epilepsy in children complaining of nothing more than vague autonomic symptoms. Some children may report intermittent autonomic disturbances unaccompanied by any other symptom or sign of central nervous system dysfunction. In almost all such instances, the early symptoms are unassociated with any serious disorder in later life (163).

### Cardiovascular Functions

Early research established the extensive cardiovascular influences of the HT, and more recent studies have emphasized the importance of these influences in human disease. Several investigators have demonstrated that the HT not only receives direct baroreceptor (103,134) and chemoreceptor (251) inputs but directly influences these reflexes under normal conditions (153). Stimulation of the sympathetic zone of the HT in animals and man causes hypertension, tachycardia, vasoconstriction of visceral beds, vasodilation of muscular beds, increased cardiac output, and increased cardiac contractility in association with the expression of rage or fear (274). Such stimulation causes various EKG changes (182,262) and can reduce the cardiac threshold for ventricular fibrillation (255). Stimulation of parasympathetic structures in the posterior HT tends to increase blood flow in the bowel and to decrease blood flow in skeletal muscle with little or no accompanying change in blood pressure or pulse rate. Stimulation of the anterior HT, on the other hand, creates profound hypotension and bradycardia (274).

An extensive literature describes the changes in cardiovascular function that can accompany disease of the brainstem and hypothalamus in man. The abnormalities include hypertension, cardiac arrhythmias of various types, and a number of EKG abnormalities, some of which can simulate acute myocardial infarction (122,236). Many such patients have no evidence of previous cardiovascular disease. It is well documented, however, that subarachnoid hemorrhage and, by inference, other acute diseases of the brainstem can induce myocardial necrosis in a nonvascular pattern, possibly due to the outpouring of catecholamines that accompanies the acute event (248). Since a high incidence of cardiovascular abnormalities especially accompanies acute intraventricular hemorrhage or acute rupture of anterior communicating artery aneurysms, both of which frequently damage the anterior HT (52), many physicians have assumed that HT damage was central to the cardiac changes. Hawkins and Clower (127), however, showed in the experimental animal that one can obtain identical changes by injecting blood into any of several subarachnoid sites. Similarly, EKG changes in diseases such as poliomyelitis were especially prominent in patients with lesions in the medulla oblongata (15). The findings imply that although the release of catecholamines and the cardiovascular changes represent damage to descending autonomic pathways, they in no way specifically reflect the presence of HT disease.

*Respiratory Functions*

Stimulation of the limbic system elicits prominent respiratory responses in both man and lower animals; in lower animals, at least, the anterior HT itself receives direct afferent stimulation from chemoreceptor (251) pathways and contains neurons with discharge patterns that relate to the respiratory cycle (150). Whether the hypothalamus specifically relates to respiratory control mechanisms in man is a more speculative matter. Two case reports (89,188) have linked neurogenic hypoventilation to hypothalamic dysfunction; both the reported children were bulemic and obese but neither had anatomical, radiographic, or biochemical evidence of HT damage.

Pulmonary edema and hemorrhage can be a sudden and striking accompaniment to acute hypothalamic injury in animals (47,98,173), and many observers believe a similar process occurs in man, perhaps causing the "wet lungs" often observed after acute intracranial injury, infection, or hemorrhage. Most recent experimental studies, however, have indicated that compression of any one of a number of CNS structures, including cerebrum (161), brainstem, and cervical spinal cord (43), can produce pulmonary edema. In keeping with this observation is the fact that although some patients develop pulmonary edema in association with anterior HT compression, intraventricular hemorrhage, or rupture of an anterior communicating arterial aneurysm (46,75,263), others develop similar lung changes in association with lesions involving intracranial regions remote from the hypothalamus. Illustrating this, Richards (218) describes a series of patients dying of acute intracranial abnormalities: one-third of those with anterior HT lesions had pulmonary edema, but so did one-third of the patients with lesions in other loci sparing the HT.

Theories of the pathogenesis of "neurogenic" pulmonary edema have centered around an adrenergic cardiovascular mechanism. Maire and Patton (173) conjectured that destruction of the parasympathetic center in the POAHT released a sympathetic, edemagenic center in the posterior HT causing hypertension and a shift of blood from systemic to pulmonary vascular beds. Other investigators similarly concluded that the critical mechanism reflects either direct sympathetic stimulation (217) or direct and abrupt loss of parasympathetic function with subsequent bradycardia and congestive heart failure (244). Studies in humans with acute neurogenic pulmonary edema, however, have failed to demonstrate either an elevated blood pressure or an elevated central venous pressure (46,75). Despite some imaginative speculations to the contrary, one would have to say from present evidence that although cardiovascular effects may exacerbate the edema, by themselves they fail to account for its genesis. Park and Sutnick (198) have proposed that the demonstrated loss of pulmonary surfactant that accompanies parasympathetic lesions may be decisive in initiating pulmonary edema, but no theory so far fulfills all the observed facts.

*Gastrointestinal Functions*

Probably the earliest allusion to the autonomic functions of the hypothalamus came from Rokitansky's 1841 observation (222) that acute gastric perforations were associated with lesions of the base of the brain. A host of subsequent and now classic experiments in the physiology of animals have noted the influence of the HT over GI function. Stimulation of both the POAHT and the more posterior, dorsolateral regions results in increased mobility and secretion of the stomach, small bowel, and colon (88,177,224). Sympathetic stimulation in the ventromedial HT and especially along the MFB depresses bowel motility. These GI functions provide the autonomic correlates of various aspects of feeding behavior.

Hemorrhage and acute ulceration are the major gastrointestinal complications of central nervous system dysfunction in man, and they can affect the GI tract anywhere from the lower esophagus to the large bowel. In animals, acute lesions placed anywhere from the level of the POAHT anteriorly (173) to the region of the vagal nuclei in the medulla caudally (65) can result in GI hemorrhage. Posterior HT lesions can result in ulceration, but anterior HT lesions seem to confine their effects to producing hemorrhagic gastromalacia (169).

Cushing (59) in 1932 refocused on Rokitansky's early observation and associated the presence of acute gastrointestinal ulcerations with diencephalic tumors. Subsequently, many clinicans have observed that intracranial surgery, head trauma, encephalitis, cerebral hemorrhage and infarction, acute multiple sclerosis, brain abscess, meningitis, and a variety of tumors all may be found in association with acute GI ulceration or hemorrhage (62,63, 81,245,260). The most frequent specific GI abnormality seems to be lower esophageal ulceration, an otherwise rare lesion (81). Quantitatively, gastric and duodenal ulcerations chalk up an even greater frequency, but much of this reflects their high nonneurogenic incidence (63).

Over the years, Cushing's initial emphasis on the diencephalic location of lesions causing acute GI ulceration or hemorrhage has yielded to the recognition that in man, as in animals, lesions of many intracranial sites can cause this disorder. Abnormality may be generalized or localized and does not need to damage the HT directly (63,106,260). Pernet (203), for example, noted a 20% incidence of gastrointestinal ulceration associated with acute trauma to the cervical spinal cord. Furthermore, neither the type nor site of the brain lesion correlates with either the site or the specific type of the GI lesion (63,260). Neurogenic ulcers occur in all age groups from the neonate (207) to those over 80 years old (62), and the ulcers themselves are morphologically identical to those that accompany the acute stress of other illness, such as burns, pneumonia, and general body trauma (81).

True neurogenic ulcers always accompany some acute central nervous system crisis, whether infection, trauma, or surgical operation. Those associated with tumors have almost always arisen consequent to surgery on the neoplasm or to acute hemorrhage or necrosis into it (62). All the evidence, therefore, points to acute gastrointestinal ulceration and hemorrhage in association with acute intracranial disease as a manifestation of nonspecific damage to descending central autonomic pathways rather than any specific hypothalamic effect. Sympathetic hyperactivity elevating the systemic levels of catecholamines and parasympathetic hyperactivity producing elevated gastrin levels have been noted in affected patients (154). No one knows which of these is more important, but both sympathectomy and vagotomy are reported to prevent ulceration (81).

Clinical reports sometimes link chronic upper GI ulceration to the presence of intracranial neoplasms (39), but it is an uncertain association. Most autopsy series of patients with intracranial tumors describe an incidence of chronic GI ulceration no greater than in the general population (62).

### Epitome

This brief review emphasizes that all of the well-documented disorders of HT autonomic function in man present themselves strictly in association with acute intracranial lesions. Although the HT normally exerts a large degree of control over autonomic functions, both the experimental and the human evidence affirms that the brain quickly adapts to disturbances of the HT control of these life-sustaining functions. Consequently, chronic lesions of the HT do not produce chronic or sustained autonomic disorders.

Most hypothalamic-engendered autonomic abnormalities result from the uncoordinated activity of sympathetic and parasympathetic neurons distributed in more distal portions of the body. Since such abnormalities also occur with lesions involving other CNS structures, and sometimes even with those involving the peripheral nervous system, they are not specifically useful in localizing disease to the HT. In contrast, the recurrent paroxysmal autonomic and affective behavioral events of diencephalic epilepsy are well coordinated and do provide an indication of disease that invades or uncontrollably captures the functions of the hypothalamus.

## FEEDING BEHAVIOR AND CALORIC BALANCE

Mammals demonstrate a remarkably strict control over caloric intake, and the body weight of most healthy human beings varies hardly more than a kilogram or so over periods of time lasting many years. This constancy implies some central setpoint function that for most individuals governs appetite and food seeking behavior in relating to weight so that those who are fat remain constantly fat and those lean, constantly lean (130). Telencephalic and, especially, limbic influences assume such an importance in feeding behavior in man that primary hypothalamic error accounts for no more than a tiny fraction of human obesity or emaciation. Nevertheless, a better understanding of those few patients with specific abnormalities and especially of the underlying hypothalamic physiology and pharmacology of caloric balance represents a promising lead when searching for treatment of the frequent and serious problems of idiopathic obesity and cachexia.

Current research on the functions of the hypothalamus in caloric balance provides general agreement on the main aspects of the functional anatomy, but one finds differences of opinion on several key points.

An attractive but unproved hypothesis postulates that the HT or its closely connected limbic areas contain specific receptors controlling feeding. Some researchers believe that a central receptor for lipid stores, possibly activated by prostaglandins produced in adipose tissue, controls long-term changes in body weight (227). The receptors that control day-to-day caloric balance also remain a mystery. Mayer (178) demonstrated that the ventral hypothalamus contains cells that take up glucose preferentially, and for several years these received special attention as possibly providing the signal for short-term caloric control. Unfortunately, it now appears that these glucoreceptors do not directly influence feeding behavior (85). The only glucoreceptors so far shown to increase feeding behavior lie rostral to the anterior HT and stimulate feeding only during profound hypoglycemia (29), removing them as candidates for controlling short-term caloric balance. Other investigators posit that still undiscovered hepatic glucoreceptors (227), or even HT insulin (69) or glucagon receptors, regulate short-term diet control. Still other suggestions are that gastric and/or intestinal mechano- or chemoreceptors mediate the sense of preabsorbtive satiety. Many adjunctive feeding stimuli exist. Those set in motion by hypothermia have particular importance for the neonatal animal and may act similarly in human babies (152).

The ventromedial nuclear region of the HT, stimulation of which inhibits feeding, comprises the "satiety center" of classic physiology (132). Lesions destroying part or all of the VM in animals usually produce immediate hyperphagia and a weight gain which later stabilizes at a new elevated setpoint, at which time the hyperphagia subsides. Hoebel and Tietelbaum (139) have shown that the larger the VM lesion, the higher is the new weight setpoint. Some investigators disagree that VM lesions *per se* cause hyperphagia since their experiments indicate that the critical damage influencing feeding may be to the catecholaminergic pathways that lie in close proximity to the VM nucleus as they ascend to the limbic forebrain, but do not synapse in the VM (2,108).

Despite induced hyperphagia, animals with VM lesions develop finicky eating habits which lead them not only to confine their eating to selectively palatable foods, but to ingest food only when it is readily available. Such animals are hypoactive and will not work to obtain food, raising the question whether VM lesions cause true bulimia (i.e., a true increase in appetite). The phenomenon has led some to conclude that hyperphagia with VM hypothalamic injury represents primarily an overresponse to external stimuli, in this case the sight of the food. Whatever the explanation, HT obesity in animals results from a combination of hyperphagia and a decrease in general level of activity.

Stimulation of the lateral HT in animals induces feeding behavior leading to the classic concept of this area as the "feeding center." Lesions damaging the LH result in a temporarily severe aphagia that slowly recovers first through a pattern of eating only palatable wet foods (i.e., finickiness), then eating all palatable foods, and finally to eating a normal diet but at a lowered body weight that represents a new setpoint (249). In this instance again, experiments suggest that the critical site affecting feeding behavior may actually be the paths traversing this LH region, e.g., the pallidofugal system (186), the internal capsule (107), or the nigrostriatal tracts (197), and not the intrinsic HT neurons themselves.

Both human and animal newborns have immature caloric control systems. Adipocytes increase rapidly in number in rats up to the age of 15 weeks and in humans until the age of approximately 1 year (37). A slower increase in number occurs until puberty, after which acquired obesity reflects only adipocyte hypertrophy but not hyperplasia (136). Neonatal rats portray features of an inactive VM with hyperphagia and finicky feeding behavior, features which Kurtz et al. (160) attribute to suppression of VM activity by high growth hormone levels. Interestingly, both infant rats and human neonates develop their adult feeding habits through the same steps that one observes within the adult during recovery from an LH lesion (44). Taken together, these observations suggest that the neonate's HT feeding mechanism is both hypoactive and unresponsive to lipid feedback signals. A suppressed HT function would ensure the developmentally important proliferation of adipocytes as well as the selective hyperphagia of foodstuffs readily digested by the immature gut.

## Hypothalamic Obesity

Mohr in 1840 (185) first described HT lesions causing obesity, and Hogner (141) further localized the critical zone to the basal HT. Most patients with hypothalamic obesity, including those with Fröhlich's syndrome (93) (see below), have shown at autopsy large lesions involving the greater part of the structure, although frequently sparing the posterior HT (26,68,153,252). Occasional examples have been due to leukemic infiltration of the periventricular HT with the most severe damage localized in the bilateral VM region (17,22).

The remarkable patient reported by Reeves and Plum (214) provides a particularly well-defined localization for the site critical to the production of obesity in man (Fig. 4). Over the course of several years, the young woman developed marked obesity and hyperphagia associated with aggressive behavior placated only by giving an 8,000 calorie per day diet. At autopsy, she was found to have a hamartoma precisely and completely destroying the VM HT, but sparing lateral, anterior, and posterior structures (Fig. 5). In contrast, a young man studied by Anderson et al. (7) was found to have only partial VM destruction due to encephalitis, and he had developed mild obesity. Apparently the degree of VM destruction influences the degree of obesity in man as in animals.

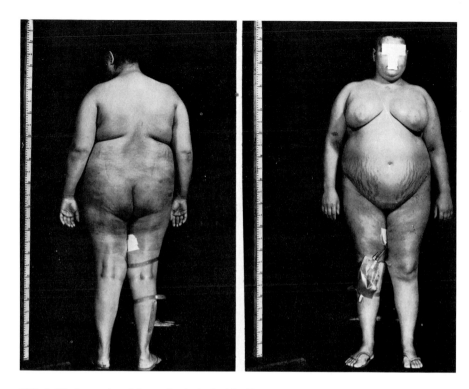

**FIG. 4.** Photographs of the patient studied by Reeves and Plum (214).

These lesions in patients with hypothalamic obesity provide the most discrete localization of any disorder of the human HT. The cases leave little doubt that damage of the medial tuberal region of the human HT produces a syndrome of hyperphagia leading to obesity, finickiness, and decreased activity just as in the experimental animal. The relative importance of the VM nucleus versus the surrounding structures in producing this disorder cannot be assessed from the human findings, however.

Bray and Gallagher (34) determined that patients with HT obesity differ metabolically from those with essential obesity or from normals only in that the HT obese patient consumes more than he expends each day throughout the period of weight gain. The findings affirm that hyperphagia, probably due to an elevated setpoint of body weight, accounts for the obesity and that general body hypoactivity accentuates it. What establishes the new setpoint remains unexplained.

When one encounters a patient with HT obesity, the presence or absence of hyperphagia at that particular point in time depends on whether or not the subject has attained his new setpoint body weight. Once the setpoint is attained, the excessive appetite subsides unless the offending lesion progresses. Cases of HT trauma resulting in obesity illustrate this

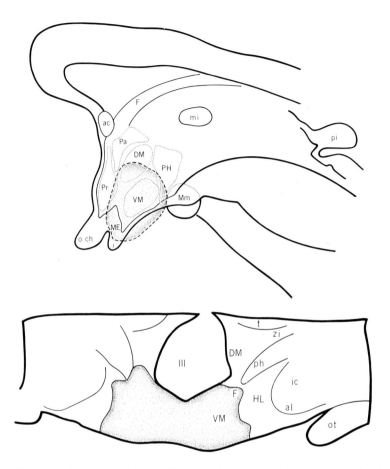

**FIG. 5.** Diagrammatic representations of the neoplasm in frontal section at the level of VM nucleus and projected on midsagittal plane. ac, anterior commissure; al, ansa lenticularis; DM, dorsomedial nuclear region; F, fornix; HL, lateral hypothalamus; i, infundibular stalk; ic, internal capsule; mi, massa intermedia; ME, median eminence; Mm, mammillary body; o ch, optic chiasm; ot, optic tract; Pa, paraventricular nucleus; ph, pallido-hypothalamic tract; PH, posterior hypothalamus; pi, pineal body; Pr, preoptic region; t, thalamus; VM, ventromedial nuclear region; zi, zona incerta; III, third ventricle. (Reprinted from ref. 214 with permission.)

principle well (34,257). Characteristically, such individuals initially display a ravenous hyperphagia and gain weight quickly until they reach their new setpoint, at which point they become normophagic and their weight stabilizes for 6 months or so. Many so injured individuals then become hypophagic and their weight declines to reapproach the pretrauma level.

Specifically engendered hypothalamic obesity has not always been differentiated from nonspecific obesity in patients with CNS disease, and perhaps for this reason, the clinical appearance of most reported cases of HT obesity has not been described. Several investigators, however, have commented on a tendency toward centripetal fat accumulation with true HT obesity (34,214).

### Essential Obesity

As noted, HT damage-induced obesity is not readily distinguished from essential obesity. Accordingly, Schacter (231) has suggested that there is a HT component in the common variety of obesity. Compared to normals, both essentially obese humans and animals with VM lesions display finicky eating behavior, eat larger and fewer meals each day with only slightly more calories per meal, and are less active but react more strongly to external stimuli. Furthermore, they tend to overeat only when food is easily obtained. It is postulated that these findings reflect poor control of internal receptor signals by the HT of the essentially obese patient.

### Disorders Suggesting Hypothalamic Dysfunction of Caloric Intake

#### *Babinski-Fröhlich Syndrome*

Originally described by Babinski (14) in 1900 and Fröhlich in 1901 (93), this disorder consists of the association of hypogenitalism with obesity. Early authors conjectured that the syndrome had a pituitary origin, but it is now generally recognized to be due to a lesion involving both the basal HT and median eminence (23). Frohlich's case was due to a craniopharyngioma, but the same syndrome can be caused by many other kinds of disorders. Damage to the VM region results in obesity, whereas destruction of the median eminence causes gonadotropin deficiency by interrupting the LHRH secretory pathways.

#### *Kleine-Levin Syndrome*

This is a disorder mainly of adolescents, most frequently of boys but occasionally of girls (27,77), that causes repeated episodes of hypersomnia, hyperphagia, hyperactivity when awake, and behavioral disturbances, especially hypersexuality. According to Critchley (51), the compulsively increased eating need not be accompanied by bulimia (excess appetite). The episodes last from days to weeks but invariably resolve without sequelae (51,156,164). No cause has been established, although a few such patients report a preceding systemic viral infection. The disorder usually disappears by the time the patient reaches his mid-twenties. The symptoms suggest paroxysmal limbic system or HT disease, but due to their self-limited nature, no cases have been examined at autopsy nor does other proof exist of intrinsic HT disease.

## Laurence-Moon-Bardet-Biedl Syndrome

The symptoms of this disorder include tapetoretinal degeneration, polydactyly, mental deficiency, hypogonadism or hypogenitalism, and obesity (155,246). The condition is hereditary and transmitted as an autosomal dominant trait. Many of the symptoms are reminiscent of HT disease, but pathologists have found neither gross nor microscopic lesions of the HT or pituitary at autopsy examination (246).

## Alström-Hallgren and Related Syndromes

These disorders somewhat resemble the Laurence-Moon-Bardet-Biedl syndrome but occur in different families with clinically distinct features. Manifestations include retinal dysplasia, deafness, obesity, adult-onset diabetes mellitus, and hypogenitalism. The few autopsied cases have shown no morphologic evidence of HT disease (5).

## Prader-Labhart-Willi Syndrome

This syndrome in the adult also is characterized by hypogenitalism, obesity, short stature, and a tendency to diabetes mellitus, but is probably not hereditary (209). Affected infants may have acromicria and tend to be aphagic, anorexic, hypotonic, somnolent, and have disturbances of temperature regulation. Anorexia changes to hyperphagia and obesity between 6 months and 2 years of age (87). No morphologic evidence of a HT lesion has been found despite the suggestive nature of the symptoms, but the patients do have abnormal diurnal variations of 11-hydroxycorticosteroid blood levels (78). One case has been reported with deficient thirst (6).

## Diencephalic Syndrome of Infancy

Russell (228) and Dods (74) first described the remarkable combination of severe emaciation, nystagmoid eye movements, and inappropriate affective behavior in infants. The condition affects children between the age of 12 and 18 months in association with an invasive tumor, usually an anterior HT or optic glioma (199). Although descriptions occasionally report an apathetic or irritable patient (268), most such infants are remarkably alert, hyperactive, and cheerful in spite of the severe wasting (60,63,74,228). These children usually maintain a surprisingly good appetite and grow to a normal height for their age, factors that have led several observers to discount a general failure of growth or hypophagia as causing their body wasting. Endocrine studies have been normal except for finding an occasional elevated growth hormone level.

The entire HT is usually destroyed by the time the affected children die, but radiographic studies performed during the early stages of emaciation indicate that the tumors all arise in the optic chiasm or anterior hypothalamus and remain confined to the anterior and tuberal region at the time that symptoms begin. Most of the children die as a result of wasting and accompanying complications before their second birthday. In those few that survive beyond 2 years of age, however, the illness often undergoes a striking qualitative transformation. With a continued good appetite, emaciation ceases and the child becomes HT obese (100), a condition rarely noted in infants less than 2 years old. At the same time, irritability and even rage tend to replace the euphoric behavior and hyperactivity of the earlier months. Pelc (199) reports that children who survive beyond 2 years old also have a greatly improved prognosis, but whether this is due to the metabolic changes or to growth characteristics of the tumor is not known.

The emaciation characteristic of this syndrome has never been adequately analyzed with detailed caloric balance or metabolic studies. Appetite has usually been adequate for normal or even increased caloric intake. Hyperactivity and vomiting due to an intracranial mass lesion have been considered to cause the negative caloric balance in some infants (63), although other patients have lacked the symptoms.

No one knows precisely why the immature HT of the infant produces symptoms different from those of the older child and adult HT. As mentioned above, HT functions governing caloric intake and energy metabolism may normally be suppressed in the neonate, possibly due to the high growth hormone levels of this age group. The HT may not control energy balance fully until after the age of 2 years. A lesion of an already nonfunctioning ventromedian region in an infant under 2 years old would not be likely to produce HT obesity. On the contrary, an HT lesion in this age group appears to disrupt normal adipose proliferation, as abnormalities of the structure may lead to insufficient appetite to compensate for the rapid growth of the period. This may be complicated by the high growth hormone levels in infants which, as Kurtz et al. (160) postulate, may shunt energy from lipid storage depots to building protein, but evidence is lacking to support this speculation.

### Emaciation in Adults and Children

Only a few adults have shown the association of emaciation with hypothalamic disease, and those instances which suggest a cause-effect association have exhibited anorexia and hypophagia as part of the syndrome. When emaciation has accompanied hypothalamic abnormalities in man, the responsible lesion usually has destroyed the entire HT by the time autopsy examination has been performed (64,105,254). A singular patient reported by White and Hain (267), however, with an emaciating illness diagnosed as anorexia nervosa during life, demonstrated an isolated cystic lesion of the right LH and depopulation of neurons in the contralateral LH. Unfortunately, even the post-mortem studies failed to distinguish whether the degeneration was a result rather than a cause of the emaciation. Hart's (126) description of hypertrophy of the subventricular HT nucleus following death by starvation has not been duplicated in other patients.

Kamalian et al. (149) described possibly the best adult example of apparent starvation in the midst of caloric plenty. The woman had a progressive wasting illness accompanied by hypophagia, but in spite of high caloric tube feedings she continued to lose weight. Metabolic balance studies were not performed, however, so that the pathogenesis remains in doubt. At autopsy, the underlying illness of multiple sclerosis had produced both new and old plaques involving the lateral hypothalamic regions but sparing the VMHT.

The localization of HT damage to LH in these two patients accords with evidence from animal experiments producing anorexia. The findings in a patient described by Heuser (133) provide another correlation with animal experiments. The subject sustained HT trauma during surgery for craniopharyngioma. Immediately postoperatively, he was anorexic and hypophagic but recovered to a stage of finicky eating habits prior to a full recovery of normal eating. The pattern resembles that which happens in animals during recovery from acute LH lesions.

## Anorexia Nervosa

Although usually assumed to be a psychogenic disorder, full-blown anorexia nervosa demonstrates several defects of HT function. As is well known, the characteristic symptoms include anorexia, weight loss, and amenorrhea occurring in an otherwise endocrinologically normal young woman. The amenorrhea reflects depressed LH-FSH levels (24), which in turn result from decreased LH-RH release from the HT. Other endocrine dysfunctions of HT origin include the occasional occurrence of hyperprolactinemia with galactorrhea and partial diabetes insipidus. Some patients show evidence of nonendocrine HT disturbances, especially poikilothermia (181).

The pathogenesis of the apparent HT dysfunction in anorexia nervosa is uncertain. Most patients examined at autopsy have had morphologically normal brains, although a few rare cases of poorly localized HT tumors have been described (165). Certainly, the functional HT abnormalities could result from starvation, and similar changes have been reported in emotionally deprived children (208). Ammenorrhea may precede any weight loss in anorexia nervosa, however, and the other HT dysfunctions sometimes persist after the anorexia resolves. Those who regard anorexia nervosa as a psychogenic disorder take these changes as reflecting the functionally abnormal influences of psychic stress on the limbic system and the HT. Their presence, however, as well as the consistent pattern of dysfunction, patient age, and sex, suggests that perhaps the neurobiologist should regard the pathogenesis of anorexia nervosa with open-minded curiosity toward a possible intrinsic HT etiology of the disease.

In summary, the functions of the HT in the control of caloric balance in man and other animals are as follows: the HT determines the set weight of the organism and maintains this weight by coordinating the affective and behavioral responses to external and internal cues for feeding. In human beings as well as in animals bilateral VM lesions cause emaciation. Many clinical disorders include at least some abnormalities in HT caloric regulation, but the precise explanation for these hypothalamic-like dysfunctions remains almost completely lacking. The diencephalic syndrome of infancy suggests that in relation to feeding behavior the immature hypothalamus functions very differently from that of the older child or adult.

## WATER BALANCE

Disturbances of water balance in man are better understood than other hypothalamic-related disorders. The relationships of the supraoptic (SO) and paraventricular (PV) nuclei to the neurohypophysis and to antidiuretic hormone (ADH) release were relatively well established even by the 1939 conference of this Association. Furthermore, disorders of water balance provide frequent problems in clinical medicine and sometimes endanger life. Recent experimental research has added important new details to the understanding of the supraoptico-neurohypophysial system and has clarified several other previously puzzling aspects of water balance control.

Homeostatic control of body water and osmolarity depends on the interactions among thirst, drinking behavior, and ADH release mechanisms. As with the ingestive behaviors that maintain caloric balance, influences mediated by the limbic system greatly alter drinking behavior. Accordingly, the HT functions to integrate the telencephalic and sensory inputs to effect a coordinated response. Water control by the HT differs from caloric control, however, in several ways. For water control, one can identify specific peripheral and local osmoreceptors; furthermore, neurosecretory cells producing a specific hormone assure that the HT maintains serum osmolality even when the HT is isolated from the rest of the brain (176).

The osmoreceptors that most influence drinking appear to lie within the POAHT itself. Andersson and Eriksson (8) have postulated that periventricular receptors in the medial POAHT respond to both intracellular dehydration, as represented by increased intraventricular sodium concentration, and extracellular dehydration. The extracellular dehydration receptors respond to the local HT blood angiotensin II concentration and elicit drinking as a response to blood volume changes and other stresses that elevate the circulating levels of that hormone. Blass (28), however, has been unable to confirm the presence of the medially placed receptors. Furthermore, he reports finding only sensors to intracellular dehydration, and that these lie in the lateral POAHT. Other HT regions also contain osmoreceptors possessing more limited functions. Those in the vicinity of the supraoptic and paraventricular nuclei influence only ADH release (11). Osmoreceptors lying in the lateral hypothalamic region appear to stimulate general excitability of the region rather than specific drinking behavior (28).

Peripheral receptors also can influence water balance (82) but less strongly than the HT receptors. Stimulation of peripheral osmoreceptors lying in the interstitial fluid and excitation of those responsible for the sensation of mouth dryness have been considered to be important influences on thirst. Excitation of volume receptors in the left atrium and in the large capacitance veins by an increased blood volume tends to inhibit drinking behavior and ADH release. Thermal stress, eating, psychic stress, and other influences all can activate drinking and ADH release so as to influence water balance.

The magnocellular neurons in the supraoptic and paraventricular nuclei of higher animals produce ADH and transport it to the median eminence and neurohypophysis, whence it is released into the blood. The neurosecretory cells maintain a low base-line firing rate that is increased by inputs from excitatory osmoreceptors or decreased by inputs from inhibitory volume receptors (10). During stimulation, the neurosecretory cells also limit their ADH release by the process of recurrent collateral inhibition in a manner similar to that found in the spinal cord's Renshaw cell (192). All evidence indicates that the threshold for ADH in response to osmotic stimuli depends entirely on the setpoint of the associated osmoreceptor. Above threshold, however, further changes in ADH release in response to changes of osmolarity appear to depend on the response of the firing rates of the neurosecretory cells to the integrated sum of all the various inputs. Under normal conditions, osmoreceptors are the most influential determinant of ADH release.

Andersson et al. (10) have produced several syndromes of abnormal ADH secretion by appropriate lesions placed in the POAHT regions of goats. Diabetes insipidus is the most easily obtained and requires at least a 90% loss of the SO and PV magnocellular neurons. Such a high percentage of destruction to ADH-secreting cells can result only from direct hypothalamic damage, which explains why hypophysectomy (which spares the neurons secreting into median eminence vessels) causes only a temporary diabetes insipidus.

Lesions that impinge on the SO nucleus but do not destroy it result in an inappropriately high secretion of ADH (10). Presumably, such damage interrupts the input to the SO nucleus from peripheral volume receptors and from recurrent inhibitory collaterals without destroying the spontaneously firing neurosecretory cells themselves. As one would expect, destruction of the receptor-containing POAHT region in animals produces a permanent adipsia, but such lesions also interfere with the ADH response to osmotic stress and result in hypovolemia and hypernatremia.

The LH, classically considered as the "drinking center," contains interneurons that respond to both osmotic and nonosmotic stimuli. In the experimental animal, electrical stimulation of this region results in polydipsia (238), whereas destructive lesions cause a temporary adipsia with recovery to habitual drinking sufficient to maintain water balance. Specific stimuli applied to the lateral hypothalamus, such as those generated by osmoreceptors, stimulate thirst and drinking. Drinking also occurs as an adjunct to nonspecific stimuli to the region such as increased arousal and general motivation (261). The LH structures critical for thirst and drinking are unknown. It is possible that alterations in drinking behavior, as in the case of eating behavior, may be related to paths traversing but not synapsing in the LH.

Stimulation of higher limbic structures in animals tends most often to inhibit drinking and ADH release. Likewise, VM stimulation inhibits drinking whereas lesions of the VM region may produce hyperdipsia (159).

### Diseases of Water Balance in Man

Disorders of water balance in man may result from abnormal ADH secretion, abnormal thirst with consequent abnormalities of drinking, or a combination of the two.

*Decreased or Absent ADH Release with Osmogenically Normal Thirst — the Syndrome of Diabetes Insipidus*

Diabetes insipidus (DI) has been the subject of many excellent reviews (176). One point deserves emphasis. The lack of ADH allows the kidney to excrete a large volume of dilute urine, but an intact thirst mechanism responds to this dehydration with a compensatory polydipsia that maintains the serum sodium concentration within or only slightly higher than the normal range (237).

Diabetes insipidus in man may be temporary if due to an intrasellar or low pituitary stalk lesion or permanent if due to damage to the median eminence or anterior hypothalamus. The structures critical to the production of DI are more precisely localized by examining the rare examples of the familial or idiopathic condition. This disease fairly uniformly presents early in life with DI responsive to exogenous ADH, but such patients lack other symptoms of hypothalamic or pituitary dysfunction (30,33,102,115,123). The serum of patients with familial DI contains a marked decrease or absence of immunoreactive ADH (142). Morphological examination of the hypothalamus at autopsy shows an absence of at least 90% of the magnocellular neurons of the SO and PV nuclei (115), but no other abnormalities. The findings beautifully confirm the experimental evidence that the magnocellular neurons of the SO and PV nuclei are the only cells that produce and secrete ADH. Thus the clinical disorder, diabetes insipidus, results from destruction of the final common pathway for ADH secretion. Its high frequency with HT disease reflects the nervous system's relative inability to compensate for a lesion of the final common path.

*Decreased or Absent ADH Release with Absence of Thirst—the Syndrome of Essential Hypernatremia*

Four features characterize essential hypernatremia: (a) hypernatremia unaccompanied by a corresponding fluid deficiency; (b) preserved renal tubular responsiveness to ADH; (c) inadequate secretion of ADH in response to osmotic stimuli; and (d) absence or deficiency of thirst despite preserved conscious behavior. True essential hypernatremia is rare; most cases of serum hyperosmolality accompanying HT disease do not meet the above four criteria but result from other causes, especially a combination of dehydration and stupor. Many examples of hypernatremia accompany bilateral anterior cerebral artery occlusion or frontal lobe disease; most of these patients have been stuporous and insufficiently well studied to know whether they have partial ADH defects akin to essential hypernatremia.

Essential hypernatremia in its milder and more chronic forms often elevates the serum sodium only modestly, and such patients characteristically have no symptoms except a remarkable lack of thirst (113,150). When sodium levels climb to the 160 to 170 mEq/dl range, however, most such patients develop symptoms including weakness and fever, as well as muscle tenderness and cramping that may progress to fatigue, ataxia, and even paralysis. Mental symptoms with mild to moderate hypernatremia include lethargy, anorexia, depression, paranoia, and irritability. With severe elevations of sodium to above 180 mEq/dl, most patients become confused or stuporous and some die. Although all patients with essential hypernatremia fail to drink in response to a deficit signal, i.e., fail to demonstrate thirst, many retain persistent habitual drinking, although the volume of water they ingest falls short of that required to maintain normal osmolality (112,254). A few patients with a lack of thirst have even given the history of initially being polydipsic, possibly indicating a preexisting state of DI (13,84,92,270).

Only mild hypovolemia or even normovolemia has been reported in patients with essential hypernatremic levels as high as 216 mEq/dl (172). Accordingly, such patients lack clinical evidence of dehydration. Hypokalemia usually is present and contributes to the muscular symptoms. Urine volumes may be low or normal but are always more dilute than would be expected from the serum hyperosmolality. An overnight fast usually produces a moderate increase in the urine osmolality, but the administration of 5 units of exogenous ADH induces a much more concentrated urine, proving the existence of at least a partial diabetes insipidus in these patients (243). Further evidence of abnormal ADH release to naturally occurring stimuli comes from the observation that either an acute water load or a hypertonic saline load can produce an increase in free water clearance.

The HT defect necessary to produce essential hypernatremia in man is less well localized than are the POAHT lesions that produce a similar picture in goats. Many patients with chronic essential hypernatremia lack radiographic evidence of CNS abnormalities (3,111,172), and in those patients with neurological lesions the entire hypothalamic region has usually been destroyed by the time autopsy occurred (13,153,162,254). A few cases have been studied, however, in which restricted lesions involved the POAHT and tuberal region (151,252,264).

Different workers have proposed alternate mechanisms to explain essential hypernatremia. One proposal maintains that HT lesions elevate the setpoint around which osmoreceptors attempt to maintain the serum osmolality, i.e., they alter the ADH release threshold producing a new high level below which ADH is no longer secreted (13,265). In contradiction to this theory, however, it has been found in these patients that a water load will result in urine dilution, implying that at least some ADH secretion must already be present in the base-line hyperosmolar state. Indeed, Sridhar et al. (243) have shown that the ADH threshold is considerably lower than 295 mosm/liter in essential hypernatremia and probably lies within normal limits. These considerations lead to the other explanation for essential hypernatremia, which argues that the defect lies not in the setpoint of osmoregulation but in the capacity of ADH secretion to respond to osmoreceptor stimulation. A patient recently studied but as yet unpublished by M. New, L. Levine, and ourselves provides strong evidence for this explanation. This 12-year-old child has evidence of several defects in hypothalamic regulation involving growth hormone release, cognition, appetite, and heat regulation and also has profound adipsia. Her serum sodium has ranged as high as 187 and as low as 134, but ADH output remains low and fixed over the entire range as well as in response to stimulation of volume receptors.

The patient illustrates that in severe cases of essential hypernatremia, osmo- and volume receptor function can be completely lost or at least uncoupled from the behavioral and hormonal control of water balance. In most instances, however, the base-line firing of neurosecretory cells maintains a low level of ADH secretion whose rate is modulated only by inhibitory volume receptors, accounting for the increase of free water clearance with hypertonic saline infusion (71). Some cases even lack this latter response. Milder cases retain some, but less than normal, ADH responsiveness to osmoreceptors. Such patients may respond to hypertonic saline infusion by concentrating their urine (243). The presence of normovolemia accompanying the hypernatremia remains difficult to explain, but may reflect the continuation of an ADH response to extracellular dehydration receptors accompanied by a selective loss of responsiveness to intracellular dehydration receptors.

Both of the above theories agree that absent thirst is critical to the production of essential hypernatremia. With thirst mechanisms intact, DI would result. The loss of thirst allows the patient to tolerate the hyperosmolar state and prevents the behavior that normally would replace the water lost as excess in the urine.

The treatment of essential hypernatremia consists of lowering the serum sodium and raising the potassium by manipulating the diet and insisting that the patient drink several liters of water each day regardless of thirst. Spironolactone, chlorpropamide, and the thiazide diuretics contribute to achieving these ends. More vigorous treatment must be initiated slowly and carefully. Because their intracellular osmolality has climbed to a new high steady state, patients with essential hypernatremia can develop symptoms of water intoxication if rapidly hydrated, even when the serum continues to be hyperosmolar when compared to normal.

### Normal ADH Release with Absent Thirst

This combination has been proposed to explain certain examples of the syndrome of essential hypernatremia, but only on the basis of patients receiving limited study. Sridhar et al. (243) point out that in none of the reported instances were the patients tested sufficiently to establish the presence of a normal release of ADH. Furthermore, some of the patients responded to exogenous ADH with a more concentrated urine than they produced during an overnight fast. Thus the reported instances of the disorder may be variants of essential hypernatremia (71,128,252).

### Elevated ADH Release with Normal Thirst—the Syndrome of Inappropriate Secretion of ADH (SIADH)

Schwartz et al. (234) first suggested SIADH in a patient with an oat cell carcinoma of the lung, and many other diseases since have been linked to it, including head trauma, subarachnoid hemorrhage, hydrocephalus, intracranial masses, meningitis, encephalitis, porphyria, peripheral neuropathy, central pontine myelinolysis, myxedema, neoplasms, cardiac failure, nonneoplastic pulmonary diseases, and drug toxicity (116). The cardinal features of SIADH consist of serum hypoosmolarity and hyponatremia, normal renal excretion of sodium, and an inappropriately high urine osmolality without clinical evidence of body fluid depletion. Serum levels of immunoreactive ADH are abnormally elevated. Several excellent reviews discuss the symptoms and treatment (21,176).

Many of the intracranial diseases causing SIADH have been regarded as compressing or irritating the HT. Lesions isolated to the HT area also can cause SIADH, but the damage in most cases comes from neoplasms, infections, and hemorrhages that are too large and imprecise to confer localizing value (36,116,175). Even the claim for strict localization within the HT in a patient described by Perlroth et al. (202) can be criticized. The patient suffered from acute intermittent porphyria with severe peripheral neuropathy. After several weeks with SIADH she died, and an autopsy disclosed an inflammatory process surrounding the SO and PV nuclei without destroying a large number of neurosecretory cells. Perlroth and co-workers postulated that either of two mechanisms could have produced the SIADH. One was irritation of neurosecretory neurons causing excessive ADH release, the other was possible destruction of inhibitory axons in the region of the SO and PV nuclei such as those carrying afferent information from volume receptors. Alternatively, since SIADH almost always accompanies the peripheral neuropathy of acute intermittent porphyria (and most other acute peripheral neuropathies as well), the findings in Perlroth's patient may have been the result of the associated neuropathy. Serial sections of the hypothalamus in two patients who died in this institution from porphyria causing peripheral neuropathy and SIADH failed to show morphological abnormalities either in the hypothalamus or elsewhere in the brain, leading us to suspect that SIADH in porphyria reflects an abnormality involving peripheral volume receptors secondary to the peripheral nerve impairment. The hypothalamic lesions observed in Perlroth's patient appear to be unique.

Impairment of receptors or interference with afferent inhibitory pathways carrying information from the volume receptors to the HT has been proposed as the mechanism for most neurogenic examples of SIADH. Patients with polyneuritis (50,61,219) resulting in excess ADH release have been regarded as having abnormalities of peripherally placed receptors or pathways. SIADH with brainstem disease (49) has been attributed to interruption of central ascending paths. SIADH can accompany the limbic system disease of herpes simplex encephalitis (225), possibly as the result of damage to telencephalic structures that normally inhibit drinking and ADH release. Unfortunately, the large and heterogeneous group of illnesses that can result in SIADH allow for many explanations but few proofs.

*Hyperdipsia*

Excessive water drinking in the absence of either hypovolemia or hyperosmolality of intracellular or extracellular fluids is termed primary hyperdipsia and must be differentiated from the compensatory hyperdipsias that occur in conditions such as diabetes insipidus, polyuric renal failure, or

TABLE 5. *The primary hyperdipsias*

| | |
|---|---|
| Beer drinker's hyponatremia | Hyperangiotensinemia |
| Infant formula dilution | Chronic renal failure |
| Situational (e.g., tea party epilepsy) | Renal artery stenosis |
| | Hypokalemia |
| Other (e.g., soothing gastric irritation or toothache) | Psychogenic |
| | Hypothalamic disease |

diabetes mellitus. Although asymptomatic hyperdipsia probably occurs commonly, the symptoms of severe hyperdipsia, i.e., hypervolemia, hyponatremia, and water intoxication, are infrequently observed. These latter develop only when the patient has either renal dysfunction or imbibes greater than 15 $l/m^2/day$ of water, the quantity necessary to overcome the diluting capacity of the normal kidney (205). Due to the ability of the kidney to excrete large amounts of free water, almost all patients with water intoxication recover with fluid restriction and proper supportive care.

Patients with primary hyperdipsia can be divided into those who imbibe fluids in response to conditioned behavior, i.e., habitual drinking, and those who drink in response to excessive thirst (Table 5). Of the habitual hyperdipsias, beer drinker's hyponatremia has been the most thoroughly studied. Such subjects drink up to 6 liters of beer per day, but this fluid volume alone is insufficient to produce water intoxication. Demanet et al. (70) postulated that chronic hyponatremia results from constant consumption of beer with a low sodium content and predisposes the patient to symptoms from the acute water load associated with a binge. Alternatively, Gwinup et al. (120) have suggested that sustained and heavy beer drinking can cause a transient inappropriate ADH release which, when coupled with a large fluid intake, results in water intoxication. The other examples of habitual hyperdipsia appear to produce their symptoms simply due to a tremendous ingestion of water over a brief period (205); incomplete renal function may add to the fluid intolerance in cases of infant formula dilution (55).

The presence of abnormal agents that stimulate hypothalamic thirst mechanisms may result in hyperdipsia. As noted previously, angiotensin II elicits thirst by stimulating extracellular dehydration receptors in animals. Brown et al. (38) and Rogers and Kurtzman (221) have postulated that this mechanism explains why some patients suffering from long-standing renal disease have constant, excessive thirst even though hemodialysis maintains them in normal electrolyte balance. Bilateral nephrectomy simultaneously corrects the markedly elevated renin levels and excessive thirst in these subjects. Other examples of excessive thirst secondary to hyperangiotensinemia have been described in patients with renal artery stenosis (110) and with hypokalemia and hyponatremia due to anorexia and vomiting (271).

One encounters hyperdipsia most often in patients with acute psychiatric disorders. The subjects complain of excessive thirst which presumably is delusional and not physiologic (45). Patients suffering symptoms from the fluid overload often have a poor urine-concentrating ability (72,137), and some have had elevated serum ADH predisposing them to water intoxication (167). Since other such patients have had normal or even depressed ADH release (20), however, SIADH cannot be a necessary factor.

The literature describes few cases of hyperdipsia due to intrinsic HT disease, and the available reports seldom differentiate primary hyperdipsia from polydipsia compensating for diabetes insipidus. Martin, Reichlin, and Brown (176) describe a patient who experienced hyperdipsia for several years prior to developing diabetes insipidus and other signs that led to the discovery of an HT tumor. Either stimulation of an HT "thirst center" or destruction of inhibitory influences on thirst in the HT has been proposed as a possible pathophysiologic mechanism, but little evidence substantiates either claim.

## CONCLUSION

The preceding chapters in this volume have described specialized peptides and transmitters of the hypothalamus that regulate the endocrine system and other parts of the body in ways no one suspected when the Association for Research in Nervous and Mental Disease discussed the HT in 1939. In the nonendocrine areas, however, the growth of knowledge more closely follows the discussions of that meeting and fulfills Fulton's remark that "no area of medicine illustrates more strikingly the wisdom of close cooperation between clinical and experimental study than the history of knowledge of hypothalamic function" (95). This chapter emphasizes that to a considerable degree one still understands the regulations of the nonendocrine HT only in descriptive or nonquantitative terms, and neurobiology and medicine still lack any precise knowledge of the chemical molecular transactions that go into how the HT integrates consciousness, cognition and affect, autonomic balance, caloric control, and much of water balance. Human disease suggests some of the leads to these answers, the laboratory bench suggests others. The hints from both indicate that when our successors meet to discuss the HT 25 years hence, the discussion will focus as heavily on the cellular basis of the nonendocrine hypothalamic regulation of the internal and external world as this conference has focused on the cellular basis of hypothalamic neuroendocrinology.

## ACKNOWLEDGMENT

The Cerebral Vascular Disease Research Center is supported by Grant NS #03346 from the USPHS.

## REFERENCES

1. Adair, E. R. (1974): Hypothalamic control of thermoregulatory behavior. In: *International Symposium on Recent Studies of Hypothalamic Function, Calgary, 1973*, edited by K. Lederis and K. E. Cooper, pp. 341–358. S. Karger, Basel.
2. Ahlskog, J. E., and Hoebel, B. G. (1973): Hyperphagia resulting from selective destruction of an ascending noradrenergic pathway in the rat brain. *Fed. Proc.*, 31:397.
3. Alford, F. P., Scoggins, B. A., and Wharton, C. (1973): Symptomatic normovolemic essential hypernatremia. *Am. J. Med.*, 54:359–370.
4. Alpers, B. J. (1940): Personality and emotional disorders associated with hypothalamic lesions. *Res. Publ. Assoc. Res. Nerv. Ment. Dis.*, 20:725–752.
5. Alström, C. H. (1972): The Lindenov-Hallgren, Alström-Hallgren and Weiss syndromes. In: *Handbook of Clinical Neurology, Vol. 13*, edited by P. J. Vinken and G. W. Bruyn, pp. 451–467. North-Holland, Amsterdam.
6. Anand, S. K., Kogut, M. D., and Lieberman, E. (1972): Persistent hypernatremia due to abnormal thirst mechanism in a 13 year old child with nephrogenic diabetes insipidus. *J. Pediatr.*, 81:1097–1105.
7. Anderson, E., Haymaker, W., and Rappaport, H. (1950): Seminiferous tubule failure associated with degenerative change in the hypothalamus. *Am. Pract.*, 1:40–45.
8. Andersson, B., and Eriksson, L. (1971): Conjoint action of sodium and angiotensin on brain mechanisms controlling water and salt balances. *Acta Physiol. Scand.*, 81:18–29.
9. Andersson, B., Gale, C., Hökfelt, B., and Larsson, B. (1965): Acute and chronic effects of preoptic lesions. *Acta Physiol. Scand.*, 65:45–60.
10. Andersson, B., Leksell, L. G., and Lishaiko, F. (1975): Perturbations in fluid balance induced by medially placed forebrain lesions. *Brain Res.*, 99:261–275.
11. Andersson, B., Olsson, K., and Warner, A. G. (1967): Dissimilarities between the central control of thirst and the release of antidiuretic hormone (ADH). *Acta Physiol. Scand.*, 71:57–64.
12. Appenzeller, O., and Snyder, R. D. (1969): Autonomic failure with persistent fever in cerebral gigantism. *J. Neurol. Neurosurg. Psychiatry*, 32:123–128.
13. Avioli, L. V., Earley, L. E., and Kashima, H. K. (1962): Chronic and sustained hypernatremia, absence of thirst, diabetes insipidus, and adrenocorticotrophin insufficiency resulting from widespread destruction of the hypothalamus. *Ann. Intern. Med.*, 56:131–140.
14. Babinski, J. (1900): Tumeur du corps pituitaire sans acromégalie et avec arrêt de développement des organes génitaux. *Rev. Neurol. (Paris)*, 8:531–533.
15. Baker, A. B., Matzke, H. A., and Brown, J. R. (1950): Poliomyelitis III: Bulbar poliomyelitis; a study of medullary function. *Arch. Neurol. Psychiatry*, 63:257–281.
16. Ban, T. (1966): The septo-preoptico-hypothalamic system and its autonomic function. In: *Progress in Brain Research, 21A Correlative Neurosciences Part A*, edited by T. Tokizane and J. P. Schadé, pp. 1–43. Elsevier, Amsterdam.
17. Barak, Y., and Liban, E. (1968): Hypothalamic hyperphagia, obesity and disturbed behavior in acute leukemia. *Acta Paediatr. Scand.*, 57:153–156.
18. Barbizet, J. (1963): Defect of memorizing of hippocampal-mammillary origin: A review. *J. Neurol. Neurosurg. Psychiatry*, 26:127–135.
19. Bard, P. (1928): A diencephalic mechanism for the expression of rage with special reference to the sympathetic nervous system. *Am. J. Physiol.*, 84:490–515.
20. Barlow, E. D., and DeWardener, H. E. (1959): Compulsive water drinking. *Q. J. Med.*, 28:235–258.
21. Bartter, F. C., and Schwartz, W. B. (1967): The syndrome of inappropriate secretion of antidiuretic hormone. *Am. J. Med.*, 42:790–806.
22. Bastrup-Madsen, P., and Greisen, O. (1963): Hypothalamic obesity in acute leukemia. *Acta Hematol.*, 29:109–116.
23. Bauer, H. G. (1959): Endocrine and metabolic conditions related to pathology in the hypothalamus: A review. *J. Nerv. Ment. Dis.*, 128:323–338.
24. Bell, E. T., Harkness, R. A., and Loraine, J. A. (1965): Gonadotrophin and oestrogen excretion in patients with anorexia nervosa. *J. Psychosom. Res.*, 9:79.
25. Benzinger, T. H. (1969): Heat regulation: Homeostasis of central temperature in man. *Physiol. Rev.*, 49:671–759.

26. Bernard, J. D., and Aguilar, M. J. (1969): Localized hypothalamic histiocytosis X. *Arch. Neurol.*, 20:368–372.
27. Billiard, M., Guilleminault, C., and Dement, W. C. (1975): A menstruation-linked periodic hypersomnia. *Neurology (Minneap.)*, 25:436–443.
28. Blass, E. (1974): Evidence for basal forebrain thirst osmoreceptors in rat. *Brain Res.*, 82:69–76.
29. Blass, E. M., and Kraly, F. S. (1974): Medial forebrain bundle lesions: Specific loss of feeding to decreased glucose utilization in rats. *J. Comp. Physiol. Psychol.*, 86:679–692.
30. Blotner, H. (1958): Primary or idiopathic diabetes insipidus—a system disease. *Metabolism*, 7:191–200.
31. Branch, E. F., Burger, P. C., and Brewer, D. L. (1971): Hypothermia in a case of hypothalamic infarction and sarcoidosis. *Arch. Neurol.*, 25:245–255.
32. Braun, F. C., Jr., and Forney, W. R. (1959): Diencephalic syndrome of early infancy associated with brain tumor. *Pediatrics*, 24:609–615.
33. Braverman, L. E., Mancini, J. P., and McGoldrick, D. M. (1965): Hereditary idiopathic diabetes insipidus. *Ann. Intern. Med.*, 63:503–508.
34. Bray, G. A., and Gallagher, T. F., Jr. (1975): Manifestations of hypothalamic obesity in man: A comprehensive investigation of eight patients and a review of the literature. *Medicine (Baltimore)*, 54:301–330.
35. Brierley, J. B. (1966): The neuropathology of amnesic states. In: *Amnesia*, edited by C. W. M. Whitty and O. L. Zangwill, pp. 150–180. Butterworths, London.
36. Brisman, R., and Chutorian, A. M. (1970): Inappropriate antidiuretic hormone secretion. Hypothalamic glioma in a child. *Arch. Neurol.*, 23:63–69.
37. Brook, C. G. D., Lloyd, J. K., and Wolf, O. H. (1972): Relation between age of onset of obesity and size and number of adipose cells. *Br. Med. J.*, 2:25–27.
38. Brown, J. J., Curtis, J. R., Lever, A. F., Robertson, J. I. S., DeWardener, H. E., and Wing, A. J. (1969): Plasma renin concentration and the control of blood pressure in patients on maintenance hemodialysis. *Nephron*, 6:329–349.
39. Bsteh, F. (1951): Des peptische Ulkus und intrakranielle Prozesse. *Wein. Klin. Wochenschr.*, 63:310–313.
40. Bustamente, M., Spatz, H., and Weisschedel, E. (1942): Die Bedeutung des Tuber cinereum. *Dtsch. Med. Wochenschr.*, 68:289.
41. Cabanac, M. (1972): Thermoregulatory behavior. In: *Essays on Temperature Regulation*, edited by J. Bligh and R. E. Moore, pp. 19–36. North-Holland, Amsterdam.
42. Castaigne, P., and Escourolle, R. (1967): Étude topographique des lésions anatomiques dans les hypersomnies. *Rev. Neurol. (Paris)*, 116:547–584.
43. Chen, H. I., Sun, S. C., and Chai, C. Y. (1973): Pulmonary edema and hemorrhage resulting from cerebral compression. *Am. J. Physiol.*, 224:223–229.
44. Cheng, M. F., Rozin, P., and Teitelbaum, P. (1971): Starvation retards development of food and water regulations. *J. Comp. Physiol. Psychol.*, 76:206–218.
45. Chinn, T. A. (1974): Compulsive water drinking. *J. Nerv. Ment. Dis.*, 158:78–80.
46. Ciongoli, A. K., and Poser, C. M. (1972): Pulmonary edema secondary to subarachnoid hemorrhage. *Neurology (Minneap.)*, 22:867–870.
47. Clark, G., Magoun, H. W., and Ranson, S. W. (1939): Hypothalamic regulation of body temperature. *J. Neurophysiol.*, 2:61–80.
48. Claude, H., and L'Hermitte, J. (1917): Le syndrome infundibulaire dans un cas de tumor du III$^e$ ventricule. *Presse Med.*, 25:417–418.
49. Conger, J. D., McIntyre, J. A., and Jacoby, W. J., Jr. (1967): Central pontine myelinolysis associated with inappropriate ADH secretion. *Am. J. Med.*, 47:813–817.
50. Cooper, W. C., Green, I. J., and Wang, S. (1965): Cerebral salt-wasting associated with the Guillain-Barré syndrome. *Arch. Intern. Med.*, 116:113–119.
51. Critchley, M. (1962): Periodic hypersomnia and megaphagia in adolescent males. *Brain*, 85:627–656.
52. Crompton, M. R. (1963): Hypothalamic lesions following the rupture of cerebral berry aneurysms. *Brain*, 86:301–314.
53. Crosby, E. C., Humphrey, T., and Lauer, E. W. (1962): *Correlative Anatomy of the Nervous System*, pp. 309–342. Macmillan, New York.
54. Cross, K. W., Hey, E. N., Kennard, D. C., Lewis, S. R., and Urich, H. (1971): Lack of

temperature control in infants with abnormalities of the central nervous system. *Arch. Dis. Child.*, 46:437–443.
55. Crumpacker, R. W., and Kriel, R. L. (1973): Voluntary water intoxication in normal infants. *Neurology (Minneap.)*, 23:1251–1255.
56. Cunliffe, W. J., Johnson, C. E., and Burton, J. L. (1971): Generalized hyperhidrosis following epilepsy. *Br. J. Dermatol.*, 85:186–191.
57. Cushing, H. (1929): *The Pituitary Body and Hypothalamus*. Charles C Thomas, Springfield, Ill.
58. Cushing, H. (1932): *Papers Relating to the Pituitary Body, Hypothalamus, and Parasympathetic System*. Charles C Thomas, Springfield, Ill.
59. Cushing, H. (1932): Peptic ulcer and the interbrain. *Surg. Gynecol. Obstet.*, 55:1–34.
60. Cussen, L. J. (1964): Diencephalic syndrome of early infancy—case report. *Med. J. Aust.*, 1:881–882.
61. Cutting, H. O. (1971): Inappropriate secretion of ADH secondary to vincristine therapy. *Am. J. Med.*, 51:269–271.
62. Dalgaard, J. B. (1958): Intracranial tumours and peptic ulcerations. *Acta Pathol. Microbiol. Scand.*, 42:313–323.
63. Dalgaard, J. B. (1959): Cerebral lesions as a cause of peptic ulceration. *Proc. World Congr. Gastroenterol.*, 1:387.
64. Daly, J. J., and Nabarro, J. D. N. (1973): Clinicopathological Conference. A case of anorexia. *Br. Med. J.*, 2:158–163.
65. Davey, L. M., Kaada, B. R., and Fulton, J. F. (1950): Effects on gastric secretion of frontal lobe stimulation. *Res. Publ. Assoc. Res. Nerv. Ment. Dis.*, 29:617–627.
66. Davison, C. (1940): Disturbances of temperature regulation in man. *Res. Publ. Assoc. Res. Nerv. Ment. Dis.*, 20:774–823.
67. Davison, C., and Demuth, E. L. (1946): Disturbances in sleep mechanisms: A clinicopathologic study: III. Lesions at the diencephalic level (hypothalamus). *Arch. Neurol. Psychiatry*, 55:111–125.
68. Davison, C., and Selby, N. E. (1935): Hypothermia in cases of hypothalamic lesions. *Arch. Neurol. Psychiatry*, 33:570–591.
69. Debons, A. F., Krimsky, I., and From, A. (1970): A direct action of insulin on the hypothalamic satiety center. *Am. J. Physiol.*, 219:938–943.
70. Demanet, J. C., Bonnyns, M., Bleiberg, H., and Stevens-Rocmans, C. (1971): Coma due to water intoxication in beer drinkers. *Lancet*, 2:1115–1117.
71. DeRubertis, F. R., Michelis, M. F., and Davis, B. B. (1974): "Essential" hypernatremia. Report of three cases and review of the literature. *Arch. Intern. Med.*, 134:889–895.
72. Devereaux, M. W., and McCormick, R. A. (1972): Psychogenic water intoxication: A case report. *Am. J. Psychiatry*, 129:628–630.
73. Diamond, E. F., and Averick, N. A. (1966): Marasmus and the diencephalic syndrome. *Arch. Neurol.*, 14:270–272.
74. Dods, L. (1957): A diencephalic syndrome of early infancy. *Med. J. Aust.*, 2:689–691.
75. Ducker, T. B. (1968): Increased intracranial pressure and pulmonary edema. *J. Neurosurg.*, 28:112–117.
76. Duff, R. S., Farrant, P. C., Leveaux, V. M., and Wray, S. M. (1961): Spontaneous periodic hypothermia. *Q. J. Med.*, 30:329–338.
77. Duffy, J. P., and Davison, K. (1969): A female case of the Kleine-Levin syndrome. *Br. J. Psychiatry*, 114:77–84.
78. Dunn, H. G. (1968): The Prader-Labhart-Willi syndrome: Review of the literature and report of nine cases. *Acta Paediatr. Scand. [Suppl.]*, 186:1–38.
79. Economo, C. von (1918): *Die Encephalitis Lethargica*. Deuticke, Vienna.
80. Economo, C. von (1930): Sleep as a problem of localization. *J. Nerv. Ment. Dis.*, 71:249–259.
81. Edmondson, H. T., Gindin, R. A., and Peebles, G. C. (1968): Cerebral lesions as a cause of esophageal ulceration. *Surgery*, 64:720–727.
82. Emmers, R. (1973): Interaction of neural systems which control body water. *Brain Res.*, 49:323–347.
83. Engel, G. L., and Aring, C. D. (1945): Hypothalamic attacks with thalamic lesion. *Arch. Neurol. Psychiatry*, 54:37–50.
84. Engstrom, W. W., and Liebman, A. (1953): Chronic hyperosmolality of the body fluids

with a cerebral lesion causing diabetes insipidus and anterior pituitary insufficiency. *Am. J. Med.*, 15:180-186.
85. Epstein, A. N., Nicolaidis, S., and Miselis, R. (1975): The glucoprivic control of food intake and the glucostatic theory of feeding behavior. In: *Neural Integration of Physiological Mechanisms and Behavior*, edited by C. J. Mogenson and F. R. Calaresu, pp. 148-168. University of Toronto Press, Toronto.
86. Erickson, T. C. (1939): Neurogenic hyperthermia: A clinical syndrome and its treatment. *Brain*, 62:172-190.
87. Evans, P. R. (1964): Hypogenital dystrophy with diabetic tendency. *Guy's Hosp. Rep.*, 113:207-227.
88. Fennegan, F. M., and Puiggari, M. J. (1966): Hypothalamus and amygdaloid influence in gastric motility in dogs. *J. Neurosurg.*, 24:497-504.
89. Fishman, L. S., Samson, J. H., and Sperling, D. R. (1965): Primary alveolar hypoventilation syndrome (Ondine's curse). *Am. J. Dis. Child.*, 110:155-161.
90. Fox, R. H., Davies, T. W., Marsh, F. P., and Urich, H. (1970): Hypothermia in a young man with an anterior hypothalamic lesion. *Lancet*, 2:185-188.
91. Fox, R. H., Wilkins, D. C., Bell, J. A., Bradley, R. D., Brouse, N. L., Cranston, W. I., Foley, T. H., Gilby, E. D., Hebden, A., Jenkins, B. S., and Rawlins, M. D. (1973): Spontaneous periodic hypothermia: Diencephalic epilepsy. *Br. Med. J.*, 2:693-695.
92. Fraser, R., Doyle, F. H., Burke, J. B., Lewis, P. D., Henry, K., Calne, D. B., and Joplin, G. F. (1973): Clinicopathological conference. A case of abnormal thirst. *Br. Med. J.*, 3:214-219.
93. Fröhlich, A. (1901): Ein Fall von Tumor der Hypophysis cerebri ohne Akromegalie. *Wien. Klin. Rdsch.*, 15:883-906.
94. Fuller, C. A., Horwitz, B. A., and Horowitz, J. M. (1975): Shivering and nonshivering thermogenic responses of cold-exposed rats to hypothalamic warming. *Am. J. Physiol.*, 228:1519-1524.
95. Fulton, J. F. (1940): Introduction: Historical resume. *Res. Publ. Assoc. Res. Nerv. Ment. Dis.*, 20:xiii-xvi.
96. Fulton, J. F., and Bailey, P. (1929): Tumors in the region of the third ventricle: Their diagnosis and relation to pathological sleep. *J. Nerv. Ment. Dis.*, 69:1-25.
97. Gale, C. C. (1973): Neuroendocrine aspects of thermoregulation. *Annu. Rev. Physiol.*, 35:391-430.
98. Gamble, J. E., and Patton, H. D. (1953): Pulmonary edema and hemorrhage from preoptic lesions in rats. *Am. J. Physiol.*, 172:623-631.
99. Gamper, E. (1928): Zur Frage der Polioencephalitis haemorrhagica der chronischen Alkoholiker. Anatomische Befunde beim alkoholischen Korsakow und ihre Bexiehungen zum klinischen bild. *Dtsch. Z. Nervenh.*, 102:122-129.
100. Gamstorp, I., Kjellman, B., and Palmgren, B. (1967): Diencephalic syndromes of infancy. *J. Pediatr.*, 70:383-390.
101. Garcia-Bengochea, F., DelaTorre, O., Esquivel, O., Vieta, R., and Fernandez, C. (1954): The section of the fornix in the surgical treatment of certain epilepsies. *Trans. Am. Neurol. Assoc.*, 79:176-178.
102. Gaupp, R., Jr. (1941): Uber der diabetes insipidus. *Z. Ges. Neurol. Psychiat.*, 171:514.
103. Gebber, G. L., and Snyder, D. W. (1970): Hypothalamic control of baroreceptor reflexes. *Am. J. Physiol.*, 218:124-131.
104. Genovesi, U., Moruzzi, G., Palestini, M., Rossi, G. F., and Zanchetti, A. (1956): EEG and behavioral patterns following lesions of the mesencephalic reticular formation in chronic cats with implanted electrodes. Abstr. Comm. 20th Int. Physiol. Congr., Bruxelles, pp. 335-336.
105. Globus, J. H., and Gang, K. M. (1945): Craniopharyngioma and suprasellar adamantinoma. *Mt. Sinai J. Med. N.Y.*, 12:220-276.
106. Globus, J. H., and Ralston, B. L. (1951): Multiple erosions and acute perforations of esophagus, stomach and duodenum in relation to disorders of nervous system. *Mt. Sinai J. Med. N.Y.*, 17:817-842.
107. Gold, R. M. (1967): Aphagia and adipsia following unilateral and bilaterally asymmetrical lesions in rats. *Physiol. Behav.*, 2:211-220.
108. Gold, R. M. (1973): Hypothalamic obesity: Myth of the VM nucleus. *Science*, 182:488-490.

109. Gold, R. M., and Proulx. D. M. (1972): Bait-shyness impaired by ventromedial hypothalamic lesions. *J. Comp. Physiol. Psychol.*, 79:201-209.
110. Goldberg, M., and McCurdy, D. K. (1963): Hyperaldosteronism and hypergranularity of juxtaglomerular cells in renal hypertension. *Ann. Intern. Med.*, 59:24-36.
111. Goldberg, M., Weinstein, G., Adesman, J., and Blescher, S. J. (1967): Asymptomatic hypovolemic hypernatremia. A variant of essential hypernatremia. *Am. J. Med.*, 43:804-810.
112. Golonka, J. E., and Richardson, J. A. (1970): Postconcussive hyperosmolality and deficient thirst. *Am. J. Med.*, 48:261-267.
113. Gordy, P. D., Peet, M. M., and Kahn, E. A. (1949): The surgery of the craniopharyngioma. *J. Neurosurg.*, 6:503-517.
114. Gowers, W. (1907): *The Borderland of Epilepsy*, P. Blakiston & Co., Philadelphia.
115. Green, J. P., Buchanan, G. C., Alvord. E. C., and Swanson, A. G. (1967): Hereditary and idiopathic types of diabetes insipidus. *Brain*, 90:707-714.
116. Greiss, K. C., Moses, A. M., and Krieger, D. T. (1974): Pituitary adenoma associated with inappropriate anti-diuretic hormone secretion. *Acta Endocrinol.*, 76:59-66.
117. Grinker, R. R. (1939): Hypothalamic functions in psychosomatic interrelations. *Psychosom. Med.*, 1:19-47.
118. Grünthal, E. (1929): Der Zellanfban des Hypothalamus bein Hunde. *Z. Ges. Neurol. Psychiat.*, 120:157.
119. Guihard, J., Velot-Lerou, A., Poitrat, C., Laloum, D., and L'Hirondel, J. (1971): Hypothermie spontanée récidivante avec agénésie du corps calleaux. *Ann. Pédiat. (Paris)*, 18:645-656.
120. Gwinup, G. Chelvam, R., Jabola, R., and Meister, L. (1972): Beer-drinker's hyponatremia. *Calif. Med.*, 116 (3):78-81.
121. Hammel, H. T. (1968): Regulation of internal body temperature. *Annu. Rev. Physiol.*, 30:641-710.
122. Hammermeister, K. E., and Reichenbach, D. D. (1969): QRS changes, pulmonary edema, and myocardial necrosis associated with subarachnoid hemorrhage. *Am. Heart J.*, 78:94-100.
123. Hanhart, E. (1940): Die erbpathologie des diabetes insipidus. In: *Handbuch der Erbbiologie des Menschen.*, Vol. 4, Part II, p. 801. Julius Springer, Berlin.
124. Hardy, J. D. (1970): Thermal comfort: Skin temperature and physiological thermoregulation. In: *Physiological and Behavioral Temperature Regulation*, edited by J. D. Hardy, A. P. Gagge, and J. A. J. Stolwijk, pp. 856-873. Charles C Thomas, Springfield, Ill.
125. Hardy, J. D. (1973): Posterior hypothalamus and the regulation of body temperature. *Fed. Proc.*, 32:1564-1571.
126. Hart, M. N. (1971): Hypertrophy of human subventricular hypothalamic nucleus in starvation. *Arch. Pathol.*, 91:493-496.
127. Hawkins, W. E., and Clower, B. R. (1971): Myocardial damage after head trauma and simulated intracranial hemorrhage in mice: The role of the autonomic nervous system. *Cardiovasc. Res.*, 5:524-529.
128. Hays, R. M., McHugh, P. R., and Williams, H. E. (1963): Absence of thirst in association with hydrocephalus. *N. Engl. J. Med.*, 269:227-231.
129. Hensel, H. (1973): Neural processes in thermoregulation. *Physiol. Rev.*, 53:948-1017.
130. Hervey, G. R. (1975): The problem of energy balance in the light of control theory. In: *Neural Integration of Physiological Mechanisms and Behavior*, edited by C. J. Mogenson and F. R. Caleresu, pp. 109-127. University of Toronto Press, Toronto.
131. Hess, W. R. (1969): *Hypothalamus and Thalamus. Experimental Documentation.* Georg Thieme Verlag, Stuttgart.
132. Hetherington, A. W., and Ranson, S. W. (1939): Experimental hypothalamico-hypophyseal obesity in the rat. *Proc. Soc. Exp. Biol. Med.*, 41:465-466.
133. Heuser, G. (1970): Trends in clinical neuroendocrinology, *Ann. Intern. Med.*, 73:783-807.
134. Hilton, S. M., and Spyer, K. M. (1971): Participation of the anterior hypothalamus in the barorecepter reflex. *J. Physiol.*, 218:271-293.
135. Hines, E. A., Jr., and Bannick, E. G. (1934): Intermittent hypothermia with disabling hyperhidrosis: Report of a case with successful treatment. *Mayo Clin. Proc.*, 9:705-708.

136. Hirsch, J., and Knittle, J. T. (1963): Cellularity of obese and nonobese human adipose tissue. *Fed. Proc.*, 29:1516–1521.
137. Hobson, J. A., and English, J. T. (1963): Self-induced water intoxication. *Ann. Intern. Med.*, 58:324–332.
138. Hockaday, T. D. R., Cranston, W. I., Cooper, K. E., and Mottram, R. F. (1962): Temperature regulation in chronic hypothermia. *Lancet*, 2:428–432.
139. Hoebel, B. G., and Teitelbaum, P. (1966): Weight regulation in normal and hypothalamic hyperphagic rats. *J. Comp. Physiol. Psychol.*, 61:189–193.
140. Hoffman, A. M., and Pobirs, F. W. (1942): Intermittent hypothermia with disabling hyperhidrosis. *J. A. M. A.*, 120:445–447.
141. Hogner, P. (1929): Clinical manifestations in diseases of the third ventricle and its walls. *Dtsch. Z. Nervenheil.*, 97:238–266.
142. Husain, M. K., Fernando, N., Shapiro, M., Kagan, A., and Glick, S. M. (1973): Radioimmunoassay of arginine vassopressin in human plasma. *J. Clin. Endocrinol. Metab.*, 37:616–625.
143. Johnson, R. H., and Spalding, J. M. K. (1974): *Disorders of the Autonomic Nervous System*, pp. 129–179. F. A. Davis, Philadelphia.
144. Jouvet, M. (1972): The role of monoamines and acetylcholine-containing neurons in the regulation of the sleep-waking cycle. *Ergeb. Physiol.*, 64:166–307.
145. Jürgens, U. (1974): The hypothalamus and behavioral patterns. *Prog. Brain Res.*, 41:445–463.
146. Kabat, H., Magoun, H. W., and Ranson, S. W. (1935): Electrical stimulation of points in the forebrain and midbrain. The resultant alteration in blood pressure. *Arch. Neurol. Psychiatry*, 34:931–955.
147. Kahn, E., and Crosby, E. C. (1972): Korsakoff's syndrome associated with surgical lesions involving the mammillary bodies. *Neurology (Minneap.)*, 22:117–125.
148. Kahn, E. A., Crosby, E. C., and DeJong, B. (1969): Tumors of the hypothalamus region. In: *Correlative Neurosurgery*, edited by E. A. Kahn, E. C. Crosby, R. C. Schneider, and J. A. Taren, pp. 94–130. Charles C Thomas, Springfield, Ill.
149. Kamalian, N., Keesey, R. E., and ZuRhein, G. M. (1975): Lateral hypothalamic demyelination and cachexia in a case of "malignant" multiple sclerosis. *Neurology (Minneap.)*, 25:25–30.
150. Kastella, K. G., Spurgeon, H. A., and Weiss, G. K. (1974): Respiratory-related neurons in anterior hypothalamus of the cat. *Am. J. Physiol.*, 227:710–713.
151. Kastin, A. J., Lipsett, M. B., and Ommaya, A. K. (1965): Asymptomatic hypernatremia. Physiological and clinical study. *Am. J. Med.*, 38:306–315.
152. Kennedy, G. C. (1975): Some aspects of the relation between appetite and endocrine development on the growing animal. In: *Neural Integration of Physiological Mechanisms and Behavior*, edited by C. J. Mogenson and F. R. Calaresu, pp. 326–338. University of Toronto Press, Toronto.
153. Killeffer, F. A., and Stern, W. E. (1970): Chronic effects of hypothalamic injury. Report of a case of near total hypothalamic destruction resulting from removal of a craniopharyngioma. *Arch. Neurol.*, 22:419–429.
154. Kitamura, T., and Ito, K. (1976): Acute gastric changes in patients with acute stroke. Part 2: Gastroendoscopic findings and biochemical observation of urinary noradrenalin, adrenalin, 17-OHCS and serum gastrin. *Stroke*, 7:464–468.
155. Klein, D. (1968): Sur quelques variétés cliniques et génétiques du syndrome de Bardet-Biedl. *Rev. Otoneuroophtalmol.*, 40:125–144.
156. Kleine, W. (1925): Periotisch Schlafsucht. *Mschr. Psychiat. Neurol.*, 57:285–298.
157. Koeppen, A. H., Daniels, J. C., and Barron, K. D. (1969): Subnormal body temperature in Wernicke's encephalopathy. *Arch. Neurol.*, 21:493–498.
158. Korner, P. L. (1971): Integrative neural cardiovascular control. *Physiol. Rev.*, 51:312–367.
159. Kucharczyk, J., and Mogenson, G. J. (1975): Separate lateral hypothalamic pathways for extracellular and intracellular thirst. *Am. J. Physiol.*, 228:295–301.
160. Kurtz, R. G., Rozin, P., and Teitelbaum, P. (1972): Ventromedial hypothalamic hyperphagia in the hypophysectomized weanling rat. *J. Comp. Physiol. Psychol.*, 80:19–25.
161. Kusajima, K., Wax, S., and Webb, W. R. (1974): Cerebral hypotension shock lung syndrome. *J. Thorac. Cardiovasc. Surg.*, 67:969–975.

162. Lascelles, P. T., and Lewis, P. D. (1972): Hypodipsia and hypernatremia associated with hypothalamic and suprasellar lesions. *Brain*, 95:249-264.
163. Lennox, W. G. (1960): *Epilepsy and Related Disorders, Vol. 1*, p. 332. Little, Brown, & Co., Boston.
164. Levin, M. (1929): Narcolepsy (Gélineau's syndrome) and other varieties of morbid somnolence. *Arch. Neurol. Psychiatry*, 22:1172-1200.
165. Lewin, K., Mattingly, D., and Mills, R. R. (1972): Anorexia nervosa associated with hypothalamic tumour. *Br. Med. J.*, 2:629-630.
166. Lindsley, D. B., Bowden, J. W., and Magoun, H. W. (1949): Effect upon the EEG of acute injury to the brain stem activating system. *Electroencephalogr. Clin. Neurophysiol.*, 1:475-486.
167. Linquette, M., Fossati, P., Lefebure, J., and Cappoen, J. P. (1973): Acute water intoxication from compulsive drinking. *Br. Med. J.*, 2:365.
168. Lipsett, M. B., Dreifuss, F. E., and Thomas, L. B. (1962): Hypothalamic syndrome following varicella. *Am. J. Med.*, 32:471-475.
169. Long, D. M., Leonard, A. S., Chou, S. N., and French, L. A. (1962): Hypothalamus and gastric ulceration. I. Gastric effects of hypothalamic lesions. *Arch. Neurol.*, 7:167-175.
170. MacLean, A. J. (1934): Autonomic epilepsy: Report of a case with observation at necropsy. *Arch. Neurol. Psychiatry*, 32:189-197.
171. MacMillan, A. L., Corbett, J. L., Johnson, R. H., Smith, A. C., Spalding, J. M. K., and Wollner, L. (1967): Temperature regulation in survivors of accidental hypothermia of the elderly. *Lancet*, 2:165-169.
172. Maddy, J. A., and Winternitz, W. W. (1971): Hypothalamic syndrome with hypernatremia and muscular paralysis. *Am. J. Med.*, 51:394-402.
173. Maire, F. W., and Patton, H. D. (1956): Neural structures involved in the genesis of "preoptic pulmonary edema," gastric erosion and behavior changes. *Am. J. Physiol.*, 184:345-350.
174. Malamud, N. (1967): Psychiatric disorder with intracranial tumors of the limbic system. *Arch. Neurol.*, 17:113-123.
175. Mangos, J. A., and Lobeck, C. C. (1964): Studies on sustained hyponatremia due to CNS infections. *Pediatrics*, 34:503-510.
176. Martin, J. B., Brown, G. M., Reichlin, S. (1977): *Clinical Neuroendocrinology. Contemporary Neurology Series*. F. A. Davis, Philadelphia.
177. Masserman, J. H., and Haertig, E. W. (1938): The influence of hypothalamic stimulation on intestinal activity. *J. Neurophysiol.*, 1:350-356.
178. Mayer, J. (1955): Regulation of energy intake and the body weight: The glucostatic theory and the lipostatic hypothesis. *Ann. N.Y. Acad. Sci.*, 63:15-43.
179. McGinty, D. J. (1969): Somnolence, recovery, and hyposomnia following ventro-medial diencephalic lesions in the rat. *Electroencephalogr. Clin. Neurophysiol.*, 26:70-79.
180. McGinty, D. J., and Sterman, M. B. (1968): Sleep suppression after basal forebrain lesions in the cat. *Science*, 160:1253-1255.
181. Mecklenburg, R. S., Loriaux, D. L., Thompson, R. H., Andersen, A. E., and Lipsett, M. B. (1974): Hypothalamic dysfunction in patients with anorexia nervosa. *Medicine (Baltimore)*, 53:147-159.
182. Melvill, K. I., Blum, B., Shister, H., and Silver, M. D. (1963): Cardiac ischemic changes and arrhythmias induced by hypothalamic stimulation. *Am. J. Cardiol.*, 12:781-791.
183. Meyer, H. H. (1913): Theorie des Fiebers und seiner Behandlung. *Verhandl. deut. bes. inn. Med.*, 30:15-25.
184. Miyasaki, K., Miyachi, Y., Arimitsu, K., Kita, E., and Yoshida, M. (1972): Post-traumatic hypothalamic obesity—an autopsy case. *Acta. Pathol. Jpn.*, 22:779-802.
185. Mohr, B. (1840): Hypertophie der hypophysis cerebri und dadurch bedingter druck afu die hirngrundflache, insbesondere auf die sehnerven, das chiasma derselben und den linkseitigen hirnschenkel. *Wochenschr. Ges. Heilk*, 6:565-571.
186. Morgane, P. J. (1961): Medial forebrain bundle and "feeding centers" of the hypothalamus. *J. Comp. Neurol.*, 117:1-25.
187. Moruzzi, G. (1972): The sleep-waking cycle. *Ergeb. Physiol.*, 64:1-165.
188. Moskowitz, M. A., Fisher, J. N., Simpser, M. D., and Strieder, D. J. (1976): Periodic apneas, exercise hypoventilation, and hypothalamic dysfunction. *Ann. Intern. Med.*, 84:171-173.

189. Myers, R. D. (1974): Fever secondary to endotoxin injection in the hypothalamus. *J. Physiol.*, 243:167-193.
190. Myers, R. D. (1974): Ionic concepts of the set-point for body temperature. In: *International Symposium on Recent Studies of Hypothalamic Function, Calgary, 1973*, edited by K. Lederis and K. E. Cooper, pp. 371-390. S. Karger, Basel.
191. Nauta, W. J. H. (1946): Hypothalamic regulation of sleep in rats. Experimental study. *J. Neurophysiol.*, 9:285-316.
192. Nicoll, R. A., and Barker, J. L. (1971): The pharmacology of recurrent inhibition in the supraoptic neurosecretory system. *Brain Res.*, 35:501-511.
193. Nisbett, R. E., Braver, A., Jusela, G., and Kezur, D. (1975): Age and sex differences in behaviors mediated by the ventromedial hypothalamus. *J. Comp. Physiol. Psychol.*, 88:735-746.
194. Noël, P., Hubert, J. P., Ectors, M., Franken, L., and Flament-Durand, J. (1973): Agenesis of the corpus callosum associated with relapsing hypothermia. A clinico-pathological report. *Brain*, 96:359-368.
195. Obrador, S., and Dierssen, G. (1967): Mental changes induced by subcortical stimulation and therapeutic lesions. *Confin. Neurol.*, 29:168.
196. Olds, M. E., and Olds, J. (1969): Effects of lesions in the medial forebrain bundle on self-stimulation behavior. *Am. J. Physiol.*, 217:1253-1264.
197. Oltmans, G. A., and Harvey, J. A. (1972): LH syndrome and brain catecholamine levels after lesions of the nigrostriatal bundle. *Physiol. Behav.*, 8:69-78.
198. Park, C. D., and Sutnic, A. J. (1973): Pulmonary surface activity alterations associated with pulmonary edema following preoptic hypothalamic lesion in rats. *Proc. Soc. Exp. Biol. Med.*, 142:1025-1030.
199. Pelc, S. (1972): The diencephalic syndrome in infants. A review in relation to optic nerve glioma. *Eur. Neurol.*, 7:321-334.
200. Penfield, W. (1929): Diencephalic autonomic epilepsy. *Arch. Neurol. Psychiatry*, 22:358-374.
201. Penfield, W., and Milner, B. (1958): Memory deficit produced by bilateral lesions in the hippocampal zone. *Arch. Neurol. Psychiatry*, 79:475-497.
202. Perlroth, M. G., Tschudy, D. P., Marver, H. S., Berard, C. W., Zeigel, R. F., Rechcigl, M., and Collins, A. (1966): Acute intermittent porphyria. New morphologic and biochemical findings. *Am. J. Med.*, 41:149-162.
203. Pernet, G. E. (1964): Stress ulcers and the neurosurgeon. *J. Iowa Med. Soc.*, 54:583-585.
204. Philip, G., and Smith, J. F. (1973): Hypothermia and Wernicke's encephalopathy. *Lancet*, 2:122-124.
205. Pickering, L. K., and Hogan, G. R. (1971): Voluntary water intoxication in a normal child. *J. Pediatr.*, 78:316-318.
206. Poeck, K., and Pilleri, G. (1965): Release of hypersexual behaviour due to lesion in the limbic system. *Acta Neurol. Scand.*, 41:233-244.
207. Pomorski, J. (1891): Experimentelles Zur Aetiologie der Melaena neonatorum. *Arch. Kinderh.*, 14:165-193.
208. Powell, G. F., Brasel, J. A., and Blizzard, R. M. (1967): Emotional deprivation and growth retardation simulating idiopathic hypopituitarism. *N. Engl. J. Med.*, 276:1271-1283.
209. Prader, A., and Willi, H. (1961): Das Syndrom von Imbezillitat, Adipositas, Muskelhypotnie, Hypogenitalismus, Hypogonadismus und Diabetes mellitus mit 'Myatonie'- Anamnese. Second International Congress on Mental Retardation, Part I:353-357.
210. Ranson, S. W. (1939): Somnolence caused by hypothalamic lesions in the monkey. *Arch. Neurol. Psychiatry*, 41:1-23.
211. Ranson, S. W. (1940): Regulation of body temperature. *Res. Publ. Assoc. Res. Nerv. Ment. Dis.*, 20:342-399.
212. Ranson, S. W., Fisher, C., and Ingram, W. R. (1937): Hypothalamic regulation of temperature in the monkey. *Arch. Neurol. Psychiatry*, 38:445-466.
213. Ranstrom, S. (1947): The hypothalamus and sleep regulation. An experimental and morphological study, *Acta. Pathol. Microbiol. Scand.* [*Suppl.*], 70:1-90.
214. Reeves, A. G., and Plum, F. (1969): Hyperphagia, rage, and dementia accompanying a ventromedial hypothalamic neoplasm. *Arch. Neurol.*, 20:616-624.

215. Reichlin, S. (1974): Neuroendocrinology. In: *Textbook of Endocrinology*, edited by R. H. Williams, pp. 774–831. W. B. Saunders Co., Philadelphia.
216. Remy, M. (1942): Contribution a l'étude de la maladie de Korsakow. *Mschrft. Psychiat. Neurol.*, 106:128–144.
217. Reynolds, R. W. (1963): Pulmonary edema as a consequence of hypothalamic lesions in rats. *Science*, 141:930–932.
218. Richards, P. (1963): Pulmonary edema and intracranial lesions. *Br. Med. J.*, 2:83–86.
219. Robertson, G. L., Bhoopalam, N., and Zekowitz, L. (1973): Vincristine neurotoxicity and abnormal secretion of antidiuretic hormone. *Arch. Intern. Med.*, 132:717–720.
220. Roeder, F., and Müller, D. (1969): The stereotaxic treatment of paedophilic homosexuality. *Ger. Med.*, 14:265–271.
221. Rogers, P. W., and Kurtzman, N. A. (1973): Renal failure, uncontrollable thirst, and hyperreninemia. *J. A. M. A.*, 225:1236–1238.
222. Rokitansky, C. (1841): Handbuch der pathologischen Anatomie, quoted by Cushing, H. (1932). *Surg. Gynecol. Obstet.*, 55:21.
223. Rosenblum, W. I., and Feigin, I. (1965): The hemorrhagic component of Wernicke's encephalopathy. *Arch. Neurol.*, 13:627–632.
224. Rostad, H. (1973): Colonic motility in the cat: III. Influence of hypothalamic and mesencephalic stimulation. *Acta Physiol. Scand.*, 89:104–115.
225. Rovit, R. L., and Sigler, M. H. (1964): Hypothalamus with herpes simplex encephalitis. *Arch. Neurol.*, 10:559–603.
226. Runnels, P., and Thompson, R. (1967): Hypothalamic structures critical for the performance of a locomotor escape response in the rat. *Brain Res.*, 13:328–337.
227. Russek, M. (1975): Current hypotheses on the control of feeding behavior. In: *Neural Integration of Physiological Mechanisms and Behavior*, edited by C. J. Mogenson and F. R. Calaresu, pp. 128–147. University of Toronto Press, Toronto.
228. Russell, A. (1951): A diencephalic syndrome of emaciation in infancy and childhood. *Arch. Dis. Child.*, 26:274.
229. Sadowsky, C., and Reeves, A. G. (1975): Agenesis of the corpus callosum with hypothermia. *Arch. Neurol.*, 32:774–776.
230. Sano, K., Mayanagi, Y., Sekino, H., Ogashiwa, M., and Ishijima, B. (1970): Results of stimulation and destruction of the posterior hypothalamus in man. *J. Neurosurg.*, 33:689–707.
231. Schachter, S. (1971): Some extraordinary facts about obese humans and rats. *Am. Psychol.*, 26:129–144.
232. Schvarcz, J. R., Driollet, R., Rios, E., and Betti, O. (1972): Stereotactic hypothalamotomy for behavior disorders. *J. Neurol. Neurosurg. Psychiatry*, 35:356–359.
233. Schwartz, M., and Teitelbaum, P. (1974): Dissociation between learning and remembering in rats with lesion in the lateral hypothalamus. *J. Comp. Physiol. Psychol.*, 87:384–398.
234. Schwartz, W. B., Bennett, W., Curelop, S., and Bartter, F. C. (1957): A syndrome of renal sodium loss and hyponatremia probably resulting from inappropriate secretion of antidiuretic hormone. *Am. J. Med.*, 23:529–542.
235. Shapiro, W. R., Williams, G. H., and Plum, F. (1969): Spontaneous recurrent hypothermia accompanying agenesis of the corpus callosum. *Brain*, 92:423–436.
236. Shuster, S. (1960): The EKG in subarachnoid hemorrhage. *Br. Heart J.*, 22:316–320.
237. Simons, B. J. (1968): Cause of excessive drinking in diabetes insipidus. *Nature*, 219:1061–1062.
238. Smith, R. W., and McCann, S. M. (1962): Alterations in food and water intake after hypothalamic lesions in the rat. *Am. J. Physiol.*, 203:366–370.
239. Solomon, G. E. (1973): Diencephalic autonomic epilepsy caused by a neoplasm. *J. Pediatr.*, 83:277–280.
240. Souques, A., Baruk, H., and Bertrand, I. (1926): Tumeur de l'infundibulum avec léthargie isolée. *Rev. Neurol. (Paris)*, 33:532–540.
241. Spiegel, E. A., and Wycis, H. T. (1968): Multiplicity of subcortical localization of various functions. *J. Nerv. Ment. Dis.*, 147:45–48.
242. Squires, R. D., and Jacobson, F. H. (1968): Chronic deficits of temperature regulation produced in cats by preoptic lesions. *Am. J. Physiol.*, 214:549–560.

243. Sridhar, C. B., Calvert, G. D., and Ibbertson, H. K. (1974): Syndrome of hypernatremia, hypodipsia and partial diabetes insipidus: A new interpretation. *J. Clin. Endocrinol. Metab.*, 38:890–901.
244. Staub, N. C., and Sagawa, Y. (1964): Pulmonary edema induced by vagotomy in the guinea pig. *Fed. Proc.*, 23:417 (Abst.).
245. Stevens, H., Guin, G. H., and Gilbert, E. F. (1963): Gastrointestinal ulceration and central nervous system lesions. *Am. J. Dis. Child.*, 106:613–619.
246. Stiggelbout, W. (1972): The Bardet-Biedl syndrome. Including Hutchinson-Laurence-Moon syndrome. In: *Handbook of Clinical Neurology, Vol. 13*, edited by P. J. Vinken and G. W. Bruyn, pp. 380–413. North-Holland, Amsterdam.
247. Sunderman, F. W., and Haymaker, W. (1947): Hypothermia and elevated serum magnesium in a patient with facial hemangioma extending into the hypothalamus. *Am. J. Med. Sci.*, 213:562–571.
248. Szakacs, J. E., and Cannon, A. (1958): L-Norepinephrine myocarditis. *Am. J. Clin. Pathol.*, 30:425–434.
249. Teitelbaum, P., and Epstein, A. N. (1962): The lateral hypothalamic syndrome: Recovery of feeding and drinking after lateral hypothalamic lesions. *Psychol. Rev.*, 69:74–90.
250. Thomas, D. J., and Green, I. D. (1973): Periodic hypothermia. *Br. Med. J.*, 2:696–697.
251. Thomas, M. R., and Calaresu, F. R. (1973): Hypothalamic inhibition of the chemoreceptor-induced bradycardia in the cat. *Am. J. Physiol.*, 225:201–208.
252. Travis, L. B., Dodge, W. F., Waggener, J. D., and Kashemsant, C. (1967): Defective thirst mechanism secondary to a hypothalamic lesion: Studies in a child with adipsia, polyphagia, obesity, and persistent hyperosmolality. *J. Pediatr.*, 70:915–926.
253. Veale, W. L., and Cooper, K. E. (1974): Evidence for the involvement of prostaglandins in fever. In: *International Symposium on Recent Studies of Hypothalamic Function, Calgary, 1973*, edited by K. Lederis and K. E. Cooper, pp. 359–370. S. Karger, Basel.
254. Vejjajiva, A., Sitprija, V., and Shuangshoti, S. (1969): Chronic sustained hypernatremia and hypovolemia in hypothalamic tumor. *Neurology (Minneap.)*, 19:161–166.
255. Verrier, R. L., Calvert, A., and Lown, B. (1975): Effect of posterior hypothalamic stimulation on ventricular fibrillation threshold. *Am. J. Physiol.*, 228:923–927.
256. Victor, M., Adams, R. D., and Collins, G. H. (1971): *The Wernicke-Korsakoff Syndrome*, pp. 166–170. F. A. Davis, Philadelphia.
257. Von Wowern, F. (1966): Obesity as a sequel of traumatic injury to the hypothalamus. *Dan. Med. Bull.*, 13:11–13.
258. Walter, W. G., Griffiths, G. M., and Nervin, S. (1939): The electroencephalogram in a case of pathological sleep due to hypothalamic tumour. *Br. Med. J.*, 1:107–109.
259. Wang, S. C., and Ranson, S. W. (1941): The role of the hypothalamus and preoptic region in the regulation of heart rate. *Am. J. Physiol.*, 132:5–8.
260. Watson, J. M., and Netsky, M. G. (1954): Ulceration and malacia of the upper alimentary tract in neurologic disorders. *Arch. Neurol. Psychiatry*, 72:426–439.
261. Wayner, M. J. (1974): The lateral hypothalamus and adjunctive drinking. *Prog. Brain Res.*, 41:371–394.
262. Weinberg, S. J., and Fuster, J. M. (1960): EKG changes with localized hypothalamic stimulation. *Ann. Intern. Med.*, 53:332–341.
263. Weissman, S. J. (1939): Edema and congestion of the lungs from intracranial hemorrhage. *Surgery*, 6:722–729.
264. Weitzman, E. D., and Triedman, M. H. (1960): Hyperosmolarity associated with a hypothalamic lesion. *Neurology (Minneap.)*, 10:584–590.
265. Welt, L. G. (1962): Hypo-and hypernatremia. *Ann. Intern. Med.*, 56:161–164.
266. White, J. C. (1940): Autonomic discharge from stimulation of the hypothalamus in man. *Res. Publ. Assoc. Res. Nerv. Ment. Dis.*, 20:854–863.
267. White, L. E., and Hain, R. F. (1959): Anorexia in association with a destructive lesion of the hypothalamus. *Arch. Pathol.*, 68:275–281.
268. White, P. T., and Ross, A. T. (1963): Inanition syndrome in infants with anterior hypothalamic neoplasms. *Neurology (Minneap.)*, 13:974–981.
269. Williams, M., and Pennybacker, J. (1954): Memory disturbances in third ventricle tumors. *J. Neurol. Neurosurg. Psychiatry*, 17:115–123.
270. Wise, B. L. (1962): Neurogenic hyperosmolarity (hypernatremia). *Neurology (Minneap.)*, 12:453–459.

271. Wolff, H. P., Vecsei, P., Kruck, F., Roschev, S., Broun, J. J., Dusterdieck, G. O., Lever, A. F., and Robertson, J. I. S. (1968): Psychiatric disturbance leading to potassium depletion, sodium depletion, raised plasma renin concentration and secondary hyperaldosteronism. *Lancet,* 1:257–261.
272. Wolff, S. M., Adler, R. C., Buskirk, E. R., and Thompson, R. H. (1964): A syndrome of periodic hypothalamic discharge. *Am. J. Med.,* 36:956–967.
273. Wortis, H., and Maurer, W. S. (1942): "Sham rage" in man. *Am. J. Psychol.,* 98:638–644.
274. Zanchetti, A. (1970): Control of the cardiovascular system. In: *The Hypothalamus,* edited by L. Martini, M. Motta, and F. Fraschini, p. 233. Academic Press, New York.
275. Zazgornik, J., Jellinger, K., Waldhausl, W., and Schmidt, P. (1974): Excessive hypernatremia and hyperosmolality associated with germinoma in the hypothalamic and pituitary region. *Eurp. Neurol.,* 12:38–46.
276. Zimmerman, H. M. (1940): Temperature disturbances and the hypothalamus. *Res. Publ. Assoc. Res. Nerv. Ment. Dis.,* 20:824–840.

## Subject Index

Acetyl-coenzyme A Co-A), 316, 326
Acetylcholine (ACh), 70, 88, 156, 159, 176, 214, 281–282, 334
Acetylcholinesterase (AChE), 88, 89f, 184, 281
N-Acetylserotonin, 306, 315
N-Acetyltransferase, 303, 306–320
ACh, see Acetylcholine
AChE, see Acetylcholinesterase
Acromegaly, 159, 332–333, 349, 356–357, 395, 396f, 404
ACTH, see Corticotropin
Actinomycin D, 260
Action potential, 182–184
  antidromic, 270–272
  pituitary hormones and, 294
Activational steroid effects, 256, 260–262
Adenohypophysis, see also Neurohypophysis, Pituitary gland
  hypothalamic control of, 283–284
  origin of, 188–189
ADH, see Antidiuretic hormone
Adipocytes, 447
Adipsia, 456
Adrenaline synthesis, 78f
α-Adrenergic
  blocking agents, 313, 361
  receptors and GH secretion, 346–347
$\beta_1$-Adrenergic receptors, 314
Afferent pathways, 275
  of tuberoinfundibular neurons, 291
Aging
  growth hormone secretion and, 355, 376
  hypothalamic function and, 418
AHA-MPOA, see Hypothalamic-medial preoptic area, anterior
Alström-Hallgren syndrome, 451
Amenorrhea, 453
  -galactorrhea syndrome, 403
Amine precursor uptake and decarboxylation (APUD) concept, 188
Aminergic pathways in nervous system, 69–119
Amino acid sequence of β-lipotropin, 239t, 335f
γ-Aminobutyric acid (GABA) 8n, 70, 88–89, 90f, 244, 282, 287
  pathway, 301
  in pineal gland, 313
  prolactin, ACTH, and FSH release and, 115
AMP, cyclic, 315–317
Amphetamine and prolactin secretion, 385
Amygdala stimulation, 275, 291–292
  GH release and, 344
Androgens, 256
Angiotensin (ANG), 36, 73t
  distribution, 76f
  as neurotransmitter in brain, 236–238
Angiotensin II, 5, 6, 461
  distribution, 104–105, 106f
  iontophoretic application of, 282–283
  structure, 9t
Anorexia nervosa, 386, 395, 396f, 397, 400, 401f, 453–454
Antidiuretic hormone (ADH), see also Vasopressin
  decreased or absent, 456–459
  elevated, 459–460
  magnocellular neurons and, 455, 457
  release, 454–462
  syndrome of inappropriate secretion of (SIADH), 459–460, 462
Antidromic action potential, 270–272
Antisera, characteristics of enzyme, 71t
Antiserum F, 61
Apathy, 428–429

475

# SUBJECT INDEX

Aphagia, 446
Apomorphine and GH secretion, 346, 347, 349, 350
Appetite control, 19
APUD (amine precursor uptake and decarboxylation) concept, 188
Arcuate nucleus, 24f, 366, 371
Arginine, 30
Aspartate, 90
Astrocyte, 319
Autonomic function disorders, 429–445
Axoaxonic interactions, 111
Axon collaterals, 272–274
  in the tuberoinfundibular system, 288f, 289–291
Axons, 182–184
  neurohormones in, 59
  ventromedial nucleus, 248
Axoplasmic flow, 15, 204

Babinski-Fröhlich syndrome, 450; see also Fröhlich's syndrome
Bacitracin, 201
Band of Broca, 51
Behavior disorders, 428
Bicuculline, 282, 301
Biosynthesis
  brain peptide, 195–209
  of releasing factors, 197–204
Body fluids
  LRH in, 225
  TRH in, 222
Bovine serum albumin (BSA), 59
  as conjugate, 73t
Brain
  cells, low-voltage processing of information by, 182–184
  peptide biosynthesis, 194–209
  rhythms, neuroendocrine regulation and, 356
  sexual differentiation of, 256
Brainstem neuron, opioid substance effect on, 174f
Bromocryptine, 356
Bromoergocryptine, 349
BSA, see Bovine serum albumin
Bulimia, 446, 450
Butaclamol and GH secretion, 347

CA, see Catecholamine
Calcitonin secretion, somatostatin effect on, 340
Caloric balance disorders, 445–454
Cardiovascular function disorders, 441–442
Carnosine, 204, 207
CAT, see Choline acetyltransferase
Catatonia, 213–214
Catecholamine (CA)
  distribution, 74–88
  nerve terminals, 84–88
  pathways, 87–88
  synthesis, 70, 78f
  systems, peptide-containing neurons and, 111–112
  turnover, 26f
Cell bodies
  enkephalin-positive, 104
  hypothalamic DA-positive, 77–84
  LHRH-positive, 92
  somatostatin-positive, 98
  substance P-positive, 100–101
  TRH-positive, 94
Cell firing pattern, 281
Central nervous system
  -mediated actions of TRF, 185t
  somatostatin in, 225–226
Cerebral cortex, opioid substance effect on, 175f
Cerebrospinal fluid, 21–23, 154
  tanycytes and, 30
ChAc, see Choline acetyltransferase
p-Chlorophenylalanine and GH secretion, 348, 349
Chlorpromazine and GH secretion, 347, 349
Cholera enterotoxin GRH, 336–337
Choline acetyltransferase (CAT) (ChAc), 23, 88, 262, 264, 281
Cholinergic systems, 88
CI628, 261, 264
Circadian
  activity, 17
  oscillator, 303
  rhythm, ACTH-cortisol, 382–383, 384
Clonidine and GH secretion, 346, 347, 356
Cocaine, 313

Cold receptors, 430
Collaterals, axon, see Axon collaterals
Coma, 420, 424t
Conduction velocity, 285
Consciousness, disorders of, 420–426
Corticotropin (ACTH)
  amino acid sequence 4–10 of, 239t, 335f
  -cortisol circadian rhythm, 382–383, 384
  GABA and, 115
  in GH secretion, 333
  regulation of, 19
  release of, 8
  releasing factor (CRF), 4, 262
    extrahypothalamic, 187, 227
    function, 8t
    VP and, 147
  somatostatin and, 340, 407
  TRH and, 397
  VP and, 115, 147
Cortisol, 259, 262
  ACTH-, circadian rhythm, 382–383, 384
  levels and TRH, 397
Craniopharyngiomas, 418
CRF, see Corticotropin releasing factor
Cross-reactions, 74
Cushing's syndrome, 397, 407
Cybernin, 189
Cycloheximide, 196, 203, 260
Cyproheptadine and GH secretion, 347, 377

**DA**, see Dopamine
DBH, see Dopamine-β-hydroxylase
DDC, see L-DOPA-decarboxylase
Dehydration, 455
Dendrites, 183–186
Deoxycholate, 206
Depression, 395
Derivatization methods, 199
Desmethylimipramine, 313, 317
Dexamethasone, 260
  phosphate, 262
Diabetes
  insipidus, 455, 456–457

  mellitus, 161
  treatment of, 159–61
Diencephalic
  epilepsy, 440–441
  gland, 4, 66, 137
  syndrome of infancy, 451–452
Dihydroxyphenylserine and GH secretion, 347
Diphenhydramine, 334
L-DOPA
  in catecholamine synthesis, 78f
  -decarboxylase (DDC), 71
    in catecholamine synthesis, 78f
  GH secretion and, 346, 349, 350, 397
Dopamine (DA), $8n$, 9, 25, 71
  in catecholamine synthesis, 78f
  cell body
    hypothalamic, 77–84
    system, ventral, 79–84
  GH secretion and, 346, 355
  nerve terminals, 87, 115
    estimation of, 116f
  neurons, 20
    SOM neurons and, 111–112
  in pineal gland, 313
  prolactin
    inhibiting factor and, 115
    secretion and, 394
  tuberoinfundibular, system, 26
Dopamine-β-hydroxylase (DBH), 71
  in catecholamine synthesis, 78f
  distribution, 75f, 86f
Drinking center, 456

**E** antibody, 153
EB, see Estradiol benzoate
Electron microscopic identification of central neurons, 107–111
Electrophysiological identification, 271–272
Emaciation in adults and children, 452–453
Emotion disorders, 428
Endocrine
  cells, peptidergic
    neural origin of, 188–190
      peptidergic neurons and, 184–186

Endocrine (*contd.*)
  cells, peptidergic (*contd.*)
    gland function, neural control over, 5f
  hypothalamus, *see* Hypothalamus, endocrine
  rhythms, sleep-related, 373–386
  tumors, 160
Endocrinology of neuron, 155–190
  endorphins, 161–182
  luteinizing hormone releasing factor, 157–158
  paracrinology versus, 186–188
  somatostatin, 158–161
  thyrotropin releasing factor, 157–158
α-Endorphin, 9t, 103, 162f–163f, 236
β-Endorphin, 9t, 103, 166–168, 236
γ-Endorphin, 9t, 103, 165f
δ-Endorphin, 166
Endorphins
  amino acid sequence of, 239t, 335f
  behavioral effects of, 178–182, 213–214
  enkephalin versus, 234n
  in GH secretion, 334
  immunocytochemical localization of, 168–169
  isolation of, 164–168
  β-lipotropin and, 170–172
  in neuron endocrinology, 161–182
  neuronal actions of, 172–178
  pain perception and, 244
  pituitary hormone release by, 172
Enkephalin (ENK), 71
  distribution, 76f, 102f, 103–104, 105f
  endorphin versus, 234n
  immunoactivity, 206–207
  leucine-, *see* Leucine-enkephalin
  mapping of, 235–236
  methionine-, *see* Methionine-enkephalin
  naloxone and, 176
  neuronal actions of, 172–178
  as neurotransmitter in brain, 233–236
  PROL release and, 115
  structure, 9t

Enzyme
  activities, estrogen effects on, 261t
  antisera, characteristics of, 71t
Ependymal cells at arcuate nucleus, 24f
Epilepsy, diencephalic, 440–441
Epinephrine, 1
  nerve terminals, 84–86
ESN, *see* Estrogen-stimulated neurophysin
Estradiol, 248, 256
  accumulation, 250
  benzoate (EB), 401f
  feedback action of, 359
  gonadotropin secretion and, 367, 369
  hemisuccinate, 262
  serum concentrations in rhesus monkey, 360f
  sleep and, 382
Estrogen, 260
  effects on enzyme activities, 261t
  LRH and, 398
  pituitary gland and, 368
  -stimulated neurophysin (ESN), 139–141, 144–145
Estrone, 258
Ether anesthesia, 336f
Euphoria, 429
Exocrine gland function, neural control over, 5f
Exocytosis and neurophysins, 209
Extrahypothalamic
  distribution of hypothalamic peptides, 217–228
  regulation of GH secretion, 344–346
Eye movement activity, 373


F antibody, 61
Facilitation, recurrent, 274, 289
Fear, 428
Feeding
  behavior disorders, 445–454
  center, 446
  ventromedial nucleus and, 446
Female reproductive behavior, 235–252
Ferritin, 30

Fields I and II, 51
  antiserum F and, 61
  LRF activity in, 62
  LRF neurons of, 59–64
  neurohormones in, 59t
Fluorescein-isothiocyanate (FITC)-conjugated antibodies, 72
Follicle-stimulating hormone (FSH)
  releasing factor, 63
  secretion
    arcuate nucleus and, 366
    GABA and, 115
    LRH injections and, 398–404
    during puberty, 380
    in rhesus monkey, 359–369
    serum concentrations in rhesus monkey, 360f
Fractionation, subcellular, 205
Fröhlich's syndrome, 1, 447; see also Babinski-Fröhlich syndrome
FSH, see Follicle-stimulating hormone

GABA, see γ-Aminobutyric acid
GAD, see Glutamate decarboxylase
Galactorrhea-amenorrhea, 403f, 453
Gastric inhibitory peptide (GIP), 156n
  extrahypothalamic, 187
Gastrin (GAS), 73t, 156, 159
  distribution, 76f, 105f, 107
  secretion, somatostatin effect on, 340
Gastrointestinal
  function disorders, 443–445
  tract, somatostatin in, 226
Genome, target cell, steroid hormone action and, 255–264
GH, see Growth hormone
GIF, see Somatostatin
Gigantism, see Acromegaly
GIP, see Gastric inhibitory peptide
Glia cells, 320
Glucagon, 156, 159–160
  in GH secretion, 333
  receptors, 446
  regulation of, 19
  secretion, somatostatin effect on, 340
Glucocorticoids, 256

Glucoreceptors, 446
Glutamate, 90, 301
  decarboxylase (GAD), 70, 71, 88–89, 90f
  distribution, 75f
Glycine, 90, 282
GnRH (gonadotropin-releasing hormone), see Luteinizing hormone releasing hormone
Gonadotropin secretion, 294
  arcuate nucleus and, 366, 371
  estradiol and, 367, 369
  LHRH in, 360–369
  during puberty, 380–382
  in rhesus monkey, 359–369
Gonadotropins, regulation of, 19
GRF, see Growth hormone releasing factor
Growth hormone (GH), 156–157
  deficiency, idiopathic, 394
  levels
    in diencephalic syndrome of infancy, 451–452
    during sleep, 374–377
  regulation, 19
    monoamines and, 346–350
  release
    cholera enterotoxin GRH and, 336–337
    cyproheptadine and, 377
    L-DOPA and, 397
    by endorphins, 172
    inhibiting factor, see Somatostatin
    TRH and, 395
  releasing factor (GRF), 228, 329
    activity, 336–337
    function, 8t
    in GH secretion, 332–333
  secretion
    aging and, 355, 376
    episodic, 330–332
    extrahypothalamic regulation of, 344–346
    hyperglycemia and, 331
    hypothalamic peptides and, 332–344
    hypothalamic regulation of, 329–332
    methysergide and, 377

Growth hormone (GH) (contd.)
secretion (contd.)
neural control of, 329–351
peptides that inhibit, 337–344
peptides that stimulate, 332–337
somatostatin and, 404–406

Haloperidol, 213–214
GH secretion and, 347
Heat
production, basal, 432
regulation, 429–439
Hippocampus
GH release and, 344
short-term memory and, 427
stimulation, 293–294
Histamine, 23, 84, 90–91, 301, 334
in pineal gland, 313
Histidine decarboxylase, 90–91
Homocarnosine, 207
Hormones
redefined, 189–90, 213
role of, 387
synthetic hypothalamic, 387–409
Horseradish peroxidase (HRP), 29–30, 61
HT, see Hypothalamus, endocrine
5-HT, see Serotonin
Hydroxyindole-O-methyltransferase, 306–309, 311, 329
5-Hydroxytryptamine, see Serotonin
5-Hydroxytryptophan
decarboxylase, 91
GH secretion and, 347
Hyperactivity, autonomic, 440–441
Hyperangiotensinemia, 461
Hyperdipsia, 460–462
Hyperglycemia, 160, 161
GH secretion and, 331
Hypernatremia, essential, 457–459
treatment of, 459
Hyperphagia, 446, 448
Hyperprolactinemia, 394
Hyperpyrexia, 437
Hypersomnolence, 421, 423t
temporary, 423
Hyperthermia
paroxysmal, 437–438
sustained, 434, 437

Hypoglycemia in rat, 356
Hypogonadism, 1–2, 3f, 11
Hypokalemia, 457
Hyponatremia, beer drinker's, 461
Hypophysial-portal blood supply, 3
Hypophysiotropic hormones, 7t, 8t
Hypophysis, origin of, 188–189
Hypopituitarism, idiopathic, 390–392
Hypothalamic
control
of adenohypophysis, 283–284
of reproductive behavior, 247–249
function, aging and, 418
hormones, synthetic, 387–409
hypothyroidism, 392
lesions, see Lesions, hypothalamic
-medial preoptic area, anterior (AHA-MPOA), 292–293
monoamines, 20–23
nuclei, 141–144
obesity, 447–450
peptides, see Peptides, hypothalamic
projections, peptidergic hypothalamo-, 65t
regions, functions and related disorders of, 416t
regulation
of FSH and LH secretion, 359–369
of GH secretion, 329–332
research, history of, 1–12
Hypothalamus (HT), endocrine
anatomy of, 15–36
hypothalamic monoamines, 20–23
median eminence, 23–35
organum vasculosum lamina terminalis, 36, 37f
preoptic and septal areas, 17–19
retinohypothalamic projections, 16–17
suprachiasmatic nucleus, 16–17
ventromedial nucleus, 19–20
anterior, see also Hypothalamus, preoptic anterior
sleep center, 420, 422
stimulation of, 439–440
basomedial, 248; see also Hypothalamus, mediobasal

Hypothalamus (HT) (contd.)
  basomedial (contd.)
    definition, 66
    functions of, 420
    mediobasal (MBH), 224, 361; see also Hypothalamus, basomedial
      "islands," 362–363
      "peninsulae," 364
      somatostatin in, 225
  neurophysiological organization of, 269–295
    neurohypophysial pathways, 270f, 270–283
    parvicellular tuberoinfundibular pathways, 270f, 283–294
  nonendocrine diseases and disorders of, 415–462
    apathy, 428–429
    autonomic function disorders, 429–445
    behavior disorders, 428
    caloric balance disorders, 445–454
    causes of, 418–420
    consciousness disorders, 420–426
    emotion disorders, 428
    euphoria, 429
    expression of, 415–418
    fear, 428
    feeding behavior disorders, 445–454
    heat regulation, 429–439
    memory disorders, 426–427
    parasympathetic function disorders, 439–445
    rage, 428
    sexual behavior disorders, 429
    sleep-wake behavior disorders, 420–426
    sympathetic function disorders, 439–445
    water balance disorders, 454–462
  origin of, 188–189
  pineal versus, 327
  posterior
    stimulation of, 439–440
    wake center, 420, 421
  preoptic anterior (POA-AHA) (POAHT), 363–365, 416; see also Hypothalamus, anterior
    in thermoregulation, 430–432
    water balance disorders and, 454–456
Hypothermia
  paroxysmal, 434, 435f–436f
  sustained, 432–434
Hypothyroidism, 395
  hypothalamic, 392
Hypovolemia, 457

Imipramine, 313
Immunocytochemical localization of endorphins, 168–169
Immunofluorescence, 21, 50
Immunohistochemistry, 70, 72
Inactivation phase, rapid, 326
INC (internuclear cells), 138, 143
Indole metabolism in pineal gland, 304f, 305–308
Infundibular recess, primate, 34f
Infundibulum, see Median eminence
Inhibin, 400
Inhibition, recurrent, 273–274, 286–287
Inhibitors of the peptidases, 201
Insulin, 156, 159
  receptors, 446
  regulation of, 19
  secretion, somatostatin effect on, 340
Intelligence, 427
Interneurons, inhibitory, 273, 287
Internuclear cells (INC), 138, 143
Interstitial cells, 319
IR-LRH (immunoassayable LRH), 223
IR-TRH (immunoassayable TRH), 218
Isoproterenol and GH secretion, 347

K complexes, 373
Ketoacidosis, diabetic, 160, 161
Kleine-Levin syndrome, 450

L-enk, see Leucine-enkephalin
Lateral hypothalamic (LH) lesions, 418, 426

Laurence-Moon-Bardet-Biedl syndrome, 451
Lesions, hypothalamic
  anterior, 433f
  bilateral, of poikilothermia, 439
  bilaterality of, 417–418
  chronic, 445
  examples of, 423t–425t
  large, slowly progressing, 425t, 426
  lateral (LH), 418, 426
  rate of progression of, 416–417
  site of, 415–416
Leucine-enkephalin (l-enk), 73t, 172, 178, 234; see also Enkephalin
  structure, 9t
LH (lateral hypothalamic) lesions, 418, 426
LH, see Luteinizing hormone
LHRH, see Luteinizing hormone releasing hormone
Libido, loss of, 429
$\beta$-Lipotropin, 8n, 164, 166, 168, 236
  amino acid sequence of, 239t, 335f
  endorphins and, 170–172
Lordosis reflex in female rodents, 245–252, 259
  circuitry for, 245–250
  peptide hormone influence on, 251–252
  steroid hormone control of, 250–251
$\beta$-LPH, see $\beta$-Lipotropin
LRF, see Luteinizing hormone releasing factor
LRH, see Luteinizing hormone releasing hormone
Luteinizing hormone releasing factor (LRF), 156–158
  activity in fields I and II, 62
  analogues, 153
  extrahypothalamic, 187
  lordosis behavior and, 251
  neurons, organization of, 49–66
  neurons and their projections, 50–59
  neurons of field I, 59–64
  in OVLT, 36
  in peptidergic neurohormone system, 65t
  in perikarya, 50

super-, 157–158
in tanycytes, 33
releasing hormone (GnRH) (LHRH) (LRH), 1, 3f, 4, 6, 73t, 387
  behavior of, 138
  behavioral effects of, 227
  biosynthesis, 195, 198, 205–206
  in body fluids, 225
  depressant effects associated with, 293f
  distribution, 75f, 91–93, 94f–96f, 108–110, 153–154
  estrogen and, 398
  extrahypothalamic, 217, 223–227
  in GH secretion, 333
  in gonadotropin secretion, 360–369
  immunoassayable (IR-LRH), 223
  injections, 397–404
  in median eminence, 23, 224, 225
  in OVLT, 363
  phylogenetic distribution of, 224–225, 231
  in plasma, 388
  sleep and, 381
  structure and function of, 7t
  tissue levels of, 260
secretion
  arcuate nucleus and, 366
  LRH injections and, 398–404
  during puberty, 380
  pulsatile, 367–368
  in rhesus monkey, 359–369
  serum concentrations in rhesus monkey, 360f
  sleep and, 379f
  tissue levels of, 260

M-enk, see Methionine-enkephalin
Magnocellular
  neurons, ADH and, 455, 457
  neurosecretory system (MNS), 137–150
  nerve cells of, 143–144
MAO, see Monoamine oxidase
MBH, see Hypothalamus, mediobasal

Median eminence (ME), 23–35, 138
  effects of, 114f
  fields I and II, 51
  LRH in, 224, 225
  neurohormones in, 59t
  neurohumoral substances in, 113–118
  projections, peptidergic hypothalamo-, 65t
  rat, 27f, 35f
  SRIF input and, 56–57
  stimulation, 286–287
Melanocyte stimulating hormone (MSH), 8n
  amino acid sequence of, 239t, 335f
  extrahypothalamic, 188, 227
  in GH secretion, 333
  immunoactivity, 207
  inhibiting factor (MIF) (MSH-IF) function, 8t
  releasing factor (MSH-RF) function, 8t
Melatonin, 303
  conversion of serotonin to, 305–306
  GH secretion and, 347
  placental transport of, 318
Memory
  disturbances, 426–427
  short-term
    hippocampus and, 427
    loss, 426
Mesencephalic projections, peptidergic hypothalamo-, 65t
Methionine-enkephalin (m-enk), 73t, 164, 170, 178, 180, 234; see also Enkephalin
  amino acid sequence of, 239t, 335f
  in GH secretion, 334
  structure, 9t
α-Methyl-p-tyrosine (α-MT) and GH secretion, 347, 348f, 349, 350f, 356
Methysergide and GH secretion, 346, 347, 348, 377
Microiontophoresis, 281
Microperoxidase, 30
MIF (MSH inhibiting factor), 8t
Milk-ejection reflex, 277–279
Mineralocorticoids, 256

MNS, see Magnocellular neurosecretory system
Monoamine
  neurons, distribution of, 20–23
  oxidase (MAO), 261
Monoamines, hypothalamic, 20–23
  GH regulation and, 346–350
Morphine, 161, 164
  sulfate in GH secretion, 334
Morphinomimetic activity, 166
MSH, see Melanocyte stimulating hormone
α-MT, see α-Methyl-p-tyrosine

Nafoxidine, 261
Naloxone, 214
  antagonism of peptide effects, 175–178
  enkephalins and, 176
NE, see Norepinephrine
Nelson's syndrome, 397, 407, 408f
Nerve
  cell activity, 10, 143–144
  terminals
    catecholamine, 84–88
    dopamine, 87, 115, 116f
    enkephalin-positive, 104
    epinephrine, 84–86
    LHRH-positive, 92
    norepinephrine, 87
    peptidergic, 209
    somatostatin-positive, 98
    substance P-positive, 101–103
    TRH-positive, 94, 218
Neural
  lobe projections, peptidergic hypothalamo-, 65t
  origin of peptidergic endocrine cells, 188–190
  pathways, biogenic amine, 8, 10f
  regulation of pineal gland, 312–314
Neural control
  over exocrine and endocrine gland function, 5f
  of growth hormone secretion, 329–351
  events related to
    oxytocin release, 276–279
    vasopressin release, 275, 279–281

Neuroendocrine
  cells, defined, 5f
  regulation
    brain rhythms and, 356
    of menstrual cycle, 359–369
    tissues, interactions of steroid hormones with, 256–259
  transducer cells, 4, 5f
Neuroendocrinology, 182, 189
Neurogenic ulcers, 444
Neurohormone, use of term, 65–66, 213
Neurohormone systems, peptidergic, 64–66
Neurohormones in perikarya, axons, and terminals, 59
Neurohumoral substances in median eminence, 113–118
Neurohypophysial
  hormone, biosynthesis, 7t, 196–197
  pathways, 270f, 270–283
  peptides, distribution of, 144–145
Neurohypophysis, 189; see Adenohypophysis, Pituitary gland
Neuron
  endocrinology of, see Endocrinology of neuron
  single-, recording of nerve cell activity, 10
Neurons
  arrangement of, at arcuate nucleus, 24f
  monoamine, distribution of, 20–23
  peptidergic, and peptidergic endocrine cells, 184–186
  preoptic, 28
  release of peptides from, 208–209
  tuberoinfundibular, see Tuberoinfundibular neurons
Neuropharmacology
  of supraoptic and paraventricular neurons, 281–283
  of tuberoinfundibular neurons, 294
Neurophysin (NF) (NP), 138; see also Vasopressin
  biosynthesis, 196–197
  distribution, 75f, 98–99, 109
  estrogen-stimulated (ESN), 139–141, 144–145
  exocytosis and, 209

  in monkey OVLT, 148
  nicotine-stimulated (NSN), 139–141, 145
  in tanycytes, 147f, 148, 154
  in zona externa of median eminence, 146
Neurosecretion, 3, 49, 66, 137
Neurosecretory
  granules, movements of, 25f
  system, magnocellular, see Magnocellular neurosecretory system
Neurotensin, 5
  in GH secretion, 334
  as neurotransmitter in brain, 238–242
  structure, 9t
  uptake, 243
Neurotransmitter, 213, 214
  peptide, 233–242
NF, see Neurophysin
Nicotine-stimulated neurophysin (NSN), 139–141, 145
Noradrenaline in catecholamine synthesis, 78f
Norepinephrine (NE), 9, 23, 25, 69, 282
  cyclic AMP and, 315–316
  GH secretion and, 346, 348, 355
  nerve terminals, 87
  in pineal gland, 313
Normorphine, 170–171, 173t
NP, see Neurophysin
Nuclear cells, 3


Obesity, 397, 447–450
Octopamine in pineal gland, 313
Oligospermia, 404
Opiate
  analgesia, 244
  receptor
    immunoactivity, 206–207
    sites, 233
  receptors, 161, 164
    mapping of, 235–236
Opioid
  peptides in GH secretion, 334–336
  substances, effect of, 174f, 175f
Optic fibers, terminations of, 16

# SUBJECT INDEX

Organum vasculosum lamina terminalis (OVLT), 36, 37f, 51
  LHRH in, 363
  neurohormones in, 59t
  peptide distribution in, 96f
  projections, peptidergic hypothalamo-, 65t
  vasopressin and neurophysin in monkey, 148
Osmoreceptors, 454–455
OT, *see* Oxytocin
OVLT, *see* Organum vasculosum lamina terminalis
Oxytocin (OT) (OXY), 71, 73t
  distribution, 76f, 98–99, 109
  as inhibitory neurotransmitter, 274
  in magnocellular neurosecretory system, 139–141, 144–145
  release, neural events related to, 276–279
  structure of, 4, 7t
  in zona externa of median eminence, 146

Pain perception, 5
  endorphins and, 244
PAP, *see* Peroxidase-antiperoxidase
Parachlorophenylalanine (PCPA), 377
Paracrine secretion, 189
Paracrinology versus endocrinology, 186–188
Parasympathetic function disorders, 439–445
Parathormone secretion, somatostatin effect on, 340
Paraventricular
  neurons, 272, 281–283
  nuclei (PVN), 137, 141–145, 147, 149
Parvicellular tuberoinfundibular pathways, 270f, 283–294
Pathway(s)
  afferent, 275
  aminergic and peptidergic, in nervous system, 69–119
  biogenic amine neural, 8, 10f
  catecholamine, 87–88
  enkephalin, 104
  GABA, 301
  inhibitory feedback, 271–274
  LHRH, 92
  neurohypophysial, 270f, 270–283
  parvicellular tuberoinfundibular, 270f, 283–294
  serotonin, 262
  somatostatin, 98
  somatostatinergic, 342–344
  substance P, 103
  TRH, 95–96
PCPA (parachlorophenylalanine), 377
Pentobarbital anesthesia, 336f, 337
Peptidases, 207–208
  inhibitors of, 201
Peptide
  biosynthesis, brain, 195–209
  -containing neurons, CA systems and, 111–112
  effects, naloxone antagonism of, 175–178
  hormones for female reproductive behavior, 235–252
  neurotransmitters, 233–242
Peptidergic
  neurons, 6f, 184–186, 195
  pathways in nervous system, 69–119
Peptides
  blood-brain barrier for, 253
  distribution of, 91–107
  as GH secretion, 332–344
  hypothalamic, 7t, 9t, 138, 217–228, 332–344
  local release of, 186–187
  mechanism of action of, 10
  neurohypophysial, distribution of, 144–145
  release of, from neurons, 208–209
Perikarya
  LRF in, 50
  neurohormones in, 59
Permeability factor, *see* Substance P
Peroxidase-antiperoxidase (PAP), 72, 110f
Phentolamine and GH secretion, 346
Phenoxybenzamine, 262
  GH secretion and, 347, 350f
Phenylalanine, 30

Phenylethanolamine-$N$-methyltransferase (PNMT), 21, 23, 71
  in catecholamine synthesis, 78f
  distribution, 75f, 83f, 85f
Phospholipid metabolism, 314
Photoillumination
  $N$-acetyltransferase and, 307–311
  TRH and, 322
Photoregulation of pineal function, 308–309
Phylogenetic distribution
  of hypothalamic peptides, 217–228
  of LHRH and LRH, 224–225, 231
  of TRH, 218–222
Picrotoxin, 282, 287, 301
PIF, see Prolactin inhibiting factor
Pimozide and GH secretion, 346, 347
Pineal cells, 308–309, 314, 319f
Pineal gland, 11
  hypothalamus versus, 327
  indole metabolism in, 304f, 305–308
  as model of neuroendocrine regulation, 303–320
  neural regulation of, 312–314
  stress and, 317
Pinealocyte, 318–320
Pituitary
  disorders, prolactin response to TRH during, 394
  gland, see also Adenohypophysis, Neurohypophysis
    estrogen and, 368
    influences on, 114f
    hormone secretory activity, sleep-wake cycle and, 374
    hormones, action potentials and, 294
    insufficiency, 389–392
    somatotroph receptors, 357
    stalk stimulation, 271–272
    tumors, somatostatin and, 406–407
Placental transport of melatonin, 318
PNMT, see Phenylethanolamine-$N$-methyltransferase
POA-AHA, see Hypothalamus, preoptic anterior
POAHT, see Hypothalamus, preoptic anterior
Poikilothermia, 438–439, 453
Polydipsia, 456

Portal vessel chemotransmitter hypothesis, 217
Portal vessels, influences on, 114f
Potassium, 313
Prader-Labhart-Willi syndrome, 451
Preoptic area, 17–19
  medial, 292
  neurons, 28, 248–249
  -septal region, 51; see also Field II
PRF, see Prolactin releasing factor
PRL, see Prolactin
Progesterone, 256–258, 398
  effect on lordosis, 251–252
  serum concentrations in rhesus monkey, 360f
Progestins, 256
Prolactin (PRL) (PROL), 73t
  distribution, 76f, 107
  inhibiting factor (PIF), 228, 394
    DA and, 115
  function, 8t
  release, 115, 172, 392, 393f, 394
  releasing factor (PRF), 8t, 227–228
  secretion, 156, 157
    amphetamine and, 385
    arcuate nucleus and, 366
    dopamine and, 394
    serotonin and, 379
    during sleep, 377–379
Proline, 199–203
Propranolol and GH secretion, 346, 349, 350f
Protein
  calorie malnutrition, 395
  template mechanism, 198
Puberty, gonadotropin secretion during, 380–382
Pulmonary edema, 443
Pulsatile patterns, 200
Purification, sequential, 199
Purification strategies, 201–204
Puromycin, 196, 203
PVN, see Paraventricular nuclei
Pyrogen receptors, 431

Rage, 428
Raphe
  neurons, 263
  nuclei, 17

## SUBJECT INDEX

Rapid eye movement (REM) sleep, 373–374, 422
RAS, see Reticular activating system
Recurrent
  facilitation, 274, 289
  inhibition, 273–274, 286–287
Refractory period, 285
Release of peptides from neurons, 208–209
Releasing factor (RF), 24, 63
  biosynthesis of, 197–204
Renal failure, 332
Renin secretion, somatostatin effect on, 340
Reproductive behavior, female, 235–252
Respiratory function disorders, 442–443
Reticular activating system (RAS), 420–422
Retinohypothalamic projections, 16–17, 308–309
RF, see Releasing factor
Ribosomal
  mechanism, 203
  synthesis, 198
RNA synthesis, 197


Saline infusion in GH secretion, 338f
Satiety center, 446
SCH, see Suprachiasmatic nucleus
Schizophrenia, 181, 214
SCN, see Suprachiasmatic nucleus
Secretin, 156, 159
  secretion, somatostatin effect on, 340
Separation of pineal cells, 319f
Septal area, 17–19
Serotonin (5-hydroxytryptamine) (5-HT), 9, 17, 26, 27, 91
  conversion of, to melatonin, 305–306
  distribution of, 21–23
  GH secretion and, 346, 347
  pathway, 262
  in pineal gland, 313
  prolactin secretion and, 379
  synthesis, 70

Sexual
  behavior, 429
  differentiation of brain, 256
  function, 10
SIADH, see Syndrome of inappropriate secretion of ADH
Sleep, 320
  center, 420, 422
  estradiol and, 382
  in GH secretion, 344, 347, 374–377
  LH and, 379f
  LHRH and, 381
  prolactin secretion during, 377–379
  rapid eye movement (REM), 373–374, 422
  -related endocrine rhythms, 373–386
  somatostatin and, 385–386
  TSH secretion and, 379–380
  -wake
    behavior, disorders of, 420–426
    cycle, pituitary hormone secretory activity and, 374
    reversal study, 374, 375f
SOHT, see Supraopticohypophysial tract
SOM, see Somatostatin
Somatomedins, 356
Somatostatin (growth hormone release inhibiting factor) (GIF) (SOM) (SRIF), 1, 3f, 4, 6, 71, 73t, 156–157, 329, 387
  ACTH levels and, 407
  analogues of, 161
  behavioral effects of, 227
  brain distribution of, 340–44
  depressant effects associated with, 293f
  distribution, 75f, 85f, 93f–95f, 96–98, 99f–101f, 108, 110
  extrahypothalamic, 187, 217, 225–226
  function of, 226–227
  functions of, 158–161
  growth hormone secretion and, 337–339, 404–406
  immunoactivity, 206
  infusion in GH secretion, 338f

Somatostatin (contd.)
  infusions, intermittent, 338
  injection, 404–408
  input, median eminence and, 56–57
  in median eminence, 23
  neurons, 49–66, 50–59, 111–112
  nonpituitary effects of, 340
  in peptidergic neurohormone system, 65t
  pituitary tumors and, 406–407
  -positive cells, tyrosine-hydroxylase-positive cells and, 153
  sleep and, 385–386
  structure and function of, 7t
  TSH secretion and, 406, 407f
Somatostatinergic pathways, 342–344
Somatotropin releasing factor (SRF) function, 8t
SON, see Supraoptic nuclei
SP, see Substance P
Spike discharges, 279
SRF (somatotropin releasing factor) function, 8t
SRIF, see Somatostatin
Starvation, 453
Steroid hormone
  action, 11
    genome involvement in, 255–264
  control of lordosis, 250–251
  effects, activational, 256, 260–262
  feedback action, 372
  for female reproductive behavior, 235–252
  interactions with neuroendocrine tissues, 256–259
  receptor sites, 263
Stress
  in GH secretion, 344
  pineal gland and, 317
Strychnine, 301
Substance P (SP), 5, 71, 73t
  distribution, 76f, 99–103
  in GH secretion, 333–334
  immunoactivity, 207
  as neurotransmitter in brain, 236–238
  structure, 9t

Suprachiasmatic nucleus (SCH) (SCN), 16–17, 138, 149, 308, 319–311
Supraoptic
  crest, 36
  neurons
    action potentials of, 271
    neuropharmacology of, 281–283
  nuclei (SON), 137, 141–145, 147, 149
Supraopticohypophysial tract (SOHT), 138, 146, 147
Sympathetic function disorders, 439–445
Synaptosomes, 205, 207, 262
  somatostatin in, 340
Syndrome of inappropriate secretion of ADH (SIADH), 459–460, 462

$T_3$ and $T_4$, 389
  levels, plasma, 380
Tanycyte theory, ependymal, 217
Tanycytes, 154
  cerebrospinal fluid and, 30
  LRF in, 33
  neurophysin in, 147f, 148, 154
  role of, 30–32
  transport capacity of, 33
Template mechanism, protein, 198
Terminals, neurohormones in, 59
Testosterone, 256–259
  accumulation, 250
  secretion, 380, 381
TH, see Tyrosine hydroxylase
Thermoregulation, 429–439
Thermostat, biologic, 431
Thirst, 456–460, 462
Thyroid-stimulating hormone, see Thyrotropin, TSH secretion
Thyrotropic area, 95–96
  lesions, 218
Thyrotropin, see also TSH secretion
  deficiency, familial isolated, 391f
  regulation of, 19
  releasing factor (TRF), 156–158
    central nervous system-mediated actions of, 185t
    extrahypothalamic, 187

Thyrotropin (*contd.*)
in peptidergic neurohormone system, 65t
releasing hormone (TRH), 1, 3f, 4, 6, 8n, 9, 73t, 387
  ACTH levels and, 397
  behavioral effects of, 227
  biosynthesis, 195, 199–205
  in body fluids, 322
  cortisol levels and, 397
  depressant effects associated with, 293f
  distribution, 75f, 93f, 94–96, 97f
  extrahypothalamic, 200, 217, 218–222, 226–227
  in GH secretion, 332–333, 395
  immunoassayable (IR-TRH), 218
  injection, 388–397
  in median eminence, 23
  obesity and, 397
  photoillumination and, 322
  phylogenetic distribution of, 218–222
  in plasma, 388
  prolactin and, 392, 393f, 394
  structure and function of, 7t
1-Thyroxine, 389, 390f
TRF, *see* Thyrotropin releasing factor
TRH, *see* Thyrotropin releasing hormone
Triton X-100, 74, 206
L-Tryptophan and GH secretion, 347
5-OH-Tryptophan and GH secretion, 347, 348
Tryptophan hydroxylase, 26, 28, 91
  pineal, 307
TSH (thyroid-stimulating hormone) secretion, 388–393; *see also* Thyrotropin
  sleep and, 379–380
  somatostatin and, 406, 407f
Tuberoinfundibular
  dopamine system, 26
  neurons, 284–289, 291, 294
  pathways, parvicellular, 270f, 283–294
  system, axon collaterals in, 288f, 289–291
  tract, 25

Tyramine, 313
Tyrosine
  in catecholamine synthesis, 78f
  hydroxylase (TH), 23, 71
    activity, 263, 264
    in catecholamine synthesis, 78f
    distribution, 75f
  -positive cell bodies, distribution of, 79f–83f, 85f
  -positive cells, somatostatin positive cells and, 153

Ulcers, neurogenic, 444
Uremia, 395
Urethane anesthesia, 336f, 337, 349

Vaginal distension, 279
Varicosities in pineal gland, 313
Vasoactive inhibitory peptide (VIP), 73t
  distribution, 76f, 105–106
  extrahypothalamic, 187
  structure, 9t
Vasopressin (VP), 71, 73t, 138–141, 146–148; *see also* Antidiuretic hormone, Neurophysin
  ACTH and, 115, 147
  behavior of, 138
  biosynthesis, 196–197
  CRF and, 147
  distribution, 76f, 98–99, 109
  in GH secretion, 333
  as inhibitory neurotransmitter, 274
  in monkey OVLT, 148
  in peptidergic neurohormone system, 65t
  release, neural events related to, 275, 279–281
  structure of, 4, 7t
  in zona externa of median eminence, 146
Velocity, conduction, 285
Ventromedial nucleus (VMH) (VMN), 19–20, 329
  aging and, 418
  axons, 248

Ventromedial nucleus (*contd.*)
  destruction, obesity and, 447
  feeding and, 446
  neurons, 57
  projections from, 248
VIP, *see* Vasoactive inhibitory peptide
VMN, *see* Ventromedial nucleus
VP, *see* Vasopressin

Wake center, 420, 421
Warm receptors, 430
Water balance, 4, 454–456
Wheal reaction, $9n$

Zona externa (ZE) of median eminence, 146–149